Pigments, Pigment Cells and Pigment Patterns

Hisashi Hashimoto • Makoto Goda •
Ryo Futahashi • Robert Kelsh • Toyoko Akiyama
Editors

Pigments, Pigment Cells and Pigment Patterns

Springer

Editors
Hisashi Hashimoto
Graduate School of Science
Nagoya University
Nagoya, Aichi, Japan

Makoto Goda
Preeminent Medical Photonics Education & Research Center
Hamamatsu University School of Medicine
Hamamatsu, Shizuoka, Japan

Ryo Futahashi
Bioproduction Research Institute
National Institute of Advanced Industrial Science and Technology (AIST)
Tsukuba, Ibaraki, Japan

Robert Kelsh
Department of Biology and Biochemistry
University of Bath
Bath, UK

Toyoko Akiyama
Department of Biology
Keio University
Yokohama, Japan

ISBN 978-981-16-1489-7 ISBN 978-981-16-1490-3 (eBook)
https://doi.org/10.1007/978-981-16-1490-3

This Springer imprint is published by the registered company Springer Nature Singapore Pte Ltd.
The registered company address is: 152 Beach Road, #21-01/04 Gateway East, Singapore 189721, Singapore

Preface

Much of the beauty in the natural world comes from the astonishing diversity of animal body coloration. Living organisms on the earth display a wide variety of colors and patterns, especially, but not exclusively, in their integument. The beauty and spectacle of these pigment patterns have attracted people, especially hobbyists, birdwatchers, scuba divers, and biologists, since antiquity. Even Aristotle, who was the first person to study biology systematically, also focused his attention on the varied integumental coloration in the natural world, including octopuses.

These patterns have been generated through the creativity of evolutionary processes. Body colors in different species have many roles, some counter-intuitive. For many animals, they are cryptic, providing camouflage specific to the natural environment of the organism—the color matching of many moths to their lichen-covered resting places is often astonishing. Some cryptic coloration may be very colorful, yet by matching flowers or brightly lit vegetation it makes the animal difficult to see—crab spiders, golden orioles, parrots are all examples of this. In other cases, bold patterns take on the role of disruptive camouflage, breaking up the characteristic outline of the organism—the striking striped patterns of zebras and the blotches of snow leopards prove surprisingly effective at hiding these animals in their contrasting habitats. In these two examples, the pigment patterns enable the prey to avoid being found, and the predator to ambush wary prey, respectively. For some organisms, bold color patterns advertise the poisonous or dangerous nature of the bearer—this aposematic coloration explains the bright colors of poison dart frogs, coral snakes, and wasps. In other cases, advertising coloration, driven by sexual selection, is used to sell an individual to a prospective mate, assuring them of the genetic quality of that individual—the color patterns (often enhanced by feather adornments) of hummingbirds, birds-of-paradise, sunbirds are likely all used in this way. Perhaps the most remarkable examples of pigment patterns are in the chameleons and in the cuttlefish and octopuses, where patterns can be changed rapidly. In these cases, their role is dynamic, being in turn cryptic, advertising, and even in displaying emotion!

Scientific study of these patterns has focused initially on the components, the pigments, and the cells that produce them, but also on the microstructures (extra- and subcellular) that interfere with incident light, resulting in structural colors. Our understanding of the chemistry and biophysics that underlies these natural patterns has progressed rapidly, yet casual exploration of the biosphere continues to reveal new pigment cells and likely new pigments to be characterized. Biochromatic pigments cover the full visible and UV spectrum, matching the wavelength sensitivities of eyes. These colors result from light absorbing pigments (with wavelengths that are not absorbed determining the perceived color of the pigment). In some cases, fluorescent pigments absorb light of one wavelength and subsequently emit light of another, longer wavelength. Structural colors result from the differential absorption and reflection of different wavelengths as they interact with physical interfaces (e.g., edges of subcellular tiny crystals and air micropockets in feathers). A further remarkable mechanism for generation of color in a select group of organisms is that of bioluminescence, where light is emitted as part of a chemical reaction; familiar from experience of fireflies and glow-worms, this process must drive (all?) the visual experience of deep-water organisms.

This understanding has often been grounded in genetic studies, especially of domesticated animals. Indeed, artificial selection in domesticated species has generated a remarkable range of color variants. The domestic mouse, now a staple of genetic studies, began its career when humans noticed and cultivated different color variants (or "fancies" as they were known). This explains, in large part, why our understanding of pigmentation is dominated by knowledge of the melanocyte and the variants of melanin pigments that it produces. Domesticated mammals are often bred to have distinctive coat colors, and our knowledge of mouse genetics has driven our understanding of how these come about, although much remains to be explained.

But beyond the components, perhaps the most fascinating question of all concerns the mechanisms that generate the *patterns* themselves—how are the cells and the pigments they produce organized in such a way as to produce the different patterns we see? This is all the more remarkable when we note that these patterns can be remarkably labile—for example, in some fish genera, sister species can show radically different pigment patterns, likely as a result of strong sexual selection, linked to speciation. Our knowledge here is often still rather rudimentary, although key insights have been obtained from study of mice, and most recently in the zebrafish. As genomic techniques have advanced, exploration of patterning mechanisms in domestic and wild cats, dogs, and horses promises to open up the field of stripe and spot generation. Mathematical modelers, from Turing onwards, have been inspired by these patterns, and a considerable body of theoretical work now underpins the field of pigment pattern formation. The next decade promises to integrate these experimental and theoretical data into a coherent understanding of patterning mechanisms, at least in a few well-studied species.

Yet we should not believe that pigment cells are the same throughout the animal kingdom. The fact that the cephalopods (mollusks) and chameleons (reptiles) have evolved the most versatile and complex pigmentary systems immediately indicates that selection for pigmentary mechanisms might be powerful enough to drive convergent evolution. That seems to be the case, with distinct pigment cell-types found in the integument of vertebrates and invertebrates. However, we remain largely ignorant of the molecular characteristics of many pigment cells, or chromatophores, so that the true extent of convergent evolution and diversification remarks very poorly understood.

Best studied are the chromatophores of vertebrates, where initial understanding of mammalian melanocytes, black pigment cells, has been expanded to include a number of chromatophores in poikilothermic vertebrates. Here the zebrafish and medaka have been best studied, again driven by their genetics, but the enormous diversity of fish, reptilian, and amphibian pigment cells have been neglected. Currently, these chromatophores are mainly classified into six categories according to the colors displayed by them, namely melanophores (black), xanthophores (yellow), erythrophores (red), cyanophores (blue), leucophores (white), and iridophores (metallic or iridescent). This nomenclature is widely accepted; nevertheless, the recent identification of variants in killifish and even in adult zebrafish suggests we have much to learn about the subtleties of pigment cell classification. With the important exception of the melanized retinal pigment epithelium of the eye, vertebrate pigment cells all seem to be derived from the neural crest, a remarkably flexible and evolutionarily important group of stem cell-like cells.

In vertebrates, diversity of pigmentation mechanisms extends beyond the diversity of cell-types, in response to evolutionary adaptations in skin appendages. Thus, in fish, reptiles, and amphibia, pigment patterns are generated by controlling the position of different chromatophore types—rather in the manner of a Pointillist painting, where individual spots of different pigments together make up the full picture. In contrast, in mammals and birds, the evolution of hair and feathers has coincided with the dominance of secreted melanin, with yellow-red pheomelanin components adding to the color palette, and in some species, combined with structural coloration resulting from the organization of microstructures within integument.

Invertebrates differ greatly from vertebrates in the mechanism of coloration. For example, in insects, genetic control of the expression of enzymes involved in diverse pigment synthesis in epidermal cells is important, unlike regulation by pigment cells in vertebrates. In addition, melanin pigments and their synthetic enzymes differ from vertebrates, and pigments such as ommochrome, which are not found in vertebrates, are commonly found in insects. In contrast, mollusks have (convergently) evolved a range of chromatophores giving colors comparable to those of poikilotherms, at their most striking in the cephalopods.

Structural colors, generated by physical phenomena of incident light on histological microstructures, contribute to diversifying the integumental colorations of the animal. The physical mechanisms of structural color expression are commonly and broadly employed in invertebrates and vertebrates. These are best known in insects, fish, and birds, but structural colors are known even in mammals, e.g., the adult male mandrill, where they determine the bluish coloration of the cheeks and buttocks, a signal of his dominance.

Interestingly, dynamic changes of pigmentation patterns and colors in vertebrates and invertebrates, while under shared neural control, rely on distinct cellular mechanisms. Thus, in cephalopods, rapid redistribution (aggregation and dispersion) of pigments in chromatophores is controlled by individual muscles attached to the pigment cells. In contrast, in vertebrates, comparable dynamic changes in pigment distribution within the cells depend upon cytoskeletal control of pigment organelle distribution. In homeothermic vertebrates, the evolution of hair and feathers and thus the secretion of melanin granules into these keratin structures precludes a rapid dynamic response. Here, coloration changes are largely restricted to moult cycles (e.g., winter and summer plumages of many temperate birds).

In this book, we attempt to bring together state-of-the-art summaries of the diverse field of pigment and pigment cell biology, by recognized experts in their fields. This book results from the enthusiastic encouragement of Dr. Misato Kochi, coordinator of Springer Japan, who attended the symposium "Pigment Cell Research" organized by some of us at the annual meeting in *Niigata* of the Zoological Society of Japan (2015). In the absence of an established modern general textbook, we offer this book as a route into the diverse concepts and observations within the literature. It complements the summary tome on pigmentation (The Pigmentary System: Physiology and Pathophysiology, Second Edition. Editor (s): James J. Nordlund; Raymond E. Boissy; Vincent J. Hearing; Richard A. King; William S. Oetting; Jean-Paul Ortonne), which is a comprehensive and integrated book of the concepts, focused on mammalian studies, broadening the reach to encompass other vertebrates and invertebrates while also providing updated summaries of the latest observations and insights across all taxa. We hope that all readers will enjoy learning about the remarkable diversity and parallels of pigment cell-related mechanisms in animals while also seeing the great gaps in our knowledge—future opportunities for the new generation of researchers!

Nagoya, Aichi, Japan — Hisashi Hashimoto
Hamamatsu, Shizuoka, Japan — Makoto Goda
Tsukuba, Ibaraki, Japan — Ryo Futahashi
Bath, UK — Robert Kelsh
Yokohama, Japan — Toyoko Akiyama

Acknowledgements

We owe a debt of gratitude to Dr. Misato Kochi (already retired from Springer in 2018) and her colleagues at Springer Japan for encouraging and then cajoling the production of this work, from dream to reality! We thank Ms. Kripa Guruprasad, project coordinator of this book, for encouraging us throughout.

Contents

Part I
Pigments and Pigment Organelles

Chapter 1
Pigments in Insects

Ryo Futahashi and Mizuko Osanai-Futahashi

Abstract Insects have an amazing variety of colors and patterns. It should be noted that pigments and/or genes involved in body color formation are markedly different between insects and vertebrates. Insect pigments have been traditionally classified into the following eight classes: melanins, ommochromes, pteridines, tetrapyrroles, carotenoids, flavonoids, papiliochromes, and quinones. Among them, melanins, ommochromes, and pteridines are three major pigments that are distributed in most insects. Insect melanins are secreted into cuticles, and dark-colored melanins are predominantly derived from dopamine, whereas light-colored melanins are mostly derived from *N*-β-alanyldopamine (NBAD) and/or *N*-acetyldopamine (NADA). Ommochromes are tryptophan-derived pigments restricted to invertebrates and are ubiquitous in the compound eyes of insects. Ommochromes and pteridines are accumulated within pigment granules ommochromasomes and pterinosomes, respectively. In recent years, much has been revealed at the molecular level about pigment synthesis and pattern formation in insects. In this review, we introduce the relationship between the pigments and insect color pattern, and summarize the genes involved in the pigment synthesis, transport, and patterning. Meanwhile, there are still many insects whose pigments have not been identified, and the future progress is expected in this field.

Keywords Insects · Pigments · Melanins · Ommochromes · Pteridines · Tetrapyrroles · Carotenoids · Flavonoids · Papiliochromes · Quinones

---

R. Futahashi (✉)
National Institute of Advanced Industrial Science and Technology (AIST), Tsukuba, Ibaraki, Japan
e-mail: ryo-futahashi@aist.go.jp

M. Osanai-Futahashi
National Institute of Advanced Industrial Science and Technology (AIST), Tsukuba, Ibaraki, Japan

Ibaraki University, Mito, Ibaraki, Japan
e-mail: mizuko.osanai-futahashi.sci@vc.ibaraki.ac.jp

H. Hashimoto et al. (eds.), *Pigments, Pigment Cells and Pigment Patterns*,
https://doi.org/10.1007/978-981-16-1490-3_1

1.1 Introduction

Insects comprise over half of the described organisms on Earth, showing an extreme variety of colors and patterns associated with a variety of ecological meanings such as sexual selection, mimicry, and thermoregulation (Grimaldi and Engel 2005). The body color of living organisms can be categorized into pigments and structural colors. These two are generally classified in that the former absorbs light in the visible part of the spectrum, and the latter is caused by light scattering, interference, or diffraction. Insect pigments have been traditionally classified into the following eight classes: melanins, ommochromes, pteridines (pterins), tetrapyrroles (porphyrins and bile pigments), carotenoids, flavonoids (anthocyanins and flavones), papiliochromes, and quinones (aphins and anthraquinones) (Table 1.1). Among them, melanin, ommochromes, and pteridines are distributed in most insects (Table 1.2).

As shown in Table 1.1, the major sources of the pigments are aromatic amino acids (i.e., tyrosine, phenylalanine, and tryptophan). Although aromatic amino acids are essential for the survival of animals, excessive amounts cause toxic effects. Defect in aromatic amino acid metabolism sometimes causes serious diseases in animals. For example, in mammals, phenylketonuria due to deficiency in conversion of phenylalanine into tyrosine causes mental retardation, convulsions, and fair skin (Williams et al. 2008). Ommochromes, tryptophan-derived pigments, are common in invertebrates such as insects, but are absent in vertebrates (Figon and Casas 2019), which seems to be associated with the absence of the two tryptophan metabolic pathways, glutaric acid pathway and nicotinamide adenine dinucleotide pathway, in protostomes (Linzen 1974). It should be noted that ommochrome pigment (xanthommatin) seems to be one of the causative agents of brown cataract, suggesting that it may be produced when tryptophan metabolism is abnormal (Tomoda et al. 1990). In this review, melanins, ommochromes, pteridines, and other pigments are introduced with respect to the colors and patterns of insects.

1.2 Melanins

1.2.1 Melanin Synthesis Pathway in Insects

Melanins are polymer of phenolic substances or quinone-based substances formed by numerous cross-linked indole and/or benzothiazine derivatives. Melanin synthesis pathway varies depending on animal species, and homologs of most vertebrate melanin synthesis genes, such as tyrosinase, tyrosinase related proteins, and dopachrome tautomerase (see Chap. 2), are not present in insects (Barek et al. 2018). Notably, the majority of insect dark-colored melanins are derived from dopamine (Fig. 1.1; Hiruma et al. 1985; Arakane et al. 2016). Dopamine-melanin is involved in the black-brown coloring of the cuticle, the exoskeleton of insects, on bodies and wings (Tables 1.1 and 1.2; Fig. 1.1). Some melanins are hypothesized to

Table 1.1 Characteristics and distribution of insect pigments

Pigment groups	Basic structure	Precursor	Distribution in organisms	Distribution in insects	Solubility	Color	Characteristics
Melanins (eg, Dopa-melanin, Dopamine-melanin, Pheomelanin)	Oxidative polymerization of indole or benzothiazine derivatives	Tyrosine (Phenylalanine)	Universally found in animals, plants and fungi.	Most insect orders	Insoluble in water, lipid, and hydrochloric acid. Dissolves when heated with 1N NaOH or concentrated sulfuric acid. Phaeomelanin sometimes dissolves in dilute hydrochloric acid.	Black, Brown, Reddish-brown, Yellowish-brown	Bleached with strong oxidants such as H_2O_2.
Ommochromes (eg, Xanthommatin, Ommatin D, Ommin A)	Oxidative condensation of 3-hydroxy-kynurenine	Tryptophan	Widely found in invertebrates (e.g., insects, spiders, crustaceans, cephalopods)	Most insect orders	Soluble in acidic methanol (e.g., 2% HCl in MeOH), poorly soluble in water (except for Ommatin D, Rhodommatin), degraded with alkali.	Red, Brown, Yellow, Purple	Redox reactions produce reversible color changes from yellowish oxidized form to reddish reduced form and vice versa.
Pteridines (eg, Xanthopterin, Leucopterin, Isoxanthopterin, Erythropterin, Sepiapterin)	Pteridine ring (fused pyrimidine and pyrazine ring)	Guanosine triphosphate (GTP)	Universally found in animals. Plant folic acid is also considered a type of pteridine.	Most insect orders	Soluble in 5% aqueous ammonia. Xanthopterin is soluble in hot water. Erythropterin is converted to leucopterin with 5% aqueous ammonia and should be eluted with sodium carbonate.	White, Yellow, Orange, Red, Blue, Colorless	The oxidized form emits strong fluorescence in UV. Fluorescence color changes depending on pH and redox state.
Tetrapyrroles (eg, Biliverdin IXα, Biliverdin IXγ, Hemoglobin)	Four pyrrole-based N-heterocyclic rings	Glycine & Succinyl-CoA	Widely found in animals	Orthoptera, Phasmatodea, Mantodea, Hemiptera, Lepidoptera, Diptera	Soluble in acidic organic solvents (e.g., 2% HCl in acetone), strong bases and acids.	Blue, Green, Red, Yellow, Brown	Often bound to proteins and present in a soluble form.
Carotenoids (eg, β-carotene, α-carotene, Lutein, Zeaxanthin, Astaxanthin)	C_{40} skeleton (tetraterpenoids)	Acetyl-CoA, Isoprene	Universally found in animals and plants. Most animals cannot be synthesized and are obtained from plants.	Orthoptera, Phasmatodea, Hemiptera, Coleoptera, Lepidoptera, Diptera (also detected in many insects regardless of body color)	Soluble in organic solvents such as acetone, hexane, chloroform and benzene. Insoluble or poorly soluble in water. Carotenes are insoluble in alcohol, whereas xanthophylls are soluble in alcohol.	Yellow, Orange, Red, Red-purple, Green, Blue	Animals do not have biosynthetic pathways, but some insects, such as aphids, acquire carotenoid synthetases by horizontal transfer.
Flavonoids (eg, Anthocyanin, Flavone, Quercetin, Quercetin-β-3-O-glucoside)	C_{15} skeleton (two phenyl rings and a heterocyclic ring)	Phenylalanine	Plants, insects (derived from plants), coelenterates, mollusks	Orthoptera, Hymenoptera, Coleoptera, Lepidoptera	Soluble in alcohol and other polar organic media. Moderately soluble in water.	Red, Purple, Blue, Yellow, White	Since the color changes with pH, it is roughly detectable by the color changes on exposure of ammonia fumes. Often found as a glycoside in plants.
Papiliochromes (eg, Papiliochrome II, Papiliochrome R)	Kynurenine combined with N-β-alanyl-dopamine (NBAD)	Tryptophan, Tyrosine, and β-alanine	Insects (papilionid butterflies)	Lepidoptera (Papilionidae)	The pale-yellow pigments are soluble in water and 70% EtOH. The deep yellow pigments are soluble in 4% methanol. The red pigments are soluble in 1N NaOH.	Yellow, Red, White	Produced by joining of melanin precursor and ommochrome precursor.
Quinones (e.g., Aphines, Anthraquinones)	Quinone skeleton (having two oxygen atoms bonded to the benzene ring)	Linear polyketides	Sea urchins, sea lilies, insects (aphids, scale insects, butterflies)	Hemiptera (Aphidoidea, Coccoidea), Lepidoptera	Soluble in organic solvents such as ether, acetone and alcohol.	Green, Yellow, Red, Red-purple	The color changes due to pH and redox reaction. Some butterfly red pigments are thought to be β-alanine-containing quinone pigments.

be dopa-melanin from 3,4-dihydroxyphenylalanine (DOPA) like vertebrates (Gibert et al. 2007; Arakane et al. 2009). In addition, *N-β*-alanyldopamine (NBAD) and *N*-acetyldopamine (NADA), which are derived from dopamine, are involved in the hardening of the cuticle, as well as reddish and yellowish coloration (Fig. 1.1; Wright 1987; Arakane et al. 2016). Moreover, pheomelanin, which contains cysteine and is widely found in vertebrates, has recently been reported as light-colored

Table 1.2 Details of distribution of pigments among insects

	Melanins	Ommochromes	Pteridines	Tetrapyrroles	Carotenoids	Flavonoids	Papiliochromes	Quinones
Odonata (dragonflies & damselflies)	Cuticles (Body & Wings)	Compound eyes, Epidermis, Wings	Epidermis					
Orthoptera (grasshoppers & crickets)	Cuticles (Body & Wings)	Compound eyes, Epidermis	Epidermis	Epidermis	Epidermis	Wings		
Phasmatodea (stick insects)		Compound eyes, Epidermis		Epidermis	Epidermis			
Mantodea (praying mantids)		Compound eyes, Epidermis	Epidermis	Epidermis				
Blattodea (cockroaches & termites)	Cuticles (Body & Wings)	Compound eyes						
Hemiptera (shield bugs, aphids & scale insects)	Cuticles (Body & Wings)	Compound eyes	Epidermis	Hemolymph, Fat body	Epidermis, Hemolymph			Hemolymph
Hymenoptera (sawflies, wasps, bees & ants)	Cuticles (Body & Wings)	Compound eyes	Epidermis			Hemolymph		
Neuroptera (lacewings)		Compound eyes, Epidermis						
Coleoptera (beetles)	Cuticles (Body & Wings)	Compound eyes			Epidermis, Wings	Fat body		
Lepidoptera (butterflies & moths) [adult]	Cuticles (Body & Wings)	Compound eyes, Epidermis, Wings	Wings, Compound eyes	Wings	Hemolymph, Wings	Wings	Wings	Wings
Lepidoptera (butterflies & moths) [larva]	Cuticles (Body & Wings)	Compound eyes, Epidermis	Epidermis	Epidermis, Hemolymph	Epidermis, Hemolymph, Wings	Epidermis		
Mecoptera (scorpionflies)	Cuticles (Body & Wings)		Epidermis					
Diptera (true flies)	Cuticles (Body & Wings)	Compound eyes	Compound eyes	Hemolymph	Hemolymph			

pigments in several insects (Galván et al. 2015; Barek et al. 2018), although the relationship between pheomelanin, NBAD-melanin, and NADA-melanin has not been investigated in detail.

Insect melanin synthesis pathway can be divided into two different steps: catecholamine (melanin precursors) production in epidermal cells and pigment precursor oxidation in the cuticle. In the first step, tyrosine is converted to DOPA by tyrosine hydroxylase (TH, encoded by the *pale* gene in the fruit fly *Drosophila melanogaster*). Tyrosine can be obtained from food as well as synthesized from phenylalanine by phenylalanine hydroxylase (PAH, encoded by the *Henna* gene in *Drosophila*). In some insects such as weevils, a large amount of tyrosine is supplied by symbiotic bacteria (Vigneron et al. 2014; Anbutsu et al. 2017). It should be noted that both TH and PAH require tetrahydrobiopterin (BH_4), which is produced by pteridine synthesis enzymes, as a cofactor (Fitzpatrick 2003) (Fig. 1.1). In many cases, DOPA is subsequently converted to dopamine by dopa decarboxylase (DDC, encoded by the *Ddc* gene in *Drosophila*). Loss of function of TH or DDC causes a defect in overall melanization in many insects (True et al. 1999; Futahashi and Fujiwara 2005; Ninomiya et al. 2006; Gorman and Arakane 2010; Liu et al. 2010,

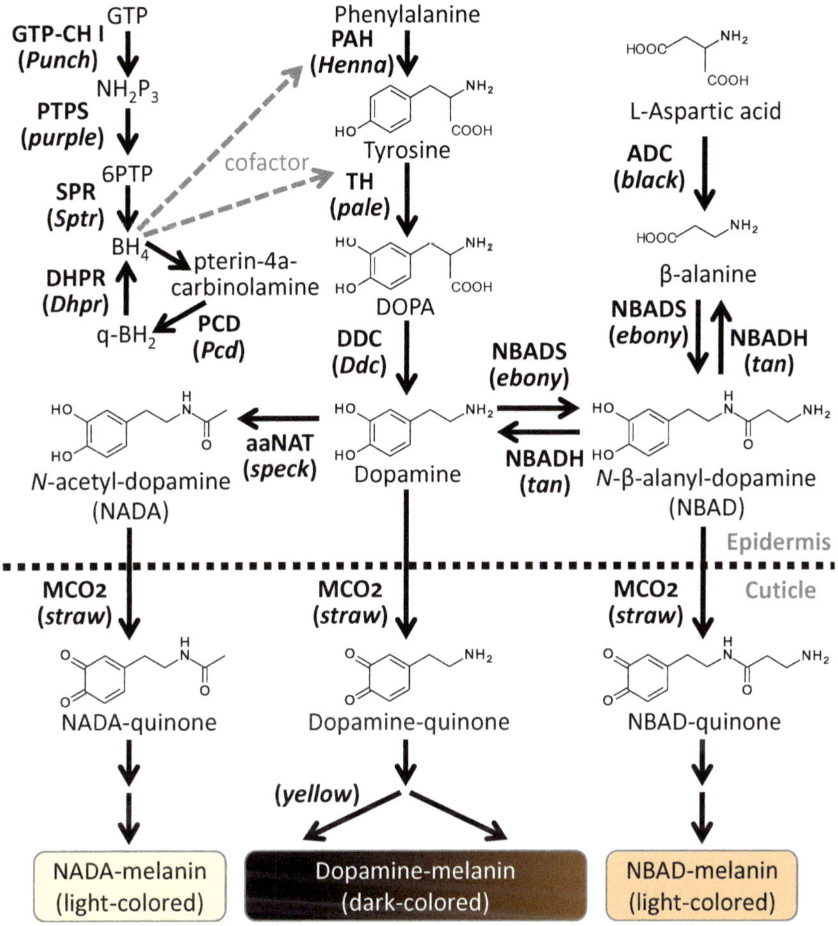

Fig. 1.1 The melanin synthesis pathway in insects. Key enzymes are shown in bold and the *Drosophila* genes that encode them are shown in parenthesis. The precise function of the *yellow* gene product remains unclear. *GTP* guanosine triphosphate, *GTP-CH I* GTP cyclohydrolase I, NH_2P_3 dihydroneopterin triphosphate, *PTPS* 6-pyruvoyl tetrahydropterin synthase, *6PTP* 6-pyruvoyl tetrahydropterin, *SPR* sepiapterin reductase, BH_4 tetrahydrobiopterin, $q\text{-}BH_2$ quinonoid dihydrobiopterin, *PCD* pterin-4a-carbinolamine dehydratase, *DHPR* dihydropteridine reductase, *PAH* phenylalanine hydroxylase, *TH* tyrosine hydroxylase, *DOPA* 3,4-dihydroxyphenylalanine, *DDC* dopa decarboxylase, *ADC* aspartate 1-decarboxylase, *aaNAT* arylalkylamine-*N*-acetyltransferase, *NBADS* *N*-β-alanyldopamine synthase, *NBADH* *N*-β-alanyldopamine hydrolase, *MCO2* multicopper oxidase 2 (also known as laccase 2)

2014; Xiao et al. 2020). NBAD and NADA are produced from dopamine by NBAD synthase (NBADS, encoded by the *ebony* gene in *Drosophila*) or arylalkylamine *N*-acetyltransferase (aaNAT, encoded by the *speck* gene in *Drosophila*) (Wright 1987; Arakane et al. 2016; Spana et al. 2020). For production of NBAD, aspartate decarboxylase (ADC, encoded by the *black* gene in *Drosophila*) converts L-aspartic acid to β-alanine. In contrast, NBAD hydrolase (NBADH, encoded by the *tan* gene in *Drosophila*) catalyzes NBAD to dopamine and β-alanine (Fig. 1.1) (True et al. 2005). In the second step, melanin precursors such as dopamine, NBAD, and NADA are secreted into cuticles and oxidized by granules containing laccase-type multicopper oxidase 2 (MCO2, also known as laccase2, encoded by the *straw* gene in *Drosophila*) (Fig. 1.1). The gene encoding MCO2 is essential for cuticular pigmentation across insects (Arakane et al. 2005; Futahashi et al. 2011; Riedel et al. 2011; Masuoka and Maekawa 2016; Okude et al. 2017; Zhang et al. 2019b; Nishide et al. 2020; Peng et al. 2020). In this step, color variation such as black and brown occurs depending on the *yellow* family gene product in the cuticle. The distribution of melanin precursors seems to differ between the cuticle layers; in the yellow mealworm *Tenebrio molitor*, loss of function of *black* and *ebony* genes (i.e., NBAD deficiency) causes darkening of the brownish outer exocuticle, whereas aaNAT deficient adults (i.e., NADA deficiency) promote pigmentation of a transparent inner endocuticle (Mun et al. 2020).

Genes involved in melanin synthesis are well conserved among insects, but the names of some genes are occasionally confused. The enzyme that converts tyrosine to DOPA in epidermal cells is TH, whereas tyrosinase-type phenol oxidase (PO) catalyzes the same reaction in insect innate immunity (Ashida 1990; Sugumaran 2002), and the term PO or tyrosinase is sometimes used when it should be described as TH. In addition, the enzyme for the oxidation of pigment precursors in the cuticle has been also previously described as PO, instead of MCO2 (laccase2).

There are several *yellow* family genes and *aaNAT* family genes in insects. Although some *yellow* family genes have been reported as dopachrome-converting enzymes for melanin biosynthesis (e.g., *yellow-f* and *yellow-f2* genes in *D. melanogaster*, Han et al. 2002), the precise function of many *yellow* family genes including typical *yellow* (also described as *yellow-y*) gene remains unclear (Drapeau 2001). In many insects, *yellow* gene is necessary for the production of black melanin (Geyer et al. 1986; Wittkopp et al. 2002a; Futahashi et al. 2008; Arakane et al. 2010), and mutant analysis suggests that the *yellow* gene functions mainly downstream of dopamine (Walter et al. 1991; Gibert et al. 2007). It has been hypothesized that a yellow protein plays a structural role in anchoring dark-colored melanin pigments or pigmentation enzymes in the cuticle (Hinaux et al. 2018). In the Asian tiger mosquito *Aedes albopictus*, *yellow-g* and *yellow-g2* genes regulate the timing of eggshell pigmentation and morphology (Noh et al. 2020). Some *yellow* family genes are associated with light coloration. For example, the *yellow-e* gene suppresses darkening in both the silkworm *Bombyx mori* and the red flour beetle *Tribolium castaneum* (Ito et al. 2010; Noh et al. 2015). It should be noted that the *yellow-e* gene is also involved in the water proofing function of the adult cuticle in *T. castaneum* (Noh et al. 2015).

Among several *aaNAT* family genes in *Drosophila*, *speck* gene (also known as *aaNAT1* or *Dat* gene) is involved in pupal coloring, but not in adult or larval body coloring or cuticle hardening excepting the wing hinge region (Brodbeck et al. 1998; Ahmed-Braimah and Sweigart 2015; Spana et al. 2020). In contrast, the mutant of the *aaNAT1* gene homologue of the silkworm *B. mori* shows an overall increase in blackness in adults, whereas only a small region such as the tail spots are affected in larvae (Dai et al. 2010; Zhan et al. 2010). In the beetles *T. castaneum* and *T. molitor*, loss of function of *aaNAT1* gene promotes darkening of adult body coloration (Noh et al. 2016; Mun et al. 2020).

1.2.2 Melanin Pigments for Insect Color Pattern

The relationship between melanin pigments and the diversity of insect body color patterns has been well characterized. Three genes, *yellow*, *ebony*, and *tan*, are often involved in the diversification of body color in the fruit flies *Drosophila* species. For example, in *D. biarmipes* with black wing markings, *yellow* gene is highly expressed in the dark marking regions where *ebony* expression is suppressed (Wittkopp et al. 2002a). The polka-dotted fruit fly *D. guttifera* has a unique black pattern on the wing, and *yellow* gene is specifically expressed in the regions that become black (Gompel et al. 2005; Werner et al. 2010, see Chap. 12). The expression pattern of the *yellow* gene is also related to the differences of the abdominal black color pattern among the three *Drosophila* species; *D. melanogaster*, *D. virilis*, and *D. subobscura* (Wittkopp et al. 2002b). In contrast, the difference in abdominal black color between *D. americana* and *D. novamexicana* is not related to the *yellow* gene but to the difference in expression of the *ebony* and *tan* genes (Wittkopp et al. 2003, 2009). Among *Drosophila* natural populations, *ebony* gene is also involved in differences in thoracic and abdominal black patterns (Takahashi et al. 2007; Rebeiz et al. 2009; Takahashi and Takano-Shimizu 2011). In addition, the difference between *D. yabuka* with dark spots on the abdomen and *D. santomea* without dark spots is mainly caused by a mutation in the promoter region of *tan* gene in *D. santomea*, which prevents the local expression of *tan* gene (Jeong et al. 2008). One of the reasons why these three genes are often involved in variations in body color and patterns is that mutations in these genes do not significantly affect survival. However, it should be noted that the mutations in *ebony* or *tan* gene affect cuticular hydrocarbon (CHC) composition (Massey et al. 2019a), whereas the mutations in *yellow* gene suppress the pigmentation of male-specific leg structures called sex combs, reducing male mating success in *D. melanogaster* (Massey et al. 2019b).

Melanin synthesis genes are also involved in the color pattern diversity of butterflies and moths. In the swallowtail butterfly *Papilio xuthus,* young larvae mimic bird dropping, whereas final instar larvae show green camouflage color with thoracic eyespots and abdominal V-shaped markings. Prior to each ecdysis, the following seven melanin synthesis or related genes; *TH*, *DDC*, *MCO2* (*laccase2*), *tan*, *yellow*, *yellow-h3*, and *GTP cyclohydrolase I* (*GTP-CH I*, a key enzyme in the synthesis of TH cofactor BH_4) are specifically expressed in the regions that become black, whereas the *ebony* gene shows specific expression in the red eyespot region (Fig. 1.2) (Futahashi and Fujiwara 2005, 2006, 2007, 2008a; Futahashi et al. 2010, 2012a). Moreover, the expression patterns of these genes are consistent with the species-specific black patterns of the three *Papilio* species, *P. xuthus*, *P. machaon*, and *P. polytes* (Fig. 1.2) (Shirataki et al. 2010). Pattern-specific expression of *TH*, *DDC*, *MCO2* (*laccase2*), *tan*, and *yellow* has been also

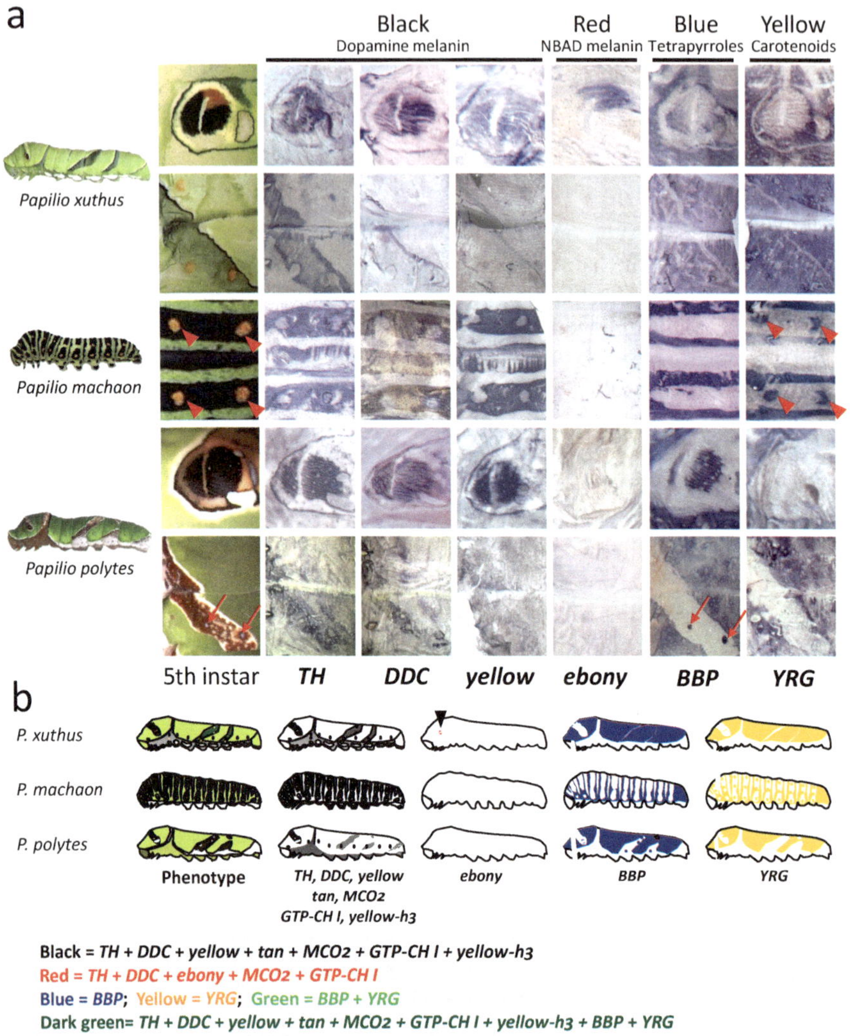

Fig. 1.2 Summary of species-specific gene expression involved in larval color pattern among three *Papilio* species. (**a**) Results of in situ hybridization. Red arrowheads indicate yellow spot regions of *P. machaon* and red arrows indicate blue spot regions of *P. polytes*. (**b**) Schematic diagram of mRNA expression. Black arrowhead indicates red region of eyespot in *P. xuthus* (Figure modified from Shirataki et al. 2010)

reported in the larvae of the silkworm *B. mori* and the armyworm *Pseudaletia separata* (Ninomiya et al. 2006; Futahashi et al. 2008, 2010). In *B. mori*, three genes *TH*, *tan*, and *yellow* are responsible for the light-colored larval mutants, *sex-linked chocolate*, *rouge*, and *chocolate*, respectively, whereas *ebony* gene is responsible for the dark-colored larval mutant *sooty* (Table 1.3) (Futahashi et al. 2008, 2010; Liu et al. 2010). Hence, the pattern-specific expression of the melanin synthesis gene is important for the formation of species-specific larval black patterns in lepidopteran insects. Notably, the two NBAD synthesis genes *black* and *ebony* are responsible for melanized pupal mutants of the silkworm *black pupa* and *sooty*, respectively (Table 1.3) (Futahashi et al. 2008; Dai et al. 2015).

In the pigmentation of adult butterfly wings, the pattern-specific expression of the *DDC* and *ebony* genes is involved in the formation of yellow and black patterns of the tiger swallowtail butterfly *Papilio glaucus* (Koch et al. 1998, 2000). In the *Heliconius* butterflies, *ebony* and *yellow-d* genes are highly expressed in the presumptive red region, whereas *tan* and *yellow-h3* genes are highly expressed in the black region (Ferguson et al. 2011a, 2011b; Hines et al. 2012). In the painted lady butterfly *Vanessa cardui*, *ebony*, *black*, and *yellow-d* genes are highly expressed in the presumptive red region, whereas *tan* gene is highly expressed in the black region (Zhang et al. 2017a). CRISPR/Cas9-based mosaic analysis revealed that *TH*, *DDC*, and *yellow* knockouts displayed reduction of dark-colored melanin pigmentation, whereas *yellow-d*, *ebony*, and *black* knockouts affected the hue of red, brown, and ochre pigmentation (Zhang et al. 2017a). Moreover, the mutation of *yellow* and *DDC* genes affects not only pigmentation but also scale structures in the mycalesine butterfly *Bicyclus anynana* (Matsuoka and Monteiro 2018).

Similarly, in coleopteran, hemipteran and blattodean insects, *TH*, *DDC*, and *yellow* genes are involved in dark-colored pigmentation, whereas *black*, *ebony*, and *aaNAT* genes are involved in light-colored pigmentation (Liu et al. 2014, 2016; Zhang et al. 2019a, b). Unlike vertebrates, there are very few reports on the relationship between melanins and eye coloration, although melanin is detected in the facet walls of the compound eyes in crane flies (Lindgren et al. 2019).

1.2.3 Application of Melanin Synthesis Gene to Transgenic Marker

With the advent of genome-editing technologies, it has become possible to produce transgenic individuals even in insects other than model species (e.g., *Drosophila*). In *Drosophila* species, recombinants are often selected by a visible genetic marker using a color mutant strain. However, many other insects do not have an appropriate mutant strain that can visually discriminate the recombinant; therefore, the recombinants are selected using fluorescent protein genes (e.g., EGFP) that is connected to an artificial promoter expressed in the compound eyes. In this case, however, a fluorescence microscope is required for screening, and it is difficult to judge the fluorescence signal when covered with dark-color pigments of the compound eyes, often present in wild type individuals.

In the silkworm *B. mori*, ubiquitous overexpression of *aaNAT* (*aaNAT1*) transgene changes the body color of the newly hatched larvae from black to light brown in a genetically dominant manner (Fig. 1.3) (Osanai-Futahashi et al. 2012b). Color changes due to overexpression of *aaNAT* gene can be observed in all larval instars and adults. In addition, overexpression of *ebony* gene suppresses the melanization of older instars, although the alteration of the pigmentation of first instar larvae is slight,

Fig. 1.3 Effect of ubiquitous overexpression of the silkworm *aaNAT* transgene together with fluorescent protein gene GFP (*IE1-NAT/3xP3EGFP*) in *B. mori*, *H. axyridis*, and *D. melanogaster*. Arrows indicate recombinants (figure modified from Osanai-Futahashi et al. 2012b)

probably due to the lack of β-alanine (Osanai-Futahashi et al. 2012b). Notably, overexpression of the silkworm *aaNAT* gene in the fruit fly *D. melanogaster* and the ladybird beetle *Harmonia axyridis* significantly reduce melanization in both species, implying the potential usefulness of this gene as a transgenic marker for diverse insect taxa (Fig. 1.3) (Osanai-Futahashi et al. 2012b).

1.3 Ommochromes

1.3.1 Types and Characteristics of Ommochromes

Ommochromes are pigments formed by oxidative condensation of 3-hydroxykynurenine, and are involved in red, brown, orange, yellow, and purple coloration of many insects. Most ommochromes are generally categorized into two types: ommatins, which are of low molecular weight, dialyzable and unstable in alkali; and ommins, which are of slightly higher molecular weight and contain sulfur element, more stable in alkali than ommatin, and cannot be dialyzed. Ommatins include xanthommatin, decarboxylated-xanthommatin, rhodommatin, and ommatin D; while ommin A is the only ommin which chemical structure has been determined (Fig. 1.4). In addition, another ommochrome pigments called ommidines, which

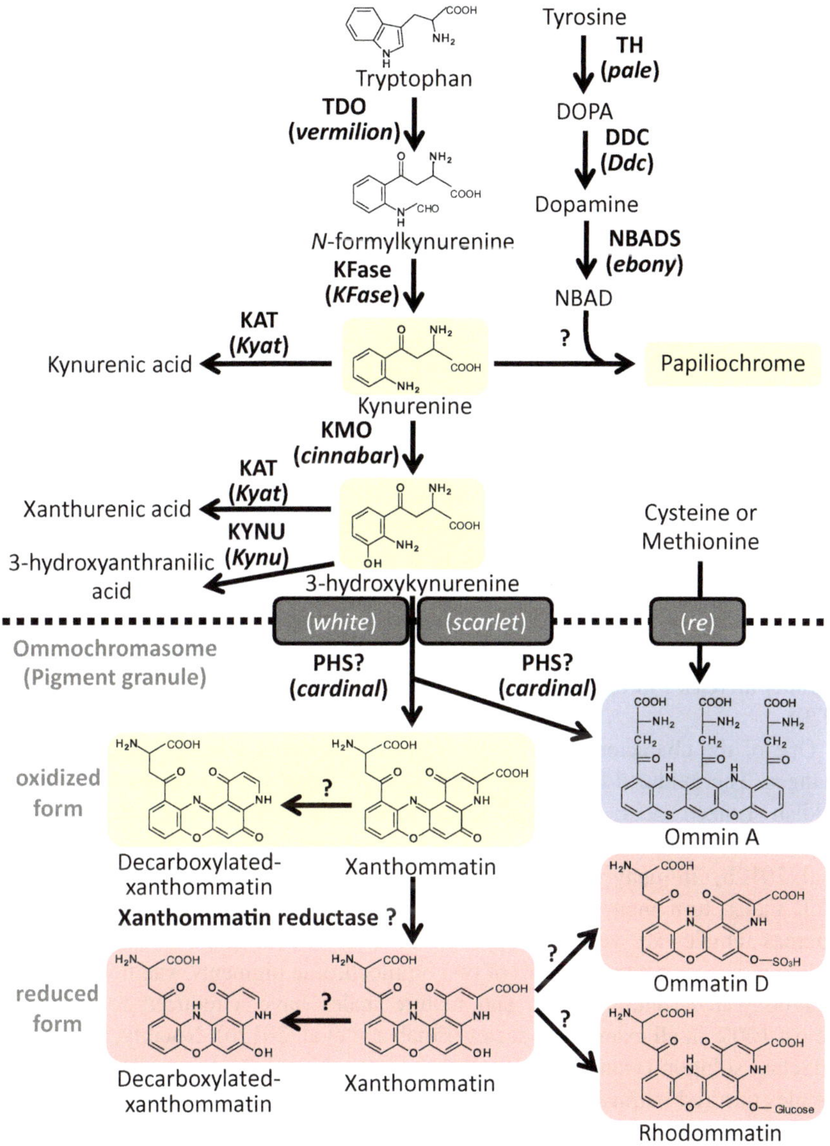

Fig. 1.4 The ommochrome synthesis pathway in insects. Key enzymes are shown in bold and *Drosophila* or *Bombyx* genes that encode them are shown in parenthesis. Three transporter genes *white*, *scarlet*, and *re* are involved in the uptake of ommochrome precursors into ommochromasomes. *Kynu* and *re* genes are absent in *Drosophila*. *TDO* tryptophan 2,3-dioxygenase, *KFase* kynurenine formamidase, *KAT* kynurenine aminotransferase, *KMO* kynurenine 3-monooxygenase, *KYNU* kynureninase, *PHS* phenoxazinone synthase, *TH* tyrosine hydroxylase, *DOPA* 3,4-dihydroxyphenylalanine, *DDC* dopa decarboxylase, *NBADS* *N*-β-alanyldopamine synthase, *NBAD* *N*-β-alanyldopamine

exhibit an intermediate property between ommatins and ommins, has been reported from the orthopteran insects (e.g., grasshoppers, locusts), although their structures have not been elucidated (Linzen 1974; Figon and Casas 2019). Ommochromes are ubiquitous in the compound eyes of arthropods including insects and are also involved in wing and body coloration in some species (Table 1.2) (e.g., nymphalid butterflies, red dragonflies). Other than insects, ommochromes are widely detected in the protostomian integuments and eyes of cephalopods such as squids and octopuses (Figon and Casas 2019; Crawford et al. 2020). By expanding or shrinking the ommochrome-containing granules, squids and octopuses can rapidly change their body color (Figon and Casas 2019).

1.3.2 Color Change by Ommochrome Pigments

Several insects change body color by migration of ommochrome-containing pigment granules (Umbers et al. 2014). For example, some dragonflies change body color by migration of ommochrome granules in a temperature-dependent manner (Veron et al. 1974; Okude and Futahashi 2021). The Indian stick insect *Carausius morosus* changes its body color by visual stimulation. When the ventral surface of the compound eye is covered with black varnish, the amount of xanthommatin and ommins increases in epidermis, and the body color turns darker (Bückmann 1977, 1979).

One of the characteristics of ommochrome pigments is redox-dependent color changes. The oxidized form is generally yellow, whereas the reduced form is more reddish (Linzen 1974; Figon and Casas 2019). This redox-based color change is responsible for the sex-specific adult color transition in red dragonflies (Futahashi et al. 2012b). In many red dragonflies of the genera *Crocothemis* and *Sympetrum*, adult males turn their body color from yellow to red upon sexual maturation, whereas females are yellowish throughout their lifetime (Fig. 1.5). Both red and yellow pigments are the mixture of two ommochrome pigments, xanthommatin and decarboxylated-xanthommatin, and mature males have a reduced form ratio of almost 100% in all examined species (Futahashi et al. 2012b). Notably, by injecting reductant solution (vitamin C) into the abdomen, both immature males and mature females turn red within several hours, like mature males (Fig. 1.5).

In the fruit fly *D. melanogaster*, compound eyes contain a high amount of reduced ommochrome pigment (xanthommatin), and a significant activity of xanthommatin reductase has been detected (Santoro and Parisi 1986). Although the substrate specificity of the enzyme and the responsible gene for the reductase are still unknown, it is plausible that male-specific expression of similar gene may be involved in the sex-specific color transition of red dragonflies. It should be noted that gynandromorphic dragonflies exhibit discontinuous yellow/red mosaicism (Fig. 1.5), suggesting that a cell-autonomous mechanism is involved in the regulation of redox status of ommochromes.

Fig. 1.5 Sex-specific adult color transition and reductant-induced color change in red dragonflies. Arrows indicate the injection sites of the reductant (figure modified from Futahashi et al. 2012b; Futahashi et al. 2019; Futahashi 2017). Photo of gynandromorphic dragonfly courtesy of Kohji Tanaka

1.3.3 Genes Involved in Ommochrome Synthesis and Transport

The compound eyes of *D. melanogaster* contain xanthommatin, an ommochrome pigment, and drosopterins, which consist of at least five kinds of pteridine compounds (Ferré et al. 1986; Kim et al. 2013). Ommochrome-related genes have been elucidated mainly in *D. melanogaster* because a defect in ommochrome synthesis or transport causes a change in the color of the compound eyes. Genes involved in *Drosophila* eye color has been classified into three groups according to their function: "pigment synthesis group," "ABC transporter group," and "granule group" (Lloyd et al. 1998) (Tables 1.3 and 1.4). In ommochrome synthesis, tryptophan is converted to an intermediate precursor 3-hydroxykynurenine via formylkynurenine and kynurenine by tryptophan 2,3-dioxygenase (TDO, encoded by the *vermilion* gene in *Drosophila*), kynurenine formamidase (encoded by the *KFase* gene in *Drosophila*), and kynurenine 3-monooxygenase (KMO, encoded by the *cinnabar* gene in *Drosophila*) (Fig. 1.4) (Searles et al. 1990; Warren et al. 1996; Han et al. 2012). Subsequently, 3-hydroxykynurenine is incorporated into pigment granules, namely ommochromasomes, which resemble vertebrate melanosomes, by heterodimers of two ABC transporters (encoded by the *white* and *scarlet* genes in *Drosophila*) and converted into final pigments within the granules (O'Hare et al. 1984; Tearle et al. 1989). The heam peroxidase-type putative phenoxazinone

Table 1.3 Responsible genes for color mutants in the fruit fly *Drosophila melanogaster* and the silkworm *Bombyx mori* (related to pigment synthesis)

	Fruit fly (*Drosophila melanogaster*) gene	Affected pigments in *Drosophila* mutant	Silkworm (*Bombyx mori*) mutant	Affected pigments in *Bombyx* mutant
Pigment synthesis (enzyme)	*pale* (*Tyrosine hydroxylase, TH*)	dark-colored melanin	*sex linked chocolate* (*sch*)	dark-colored melanin
	Dopa decarboxylase (*Ddc*)	dark-colored melanin		
	straw (*multicopper oxidase 2, MCO2, laccase2*)	dark-colored melanin		
	tan (*N-β-alanyldopamine hydrolase, NBADH*)	dark-colored melanin	*rouge* (*ro*)	dark-colored melanin
	black (*aspartate 1-decarboxylase, ADC*)	light-colored melanin	*black pupa* (*bp*)	light-colored melanin
	ebony (*N-β-alanyldopamine synthase, NBADS*)	light-colored melanin	*sooty* (*so*)	light-colored melanin
	speck (*Arylalkylamine N-acetyltransferase 1, aaNAT1, Dat*)	light-colored melanin	*melanism* (*mln*)	light-colored melanin
	vermilion (*tryptophan 2,3-dioxygenase, TDO*)	ommochrome		
	Kynurenine formamidase (*KFase, Arylformamidase, AFMID*)	(Not reported)		
	cinnabar (*kynurenine 3-monooxygenase, KMO*)	ommochrome	*white egg 1* (*w-1*)	ommochrome
	cardinal (*phenoxazinone synthase*)	ommochrome	*pink-eyed white egg* (*pe*)	ommochrome
	Kynurenine aminotransferase (*Kyat*)	(Not reported)		
	Absent		*red blood* (*rb, Kynureninase, KYNU*)	ommochrome
	Punch (*GTP cyclohydrolase I, GTP-CH I*)	pteridine, dark-colored melanin		
	purple (*6-pyruvoyl tetrahydropterin synthase, PTPS*)	pteridine	*albino* (*al*)	dark-colored melanin, pteridine
	Sepiapterin reductase (*Sptr*)	(Not reported)	*lemon* (*lem*)	pteridine, dark-colored melanin
	pterin-4α-carbinolamine dehydratase (*Pcd*)	(Not reported)		
	Dihydropteridine reductase (*Dhpr*)	(Not reported)		
	Dihydrofolate reductase (*Dhfr*)	(Not reported)		
	sepia (*pyrimidodiazepine synthase, PDAS*)	pteridine		
	clot (*Thioredoxin domain-containing protein 17*)	pteridine		
	Dihydropterin deaminase (*DhpD*)	pteridine		
	Henna (*phenylalanine hydroxylase, PAH*)	pteridine		
	rosy (*xanthine dehydrogenase, XDH*)	pteridine	*q-translucent* (*oq*)	uric acid
Pigment synthesis related (yellow family)	*yellow* (*yellow-y*)	dark-colored melanin	*chocolate* (*ch*)	dark-colored melanin
	yellow-b	(Not reported)		
	yellow-c	(Not reported)		
	yellow-d	(Not reported)		
	yellow-d2	(Not reported)	Absent	
	yellow-e	(Not reported)	*brown head and tail* (*bts*)	lighr-colored melanin
	yellow-e2	(Not reported)	Absent	
	yellow-e3	(Not reported)	Absent	
	yellow-f	(Not reported)		
	yellow-f2	(Not reported)		
	yellow-g	(Not reported)	Absent	
	yellow-g2	(Not reported)	Absent	
	yellow-h	(Not reported)		
	yellow-k	(Not reported)	Absent	
(GTP synthesis)	*Phosphoribosylamidotransferase* (*Prat*)	(Not reported)		
	Phosphoribosylformylglycinamidine synthase (*Pfas*)	pteridine		
	GART trifunctional enzyme (*Gart*)	pteridine		
	PAICS bifunctional enzyme (*Paics*)	pteridine		
	Adenylosuccinate Lyase (*AdSL*)	pteridine		
	CG11089 (*AICAR transformylase, atic*)	(Not reported)		
	raspberry (*IMP dehydrogenase*)	pteridine		
	burgundy (*GMP synthetase*)	pteridine		
(MoCo synthesis)	*Molybdenum cofactor synthesis 1* (*Mocs1*)	pteridine	*Yamamoto translucent* (*oya*)	uric acid
	Molybdenum cofactor synthesis 2A (*Mocs2A*)	(Not reported)		
	Molybdenum cofactor synthesis 2B (*Mocs2B*)	(Not reported)		
	Ubiquitin activating enzyme 4 (*Uba4, Mocs3*)	(Not reported)		
	cinnamon (*Gephyrin*)	pteridine	*Hime'nichi translucent* (*ohi*)	uric acid
	maroon-like (*Moco sulfurase*)	pteridine	*giallo ascoli translucent* (*og*) *Ozaki's translucent* (*og(Z)*)	uric acid
(Carboxypeptidase)	*silver*	dark-colored melanin		
(Phosphoesterase)	*prune*	pteridine		
(Guanylyl cyclase)	*CG31183*	(Not reported)	*quail* (*q*)	dark-colored melanin, ommochrome, pteridine

Table 1.4 Responsible genes for color mutants in the fruit fly *Drosophila melanogaster* and the silkworm *Bombyx mori* (related to pigment precursor transport and pigment granule formation)

	Fruit fly (*Drosophila melanogaster*) gene	Affected pigments in *Drosophila* mutant	Silkworm (*Bombyx mori*) mutant	Affected pigments in *Bombyx* mutant
Pigment precursor transport (ABC transporter)	*white*	ommochrome, pteridine	*white egg3* (*w-3*)	ommochrome, uric acid
	scarlet	ommochrome	*white egg 2* (*w-2*)	ommochrome
	brown	pteridine		(riboflavin)
	Absent		*kinshiryu translucent* (*ok*)	uric acid
(MFS transporter)	*karmoisin*	ommochrome		
	Absent		*red egg* (*re*)	ommochrome
	CG14439	(Not reported)	*cheek and tail spots* (*cts*)	dark-colored melanin
	monocarboxylate transporter	(Not reported)	*chinese translucent* (*oc*)	uric acid
(Amino acid transporter)	*mahogany*	ommochrome, pteridine	*brown egg 4* (*b-4*)	ommochrome
	Absent		*sex-linked translucent* (*os*)	uric acid, ommochrome
Pigment granule formation (AP-3 complex)	*ruby* (*AP-3β*)	ommochrome, pteridine		
	garnet (*AP-3δ*)	ommochrome, pteridine		
	carmine (*AP-3μ*)	ommochrome, pteridine		
	orange (*AP-3σ*)	ommochrome, pteridine		
(BLOC-1 complex)	*Biogenesis of lysosome-related organelles complex 1, subunit 1* (*blos1*)	ommochrome, pteridine		
	Biogenesis of lysosome-related organelles complex 1, subunit 2 (*blos2*)	(Not reported)	*distinct translucent* (*od*)	uric acid, ommochrome
	Biogenesis of lysosome-related organelles complex 1, subunit 3 (*blos3*)	(Not reported)		
	Biogenesis of lysosome-related organelles complex 1, subunit 4 (*blos4*)	ommochrome, pteridine	Absent	
	Dysbindin	ommochrome, pteridine	*mottled translucent of Var* (*ov*)	uric acid
	Pallidin	ommochrome, pteridine		
	Muted	(Not reported)	*Tanaka's mottled translucent* (*otm*)	uric acid
	SNAP associated protein (*Snapin*)	(Not reported)		
(BLOC-2 complex)	*pink* (*Hermansky-Pudlak Syndrome 5, HPS5*)	ommochrome, pteridine	*aojuku translucent* (*oa*)	uric acid
	CG14562 (*Hermansky-Pudlak Syndrome 3, HPS3*)	(Not reported)	Absent	
(BLOC-3 complex)	*Hermansky-Pudlak Syndrome 4* (*HPS4*)	ommochrome, pteridine		
	Hermansky-Pudlak Syndrome 1 (*HPS1*)	(Not reported)		
(HOPS complex)	*carnation* (*Vps33A*)	ommochrome, pteridine		
	deep orange (*Vps18*)	ommochrome, pteridine		
	light (*Vps41*)	ommochrome, pteridine		
	maroon (*Vacuolar protein sorting 16A, Vps16a*)	ommochrome, pteridine		
	Vacuolar protein sorting 11 (*Vps11*)	(Not reported)		
	Vacuolar protein sorting 39 (*Vps39*)	(Not reported)		
(HOPS-interacting protein)	*UV-resistance associated gene* (*UVRAG, Vps38*)	ommochrome, pteridine		
(Rab38/Rab32)	*lightoid* (*Rab32*)	ommochrome, pteridine		
(RabGEF)	*claret*	ommochrome, pteridine		
(VATPase)	*chocolate* (*VhaAC39-1-PA*)	ommochrome, pteridine		
(LysM domain protein)	*red Malpighian tubules*	ommochrome, pteridine		
(BEACH-domain containing protein)	*mauve*	ommochrome, pteridine		
(Methylene-tetrahydrofolate dehydrogenase)	*pugilist*	ommochrome, pteridine		
(Varp)	Absent		*waxy translucent* (*ow*)	uric acid
(Unannotated)	*purpleoid*	ommochrome, pteridine		

synthase (PHS, encoded by the *cardinal* gene in *Drosophila*) catalyzes the conversion of 3-hydroxykynurenin to xanthommatin (Harris et al. 2011). Ommochrome precursors kynurenine and 3-hydroxykynurenine can be converted to kynurenic acid and xanthurenic acid, respectively, by kynurenine aminotransferase (KAT, encoded by the *Kyat* gene in *Drosophila*) (Navrotskaya et al. 2018), although this process may not involve ommochromasomes.

In the silkworm *B. mori*, *cardinal* homolog gene is responsible for the egg and eye color mutant *pink-eyed white egg* (*pe*) (Fig. 1.6), which accumulates 3-hydroxykynurenin and reduces the total amount of ommins and ommatins

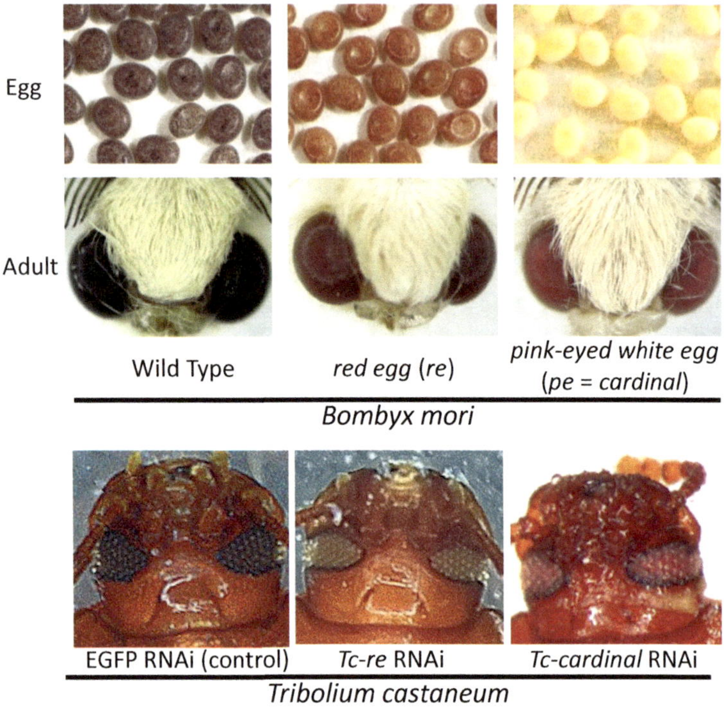

Fig. 1.6 Egg and eye color mutants in *Bombyx mori* and RNAi phenotype of their ortholog in *Tribolium castaneum* (figure modified from Osanai-Futahashi et al. 2012a; Osanai-Futahashi et al. 2016)

(xanthommatin and decarboxylated-xanthommatin) (Osanai-Futahashi et al. 2016), suggesting that *cardinal* gene is involved in the oxidation step to produce both ommins and ommatins (Fig. 1.4). Similarly, RNAi-based knockdown of *cardinal* gene causes reddish eye color instead of normal black or brown in the red flour beetle *T. castaneum* and the brown planthopper *Nilaparvata lugens* (Fig. 1.6) (Osanai-Futahashi et al. 2016; Liu et al. 2019). Within ommochromasomes, ommochrome-binding proteins are presumed to stabilize the reduced form of ommochromes (Linzen 1974; Figon and Casas 2019). It should be noted that ommochrome structures are modified (e.g., side-chain methylations) at visible light radiation and room temperature (Bolognese et al. 1988; Figon and Casas 2019).

Pigment granules containing ommochromes (ommochromasomes) and pteridines (pterinosomes) are lysosome-related organelles, and it has been elucidated that the formation of pigment granules involves genes required for lysosomal formation (e.g., Hermansky–Pudlak syndrome proteins) (Lloyd et al. 1998; Mullins et al. 2000; Falcón-Pérez et al. 2007; Cheli et al. 2010; Harris et al. 2011; Grant et al. 2016). The "granule group" genes and the ABC transporter *white* gene are involved not only in ommochrome pigments but also in pteridine pigments and uric acid (Table 1.4).

Some of the ommochrome mutants in *Drosophila* are known to have similar mutants in other insects. In particular, the analogy between the silkworm mutants and the *Drosophila* mutants has been pointed out from comparison of phenotypes and transplantation experiments (Kikkawa 1941). In the silkworm *B. mori*, ommochrome pigments are also present in the egg serosa, thus, a mutation in the ommochrome-related gene affects the color of the eggs as well as the compound eyes. In the past decades, genes responsible for several eye and/or egg color mutants of *B. mori* have been identified, and the relationships of *Drosophila* eye color mutants are becoming apparent (Tables 1.3 and 1.4). For example, the responsible genes for the white egg mutants *w-1*, *w-2*, and *w-3* of *Bombyx* are the orthologs of *Drosophila cinnabar*, *scarlet*, and *white*, respectively (Quan et al. 2002; Kômoto et al. 2009; Tatematsu et al. 2011). Similarly, the responsible genes for the eye color mutants *white*, *pearl*, and *ivory* of *Tribolium* are the orthologs of *vermilion*, *scarlet*, and *white* of *Drosophila*, respectively (Lorenzen et al. 2002; Grubbs et al. 2015). It should be noted that mutations in lysosome-related genes of *B. mori* exhibit phenotypes as "oily mutants," which display translucent epidermis instead of normal white coloration due to inability to accumulate white uric acid in the epidermis, rather than as eye and/or egg color mutants (Table 1.4) (Fujii et al. 2008, 2012; Wang et al. 2013a; Zhang et al. 2017c). In addition, some MFS transporter and amino acid transporter genes are involved in ommochrome and uric acid pigmentation (Table 1.4) (Kiuchi et al. 2011; Osanai-Futahashi et al. 2012a; Ito et al. 2012; Zhao et al. 2012; Grant et al. 2016; Wang et al. 2020).

1.3.4 Novel Ommochrome-Related Genes Discovered from the Silkworm

Several ommochrome synthesis mutants in silkworms have no analogous mutants in *Drosophila*. One example is *red blood*, a mutant that abnormally accumulates 3-hydroxykynurenine, and *Kynureninase* (*Kynu*) gene has been identified as the responsible gene for this mutant (Table 1.3; Fig. 1.4) (Meng et al. 2009a). This gene is not present in most insects other than silkworms, suggesting that it was acquired from bacteria by horizontal gene transfer.

In the compound eyes of most insects other than cyclorrhaphan flies (e.g., *Drosophila*), ommin is present in addition to xanthommatin. The silkworm mutant, namely *red egg* (*re*) is defective in ommin synthesis, exhibits reddish eye and egg color instead of black (Fig. 1.6). The gene responsible for *re* is a transporter gene belonging to MFS superfamily (Osanai-Futahashi et al. 2012a). Because ommin contains sulfur derived from cysteine or methionine, the gene responsible for *re* may function to incorporate these amino acids into ommochromasomes. The orthologs of the silkworm *re* gene are present in most insect species other than *Drosophila*, and RNAi-based gene knockdown changes the eye color of the red flour beetle *T. castaneum* from black to brown (Fig. 1.6), suggesting that this transporter gene is involved in the eye pigmentation in a wide range of insects (Osanai-Futahashi et al. 2012a). Interestingly, in *Drosophila* species, in which the ommin pigment is absent in the compound eye, the homolog of *re* gene is missing from the genome, suggesting that the loss of *re* gene may coincide with the redness of the eyes in the fruit flies. Notably, MFSD12, the *re* homolog in vertebrates, is a transporter that incorporates cysteine into melanosomes and is involved in the production of pheomelanin (Adelmann et al. 2020), implying a similarity between melanosomes and ommochromasomes.

1.3.5 Ommochrome-Related Genes Involved in Butterfly Wing Patterns

Ommochromes are also responsible for the reddish wing pigmentation of nymphalid butterflies. The buckeye butterfly *Junonia coenia* has two different seasonal wing color morphs, the pale tan summer morph (f. *linea*) and the dark reddish-brown autumn morph (f. *rosa*). The summer morph has a larger amount of oxidized xanthommatin, whereas the autumn morph contains mainly reduced xanthommatin and ommatin D (Nijhout 1997). The expression level of *cinnabar* gene in the late pupal stage is higher in the autumn morph than in the summer morph (Daniels et al. 2014). Moreover, this seasonal color difference is associated with high expression of the three genes *cortex*, *trehalase*, and *herfst* in the autumn morph (van der Burg et al. 2020). In *Elymnias hypermnestra*, presence of multiple ommochrome pigments have been suggested (Panettieri et al. 2018).

In the genus *Heliconius*, the red and yellow regions of the wings consist of xanthommatin and 3-hydroxykynurenine, respectively (Fig. 1.7), and *Kynurenine formamidase* and *cinnabar* genes are specifically expressed in the red marking regions (Reed et al. 2008; Ferguson and Jiggins 2009; Hines et al. 2012). It should be noted that no significant expression of the ABC transporter *scarlet* and *white* genes was detected in the red marking regions in *Heliconius erato*, whereas two other ABC transporter genes were expressed in a pattern-specific manner, suggesting that the incorporation of 3-hydroxykynurenine into pigment granules in the wings may be mediated by novel ABC transporter genes (Hines et al. 2012).

In the monarch butterfly *Danaus plexippus*, a recessive *nivosus* allele with white wings appears in Hawaii islands. A strong association of *nivosus* allele with the *myosin 5a* gene has been reported by population genetics, suggesting that the *myosin 5a* gene may be required for the transport of wing ommochrome pigments (Zhan et al. 2014).

Fig. 1.7 Examples of pigments on butterfly wings (Figure modified from Futahashi 2015)

1.4 Pteridines

1.4.1 Characteristics of Pteridines

Pteridines are substances with a pteridine ring, a structure similar to the purine ring of uric acid (Fig. 1.8). Pteridines are widely found in the animal and even in plant kingdoms (e.g., folic acid), and most insects possess pteridine pigments (Table 1.2). In insects, most pteridine pigments are members of pterin as follows: xanthopterin (yellow), isoxanthopterin (white), leucopterin (white), erythropterin (orange), sepiapterin (yellow), and drosopterins (at least five compounds, red) (Kim et al. 2013) (Fig. 1.8). The oxidized form of pteridine emits strong fluorescence under UV light, and UV-absorbing pteridines are sometimes involved in the difference in UV reflection (Makino et al. 1952; Henze et al. 2019). In mammals, excess pterin is excreted in urine and does not accumulate in the skin, but is widely found in pigment cells (e.g., xanthophore, leucophore) in fishes and amphibians. In the silkworm *B. mori*, ommochromes and pteridines (including uric acid) are present in different granules of larval epidermis (Kato et al. 2006). Most pteridines are stable under

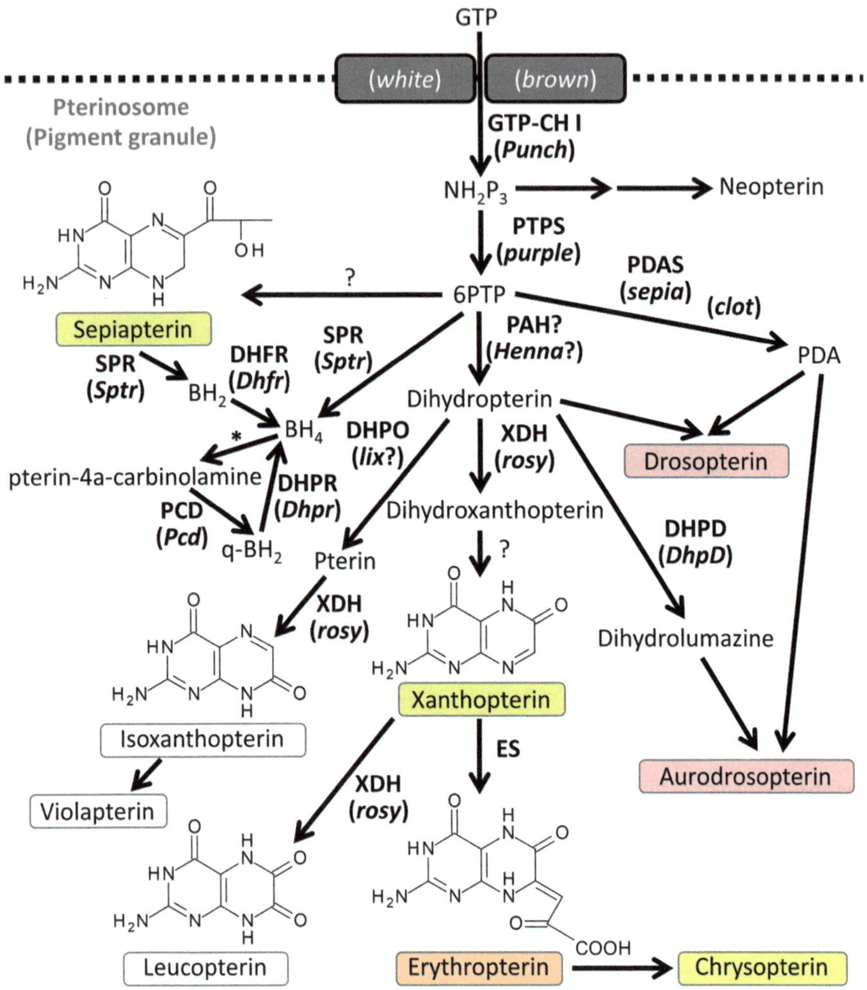

Fig. 1.8 The pteridine synthesis pathway in insects. Two transporter genes *white* and *brown* are involved in uptake of pteridine precursors into pterinosomes. Key enzymes are shown in bold and the *Drosophila* genes that encode them are shown in parenthesis. *GTP* guanosine triphosphate, *GTP-CH I* GTP cyclohydrolase I, NH_2P_3 dihydroneopterin triphosphate, *PTPS* 6-pyruvoyl tetrahydropterin synthase, *6PTP* 6-pyruvoyl tetrahydropterin, *SPR* sepiapterin reductase, *DHFR* dihydrofolate reductase, BH_2 dihydrobiopterin, BH_4 tetrahydrobiopterin, $q\text{-}BH_2$ quinonoid dihydrobiopterin, PCD, pterin-4a-carbinolamine dehydratase, *DHPR* dihydropteridine reductase, *PAH* phenylalanine hydroxylase, *DHPO* dihydropterin oxidase, *PDAS* pyrimidodiazepine synthase, *PDA* pyrimidodiazepine, *XDH* xanthine dehydrogenase, *DHPD* dihydropterin deaminase, *ES* erythropterin synthetase. The reaction indicated by an asterisk can be catalyzed by one of the following enzymes: phenylalanine hydroxylase (PAH), tyrosine hydroxylase (TH), tryptophan hydroxylase (TRH), nitric oxide synthase (NOS), and alkylglycerol monooxygenase (AGMO)

physiological conditions, although erythropterin is considered to be readily hydrolyzed to chrysopterin (Fig. 1.8) (Kayser 1985).

1.4.2 Pteridine Pigments for Insect Color Pattern

Pteridine pigments are involved in the wing coloration of pierid butterflies. The orange sulfur butterfly *Colias eurytheme* has five types of pterins: xanthopterin, isoxanthopterin, leucopterin, erythropterin, and sepiapterin (Fig. 1.8). The proportion of xanthopterin, erythropterin, and sepiapterin in white females is lower than in yellow females (Watt 1964, 1973). The amount and distribution of uric acid and pteridines in the wings are also associated with differences in UV reflection in pierid butterflies (Makino et al. 1952; Nakagoshi et al. 2002). Given that most insects can recognize ultraviolet (UV) light (Osorio and Vorobyev 2008; Futahashi et al. 2015; Feuda et al. 2016), the diversity of UV reflection may be important for partner recognition in various species.

The reddish coloration of many hemipteran insects (e.g., shield bugs, firebugs) is due to pteridines, not ommochromes (Linzen 1974; Krajíček et al. 2014). Several body color mutants have been found in the firebug *Pyrrhocoris apterus*. From pigment analysis, the amount of total pteridines is found to be lower in the white strain than in the reddish wild type. The yellow strain has lower amounts of erythropterin and higher amounts of isoxanthopterin and violapterin (hydrolysate of isoxanthopterin, also known as isoxantholumazine) than the wild type (Bel et al. 1997). Pteridines are also involved in the diversification of embryonic reddish colors in water striders (Vargas-Lowman et al. 2019), and seasonal polymorphism in body color of the Japanese scorpionfly *Panorpa japonica* (Nakagoshi et al. 1984). Adults of *P. japonica* are blackish in spring and yellowish in summer, and this is attributed to seasonal differences in the accumulation of black melanin and yellow sepiapterin (Nakagoshi et al. 1984).

1.4.3 Genes Involved in Pteridine Synthesis and Transport

Pteridine pigments are also located in the compound eyes of lepidopteran and dipteran insects (Table 1.2). Genes involved in pteridine synthesis and transport have been studied primarily using *Drosophila* eye color mutants (Ferré et al. 1986). In the compound eyes of *D. melanogaster* and the embryo of water sliders, pteridine precursors (GTP is assumed, Montell et al. 1992) are transported into pigment granules (pterinosomes) by the heterodimeric ABC transporters encoded by the *brown* and *white* genes (Fig. 1.8) (Dreesen et al. 1988; Vargas-Lowman et al. 2019). It should be noted that *brown* gene is not involved in compound eyes or body coloration in the red flour beetle *T. castaneum* and the silkworm *B. mori* (Grubbs et al. 2015; Zhang et al. 2018), but involved in riboflavin transport in

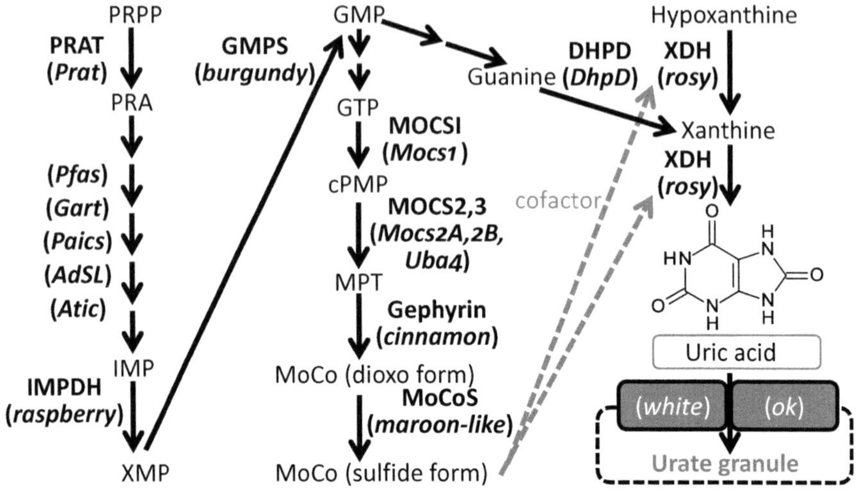

Fig. 1.9 Biosynthetic pathway of GTP, MoCo, and uric acid in insects. Two transporter genes *white* and *ok* are involved in the uptake of uric acid into urate granules. *PRPP* phosphoribosylpyrophosphate, *PRAT* phosphoribosylamidotransferase, *PRA* phosphoribosylamine, *IMP* inosine monophosphate, *IMPDH* inosine monophosphate dehydrogenase, *XMP* xanthosine monophosphate, *GMP* guanosine monophosphate, *GMPS* guanosine monophosphate synthetase, *GTP* guanosine triphosphate, *cPMP* cyclic pyranopterin monophosphate, *MPT* molybdopterin, *MoCo* molybdenum cofactor, *MOCS* molybdenum cofactor synthesis, *MoCoS* molybdenum cofactor sulfurase, *DHPD* dihydropterin deaminase, *XDH* xanthine dehydrogenase

Malpighian tubules in the silkworm *B. mori* (Zhang et al. 2018). In addition, uric acid is transported into urate granules by the heterodimeric ABC transporters encoded by the *white* and *ok* gene in *B. mori* (Fig. 1.9) (Wang et al. 2013b).

The pteridine synthesis pathway is similar in insects and vertebrates (Braasch et al. 2007, see Chap. 4) (Fig. 1.8). Defects of GTP cyclohydrolase I (GTP-CH I, encoded by the *Punch* gene in *Drosophila*) or 6-pyruvoyl tetrahydropterin synthase (PTPS, encoded by the *purple* gene in *Drosophila*) inhibit synthesis of most pteridine pigments, whereas loss of function of xanthine dehydrogenase (XDH, encoded by the *rosy* gene in *Drosophila*) reduces the amounts of isoxanthopterin, xanthopterin, leucopterin, erythropterin as well as uric acid (Reaume et al. 1991; McLean et al. 1993; Kim et al. 1996; Vargas-Lowman et al. 2019) (Figs. 1.8 and 1.9). Three *Drosophila* genes *sepia*, *clot*, and *DhpD* are involved in the synthesis of drosopterin and aurodrosopterin synthesis (Kim et al. 2013) (Fig. 1.8). Curiously, phenylalanine hydroxylase (PAH, encoded by the *Henna* gene in *Drosophila*) is involved in the conversion of 6-pyruvoyl tetrahydropterin to dihydropterin (Wang et al. 2008), and loss of function of PAH affects pteridine pigmentation, instead of melanin pigmentation (Figs. 1.1 and 1.8, Table 1.3). Sepiapterin reductase (SPR), dihydrofolate reductase (DHFR), pterin-4a-carbinolamine dehydratase (PCD), and dihydropteridine reductase (DHPR) are involved in the production of BH_4 (Figs. 1.1 and 1.8) (Park et al. 2000). In addition, genes involved in the synthesis of pteridine precursor GTP (Nash et al. 1994; O'Donnell et al. 2000) and sulfide form of the molybdenum cofactor (MoCo), which is essential for xanthine dehydrogenase (XDH) enzyme, are important in pteridine and uric acid production (Fig. 1.9, Table 1.3) (Kômoto et al. 2003; Fujii et al. 2009, 2016, 2018). It should be noted that the presence of different pigment-binding proteins depending on the type of pteridine has been suggested from silkworm epidermis (Tsujita and Sakurai 1965).

The expression of the *GTP-CH I* gene is associated with the white pterin-based eyespot markings of the buckeye butterfly *Junonia coenia* (Sawada et al. 2002). Since the *GTP-CH I* gene is involved in the synthesis of BH_4, a cofactor for melanin synthesis gene *TH* (Fig. 1.1), black marking-specific expression pattern of *GTP-CH I* has been observed in the larvae of the swallowtail butterfly *P. xuthus*, as described above (Futahashi and Fujiwara 2006). It has also been reported that silkworm mutants of *purple* and *Sepiapterin reductase* genes, which are also involved in the synthesis of BH_4, inhibit melanin pigmentation (Table 1.3) (Meng et al. 2009b; Fujii et al. 2013). Notably, in the silkworm *B. mori*, a mutation in cGMP synthesis enzyme guanylyl cyclase affects the production of dark-colored melanins, ommochromes, and pteridines, suggesting the importance of cGMP in the synthesis of all three major insect pigments (Table 1.3) (Yuasa et al. 2016).

1.5 Other Pigments

1.5.1 Tetrapyrroles

Tetrapyrroles (bile pigments and porphyrins) contain four pyrrole-based *N*-heterocyclic rings. Bile pigments are linear tetrapyrroles, whereas porphyrins are cyclic tetrapyrroles. Bile pigments often contribute to the blue to green color of insects (Tables 1.1 and 1.2). Of insect bile pigments, biliverdin IXα and biliverdin IXγ have been widely found in plant-eating insects (Kayser 1985). Biliverdin IXα has been detected predominantly in hemimetabolous insects (e.g., grasshoppers, praying mantis), and biliverdin IXγ has been detected in holometabolous insects (e.g., butterflies, moths, water beetles). In several butterflies and moths, phorcabilin and sarpedobilin, derivatives of biliverdin IXγ, have been also detected as bluish pigments (Kayser 1985). Bile pigments are usually associated with proteins belonging to the lipocalin or hexamerin family (Riddiford et al. 1990; Schmidt and Skerra 1994; Miura et al. 1998; Shirataki et al. 2010; Futahashi et al. 2012a; Yoda et al. 2020). Meanwhile, hemoglobin, which contains porphyrin, contributes to the red color of mosquito larvae (Kayser 1985).

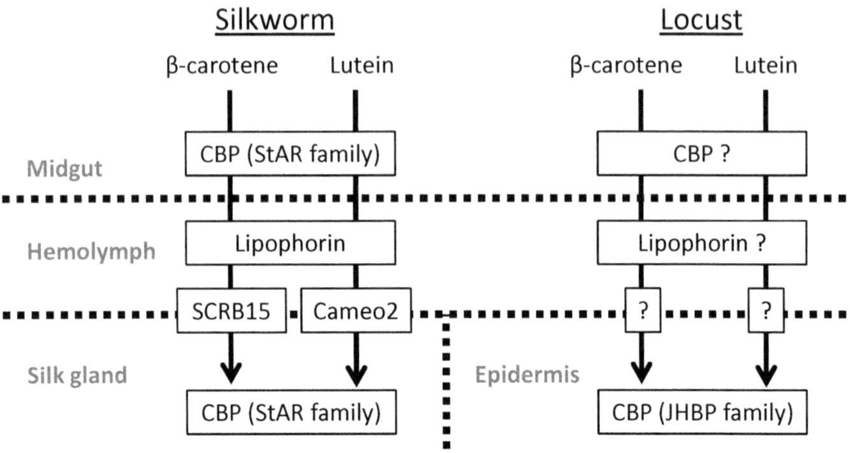

Fig. 1.10 Model of carotenoid transport in the silkworm and locusts. *CBP* carotenoid binding protein, *SCRB* scavenger receptor class B, *Cameo* C locus-associated membrane protein homologous to mammalian HDL receptor, *StAR* steroidogenic acute regulatory protein, *JHBP* juvenile hormone binding protein

1.5.2 Carotenoids

Carotenoids are yellow, orange, and red pigments produced by plants, algae, fungi, and bacteria. Carotenoids can be divided to two subclasses: carotenes (e.g., β-carotene, α-carotene) and xanthophylls (oxygen-containing carotenes, e.g., lutein, zeaxanthin). Most insects acquire carotenoids or their precursors from plants, while pea aphids and the gall midges are known to produce carotenoids by carotenoid synthetic genes acquired through lateral transfer from fungi (Moran and Jarvik 2010; Nováková and Moran 2012; Cobbs et al. 2013). In many hemipteran insects, carotenoids are assumed to be synthesized by microbial symbionts (Kayser 1985; Sloan and Moran 2012). Although the relationship with body color is unknown, carotenoids themselves have been detected in many insects (Maoka et al. 2020).

Among carotenoid pigments, β-carotene, α-carotene, lutein, and zeaxanthin are widely found across insects (Kayser 1985; Umebachi 2000). The green coloration of plant-eating insects is usually mixtures of bluish tetrapyrroles (bile pigments) and yellowish plant-derived carotenoids (Kayser 1985). Carotenoids, like tetrapyrroles, are usually associated with proteins, and many of which belong to the takeout/JHBP family (Fig. 1.10) (Futahashi et al. 2012a; Sugahara and Tanaka 2019; Sugahara et al. 2020). The lutein binding proteins are responsible for yellow coloration in the desert locust *Schistocerca gregaria* (Sugahara and Tanaka 2019; Sugahara et al. 2020), and the β-carotene binding protein is involved in body color transition between black and green in the migratory locust *Locusta migratoria* (Yang et al. 2019). Similar JHBP proteins are also associated with yellow coloration in the swallowtail butterflies, and the combination of biliverdin binding protein (BBP) genes, putative carotenoid binding protein genes and yellow related gene (YRG) correlated with stage- and species-specific larval blue, yellow, and green coloration as well as pupal green coloration in the swallowtail butterflies *Papilio* species (Fig. 1.2) (Futahashi and Fujiwara 2008b; Shirataki et al. 2010; Futahashi et al. 2012a; Yoda et al. 2020).

In the silkworm *B. mori*, hemolymph and cocoon color differ between strains depending on the existence of carotenoid pigments, which is defined genetically. The intracellular carotenoid binding protein *Yellow blood* gene belonging to the steroidogenic acute regulatory protein (StAR) family is essential for uptake of carotenoids into the intestinal mucosa and the silk gland (Fig. 1.10) (Sakudoh et al. 2005, 2007). Moreover, *Cameo2* and *SCRB15* genes, members of the cluster determinant 36 (CD36) family, are involved in the selective uptake of lutein and β-carotene into the silk gland, respectively (Fig. 1.10) (Sakudoh et al. 2010, 2013).

1.5.3 Flavonoids

Flavonoids (e.g., anthocyanins, flavones) are polyphenolic pigments derived from phenylalanine, and are common plant pigments occasionally involved in body color in plant-eating insect such as butterflies and grasshoppers (Table 1.2) (Wilson 1987; Umebachi 2000). For example, green pigments of the large hawkmoth *Euchloron megaera* are anthocyanidin derivatives (Barbier 1984). In the common blue butterfly *Polyommatus icarus*, UV-absorbing flavonoid pigments in the wings make females more attractive to males (Burghardt et al. 2000). The quantity of flavonoids is also associated with intraspecific wing color variation in the marbled white butterfly *Melanargia galathea*, with the yellow type having higher amounts of flavonoids (Wilson 1986). The major components of yellowish hindwing of the Carolina grasshopper *Dissoteira carolina* are quercetin and quercetin-β-3-O-glucoside (Hopkins and Ahmad 1991).

Flavonoids are accumulated in the cocoons of some moths and sawflies (Schopf et al. 1982; Hirayama et al. 2013). In the silkworm *B. mori*, quercetin 5-O-glucosides are the major cocoon flavonoids. The quercetin 5-O-glucosyltransferase gene is essential for greenish cocoon coloration with UV-shielding properties (Daimon et al. 2010). Deficiency of a pyrroline-5-carboxylate reductase produces prolinylflavonols, the unusual flavonoids, resulting in yellowish-green cocoon coloration (Hirayama et al. 2018).

1.5.4 Papiliochromes

The white, yellow, and reddish-brown pigments in the adult wings of papilionid butterflies are called papiliochromes, which are formed by the combination of NBAD and kynurenine (Fig. 1.4) (Umebachi 1985). For example, papiliochrome II and its related pigments (kynurenine chemically linked to the NBAD monomer, dimer, or trimer) are detected in the pale-yellow markings of the wings of the common mormon butterfly *Papilio polytes*, where expression of both melanin and ommochrome synthesis genes are enhanced (Fig. 1.7) (Nishikawa et al. 2013). The reddish-brown pigment of several papilionid butterflies is identified as papiliochrome R, which is assumed to be a compound in which kynurenine chemically linked to the NBAD polymer (Fig. 1.7) (Umebachi 1985). Both melanin and ommochrome synthesis genes are also enhanced in the reddish-brown marking regions of *P. polytes* females (Nishikawa et al. 2013). Papiliochromes and melanin pigments as well as structural coloration display a colorful pattern in papilionid butterflies (Stavenga et al. 2014). In the birdwing *Ornithoptera* butterflies, papiliochrome pigments function as spectral filters, combining structural and pigment effects to adjust the color (Wilts et al. 2015).

1.5.5 Quinones

Some pigments have a quinone skeleton with two oxygen atoms attached to the benzene ring. Anthraquinones are red, yellow, and crimson pigments found in scale insects (Tables 1.1 and 1.2) (Kayser 1985). Red cochineal pigment (carminic acid) has been widely used as a dye. Aphines, polycyclic quinone pigments, have been identified from aphids (Bowie et al. 1966). Aphines are structurally diverse and exhibit a variety of colors, such as green, yellow, and red. The composition and amount of aphins vary between different morphs within the same species as well as between different colored aphid species (Bowie et al. 1966; Tsuchida 2016). Notably, a specific facultative symbiotic bacterium *Rickettsiella viridis* can influence the production of green aphins (Tsuchida et al. 2010; Nikoh et al. 2018).

The reddish wing pigments of butterflies were classically divided into five types (A–E) based on their solubility and the responses to acid and alkali (Ford 1942). Type B, C, and D+E pigments correspond to papiliochromes, ommochromes, and pteridines, respectively. Type A pigments found in several butterflies (e.g., *Pachliopta aristolochiae*, *Luehdotlfia japonica*) have been suggested to be quinone related pigments containing β-alanine (Fig. 1.7) (Umebachi 2000).

1.6 Pigment-Associated Structural Coloration

Pigments are also essential for producing structural colors. The presence or absence of the melanin pigments contributes to the metallic blue and white pattern of the morpho butterfly *Morpho cypris* (Yoshioka and Kinoshita 2006). Some pigments themselves are responsible for structural colors. For instance, many dragonflies display a non-iridescent light blue color by coherent light scattering (Prum et al. 2004), which seems to be attributed to the light scattering granules containing pteridine pigments in the distal layer and the light absorbing granules containing ommochrome pigments in the proximal layer (Veron et al. 1974; Henze et al. 2019; Okude and Futahashi 2021). Some dragonflies are partly coated with wax, which was originally defined as a supracuticular pigment (Tillyard 1917; Futahashi 2016). Dragonfly's wax spontaneously forms light scattering scale-like fine structures that strongly reflect both ultraviolet and visible light (Futahashi 2020). Very long-chain methyl ketones and very long-chain aldehydes are the major components of the dragonfly wax, which are totally different from previously known biological waxes (Futahashi et al. 2019).

1.7 Genes Involved in the Insect Color Patterning

Insect color patterns are formed when specific pigments are synthesized or transported in a region-specific manner. The genes involved in body color patterning have been identified mainly in fruit flies and butterflies, and many of which are the genes that control the expression of other genes such as secreted signaling factors that act as morphogens, their receptors, and transcription factors. For example, the polka-dot pattern on the wings of *D. guttifera* is regulated by the expression of the *wingless* gene, a well-known morphogen (Werner et al. 2010; Koshikawa et al. 2015, see Chap. 12), and the *Drosophila* abdominal black marking pattern is regulated by the transcription factor genes *bric-a-brac* and *Abdominal-B* (Kopp et al. 2000). Some *Drosophila* species have female-limited abdominal marking polymorphisms, and many of them are regulated by the transcription factor gene *POU domain motif 3* (*pdm3*) (Yassin et al. 2016b). It should be noted that this polymorphism has evolved by at least three independent changes in the *pdm3* gene (Yassin et al. 2016b). Meanwhile, the similar female-limited polymorphism of *Drosophila erecta* is regulated by cis regulatory changes in the melanin synthesis gene *tan* (Yassin et al. 2016a).

The first identified gene involved in butterfly wing color pattern is the *Distal-less* gene, a transcription factor involved in *Drosophila* leg development and wing markings (Arnoult et al. 2013), that is expressed at the center of the eyespot in the larval wing disc of the buckeye butterfly, *Junonia coenia* (Carroll et al. 1994). Since then, dozens of genes associated with the eyespot pattern have been reported (Monteiro and Prudic 2010; Saenko et al. 2010; Zhang and Reed 2016). Moreover, the transcription factor gene *aristaless*, and the secreted signaling factor genes *wingless* (*Wnt1*), *Wnt6*, *Wnt10*, and *WntA* are associated with the symmetric stripes of butterfly wing pattern (Martin and Reed 2010, 2014).

In the past decade, the master gene that controls wing pattern polymorphism in butterflies have been identified, mainly in the genera *Heliconius* and *Papilio*. Among *Heliconius* butterflies, wing pattern polymorphism can be explained by a small number of loci (Kronforst and Papa 2015; Van Belleghem et al. 2017). The pattern of red markings based on ommochrome pigments is controlled by the transcription factor gene *optix* (Reed et al. 2011; Zhang et al. 2017b; Lewis et al. 2019), the variation in stripe patterning (mainly the shape of black melanin contours of forewings) is controlled by the secreted signaling factor gene *WntA* (Martin et al. 2012; Mazo-Vargas et al. 2017; Concha et al. 2019), and the variation between yellow pigment (based on ommochrome precursor 3-hydroxykynurenin) and white structural color pattern is controlled by the transcription factor gene *aristaless1* (Westerman et al. 2018). Notably, *optix* gene is also associated with structural iridescence in the genus *Junonia* (Zhang et al. 2017b), and *WntA* gene is also involved in the color pattern polymorphism in the red-spotted purple butterfly *Limenitis arthemis* (Timmermans et al. 2014). In addition, the *P* locus, which controls the most white and yellow color pattern elements of the entire wing of *Heliconius numata*, is regulated by two small inversions (Joron et al. 2011), among which the cell cycle regulator gene *cortex* seems to be responsible (Nadeau et al. 2016). The *cortex* gene is also involved in the industrial melanism of the peppered moth *Biston betularia* (Van't Hof et al. 2016), and the seasonal color difference of *J. coenia* (van der Burg et al. 2020). With respect to the genus *Papilio*, an inversion of the *doublesex* gene, a transcription factor involved in sexual differentiation, is responsible for the female-limited mimetic polymorphism in *P. polytes* (Kunte et al. 2014; Nishikawa et al. 2015). The *doublesex* gene is also involved in female polymorphisms in several *Papilio* species (Iijima et al. 2018, 2019; Palmer and Kronforst 2019). Meanwhile, the transcription factor gene *engrailed* is associated with female polymorphisms in *Papilio dardanus* (Timmermans et al. 2014). In the *Colias* butterflies, the variation of wing pteridine pigments between yellow and white color polymorphism is controlled by the transcription factor gene *BarH-1* (Woronik et al. 2019).

In the silkworm *B. mori*, there are numerous larval color pattern mutants. The *p* locus comprises at least 15 alleles of the larval melanin pattern that are controlled by the transcription factor *apontic-like* gene (Yoda et al. 2014). Moreover, the secreted signaling factors *Wnt1* (*wingless*) and *spätzle-3* are responsible for the variations of larval twin spot marking and black stripe in the anterior margin of each segment, respectively (Yamaguchi et al. 2013; KonDo et al. 2017). In the larva of the swallowtail butterfly *P. xuthus*, the thoracic eyespot markings are regulated by the transcription factor gene *clawless*, and the abdominal V-shaped markings are regulated by the transcription factor genes *abdominal-A* and *Abdominal-B* (Jin et al. 2019). Altered expression of *Abdominal-B* is also associated with abdominal color pattern polymorphisms in bumble bees of the genus *Bombus* (Tian et al. 2019).

In the multicolored Asian ladybird beetle, *Harmonia axyridis*, more than 200 distinct forewing color forms have been described. The transcription factor gene *pannier* is responsible for this polymorphism, and the repeated inversions within an intron causes the expression changes of *pannier* gene (Ando et al. 2018; Gautier et al. 2018).

1.8 Conclusion and Perspective

The mechanism of pigmentation by melanins and ommochromes in insects differs greatly from that in vertebrates, while that by pteridines has many similarities. With the rapid spread of next-generation sequencers and genome-editing techniques in this century, the molecular mechanisms involved in insect color pattern formation are being dramatically elucidated, particularly in flies and butterflies. Meanwhile, there are many insects that lack basic information, including the identity of pigments, and future progress is expected for topics on the molecular mechanisms involved in the formation and evolution of insect color patterns.

Acknowledgements We would like to thank Kohji Tanaka for photo of gynandromorphic dragonfly, Genta Okude for helpful comments of the manuscript. The writing was partly supported by JSPS KAKENHI (JP18H02491 and JP20H04936).

References

Adelmann CH, Traunbauer AK, Chen B, Condon KJ, Chan SH, Kunchok T, Lewis CA, Sabatini DM (2020) MFSD12 mediates the import of cysteine into melanosomes and lysosomes. Nature 588:699–704

Ahmed-Braimah YH, Sweigart AL (2015) A single gene causes an interspecific difference in pigmentation in *Drosophila*. Genetics 200(1):331–342

Anbutsu H, Moriyama M, Nikoh N, Hosokawa T, Futahashi R, Tanahashi M, Meng XY, Kuriwada T, Mori N, Oshima K, Hattori M, Fujie M, Satoh N, Maeda T, Shigenobu S, Koga R, Fukatsu T (2017) Small genome symbiont underlies cuticle hardness in beetles. Proc Natl Acad Sci U S A 114(40):E8382–E8391

Ando T, Matsuda T, Goto K, Hara K, Ito A, Hirata J, Yatomi J, Kajitani R, Okuno M, Yamaguchi K, Kobayashi M, Takano T, Minakuchi Y, Seki M, Suzuki Y, Yano K, Itoh T, Shigenobu S, Toyoda A, Niimi T (2018) Repeated inversions within a pannier intron drive diversification of intraspecific colour patterns of ladybird beetles. Nat Commun 9(1):3843

Arakane Y, Muthukrishnan S, Beeman RW, Kanost MR, Kramer KJ (2005) Laccase 2 is the phenoloxidase gene required for beetle cuticle tanning. Proc Natl Acad Sci U S A 102 (32):11337–11342

Arakane Y, Lomakin J, Beeman RW, Muthukrishnan S, Gehrke SH, Kanost MR, Kramer KJ (2009) Molecular and functional analyses of amino acid decarboxylases involved in cuticle tanning in *Tribolium castaneum*. J Biol Chem 284:16584–16594

Arakane Y, Dittmer NT, Tomoyasu Y, Kramer KJ, Muthukrishnan S, Beeman RW, Kanost MR (2010) Identification, mRNA expression and functional analysis of several yellow family genes in Tribolium castaneum. Insect Biochem Mol Biol 40(3):259–266

Arakane Y, Noh MY, Asano T, Kramer KJ (2016) Tyrosine metabolism for insect cuticle pigmentation and sclerotization. In: Cohen E, Moussian B (eds) Extracellular composite matrices in arthropods. Springer, New York, pp 165–220

Arnoult L, Su KFY, Manoel D, Minervino C, Magriña J, Gompel N, Prud'homme B (2013) Emergence and diversification of fly pigmentation through evolution of a gene regulatory module. Science 339:1423–1426

Ashida M (1990) The prophenoloxidase cascade in insect immunity. Res Immunol 141(9):908–910

Barbier M (1984) A substance which acts as a pH indicator from the moth *Euchloron megaera* L. J Chem Ecol 10(7):1109–1113

Barek H, Sugumaran M, Ito S, Wakamatsu K (2018) Insect cuticular melanins are distinctly different from those of mammalian epidermal melanins. Pigment Cell Melanoma Res 31 (3):384–392

Bel Y, Porcarm M, Socha R, Nemec V, Ferre J (1997) Analysis of pteridines in *Pyrrhocoris apterus* (L.) (heteroptera, pyrrhocoridae) during development and in body - color mutants. Arch Insect Biochem Physiol 34:83–98

Bolognese A, Liberatore R, Scherillo G (1988) Photochemistry of ommochromes and related compounds. J Heterocyclic Chem 25:979–983

Bowie JH, Cameron DW, Findlay JA, Quartey JAK (1966) Haemolymph pigments of aphids. Nature 210:395–397

Braasch I, Schartl M, Volff JN (2007) Evolution of pigment synthesis pathways by gene and genome duplication in fish. BMC Evol Biol 11(7):74

Brodbeck D, Amherd R, Callaerts P, Hintermann E, Meyer UA, Affolter M (1998) Molecular and biochemical characterization of the aaNAT1 (Dat) locus in *Drosophila melanogaster*: differential expression of two gene products. DNA Cell Biol 17(7):621–633

Bückmann D (1977) Morphological colour change (1): stage independent, optically induced ommochrome synthesis in larvae of the stick insect, *Carausius morosus* Br. J Comp Physiol 115(2):185–193

Bückmann D (1979) Morphological colour change (2): the effects of total and partial blinding on epidermal ommochrome content in the stick insect, *Carausius morosus* Br. J Comp Physiol 130 (4):331–336

Burghardt F, Knüttel H, Becker M, Fiedler K (2000) Flavonoid wing pigments increase attractiveness of female common blue (*Polyommatus icarus*) butterflies to mate-searching males. Naturwissenschaften 87(7):304–307

Carroll SB, Gates J, Keys DN, Paddock SW, Panganiban GE, Selegue JE, Williams JA (1994) Pattern-formation and eyespot determination in butterfly wings. Science 265:109–114

Cheli VT, Daniels RW, Godoy R, Hoyle DJ, Kandachar V, Starcevic M, Martinez-Agosto JA, Poole S, DiAntonio A, Lloyd VK, Chang HC, Krantz DE, Dell'Angelica EC (2010) Genetic modifiers of abnormal organelle biogenesis in a *Drosophila* model of BLOC-1 deficiency. Hum Mol Genet 19(5):861–878

Cobbs C, Heath J, Stireman JO, Abbot P (2013) Carotenoids in unexpected places: gall midges, lateral gene transfer, and carotenoid biosynthesis in animals. Mol Phylogenet Evol 68 (2):221–228

Concha C, Wallbank RWR, Hanly JJ, Fenner J, Livraghi L, Rivera ES, Paulo DF, Arias C, Vargas M, Sanjeev M, Morrison C, Tian D, Aguirre P, Ferrara S, Foley J, Pardo-Diaz C, Salazar C, Linares M, Massardo D, Counterman BA, Scott MJ, Jiggins CD, Papa R, Martin A, McMillan WO (2019) Interplay between developmental flexibility and determinism in the evolution of mimetic *Heliconius* wing patterns. Curr Biol 29(23):3996–4009

Crawford K, Quiroz JFD, Koenig KM, Ahuja N, Albertin CB, Rosenthal JJC (2020) Highly efficient knockout of a squid pigmentation gene. Curr Biol 30(17):3484–3490

Dai FY, Qiao L, Tong XL, Cao C, Chen P, Chen J, Lu C, Xiang ZH (2010) Mutations of an arylalkylamine-N-acetyltransferase, Bm-iAANAT, are responsible for silkworm melanism mutant. J Biol Chem 285(25):19553–19560

Dai F, Qiao L, Cao C, Liu X, Tong X, He S, Hu H, Zhang L, Wu S, Tan D, Xiang Z, Lu C (2015) Aspartate decarboxylase is required for a normal pupa pigmentation pattern in the silkworm, *Bombyx mori*. Sci Rep 5:10885

Daimon T, Hirayama C, Kanai M, Ruike Y, Meng Y, Kosegawa E, Nakamura M, Tsujimoto G, Katsuma S, Shimada T (2010) The silkworm *Green b* locus encodes a quercetin 5-O-glucosyltransferase that produces green cocoons with UV-shielding properties. Proc Natl Acad Sci U S A 107(25):11471–11476

Daniels EV, Murad R, Mortazavi A, Reed RD (2014) Extensive transcriptional response associated with seasonal plasticity of butterfly wing patterns. Mol Ecol 23(24):6123–6134

Drapeau MD (2001) The family of yellow-related drosophila melanogaster proteins. Biochem Biophys Res Commun 281(3):611–613

Dreesen TD, Johnson DH, Henikoff S (1988) The brown protein of *Drosophila melanogaster* is similar to the white protein and to components of active transport complexes. Mol Cell Biol 8 (12):5206–5215

Falcón-Pérez JM, Romero-Calderón R, Brooks ES, Krantz DE, Dell'Angelica EC (2007) The *Drosophila* pigmentation gene *pink* (*p*) encodes a homologue of human Hermansky-Pudlak syndrome 5 (HPS5). Traffic 8(2):154–168

Ferguson LC, Jiggins CD (2009) Shared and divergent expression domains on mimetic *Heliconius* wings. Evol Dev 11(5):498–512

Ferguson LC, Green J, Surridge A, Jiggins CD (2011a) Evolution of the insect *yellow* gene family. Mol Biol Evol 28(1):257–272

Ferguson LC, Maroja L, Jiggins CD (2011b) Convergent, modular expression of *ebony* and *tan* in the mimetic wing patterns of Heliconius butterflies. Dev Genes Evol 221(5-6):297–308

Ferré J, Silva FJ, Real MD, Ménsua JL (1986) Pigment patterns in mutants affecting biosynthesis of pteridines and xanthommatin in *Drosophila melanogaster*. Biochem Genet 24:545–569

Feuda R, Marlétaz F, Bentley MA, Holland PW (2016) Conservation, duplication, and divergence of five opsin genes in insect evolution. Genome Biol Evol 8:579–587

Figon F, Casas J (2019) Ommochromes in invertebrates: biochemistry and cell biology. Biol Rev 94:156–183

Fitzpatrick PF (2003) Mechanism of aromatic amino acid hydroxylation. Biochemistry 42 (48):14083–14091

Ford KB (1942) Studies on the chemistry of pigments in the Lepidoptera, with reference to their bearing on systematics. 3. The red pigments of the Papilionidae. Proc R Soc Lond A 17:87–92

Fujii T, Abe H, Katsuma S, Mita K, Shimada T (2008) Mapping of sex-linked genes onto the genome sequence using various aberrations of the Z chromosome in *Bombyx mori*. Insect Biochem Mol Biol 38(12):1072–1079

Fujii T, Ozaki M, Masamoto T, Katsuma S, Abe H, Shimada T (2009) A *Bombyx mandarina* mutant exhibiting translucent larval skin is controlled by the molybdenum cofactor sulfurase gene. Genes Genet Syst 84(2):147–152

Fujii T, Banno Y, Abe H, Katsuma S, Shimada T (2012) A homolog of the human Hermansky-Pudluck syndrome-5 (HPS5) gene is responsible for the *oa* larval translucent mutants in the silkworm, *Bombyx mori*. Genetica 140(10-12):463–468

Fujii T, Abe H, Kawamoto M, Katsuma S, Banno Y, Shimada T (2013) *Albino* (*al*) is a tetrahydrobiopterin (BH_4)-deficient mutant of the silkworm *Bombyx mori*. Insect Biochem Mol Biol 43(7):594–600

Fujii T, Yamamoto K, Banno Y (2016) Molybdenum cofactor deficiency causes translucent integument, male-biased lethality, and flaccid paralysis in the silkworm *Bombyx mori*. Insect Biochem Mol Biol 73:20–26

Fujii T, Yamamoto K, Banno Y (2018) Translucent larval integument and flaccid paralysis caused by genome editing in a gene governing molybdenum cofactor biosynthesis in *Bombyx mori*. Insect Biochem Mol Biol 99:11–16

Futahashi R (2015) Molecular mechanisms underlying butterfly pattern formation. In: Yagi T (ed) Biology of butterfly pattern formation. Osaka Municipal Universities Press, Osaka, pp 35–66

Futahashi R (2016) Color vision and color formation in dragonflies. Curr Opin Insect Sci 17:32–39

Futahashi R (2017) Molecular mechanisms underlying color vision and color formation in dragonflies. In: Sekimura T, Nijhout HF (eds) Diversity and evolution of butterfly wing patterns: an integrative approach. Springer, Singapore, pp 303–321

Futahashi R (2020) Diversity of UV reflection patterns in Odonata. Front Ecol Evol 8:201

Futahashi R, Fujiwara H (2005) Melanin-synthesis enzymes coregulate stage-specific larval cuticular markings in the swallowtail butterfly, *Papilio xuthus*. Dev Genes Evol 215:519–529

Futahashi R, Fujiwara H (2006) Expression of one isoform of *GTP cyclohydrolase I* coincides with the larval black markings of the swallowtail butterfly, *Papilio xuthus*. Insect Biochem Mol Biol 36(1):63–70

Futahashi R, Fujiwara H (2007) Regulation of 20-hydroxyecdysone on the larval pigmentation and the expression of melanin synthesis enzymes and *yellow* gene of the swallowtail butterfly, *Papilio xuthus*. Insect Biochem Mol Biol 37(8):855–864

Futahashi R, Fujiwara H (2008a) Juvenile hormone regulates butterfly larval pattern switches. Science 319(5866):1061

Futahashi R, Fujiwara H (2008b) Identification of stage-specific larval camouflage associated genes in the swallowtail butterfly, *Papilio xuthus*. Dev Genes Evol 218(9):491–504

Futahashi R, Sato J, Meng Y, Okamoto S, Daimon T, Yamamoto K, Suetsugu Y, Narukawa J, Takahashi H, Banno Y, Katsuma S, Shimada T, Mita K, Fujiwara H (2008) *yellow* and *ebony* are the responsible genes for the larval color mutants of the silkworm *Bombyx mori*. Genetics 180(4):1995–2005

Futahashi R, Banno Y, Fujiwara H (2010) Caterpillar color patterns are determined by a two-phase melanin gene prepatterning process: new evidence from *tan* and *laccase2*. Evol Dev 12 (2):157–167

Futahashi R, Tanaka K, Matsuura Y, Tanahashi M, Kikuchi Y, Fukatsu T (2011) Laccase2 is required for cuticular pigmentation in stinkbugs. Insect Biochem Mol Biol 41(3):191–196

Futahashi R, Shirataki H, Narita T, Mita K, Fujiwara H (2012a) Comprehensive microarray-based analysis for stage-specific larval camouflage pattern-associated genes in the swallowtail butterfly, *Papilio xuthus*. BMC Biol 10:46

Futahashi R, Kurita R, Mano H, Fukatsu T (2012b) Redox alters yellow dragonflies into red. Proc Natl Acad Sci U S A 109(31):12626–12631

Futahashi R, Kawahara-Miki R, Kinoshita M, Yoshitake K, Yajima S, Arikawa K, Fukatsu T (2015) Extraordinary diversity of visual opsin genes in dragonflies. Proc Natl Acad Sci U S A 112(11):E1247–E1256

Futahashi R, Yamahama Y, Kawaguchi M, Mori N, Ishii D, Okude G, Hirai Y, Kawahara-Miki R, Yoshitake K, Yajima S, Hariyama T, Fukatsu T (2019) Molecular basis of wax-based color change and UV reflection in dragonflies. elife 8:e43045

Galván I, Jorge A, Edelaar P, Wakamatsu K (2015) Insects synthesize pheomelanin. Pigment Cell Melanoma Res 28(5):599–602

Gautier M, Yamaguchi J, Foucaud J, Loiseau A, Ausset A, Facon B, Gschloessl B, Lagnel J, Loire E, Parrinello H, Severac D, Lopez-Roques C, Donnadieu C, Manno M, Berges H, Gharbi K, Lawson-Handley L, Zang LS, Vogel H, Estoup A, Prud'homme B (2018) The genomic basis of color pattern polymorphism in the harlequin ladybird. Curr Biol 28 (20):3296–3302

Geyer PK, Spana C, Corces VG (1986) On the molecular mechanism of gypsy-induced mutations at the *yellow* locus of *Drosophila melanogaster*. EMBO J 5(10):2657–2662

Gibert JM, Peronnet F, Schlötterer C (2007) Phenotypic plasticity in *Drosophila* pigmentation caused by temperature sensitivity of a chromatin regulator network. PLoS Genet 3:e30

Gompel N, Prud'homme B, Wittkopp PJ, Kassner VA, Carroll SB (2005) Chance caught on the wing: cis-regulatory evolution and the origin of pigment patterns in *Drosophila*. Nature 433 (7025):481–487

Gorman MJ, Arakane Y (2010) Tyrosine hydroxylase is required for cuticle sclerotization and pigmentation in *Tribolium castaneum*. Insect Biochem Mol Biol 40(3):267–273

Grant P, Maga T, Loshakov A, Singhal R, Wali A, Nwankwo J, Baron K, Johnson D (2016) An eye on trafficking genes: identification of four eye color mutations in *Drosophila*. G3 (Bethesda) 6 (10):3185–3196

Grimaldi D, Engel MS (2005) Evolution of the insects. Cambridge University Press, New York

Grubbs N, Haas S, Beeman RW, Lorenzen MD (2015) The ABCs of eye color in *Tribolium castaneum*: orthologs of the *Drosophila white*, *scarlet*, and *brown* genes. Genetics 199 (3):749–759

Han Q, Fang J, Ding H, Johnson JK, Christensen BM, Li J (2002) Identification of *Drosophila melanogaster* yellow-f and yellow-f2 proteins as dopachrome conversion enzymes. Biochem J 368(Pt 1):333–340

Han Q, Robinson H, Li J (2012) Biochemical identification and crystal structure of kynurenine formamidase from *Drosophila melanogaster*. Biochem J 446(2):253–260

Harris DA, Kim K, Nakahara K, Vasquez-Doorman C, Carthew RW (2011) Cargo sorting to lysosome-related organelles regulates siRNA-mediated gene silencing. J Cell Biol 194:77–87

Henze MJ, Lind O, Wilts BD, Kelber A (2019) Pterin-pigmented nanospheres create the colours of the polymorphic damselfly *Ischnura elegans*. J R Soc Interface 16(153):20180785

Hinaux H, Bachem K, Battistara M, Rossi M, Xin Y, Jaenichen R, Le Poul Y, Arnoult L, Kobler JM, Grunwald Kadow IC, Rodermund L, Prud'homme B, Gompel N (2018) Revisiting the developmental and cellular role of the pigmentation gene yellow in Drosophila using a tagged allele. Dev Biol 438(2):111–123

Hines HM, Papa R, Ruiz M, Papanicolaou A, Wang C, Nijhout HF, McMillan WO, Reed RD (2012) Transcriptome analysis reveals novel patterning and pigmentation genes underlying *Heliconius* butterfly wing pattern variation. BMC Genomics 13:288

Hirayama C, Ono H, Meng Y, Shimada T, Daimon T (2013) Flavonoids from the cocoon of *Rondotia menciana*. Phytochemistry 94:108–112

Hirayama C, Mase K, Iizuka T, Takasu Y, Okada E, Yamamoto K (2018) Deficiency of a pyrroline-5-carboxylate reductase produces the yellowish green cocoon 'Ryokuken' of the silkworm, *Bombyx mori*. Heredity 120(5):422–436

Hiruma K, Riddiford LM, Hopkins TL, Morgan TD (1985) Roles of dopa decarboxylase and phenoloxidase in the melanization of tobacco hornworm and their control by 20-hydroxyecdysone. J Comp Physiol B 155:659–669

Hopkins TL, Ahmad SA (1991) Flavonoid wing pigments in grasshoppers. Experientia 47:1089–1091

Iijima T, Kajitani R, Komata S, Lin CP, Sota T, Itoh T, Fujiwara H (2018) Parallel evolution of Batesian mimicry supergene in two *Papilio* butterflies, *P. polytes* and *P. memnon*. Sci Adv 4(4): eaao5416

Iijima T, Yoda S, Fujiwara H (2019) The mimetic wing pattern of *Papilio polytes* butterflies is regulated by a *doublesex*-orchestrated gene network. Commun Biol 2:257

Ito K, Katsuma S, Yamamoto K, Kadono-Okuda K, Mita K, Shimada T (2010) Yellow-e determines the color pattern of larval head and tail spots of the silkworm *Bombyx mori*. J Biol Chem 285(8):5624–5629

Ito K, Kidokoro K, Katsuma S, Shimada T, Yamamoto K, Mita K, Kadono-Okuda K (2012) Positional cloning of a gene responsible for the *cts* mutation of the silkworm, *Bombyx mori*. Genome 55(7):493–504

Jeong S, Rebeiz M, Andolfatto P, Werner T, True J, Carroll SB (2008) The evolution of gene regulation underlies a morphological difference between two *Drosophila* sister species. Cell 132 (5):783–793

Jin H, Seki T, Yamaguchi J, Fujiwara H (2019) Prepatterning of *Papilio xuthus* caterpillar camouflage is controlled by three homeobox genes: *clawless*, *abdominal-A*, and *Abdominal-B*. Sci Adv 5(4):eaav7569

Joron M, Frezal L, Jones RT, Chamberlain NL, Lee SF, Haag CR, Whibley A, Becuwe M, Baxter SW, Ferguson L, Wilkinson PA, Salazar C, Davidson C, Clark R, Quail MA, Beasley H, Glithero R, Lloyd C, Sims S, Jones MC, Rogers J, Jiggins CD, French-Constant RH (2011) Chromosomal rearrangements maintain a polymorphic supergene controlling butterfly mimicry. Nature 477:203–206

Kato T, Sawada H, Yamamoto T, Mase K, Nakagoshi M (2006) Pigment pattern formation in the *quail* mutant of the silkworm, *Bombyx mori*: parallel increase of pteridine biosynthesis and pigmentation of melanin and ommochromes. Pigment Cell Res 19(4):337–345

Kayser H (1985) Pigments. Comprehensive insect physiology. In: Kerkut GA, Gilbert LI (eds) Biochemistry and pharmacology, vol. 10. Pergamon, New York, pp 367–415

Kikkawa H (1941) Mechanism of pigment formation in *Bombyx* and *Drosophila*. Genetics 26 (6):587–607

Kim N, Kim J, Park D, Rosen C, Dorsett D, Yim J (1996) Structure and expression of wild-type and suppressible alleles of the *Drosophila purple* gene. Genetics 142(4):1157–1168

Kim H, Kim K, Yim J (2013) Biosynthesis of drosopterins, the red eye pigments of *Drosophila melanogaster*. IUBMB Life 65(4):334–340

Kiuchi T, Banno Y, Katsuma S, Shimada T (2011) Mutations in an amino acid transporter gene are responsible for sex-linked translucent larval skin of the silkworm, *Bombyx mori*. Insect Biochem Mol Biol 41(9):680–687

Koch PB, Keys DN, Rocheleau T, Aronstein K, Blackburn M, Carroll SB, French-Constant RH (1998) Regulation of dopa decarboxylase expression during colour pattern formation in wild-type and melanic tiger swallowtail butterflies. Development 125(12):2303–2313

Koch PB, Behnecke B, French-Constant RH (2000) The molecular basis of melanism and mimicry in a swallowtail butterfly. Curr Biol 10(10):591–594

Kômoto N, Sezutsu H, Yukuhiro K, Banno Y, Fujii H (2003) Mutations of the silkworm molybdenum cofactor sulfurase gene, *og*, cause translucent larval skin. Insect Biochem Mol Biol 33 (4):417–427

Kômoto N, Quan GX, Sezutsu H, Tamura T (2009) A single-base deletion in an ABC transporter gene causes white eyes, white eggs, and translucent larval skin in the silkworm *w-3(oe)* mutant. Insect Biochem Mol Biol 39(2):152–156

KonDo Y, Yoda S, Mizoguchi T, Ando T, Yamaguchi J, Yamamoto K, Banno Y, Fujiwara H (2017) Toll ligand Spätzle3 controls melanization in the stripe pattern formation in caterpillars. Proc Natl Acad Sci U S A 114(31):8336–8341

Kopp A, Duncan I, Godt D, Carroll SB (2000) Genetic control and evolution of sexually dimorphic characters in *Drosophila*. Nature 408(6812):553–559

Koshikawa S, Giorgianni MW, Vaccaro K, Kassner VA, Yoder JH, Werner T, Carroll SB (2015) Gain of cis-regulatory activities underlies novel domains of wingless gene expression in *Drosophila*. Proc Natl Acad Sci U S A 112(24):7524–7529

Krajíček J, Kozlík P, Exnerová A, Stys P, Bursová M, Cabala R, Bosáková Z (2014) Capillary electrophoresis of pterin derivatives responsible for the warning coloration of Heteroptera. J Chromatogr A 1336:94–100

Kronforst MR, Papa R (2015) The functional basis of wing patterning in Heliconius butterflies: the molecules behind mimicry. Genetics 200(1):1–19

Kunte K, Zhang W, Tenger-Trolander A, Palmer DH, Martin A, Reed RD, Mullen SP, Kronforst MR (2014) *doublesex* is a mimicry supergene. Nature 507(7491):229–232

Lewis JJ, Geltman RC, Pollak PC, Rondem KE, Van Belleghem SM, Hubisz MJ, Munn PR, Zhang L, Benson C, Mazo-Vargas A, Danko CG, Counterman BA, Papa R, Reed RD (2019) Parallel evolution of ancient, pleiotropic enhancers underlies butterfly wing pattern mimicry. Proc Natl Acad Sci U S A 116(48):24174–24183

Lindgren J, Nilsson DE, Sjövall P, Jarenmark M, Ito S, Wakamatsu K, Kear BP, Schultz BP, Sylvestersen RL, Madsen H, LaFountain JR Jr, Alwmark C, Eriksson ME, Hall SA, Lindgren P, Rodríguez-Meizoso I, Ahlberg P (2019) Fossil insect eyes shed light on trilobite optics and the arthropod pigment screen. Nature 573(7772):122–125

Linzen B (1974) The Tryptophan→Ommochrome pathway in insects. Adv Insect Physiol 10:117–246

Liu C, Yamamoto K, Cheng TC, Kadono-Okuda K, Narukawa J, Liu SP, Han Y, Futahashi R, Kidokoro K, Noda H, Kobayashi I, Tamura T, Ohnuma A, Banno Y, Dai FY, Xiang ZH, Goldsmith MR, Mita K, Xia QY (2010) Repression of tyrosine hydroxylase is responsible for the *sex-linked chocolate* mutation of the silkworm, *Bombyx mori*. Proc Natl Acad Sci U S A 107 (29):12980–12985

Liu J, Lemonds TR, Popadić A (2014) The genetic control of aposematic black pigmentation in hemimetabolous insects: insights from *Oncopeltus fasciatus*. Evol Dev 16(5):270–277

Liu J, Lemonds TR, Marden JH, Popadić A (2016) A pathway analysis of melanin patterning in a hemimetabolous insect. Genetics 203(1):403–413

Liu SH, Luo J, Yang BJ, Wang AY, Tang J (2019) *karmoisin* and *cardinal* ortholog genes participate in the ommochrome synthesis of *Nilaparvata lugens* (Hemiptera: Delphacidae). Insect Sci 26(1):35–43

Lloyd V, Ramaswami M, Krämer H (1998) Not just pretty eyes: *Drosophila* eye-colour mutations and lysosomal delivery. Trends Cell Biol 8:257–259

Lorenzen MD, Brown SJ, Denell RE, Beeman RW (2002) Cloning and characterization of the *Tribolium castaneum* eye-color genes encoding tryptophan oxygenase and kynurenine 3-monooxygenase. Genetics 160(1):225–234

Makino K, Satoh K, Koike M, Ueno N (1952) Sex in *Pieris rapae* L. and the pteridine content of their wings. Nature 170:933–934

Maoka T, Kawase N, Ueda T, Nishida R (2020) Carotenoids of dragonflies, from the perspective of comparative biochemical and chemical ecological studies. Biochem Syst Ecol 89:104001

Martin A, Reed RD (2010) *wingless* and *aristaless2* define a developmental ground plan for moth and butterfly wing pattern evolution. Mol Biol Evol 27:2864–2878

Martin A, Reed RD (2014) *Wnt* signaling underlies evolution and development of the butterfly wing pattern symmetry systems. Dev Biol 395:367–378

Martin A, Papa R, Nadeau NJ, Hill RI, Counterman BA, Halder G, Jiggins CD, Kronforst MR, Long AD, McMillan WO, Reed RD (2012) Diversification of complex butterfly wing patterns by repeated regulatory evolution of a *Wnt* ligand. Proc Natl Acad Sci U S A 109:12632–12637

Massey JH, Akiyama N, Bien T, Dreisewerd K, Wittkopp PJ, Yew JY, Takahashi A (2019a) Pleiotropic effects of *ebony* and *tan* on pigmentation and cuticular hydrocarbon composition in *Drosophila melanogaster*. Front Physiol 10:518

Massey JH, Chung D, Siwanowicz I, Stern DL, Wittkopp PJ (2019b) The *yellow* gene influences *Drosophila* male mating success through sex comb melanization. elife 8:e49388

Masuoka Y, Maekawa K (2016) Gene expression changes in the tyrosine metabolic pathway regulate caste-specific cuticular pigmentation of termites. Insect Biochem Mol Biol 74:21–31

Matsuoka Y, Monteiro A (2018) Melanin pathway genes regulate color and morphology of butterfly wing scales. Cell Rep 24(1):56–65

Mazo-Vargas A, Concha C, Livraghi L, Massardo D, Wallbank RWR, Zhang L, Papador JD, Martinez-Najera D, Jiggins CD, Kronforst MR, Breuker CJ, Reed RD, Patel NH, McMillan WO, Martin A (2017) Macroevolutionary shifts of WntA function potentiate butterfly wing-pattern diversity. Proc Natl Acad Sci U S A 114(40):10701–10706

McLean JR, Krishnakumar S, O'Donnell JM (1993) Multiple mRNAs from the *Punch* locus of *Drosophila melanogaster* encode isoforms of GTP cyclohydrolase I with distinct N-terminal domains. J Biol Chem 268(36):27191–27197

Meng Y, Katsuma S, Mita K, Shimada T (2009a) Abnormal red body coloration of the silkworm, *Bombyx mori*, is caused by a mutation in a novel kynureninase. Genes Cells 14(2):129–140

Meng Y, Katsuma S, Daimon T, Banno Y, Uchino K, Sezutsu H, Tamura T, Mita K, Shimada T (2009b) The silkworm mutant *lemon* (*lemon lethal*) is a potential insect model for human sepiapterin reductase deficiency. J Biol Chem 284(17):11698–11705

Miura K, Shinoda T, Yura M, Nomura S, Kamiya K, Yuda M, Chinzei Y (1998) Two hexameric cyanoprotein subunits from an insect, Riptortus clavatus. Sequence, phylogeny and developmental and juvenile hormone regulation. Eur J Biochem 258(3):929–940

Monteiro A, Prudic KL (2010) Multiple approaches to study color pattern evolution in butterflies. Trends Evol Biol 2:e2

Montell I, Rasmuson A, Rasmuson B, Holmgren P (1992) Uptake and incorporation in pteridines of externally supplied GTP in normal and pigment-deficient eyes of *Drosophila melanogaster*. Biochem Genet 30(1-2):61–75

Moran NA, Jarvik T (2010) Lateral transfer of genes from fungi underlies carotenoid production in aphids. Science 328(5978):624–627

Mullins C, Hartnell LM, Bonifacino JS (2000) Distinct requirements for the AP-3 adaptor complex in pigment granule and synaptic vesicle biogenesis in *Drosophila melanogaster*. Mol Gen Genet 263(6):1003–1014

Mun S, Noh MY, Kramer KJ, Muthukrishnan S, Arakane Y (2020) Gene functions in adult cuticle pigmentation of the yellow mealworm, *Tenebrio molitor*. Insect Biochem Mol Biol 117:103291

Nadeau NJ, Pardo-Diaz C, Whibley A, Supple MA, Saenko SV, Wallbank RW, Wu GC, Maroja L, Ferguson L, Hanly JJ, Hines H, Salazar C, Merrill RM, Dowling AJ, French-Constant RH, Llaurens V, Joron M, McMillan WO, Jiggins CD (2016) The gene *cortex* controls mimicry and crypsis in butterflies and moths. Nature 534(7605):106–110

Nakagoshi M, Masada M, Tsutsue M (1984) The nature of the seasonal colour dimorphism in the scorpion fly, *Panorpa japonica* Thunberg. Insect Biochem 14(6):615–618

Nakagoshi M, Kondo R, Sawada H, Takikawa S, Yoshida A (2002) Pteridines and pigment granules of wing scales concerned with sexual difference in wing's capability reflecting near-UV rays in the Japanese cabbage butterfly *Pieris rapae crucibora*. In: Milstien S, Kapatos G, Levine RA, Shane B (eds) Chemistry and biology of pteridines and folates. Springer, Boston, pp 229–234

Nash D, Hu S, Leonard NJ, Tiong SY, Fillips D (1994) The *raspberry* locus of *Drosophila melanogaster* includes an inosine monophosphate dehydrogenase like coding sequence. Genome 37(2):333–244

Navrotskaya V, Wnorowski A, Turski W, Oxenkrug G (2018) Effect of kynurenic acid on pupae viability of *Drosophila melanogaster cinnabar* and *cardinal* eye color mutants with altered tryptophan-kynurenine metabolism. Neurotox Res 34(2):324–331

Nijhout HF (1997) Ommochrome pigmentation of the *linea* and *rosa* seasonal forms of *Precis coenia* (Lepidoptera: Nymphalidae). Arch Insect Biochem Physiol 36:215–222

Nikoh N, Tsuchida T, Maeda T, Yamaguchi K, Shigenobu S, Koga R, Fukatsu T (2018) Genomic insight into symbiosis-induced insect color change by a facultative bacterial endosymbiont, *Candidatus* Rickettsiella viridis. MBio 9(3):e00890

Ninomiya Y, Tanaka K, Hayakawa Y (2006) Mechanisms of black and white stripe pattern formation in the cuticles of insect larvae. J Insect Physiol 52:638–645

Nishide Y, Kageyama D, Hatakeyama M, Yokoi K, Jouraku A, Tanaka H, Koga R, Futahashi R, Fukatsu T (2020) Diversity and function of multicopper oxidase genes in the stinkbug *Plautia stali*. Sci Rep 10(1):3464

Nishikawa H, Iga M, Yamaguchi J, Saito K, Kataoka H, Suzuki Y, Sugano S, Fujiwara H (2013) Molecular basis of wing coloration in a Batesian mimic butterfly, *Papilio polytes*. Sci Rep 3:3184

Nishikawa H, Iijima T, Kajitani R, Yamaguchi J, Ando T, Suzuki Y, Sugano S, Fujiyama A, Kosugi S, Hirakawa H, Tabata S, Ozaki K, Morimoto H, Ihara K, Obara M, Hori H, Itoh T, Fujiwara H (2015) A genetic mechanism for female-limited Batesian mimicry in *Papilio* butterfly. Nat Genet 47(4):405–409

Noh MY, Kramer KJ, Muthukrishnan S, Beeman RW, Kanost MR, Arakane Y (2015) Loss of function of the *yellow-e* gene causes dehydration-induced mortality of adult *Tribolium castaneum*. Dev Biol 399(2):315–324

Noh MY, Koo B, Kramer KJ, Muthukrishnan S, Arakane Y (2016) A*rylalkylamine N-acetyltransferase 1* gene (TcAANAT1) is required for cuticle morphology and pigmentation of the adult red flour beetle, *Tribolium castaneum*. Insect Biochem Mol Biol 79:119–129

Noh MY, Kim SH, Gorman MJ, Kramer KJ, Muthukrishnan S, Arakane Y (2020) Yellow-g and yellow-g2 proteins are required for egg desiccation resistance and temporal pigmentation in the Asian tiger mosquito, *Aedes albopictus*. Insect Biochem Mol Biol 117:103386

Nováková E, Moran NA (2012) Diversification of genes for carotenoid biosynthesis in aphids following an ancient transfer from a fungus. Mol Biol Evol 29(1):313–323

O'Donnell AF, Tiong S, Nash D, Clark DV (2000) The *Drosophila melanogaster ade5* gene encodes a bifunctional enzyme for two steps in the de novo purine synthesis pathway. Genetics 154(3):1239–1253

O'Hare K, Murphy C, Levis R, Rubin GM (1984) DNA sequence of the *white* locus of *Drosophila melanogaster*. J Mol Biol 180(3):437–455

Okude G, Futahashi R (2021) Pigmentation and color pattern diversity in Odonata. Curr Opin Genet Dev 69:14–20

Okude G, Futahashi R, Kawahara-Miki R, Yoshitake K, Yajima S, Fukatsu T (2017) Electroporation-mediated RNA interference reveals a role of multicopper oxidase 2 gene in dragonfly's cuticular pigmentation. Appl Entomol Zool 52:379–387

Osanai-Futahashi M, Tatematsu K, Yamamoto K, Narukawa J, Uchino K, Kayukawa T, Shinoda T, Banno Y, Tamura T, Sezutsu H (2012a) Identification of the *Bombyx red egg* gene reveals involvement of a novel transporter family gene in late steps of the insect ommochrome biosynthesis pathway. J Biol Chem 287(21):17706–17714

Osanai-Futahashi M, Ohde T, Hirata J, Uchino K, Futahashi R, Tamura T, Niimi T, Sezutsu H (2012b) A visible dominant marker for insect transgenesis. Nat Commun 3:1295

Osanai-Futahashi M, Tatematsu KI, Futahashi R, Narukawa J, Takasu Y, Kayukawa T, Shinoda T, Ishige T, Yajima S, Tamura T, Yamamoto K, Sezutsu H (2016) Positional cloning of a *Bombyx pink-eyed white egg* locus reveals the major role of *cardinal* in ommochrome synthesis. Heredity 116(2):135–145

Osorio D, Vorobyev M (2008) A review of the evolution of animal colour vision and visual communication signals. Vis Res 48:2042–2051

Palmer DH, Kronforst MR (2019) A shared genetic basis of mimicry across swallowtail butterflies points to ancestral co-option of *doublesex*. Nat Commun 11(1):6

Panettieri S, Gjinaj E, John G, Lohman DJ (2018) Different ommochrome pigment mixtures enable sexually dimorphic Batesian mimicry in disjunct populations of the common palmfly butterfly, *Elymnias hypermnestra*. PLoS One 13(9):e0202465

Park D, Park S, Yim J (2000) Molecular characterization of *Drosophila melanogaster* dihydropteridine reductase. Biochim Biophys Acta 1492(1):247–251

Peng CL, Mazo-Vargas A, Brack BJ, Reed RD (2020) Multiple roles for laccase2 in butterfly wing pigmentation, scale development, and cuticle tanning. Evol Dev 22(4):336–341

Prum RO, Cole JA, Torres RH (2004) Blue integumentary structural colours in dragonflies (Odonata) are not produced by incoherent Tyndall scattering. J Exp Biol 207:3999–4009

Quan GX, Kim I, Kômoto N, Sezutsu H, Ote M, Shimada T, Kanda T, Mita K, Kobayashi M, Tamura T (2002) Characterization of the *kynurenine 3-monooxygenase* gene corresponding to the white egg 1 mutant in the silkworm *Bombyx mori*. Mol Gen Genomics 267(1):1–9

Reaume AG, Knecht DA, Chovnick A (1991) The *rosy* locus in *Drosophila melanogaster*: xanthine dehydrogenase and eye pigments. Genetics 129(4):1099–1109

Rebeiz M, Pool JE, Kassner VA, Aquadro CF, Carroll SB (2009) Stepwise modification of a modular enhancer underlies adaptation in a *Drosophila* population. Science 326 (5960):1663–1667

Reed RD, McMillan WO, Nagy LM (2008) Gene expression underlying adaptive variation in *Heliconius* wing patterns: non-modular regulation of overlapping *cinnabar* and *vermilion* prepatterns. Proc Biol Sci 275(1630):37–45

Reed RD, Papa R, Martin A, Hines HM, Counterman BA, Pardo-Diaz C, Jiggins CD, Chamberlain NL, Kronforst MR, Chen R, Halder G, Nijhout HF, McMillan WO (2011) *optix* drives the repeated convergent evolution of butterfly wing pattern mimicry. Science 333:1137–1141

Riddiford LM, Palli SR, Hiruma K, Li W, Green J, Hice RH, Wolfgang WJ, Webb BA (1990) Developmental expression, synthesis, and secretion of insecticyanin by the epidermis of the tobacco hornworm, *Manduca sexta*. Arch Insect Biochem Physiol 14(3):171–190

Riedel F, Vorkel D, Eaton S (2011) Megalin-dependent yellow endocytosis restricts melanization in the *Drosophila* cuticle. Development 138(1):149–158

Saenko SV, Brakefield PM, Beldade P (2010) Single locus affects embryonic segment polarity and multiple aspects of an adult evolutionary novelty. BMC Biol 8:111

Sakudoh T, Tsuchida K, Kataoka H (2005) BmStart1, a novel carotenoid-binding protein isoform from *Bombyx mori*, is orthologous to MLN64, a mammalian cholesterol transporter. Biochem Biophys Res Commun 336(4):1125–1135

Sakudoh T, Sezutsu H, Nakashima T, Kobayashi I, Fujimoto H, Uchino K, Banno Y, Iwano H, Maekawa H, Tamura T, Kataoka H, Tsuchida K (2007) Carotenoid silk coloration is controlled by a carotenoid-binding protein, a product of the *Yellow blood* gene. Proc Natl Acad Sci U S A 104(21):8941–8946

Sakudoh T, Iizuka T, Narukawa J, Sezutsu H, Kobayashi I, Kuwazaki S, Banno Y, Kitamura A, Sugiyama H, Takada N, Fujimoto H, Kadono-Okuda K, Mita K, Tamura T, Yamamoto K, Tsuchida K (2010) A CD36-related transmembrane protein is coordinated with an intracellular lipid-binding protein in selective carotenoid transport for cocoon coloration. J Biol Chem 285 (10):7739–7751

Sakudoh T, Kuwazaki S, Iizuka T, Narukawa J, Yamamoto K, Uchino K, Sezutsu H, Banno Y, Tsuchida K (2013) CD36 homolog divergence is responsible for the selectivity of carotenoid species migration to the silk gland of the silkworm *Bombyx mori*. J Lipid Res 54(2):482–495

Santoro P, Parisi G (1986) A new enzyme from *Drosophila melanogaster*: in vitro conversion of xanthommatin into its dihydroform by means of xanthommatin reductase. J Exp Zool 239:169–173

Sawada H, Nakagoshi M, Reinhardt RK, Ziegler I, Koch PB (2002) Hormonal control of *GTP cyclohydrolase I* gene expression and enzyme activity during color pattern development in wings of *Precis coenia*. Insect Biochem Mol Biol 32(6):609–615

Schmidt FS, Skerra A (1994) The bilin-binding protein of Pieris brassicae. cDNA sequence and regulation of expression reveal distinct features of this insect pigment protein. Eur J Biochem 219(3):855–863

Schopf R, Mignat C, Hedden P (1982) As to the food quality of spruce needles for forest damaging insects. 18. Resorption of secondary plant metabolites by the sawfly *Gilpinia hercyniae* Htg (Hym., Diprionidae). Z Angew Entomol 93:244–257

Searles LL, Ruth RS, Pret AM, Fridell RA, Ali AJ (1990) Structure and transcription of the *Drosophila melanogaster vermilion* gene and several mutant alleles. Mol Cell Biol 10 (4):1423–1431

Shirataki H, Futahashi R, Fujiwara H (2010) Species-specific coordinated gene expression and trans-regulation of larval color pattern in three swallowtail butterflies. Evol Dev 12(3):305–314

Sloan DB, Moran NA (2012) Endosymbiotic bacteria as a source of carotenoids in whiteflies. Biol Lett 8(6):986–989

Spana EP, Abrams AB, Ellis KT, Klein JC, Ruderman BT, Shi AH, Zhu D, Stewart A, May S (2020) *Speck*, first identified in *Drosophila melanogaster* in 1910, is encoded by the Arylalkalamine N-Acetyltransferase (AANAT1) gene. G3 10(9):3387–3398

Stavenga DG, Leertouwer HL, Wilts BD (2014) The colouration toolkit of the pipevine swallowtail butterfly, *Battus philenor*: thin films, papiliochromes, and melanin. J Comp Phys A 200:547–561

Sugahara R, Tanaka S (2019) Yellowing and YPT gene expression in the desert locust, *Schistocerca gregaria*: effects of developmental stages and fasting. Arch Insect Biochem Physiol 101(2): e21551

Sugahara R, Tsuchiya W, Yamazaki T, Tanaka S, Shiotsuki T (2020) Recombinant yellow protein of the takeout family and albino-related takeout protein specifically bind to lutein in the desert locust. Biochem Biophys Res Commun 522(4):876–880

Sugumaran M (2002) Comparative biochemistry of eumelanogenesis and the protective roles of phenoloxidase and melanin in insects. Pigment Cell Res 15(1):2–9

Takahashi A, Takano-Shimizu T (2011) Divergent enhancer haplotype of ebony on inversion In (3R)Payne associated with pigmentation variation in a tropical population of *Drosophila melanogaster*. Mol Ecol 20(20):4277–4287

Takahashi A, Takahashi K, Ueda R, Takano-Shimizu T (2007) Natural variation of *ebony* gene controlling thoracic pigmentation in *Drosophila melanogaster*. Genetics 177(2):1233–1237

Tatematsu K, Yamamoto K, Uchino K, Narukawa J, Iizuka T, Banno Y, Katsuma S, Shimada T, Tamura T, Sezutsu H, Daimon T (2011) Positional cloning of silkworm *white egg 2* (*w-2*) locus shows functional conservation and diversification of ABC transporters for pigmentation in insects. Genes Cells 16(4):331–342

Tearle RG, Belote JM, McKeown M, Baker BS, Howells AJ (1989) Cloning and characterization of the *scarlet* gene of *Drosophila melanogaster*. Genetics 122(3):595–606

Tian L, Rahman SR, Ezray BD, Franzini L, Strange JP, Lhomme P, Hines HM (2019) A homeotic shift late in development drives mimetic color variation in a bumble bee. Proc Natl Acad Sci U S A 116(24):11857–11865

Tillyard RJ (1917) The biology of dragonflies. Cambridge University Press, Cambridge

Timmermans MJ, Baxter SW, Clark R, Heckel DG, Vogel H, Collins S, Papanicolaou A, Fukova I, Joron M, Thompson MJ, Jiggins CD, French-Constant RH, Vogler AP (2014) Comparative genomics of the mimicry switch in *Papilio dardanus*. Proc Biol Sci 281(1787):20140465

Tomoda A, Yoneyama Y, Yamaguchi T, Shirao E, Kawasaki K (1990) Mechanism of coloration of human lenses induced by near-ultraviolet-photo-oxidized 3-hydroxykynurenine. Ophthalmic Res 22(3):152–159

True JR, Edwards KA, Yamamoto D, Carroll SB (1999) *Drosophila* wing melanin patterns form by vein-dependent elaboration of enzymatic prepatterns. Curr Biol 9(23):1382–1391

True JR, Yeh SD, Hovemann BT, Kemme T, Meinertzhagen IA, Edwards TN, Liou SR, Han Q, Li J (2005) *Drosophila tan* encodes a novel hydrolase required in pigmentation and vision. PLoS Genet 1(5):e63

Tsuchida T (2016) Molecular basis and ecological relevance of aphid body colors. Curr Opin Insect Sci 17:74–80

Tsuchida T, Koga R, Horikawa M, Tsunoda T, Maoka T, Matsumoto S, Simon JC, Fukatsu T (2010) Symbiotic bacterium modifies aphid body color. Science 330(6007):1102–1104

Tsujita M, Sakurai S (1965) Purification of the three specific soluble chromoproteins from chromogranules in hypodermal cells of the silkworm larva. Proc Jpn Acad 41:225–229

Umbers KD, Fabricant SA, Gawryszewski FM, Seago AE, Herberstein ME (2014) Reversible colour change in Arthropoda. Biol Rev 89(4):820–848

Umebachi Y (1985) Papiliochrome, a new pigment group of butterfly. Zool Sci 2:163–174

Umebachi Y (2000) The pigment of animals. Uchida-Rohkakuho, Tokyo

Van Belleghem SM, Rastas P, Papanicolaou A, Martin SH, Arias CF, Supple MA, Hanly JJ, Mallet J, Lewis JJ, Hines HM, Ruiz M, Salazar C, Linares M, Moreira GRP, Jiggins CD, Counterman BA, McMillan WO, Papa R (2017) Complex modular architecture around a simple toolkit of wing pattern genes. Nat Ecol Evol 1(3):52

van der Burg KRL, Lewis JJ, Brack BJ, Fandino RA, Mazo-Vargas A, Reed RD (2020) Genomic architecture of a genetically assimilated seasonal color pattern. Science 370(6517):721–725

Van't Hof AE, Campagne P, Rigden DJ, Yung CJ, Lingley J, Quail MA, Hall N, Darby AC, Saccheri IJ (2016) The industrial melanism mutation in British peppered moths is a transposable element. Nature 534(7605):102–105

Vargas-Lowman A, Armisen D, Burguez Floriano CF, da Rocha Silva Cordeiro I, Viala S, Bouchet M, Bernard M, Le Bouquin A, Santos ME, Berlioz-Barbier A, Salvador A, Figueiredo Moreira FF, Bonneton F, Khila A (2019) Cooption of the pteridine biosynthesis pathway underlies the diversification of embryonic colors in water striders. Proc Natl Acad Sci U S A 116(38):19046–19054

Veron JE, O'Farrell AF, Dixon B (1974) The fine structure of Odonata chromatophores. Tissue Cell 6(4):613–626

Vigneron A, Masson F, Vallier A, Balmand S, Rey M, Vincent-Monégat C, Aksoy E, Aubailly-Giraud E, Zaidman-Rémy A, Heddi A (2014) Insects recycle endosymbionts when the benefit is over. Curr Biol 24:2267–2273

Walter MF, Black BC, Afshar G, Kermabon AY, Wright TR, Biessmann H (1991) Temporal and spatial expression of the *yellow* gene in correlation with cuticle formation and dopa decarboxylase activity in *Drosophila* development. Dev Biol 147:32–45

Wang Q, Zhao C, Bai L, Deng X, Wu C (2008) Reduction of drosopterin content caused by a 45-nt insertion in *Henna* pre-mRNA of *Drosophila melanogaster*. Sci China Life Sci 51(8):702–710

Wang L, Kiuchi T, Fujii T, Daimon T, Li M, Banno Y, Katsuma S, Shimada T (2013a) Reduced expression of the *dysbindin-like* gene in the *Bombyx mori ov* mutant exhibiting mottled translucency of the larval skin. Genome 56(2):101–108

Wang L, Kiuchi T, Fujii T, Daimon T, Li M, Banno Y, Kikuta S, Kikawada T, Katsuma S, Shimada T (2013b) Mutation of a novel ABC transporter gene is responsible for the failure to incorporate uric acid in the epidermis of ok mutants of the silkworm, *Bombyx mori*. Insect Biochem Mol Biol 43(7):562–571

Wang L, Yin Y, Wang K, Cao J, Cheng T, Liu C, Zhang Y, Zhu Y (2020) *Bombyx mori* monocarboxylate transporter 9 (BmMCT9) is involved in the transport of uric acid in silkworm integument. Genes Cells 25(1):33–40

Warren WD, Palmer S, Howells AJ (1996) Molecular characterization of the *cinnabar* region of *Drosophila melanogaster*: identification of the *cinnabar* transcription unit. Genetica 98 (3):249–262

Watt WB (1964) Pteridine components of wing pigmentation in the butterfly *Colias eurytheme*. Nature 201:1326–1327

Watt WB (1973) Adaptive significance of pigment polymorphisms in *Colias* butterflies. III. Progress in the study of the '*alba*' variant. Evolution 27:537–548

Werner T, Koshikawa S, Williams TM, Carroll SB (2010) Generation of a novel wing colour pattern by the Wingless morphogen. Nature 464(7292):1143–1148

Westerman EL, VanKuren NW, Massardo D, Tenger-Trolander A, Zhang W, Hill RI, Perry M, Bayala E, Barr K, Chamberlain N, Douglas TE, Buerkle N, Palmer SE, Kronforst MR (2018) *Aristaless* controls butterfly wing color variation used in mimicry and mate choice. Curr Biol 28 (21):3469–3474

Williams RA, Mamotte CD, Burnett JR (2008) Phenylketonuria: an inborn error of phenylalanine metabolism. Clin Biochem Rev 29(1):31–41

Wilson A (1986) Flavonoid pigments and wing color in *Melanargia galathea*. J Chem Ecol 12 (1):49–68

Wilson A (1987) Flavonoid pigments in chalkhill blue (*Lysandra coridon* Poda) and other lycaenid butterflies. J Chem Ecol 13(3):473–493

Wilts BD, Matsushita A, Arikawa K, Stavenga DG (2015) Spectrally tuned structural and pigmentary coloration of birdwing butterfly wing scales. J R Soc Interface 12:20150717

Wittkopp PJ, True JR, Carroll SB (2002a) Reciprocal functions of the *Drosophila* Yellow and Ebony proteins in the development and evolution of pigment patterns. Development 129:1849–1858

Wittkopp PJ, Vaccaro K, Carroll SB (2002b) Evolution of yellow gene regulation and pigmentation in Drosophila. Curr Biol 12(18):1547–1556

Wittkopp PJ, Williams BL, Selegue JE, Carroll SB (2003) *Drosophila* pigmentation evolution: divergent genotypes underlying convergent phenotypes. Proc Natl Acad Sci U S A 100 (4):1808–1813

Wittkopp PJ, Stewart EE, Arnold LL, Neidert AH, Haerum BK, Thompson EM, Akhras S, Smith-Winberry G, Shefner L (2009) Intraspecific polymorphism to interspecific divergence: genetics of pigmentation in *Drosophila*. Science 326(5952):540–544

Woronik A, Tunström K, Perry MW, Neethiraj R, Stefanescu C, Celorio-Mancera MP, Brattström O, Hill J, Lehmann P, Käkelä R, Wheat CW (2019) A transposable element insertion is associated with an alternative life history strategy. Nat Commun 10(1):5757

Wright TRF (1987) The genetics of biogenic amine metabolism, sclerotization, and melanization in *Drosophila melanogaster*. Adv Genet 24:127–222

Xiao D, Chen X, Tian R, Wu M, Zhang F, Zang L, Harwood JD, Wang S (2020) Molecular and potential regulatory mechanisms of melanin synthesis in *Harmonia axyridis*. Int J Mol Sci 21 (6):E2088

Yamaguchi J, Banno Y, Mita K, Yamamoto K, Ando T, Fujiwara H (2013) Periodic Wnt1 expression in response to ecdysteroid generates twin-spot markings on caterpillars. Nat Commun 4:1857

Yang M, Wang Y, Liu Q, Liu Z, Jiang F, Wang H, Guo X, Zhang J, Kang L (2019) A β-carotene-binding protein carrying a red pigment regulates body-color transition between green and black in locusts. elife 8:e41362

Yassin A, Bastide H, Chung H, Veuille M, David JR, Pool JE (2016a) Ancient balancing selection at tan underlies female colour dimorphism in *Drosophila erecta*. Nat Commun 7:10400

Yassin A, Delaney EK, Reddiex AJ, Seher TD, Bastide H, Appleton NC, Lack JB, David JR, Chenoweth SF, Pool JE, Kopp A (2016b) The pdm3 locus is a hotspot for recurrent evolution of female-limited color dimorphism in *Drosophila*. Curr Biol 26(18):2412–2422

Yoda S, Yamaguchi J, Mita K, Yamamoto K, Banno Y, Ando T, Daimon T, Fujiwara H (2014) The transcription factor *Apontic-like* controls diverse colouration pattern in caterpillars. Nat Commun 5:4936

Yoda S, Otaguro E, Nobuta M, Fujiwara H (2020) Molecular mechanisms underlying pupal protective color switch in *Papilio polytes* butterflies. Front Ecol Evol 8:51

Yoshioka S, Kinoshita S (2006) Structural or pigmentary? Origin of the distinctive white stripe on the blue wing of a Morpho butterfly. Proc Biol Sci 273(1583):129–134

Yuasa M, Kiuchi T, Banno Y, Katsuma S, Shimada T (2016) Identification of the silkworm *quail* gene reveals a crucial role of a receptor guanylyl cyclase in larval pigmentation. Insect Biochem Mol Biol 68:33–40

Zhan S, Guo Q, Li M, Li M, Li J, Miao X, Huang Y (2010) Disruption of an N-acetyltransferase gene in the silkworm reveals a novel role in pigmentation. Development 137(23):4083–4090

Zhan S, Zhang W, Niitepõld K, Hsu J, Haeger JF, Zalucki MP, Altizer S, de Roode JC, Reppert SM, Kronforst MR (2014) The genetics of monarch butterfly migration and warning colouration. Nature 514(7522):317–321

Zhang L, Reed RD (2016) Genome editing in butterflies reveals that *spalt* promotes and *Distal-less* represses eyespot colour patterns. Nat Commun 7:11769

Zhang L, Martin A, Perry MW, van der Burg KR, Matsuoka Y, Monteiro A, Reed RD (2017a) Genetic basis of melanin pigmentation in butterfly wings. Genetics 205(4):1537–1550

Zhang L, Mazo-Vargas A, Reed RD (2017b) Single master regulatory gene coordinates the evolution and development of butterfly color and iridescence. Proc Natl Acad Sci U S A 114 (40):10707–10712

Zhang H, Kiuchi T, Wang L, Kawamoto M, Suzuki Y, Sugano S, Banno Y, Katsuma S, Shimada T (2017c) *Bm-muted*, orthologous to mouse muted and encoding a subunit of the BLOC-1 complex, is responsible for the *otm* translucent mutation of the silkworm *Bombyx mori*. Gene 629:92–100

Zhang H, Kiuchi T, Hirayama C, Katsuma S, Shimada T (2018) *Bombyx* ortholog of the *Drosophila* eye color gene *brown* controls riboflavin transport in Malpighian tubules. Insect Biochem Mol Biol 92:65–72

Zhang Y, Li H, Du J, Zhang J, Shen J, Cai W (2019a) Three melanin pathway genes, *TH*, *yellow*, and *aaNAT*, regulate pigmentation in the twin-spotted assassin bug, *Platymeris biguttatus* (Linnaeus). Int J Mol Sci 20(11):E2728

Zhang Y, Wang XX, Feng ZJ, Cong HS, Chen ZS, Li YD, Yang WM, Zhang SQ, Shen LF, Tian HG, Feng Y, Liu TX (2019b) Superficially similar adaptation within one species exhibits similar morphological specialization but different physiological regulations and origins. Front Cell Dev Biol 8:300

Zhao Y, Zhang H, Li Z, Duan J, Jiang J, Wang Y, Zhan S, Akinkurolere RO, Xu A, Qian H, Miao X, Tan A, Huang Y (2012) A major facilitator superfamily protein participates in the reddish brown pigmentation in *Bombyx mori*. J Insect Physiol 58(11):1397–1405

Chapter 2
Melanins in Vertebrates

Kazumasa Wakamatsu and Shosuke Ito

Abstract It is known that melanin pigments composed of black to dark brown eumelanin (EM) and yellow to reddish-brown pheomelanin (PM) are widely distributed in vertebrates. Melanin pigments in vertebrate are produced in melanocytes which are distributed in the epidermis, hair follicles, choroid, iris, inner ear, and other tissues. The diversity of the color in the internal and external tissues in vertebrates is mainly attributed to the quantity and ratio of EM and PM. Melanin pigments are highly oxidized and complex pigments. It is thus important to analyze these two types of melanin pigments in studies of the biochemical and genetic bases of pigmentation in vertebrates. Two microanalytical methods to perform the simultaneous measurement of eumelanin and pheomelanin were established to characterize melanin and melanogenesis.

In this chapter, we explain microanalytical application with the chemical degradation of melanin and focus on internal and external melanins produced by pigment cells in vertebrates: mammals, birds, fish, reptiles, and amphibians. By using these methods for the evaluation of the "chemical phenotype," the mutual triangular relationship among "chemical phenotype," "visual phenotype," and "genotype" is reviewed.

Keywords Melanin · Eumelanin · Pheomelanin · Neuromelanin · Vertebrates

Abbreviations

3-AHP	3-amino-4-hydroxyphenylalanine
4-AHP	4-amino-3-hydroxyphenylalanine
5SCD	5-*S*-cysteinyldopa
8-oxodG	8-oxo-7,8-dihydro-2′-deoxyguanosine
α-MSH	α-melanocyte-stimulating hormone

K. Wakamatsu (✉) · S. Ito
Institute for Melanin Chemistry, Fujita Health University, Toyoake, Japan
e-mail: kwaka@fujita-hu.ac.jp

H. Hashimoto et al. (eds.), *Pigments, Pigment Cells and Pigment Patterns*,
https://doi.org/10.1007/978-981-16-1490-3_2

A500	absorption at 500 nm
Asip	agouti signaling protein
BT	benzothiazine
BT-PM	PM with a BT moiety
BZ	benzothiazole
BZ-AA	6-(2-amino-2-carboxyethyl)-4-hydroxybenzothiazole
BZ-PM	PM with a BZ moiety
cAMP	cyclic adenosine monophosphate
CD	cysteinyldopa
CPD	cyclobutane pyrimidine dimer
CTNS	cystinosin, lysosomal cystine transporter
CYS	cysteine
DA	dopamine
DC	dopachrome
DCT	dopachrome tautomerase
DHBTCA	7-(2-amino-2-carboxyethyl)-5-hydroxy-dihydro-1,4-benzothiazine-3-carboxylic acid
DHI	5,6-dihydroxyindole
DHICA	5,6-dihydroxyindole-2-carboxylic acid
DOMA	3,4-dihydroxymandelic acid
DOPA	3,4-dihydroxyphenylalanine
DOPAC	3,4-dihydroxyphenylacetic acid
DOPET	3,4-dihydroxyphenylethanol
DOPEG	3,4-dihydroxyphenylethylene glycol
DQ	dopaquinone
EM	eumelanin
EPR	electron paramagnetic resonance
GSH	glutathione
HPLC	high-performance liquid chromatography
HPS	Hermansky–Pudlak syndrome
iNOS	inducible nitric oxide synthase
isoPTCA	pyrrole-2,3,4-tricarboxylic acid
LC	locus coeruleus
MATP	membrane-associated transporter protein
MCH	melanin-concentrating hormone
MC1R	melanocortin-1 receptor gene
MFSD12	major facilitator superfamily domain-containing protein 12
MITF	microphthalmia-associated transcription factor
MRI	magnetic resonance imaging
NADPH	nicotinamide adenine dinucleotide phosphate
NE	norepinephrine
NM	neuromelanin

NOX	reduced nicotinamide adenine dinucleotide phosphate (NADPH) oxidase
OCA	oculocutaneous albinism
PD	Parkinson's disease
PM	pheomelanin
PMEL	premelanosome protein
PDCA	pyrrole-2,3-dicarboxylic acid
PTCA	pyrrole-2,3,5-tricarboxylic acid
PTeCA	pyrrole-2,3,4,5-tetracarboxylic acid
RD	rhododendrol
RES	resveratrol
ROS	reactive oxygen species
sAC	soluble adenylyl cyclase
SN	substantia nigra
SPE	solid-phase extraction
Sox10	SRY-Box transcription factor 10
TDCA	thiazole-4,5-dicarboxylic acid
TTCA	thiazole-2,4,5-tricarboxylic acid
Tyrp	tyrosinase-related protein
UV	ultraviolet

2.1 Introduction

Melanin pigments are composed of black to dark brown insoluble eumelanin (EM) and yellow to reddish-brown alkali-soluble pheomelanin (PM) (Ito and Wakamatsu 2003, 2008) (Fig. 2.1). Melanins are widely distributed in animals and plants; in vertebrates, most melanins are present on the body surface. Melanin pigments in vertebrate are produced in melanocytes within membrane-bound

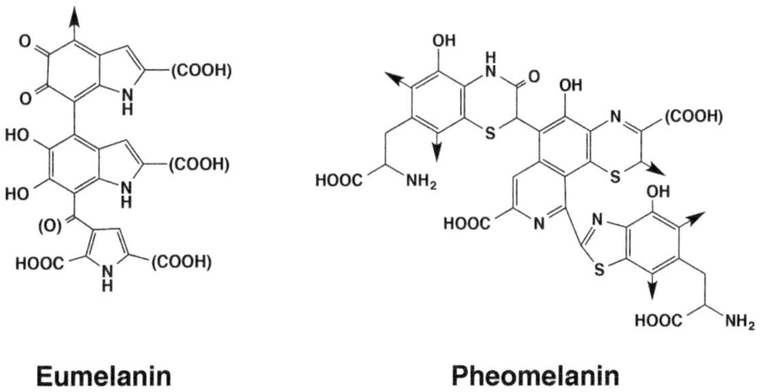

Fig. 2.1 Chemical structures of EM and PM (Adapted from Ito and Wakamatsu 2008)

organelles termed melanosomes; thereafter, the melanosomes in the hair follicles and epidermal melanocytes are transferred to the surrounding keratinocytes to afford diverse colors to the hair and a wide range of colors from light to dark in the skin (Sturm 2009). Melanocytes are distributed in the epidermis, hair follicles, choroid, iris, inner ear, and other tissues. Unlike ordinary melanins, a pigment called neuromelanin (NM) is deposited in the dopaminergic and norepinephrinergic neurons of the substantia nigra (SN) and the locus coeruleus (LC) in the midbrain of humans.

The term "melanin" was first coined by Berzelius in 1840 to refer to black animal pigments (d'Ischia et al. 2013). Since then, it has been widely used to refer to any black or dark brown organic pigment. Nicolaus suggested the classification of melanins into three groups: EM, PM, and allomelanins, which encompass a broad variety of dark non-nitrogenous pigments of plant, fungal, and bacterial origin (Nicolaus 1969). Since then, Prota proposed a more restrictive use of the term "melanin" to include only those pigments that are formed intracellularly by the oxidation of tyrosine and related metabolites (Prota 1995). Recently, d'Ischia et al. (2013) proposed the use of the term "melanin" as follows: pigments of diverse structure and origin derived from the oxidation and polymerization of tyrosine in animals or phenolic compounds in lower organisms. EM: A black-to-brown insoluble subgroup of melanin pigments derived at least in part from the oxidative polymerization of L-3,4-dihydroxyphenylalanine (DOPA) via 5,6-dihydroxyindole (DHI) intermediates. PM: a yellow-to-reddish brown, alkali-soluble, sulfur-containing subgroup of melanin pigments derived from oxidation of cysteinyldopa (CD) precursors via benzothiazine (BT) and benzothiazole (BZ) intermediates. NM: Dark pigments produced within neurons by oxidation of dopamine (DA) and other catecholamine precursors. Pyomelanins: Dark pigments produced mainly by microorganisms, but not exclusively, from homogentisate. Melanin pigments produced by oxidation of DA are universally found on the body surface of insects (see Chap. 1).

In this chapter, we focus on internal and external melanins produced by pigment cells in vertebrates: mammals, birds, fish, reptiles, amphibians, and fossils.

2.2 Biochemical Pathway for the Production of EM, PM, and NM

The biochemical pathway of melanogenesis has been described in detail in our recent reviews (Simon et al. 2009; Ito and Wakamatsu 2011a) and a summary of the scheme is shown in Fig. 2.2. Both EM and PM are derived from the common precursor dopaquinone (DQ) that is produced from tyrosine by the action of tyrosinase. This catalytic conversion of tyrosine to DQ by tyrosinase is the initial and key step of melanogenesis (Ito and Wakamatsu 2011a). DQ, an *ortho*-quinone, is a highly reactive intermediate that reacts extremely rapidly with thiol compounds (Ito et al. 2020a). In the absence of thiol compounds, DQ undergoes intramolecular cyclization of its amino group to produce dopachrome (DC) via cyclodopa (Land

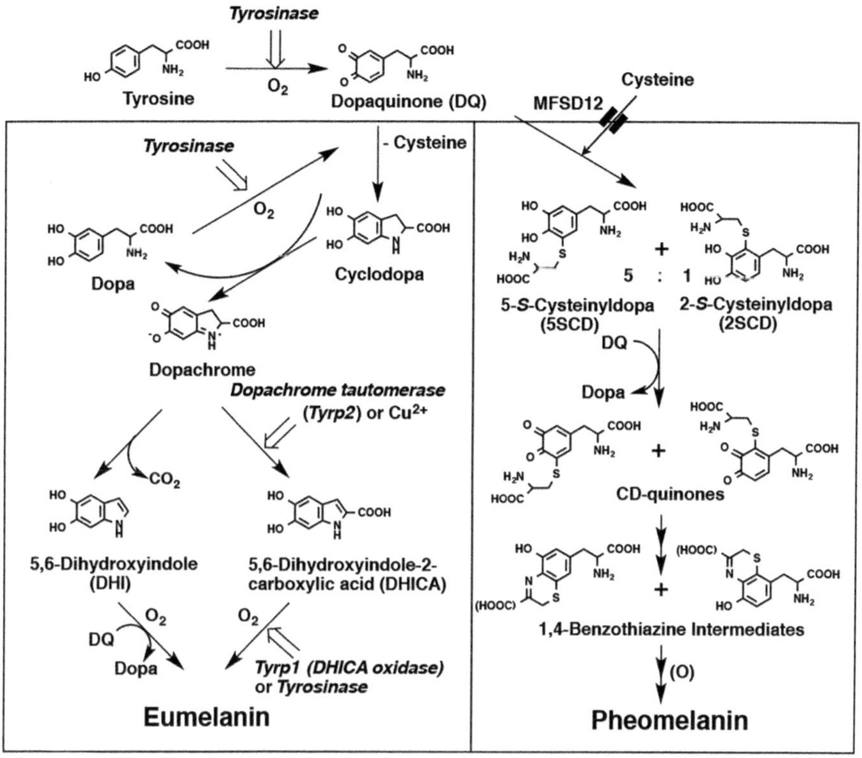

Fig. 2.2 Biosynthetic pathways leading to the production of EM and PM. Note that the activities of tyrosinase, Tyrp1 and Dct/Tyrp2, and the precursor tyrosine are involved in the production of EM, while only tyrosinase (and the precursors tyrosine and cysteine) is necessary for the production of PM. Additionally, note that human TYRP1 does not act as a DHICA oxidase (Boissy et al. 1998). (Modified from Ito and Wakamatsu 2011a)

et al. 2003). DC is then spontaneously and gradually converted mostly to DHI by decarboxylative rearrangement. A tyrosinase-related protein (Tyrp), called DC tautomerase (Dct, also known as Tyrp2), catalyzes the tautomerization of DC to 5,6-dihydroxyindole-2-carboxylic acid (DHICA) (Pawelek et al. 1980; Tsukamoto et al. 1992; Kroumpouzos et al. 1994). Copper ions can also catalyze this process (Ito et al. 2013a). Oxidative polymerization of DHI and DHICA in various ratios produces black to dark brown EM. Oxidative polymerization of DHI is catalyzed directly by tyrosinase or indirectly by DQ, while oxidation of DHICA appears to be catalyzed by Tyrp1, the *brown* locus protein, at least in mice (Jiménez-Cervantes et al. 1994; Olivares et al. 2001). However, the human homolog TYRP1 may not act in the same way as in mice (Boissy et al. 1998), and its precise enzymatic function in humans is not yet clear. In contrast, the production of PM appears to proceed spontaneously after the production of DQ as long as cysteine (CYS) is present (Land and Riley 2000; Ito and Wakamatsu 2008). 5-*S*-cysteinyldopa (5SCD) is rapidly formed as the major isomer (Ito and Prota 1977). Oxidation of CD proceeds by redox exchange with DQ to form the quinone form. Cyclization and its rearrangement afford BT intermediates that are oxidized to form PM with a BT moiety (BT-PM) and then gradually converted to PM with a BZ moiety (BZ-PM) (Wakamatsu et al. 2009).

To date, more than 250 genes have been involved directly or indirectly in the control of skin pigmentation in humans (Baxter et al. 2019). Among these genes, MITF (microphthalmia-associated transcription factor) plays a key role in these developmental processes. MITF is a transcription factor that regulates the expression of many genes involved in the proliferation, survival, and migration of melanocytes. Using combined bioinformatics analysis of public chromatin immunoprecipitation-sequencing and transcriptomic data, Gaudel et al. (2020) identified direct MITF-target genes, SLC7A5, which has not been previously reported as MITF-target genes and pigmentation gene. They described that SLC7A5 encodes an amino acid transporter and transports tyrosine which is the key substrate for melanin synthesis. Therefore, inhibition of SLC7A5 might reduce cellular tyrosine level. Using MelanoIP, a method for rapidly isolating melanosomes and profiling their labile metabolites contents, Adelmann et al. (2020) found that MFSD12 (major facilitator superfamily domain-containing protein 12) is required to maintain normal levels of cystine, the oxidized dimer of CYS, in melanosomes, and to produce CD (Fig. 2.2). MFSD12 is a part of the large major facilitator superfamily of 12-transmembrane domain proteins and causes darker pigmentation in mice and humans when suppressed (Crawford et al. 2017; Adhikari et al. 2019). Isolated melanosomes were known to import CYS, but the transporter responsible for this activity had not yet been identified (Potterf et al. 1999). How CYS enters melanosomes and lysosomes was not known, but cystine was known to efflux out of these organelles through the transported cystinosin (CTNS) (Gahl et al. 1982; Jonas et al. 1982). Comparing wild-type melanocytes with metabolites in melanosomes isolated from MFSD12-deficient melanocytes, the levels of cystine and CD were significantly reduced in MFSD12-deficient melanosomes. Their results suggested that MFSD12 is a component of melanosomal CYS import system and the pigmentation protein MFSD12 is necessary and probably sufficient for the transport of CYS into melanosomes and lysosomes, processes that have been long described (Pisoni et al. 1990; Potterf et al. 1999). Adelmann et al. suggested that MFSD12 inhibitors may represent a new therapeutic class for the treatment of cystinosis. They found that by inhibiting MFSD12 and preventing CYS from entering lysosomes, they could reverse the buildup of cystine in cells with the genetic mutation linked to cystinosis, suggesting a potential therapeutic use for MFSD12 inhibitors.

NMs are complex polymeric compounds present in the central nervous system of humans and other phylogenetically close mammals, including rats, chimpanzees, gibbons, baboons, and more distant ones, such as horses and sheep (Marsden 1961). NMs are localized in cytoplasmic organelles surrounded by a double membrane, together with lipid bodies and proteins (Sulzer et al. 2008; Zecca et al. 2008). NM is composed of different compounds such as melanin, proteins, lipids, and metal ions (Zecca et al. 2008; Engelen et al. 2012; Wakamatsu et al. 2012a; Zucca et al. 2014, 2018). NM is mainly synthesized in dopaminergic neurons of the SN and noradrenergic neurons of the LC; however, it has been demonstrated that NM is also synthesized and accumulated in the neurons of other brain areas (Zecca et al. 2008). The synthesis and accumulation of NM inside neurons occur during brain aging and there is evidence that NM is involved in the pathogenesis of neurodegenerative diseases such as Parkinson's disease (PD) (Marsden 1983; Mann and Yates 1983; Zecca et al. 2006). Notably, the SN and LC are the regions of the brain with the highest concentration of NM (Zecca et al. 2004). Synthesis of peripheral melanins (e.g., skin and hair) is mediated by tyrosinase, an enzyme also present at low levels in the brain. Whether brain tyrosinase may actually contribute to NM synthesis is currently unknown. Carballo-Carbajal et al. (2019) reported that overexpression of human tyrosinase in rat substantia nigra results in age-dependent production of human-like neuromelanin within nigral dopaminergic neurons, up to levels reached in elderly humans. We previously performed chemical analyses to elucidate the structure of NM in the SN (Wakamatsu et al. 2003, 2012a; Zecca et al. 2008), which suggested that the pigmented part of NM in the SN is derived from DA and CYS in a molar ratio of 2:1 (Zecca et al. 2008; Wakamatsu et al. 2012a). In addition, it was recently suggested that various catecholic metabolites are incorporated into NM from the SN and the LC, including DOPA, 3,4-dihydroxyphenylethanol (DOPET), 3,4-dihydroxyphenylethylene glycol (DOPEG), 3,4-dihydroxyphenylacetic acid (DOPAC), and 3,4-dihydroxymandelic acid (DOMA). These compounds are metabolites of DA and norepinephrine (NE) formed by oxidative deamination by monoamine oxidase followed by reduction/oxidation (Eisenhofer et al. 2004; Wakamatsu et al. 2014, 2015) (Fig. 2.3). The location of NM is related to the double-edged sword properties sometimes attributed to NM, as melanin formation in the cytosol of neurons could have cytotoxic properties. Once the NM polymer is formed, it can be a cytosolic scavenger for potentially neurotoxic sub-products, because Fe (III) or reactive oxygen species (ROS) oxidize catecholamine neurotransmitters. It has been reported that α-synuclein could be one of the factors responsible for the positive association between PD and melanoma via its differential roles in melanin synthesis in melanoma and in dopaminergic neuronal cells (Goedert 2001; Pan et al. 2012). NM pigment accumulates inside specific autophagic organelles, which contain NM-iron complexes, along with lipids and various proteins (Zucca et al. 2017). As NM–iron complexes are paramagnetic, they can be imaged using magnetic resonance imaging (MRI) (Sasaki et al. 2006; Trujillo et al. 2017; Sulzer et al. 2018; Cassidy et al. 2019). Cassidy et al. (2019) indicated that noninvasive NM-MRI is a promising tool that could have diverse research and clinical applications to investigate in vivo the role of DA in neuropsychiatric illness.

2.3 Microanalytical Application with the Chemical Degradation of Melanin

Most natural melanin pigments consist of both EM and PM (a concept of "mixed melanogenesis"). To characterize melanin and melanogenesis, we developed a microanalytical method to analyze EM and PM (Ito and Fujita 1985) based on the

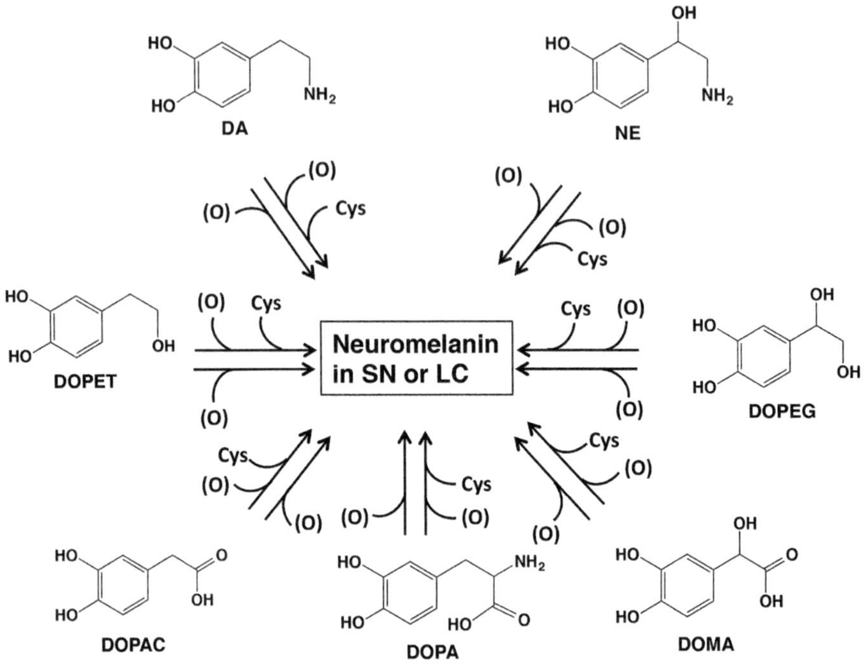

Fig. 2.3 Possible participation of various catecholic metabolites present in various regions of the brain that may be incorporated into NM. In addition to DA and NE and the corresponding Cys-derivatives, these other metabolites are also thought to be incorporated into NM. (O) represents the oxidants. (Adapted from Wakamatsu et al. 2015)

chemical degradation of melanin pigments followed by the analysis of the degradation products using high-performance liquid chromatography (HPLC). We recently established a more convenient method for the simultaneous measurement of EM and PM (Wakamatsu et al. 2002; Ito et al. 2011c, d, 2020b) (Fig. 2.4). This method is based on alkaline hydrogen peroxide (H_2O_2) oxidation to generate the specific markers, pyrrole-2,3,5-tricarboxylic acid (PTCA), pyrrole-2,3-dicarboxylic acid (PDCA), thiazole-2,4,5-tricarboxylic acid (TTCA), and thiazole-4,5-dicarboxylic acid (TDCA). PTCA is a specific biomarker of DHICA units or 2-substituted DHI units, whereas PDCA is a specific biomarker for DHI-derived units in EM, while TTCA and TDCA are specific biomarkers for BZ-derived moieties in PM. In addition to PTCA and PDCA, pyrrole-2,3,4,5-tetracarboxylic acid (PTeCA) and pyrrole-2,3,4-tricarboxylic acid (isoPTCA) have also been detected in fossil ink sacs (Glass et al. 2012; d'Ischia et al. 2013), which suggests extra cross-linking of DHI units during aging of the EM polymer. The PTeCA/PTCA ratio has been proposed as a good indicator of EM aging (Ito et al. 2013b).

PM is more readily analyzed than EM, as it is more soluble than EM. Most PM can be dissolved in alkaline media and consist of oligomers formed from sulfur-containing units, mostly BT and BZ moieties. Analysis of BT-derived moieties in

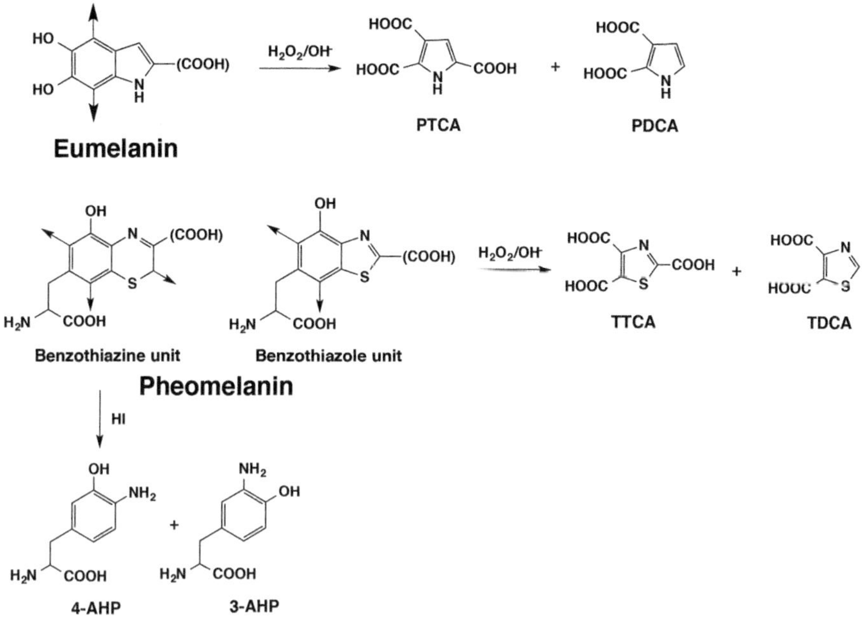

Fig. 2.4 Chemical degradation of EM and PM. Note that upon H_2O_2 oxidation, DHICA- and DHI-derived units in EM yield PTCA and PDCA, respectively. BZ moieties in PM yield TTCA and TDCA. Upon hydroiodic acid hydrolysis, 5-*S*- and 2-*S*-cysteinyldopa-derived BT moieties yield 4-AHP and 3-AHP, respectively. (Adapted from Ito et al. 2011c)

PM was performed using hydroiodic acid, yielding 4-amino-3-hydroxyphenylalanine (4-AHP) and its isomer 3-amino-4-hydroxyphenylalanine (3-AHP) (Ito and Fujita 1985; Wakamatsu et al. 2002). A spectrophotometric method was also developed to measure the total melanin content estimated by absorbance at 500 nm (A500), which measures the combined amount of EM and PM; this was performed by solubilizing the samples in hot Soluene-350 plus water (Ozeki et al. 1996). This method also provides a rough estimate of the ratio of EM to PM by analyzing the ratio of absorbance values at 650/500 nm. The EM to PM ratio was within the following ranges for various hair colors: 0.12–0.14 for red, 0.15–0.21 for blond, 0.19–0.26 for light brown, and 0.30–0.32 for brown to black (Ozeki et al. 1996). Thus, the ratio of absorbance at A650/A500 aids in estimating the relative ratio of EM to PM. Ito et al. (2018c) recently developed a novel method of melanin characterization based on acid hydrolysis of melanins followed by alkaline H_2O_2 oxidation of tissue samples. This method provides more simplified HPLC chromatograms compared with conventional H_2O_2 oxidation owing to the removal of proteins and low-molecular-weight compounds. It is useful not only to characterize melanins but also to confirm the presence of trace levels of EM and PM in various tissues and fossil samples (McNamara et al. 2018). Rioux et al. (2019) recently reported a method for measuring EM and PM in melanoma tumors and cells using solid-phase extraction (SPE) after alkaline hydrogen peroxide oxidation and

HPLC-diode array detection (HPLC-DAD) analysis. Affenzeller et al. (2019) introduced the purification method of the oxidation products by SPE (reversed-phase) followed by the selective identification using LC-MS. Electron paramagnetic resonance (EPR) spectroscopy has also been to characterize melanin. EM contains oxygen-centered free radicals of the semiquinone type, while PM contains nitrogen-centered radicals of the semiquinonimine type (Sealy et al. 1982). Unlike the singlet EPR spectrum of EM, the PM spectrum includes at least three peaks, the g-factor of the main central peak is practically the same as that of the EM singlet. The EPR method is suitable for the analysis of large numbers of samples in a short time but is even more indirect than the chemical one. Vsevolodov et al. (1991) reported that both HPLC and EPR methods fit well with the relative values of EM and PM contents in hair samples from newborn lambs. The HPLC method appears to be more sensitive for the detection of low concentrations of PM, while the EPR method is more suitable for mass selection purposes.

2.4 Hair Color

The contents of EM and PM (chemical phenotype) by the microanalytical application in human hairs of black, dark brown, brown, light brown, blond, and red color (visual phenotype) were measured using the chemical degradation methods. EM contents decrease in that order, with a trace but constant level of PM, except for red hair which contains about equal levels of PM and EM. Thus, the chemical phenotype correlates well with the visual phenotype (Ito et al. 2011c). However, we have encountered that human red hair is occasionally difficult to differentiate from light to medium brown hair, whereas the chemical degradation methods clearly differentiate between those phenotypes. This illustrates the significance of the chemical phenotype compared with visual phenotype.

Mouse coat color pigmentation has been extensively studied and is described in our earlier review (Ito and Wakamatsu 2011a). Variations in mouse coat color have a long history of genetic as well as biochemical studies, as summarized in a classical treatise by Silvers (Silvers 1979) and a review by Bennett and Lamoreux (Bennett and Lamoreux 2003). Currently, >170 mouse coat color genes have been cloned, although many coat color genes remain to be cloned (http://www.espcr.org/micemut/). However, fewer than 20 genes are directly involved in the production of melanin or in the regulation of the ratio of EM to PM (Ito and Wakamatsu 2011b). It should be stressed that congenic mice are available for the study of the pigmentary effects of many specific coat color genes. The synthesis of EM and PM in mice is regulated by a number of coat color genes. EM and PM contents were measured in cultured melanocytes and in the epidermis/dermis and hairs of C57BL/10JHir (B10) and congenic mice carrying mutations of various coat color genes (Hirobe 2012). Among the various strains of mice, C57BL/10JHir mice (B10 mice; Fig. 2.5a) possess the greatest number of epidermal melanoblasts and melanocytes (Hirobe

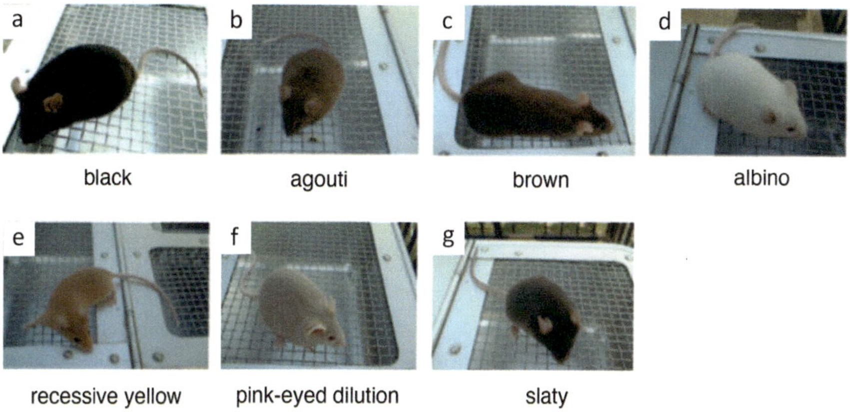

Fig. 2.5 Photos of B10 (**a**, black) and congenic strain mice. (**b**) Agouti; (**c**) brown; (**d**) albino; (**e**) recessive yellow; (**f**) pink-eyed dilution; (**g**) slaty. (Adapted from Hirobe 2011)

1992). Thus, B10 mice are useful for in vivo and in vitro studies of the mechanisms regulating melanocyte proliferation and differentiation.

Two major loci among several loci that affect melanin synthesis in mice are known to control the nature of the pigment formed. Several alleles of the *agouti* and *extension* loci are involved in the regulation of the relative amount and distribution of PM in the hair of the coat. In the coat color of wild-type (*A/A*) mice, individual hairs possess a subterminal yellow band in an otherwise black background (Fig. 2.5b). This phenotype is called an agouti pattern, which is altered by gene substitutions at the *agouti* locus (Sakurai et al. 1975). Animals homozygous for the *a* allele produce black EM only (Silvers 1979). The switch between the synthesis of EM and PM is regulated by the α-melanocyte-stimulating hormone (α-MSH) and the agouti signaling protein (Asip), the product of the *A* allele expressed in hair bulbs (Barsh 1996). A recent study showed that the loss or gain of function of β-catenin in dermal papilla cells resulted in yellow or black mice, respectively, via a mechanism that is independent of Asip (Enshell-Seijiffers et al. 2010). The EM content in agouti hair did not differ from that in black mice, but the PM content in agouti hairs increased dramatically (Table 2.1). Asip requires two accessory proteins for pigment type switching: products of the mahogany (*mg*) and mahoganoid (*md*) loci (Walker and Gunn 2010). The *mg* locus was identified as the mouse ortholog of the human attractin (*ATRN*) gene and the *md* locus encodes a novel RING-domain-containing protein. Mice homozygous for *mg* and heterozygous for lethal yellow (A^y) produce a mixed-type melanin with a low level of EM (ca. 15% of black) and have a reduced level of PM (ca. 60% of A^y). Similarly, Gunn et al. (2001) found that three Atrn mutants, either homozygous or compound heterozygous, showed a PM content five- to tenfold lower than wild-type agouti C3H/HeJ mice. Another control point in the regulation of eumelanogenesis and pheomelanogenesis is the CYS concentration in melanosomes (del Marmol et al. 1996). Chintala et al. (2005) showed that the murine subtle gray (*sut*) mutation occurred because of a mutation in the *Slc7a11* gene, which

Table 2.1 Effects of coat color genes on EM and PM synthesis in cultured melanocytes (*m*), the epidermis (*e*), dermis (*d*), and hairs (*h*) of mice

Gene	Strain	Eu (*m*)	Pheo (*m*)	Eu (*e*)	Pheo (*e*)	Eu (*d*)	Pheo (*d*)	Eu (*h*)	Pheo (*h*)	References
A	B10	→	→	→	↑↑			→	↑↑	Ozeki et al. (1995), Hirobe et al. (2004)
A	B6							→	↑↑	Lamoreux et al. (2001)
$Tyrp1^b$	B10	↓	↑					↓	↑	Ozeki et al. (1995), Hirobe et al. (1998)
$Tyrp1^b$	B6							↓	↑	Lamoreux et al. (2001)
Tyr^c	B10	0	0	0	0	0	0	0	0	Ozeki et al. (1995), Hirobe et al. (1998)
$Mc1r^e$	B10	↑	↓	↓↓	↑	↓↓	↑↑↑	↓↓	↑↑↑	Hirobe et al. (2007a, b)
$Mc1r^e$	B6							↓↓	↑↑	Ozeki et al. (1995)
$Oca2^p$	B10	↓↓	↓					↓↓↓	→	Hirobe et al. (2003, 2011a)
Dct^{slt}	B10	↓	→	↓	↓	↓↓	↓	↓	→	Hirobe et al. (2006)
Dct^{slt}	B6							↓	→	Ozeki et al. (1995)
$Hps5^{ru2\text{-}d}$	B10	↓	→	↓↓	↓	↓↓↓	↓	↓	↑	Hirobe (2011), Hirobe et al. (2011b)
A^y	B6							↓↓↓	↑↑	Lamoreux et al. (2001)

PTCA and 4-AHP (or AHP) were measured in cultured melanocytes (Eu (*m*), Pheo (*m*)), the epidermis (Eu (*e*), Pheo (*e*)), dermis (Eu (*d*), Pheo (*d*)), and hairs (Eu (*h*), Pheo (*h*)). → no effects, ↑ slightly increased (~ ×10), ↑↑ increased (~ ×100), ↑↑↑ greatly increased (~ ×1000), ↓ slightly decreased (~ ×1/10), ↓↓ decreased (~ ×1/100), ↓↓↓ greatly decreased (~ ×1/1000). The effects of coat color genes were compared with control melanocytes (B10 mice), control epidermis (B10), dermis (B10), and hairs (B10 or B6) of mice. (Adapted from Hirobe 2012)

encodes the plasma membrane cystine/glutamate exchanger xCT. The resulting low rate of extracellular cystine transport into *sut* melanocytes reduces PM synthesis with little or no effect on EM synthesis. The effect of the *sut* mutation on PM synthesis was greatly emphasized on an A^y/a background, thereby decreasing the levels of PM in the hair to one-sixth of the control level.

B (*Tyrp1*), the wild-type allele at the *brown* locus, produces black EM, while *b* ($Tyrp1^b$), the recessive allele, produces brown EM. The coat color of brown mice is lighter than that of black mice, whereas tyrosinase activity in brown mice is higher than that in black mice (Foster 1965; Hirobe 1984; Tamate et al. 1989, Fig. 2.5c).

EM content in brown hairs is lower than in black hairs, whereas PM content is increased in both the B10 (Ozeki et al. 1995) and the B6 (Lamoreux et al. 2001) backgrounds (Table 2.1). Tyrp1 functions enzymatically as DHICA oxidase in mice (Jiménez-Cervantes et al. 1994; Kobayashi et al. 1994). The *brown* mutation encodes Tyrp1, which is not properly translocated to melanosomes, resulting in no functional Tyrp1 activity and decreased tyrosinase function (Jackson et al. 1990). Brown melanocytes seem to inhibit EM synthesis (TM and PTCA values). The *brown* mutation does not significantly alter the proportion of DHICA in the EM synthesized, but rather, brown EM seems to possess a smaller molecular size compared to black EM (Ozeki et al. 1997).

Albino mice lack pigment in their coat and eyes (Fig. 2.5d, Table 2.1). The inability of albino mice to produce pigment is not because of the absence of melanoblasts, but because of the deficiency of Tyr activity (Tanaka et al. 1990; Hirobe and Abe 1999). The enzyme Tyr is encoded at the *albino/tyrosinase* (*C/Tyr*) locus in mice. C (*Tyr*), the wild-type allele of the *albino* locus, produces melanin, while *c* (Tyr^c), the recessive allele, produces no pigment in the coat or eyes (Silvers 1979; Yamamoto et al. 1989; Tanaka et al. 1990).

The phenotype that mostly produces PM is regulated by two alleles, recessive yellow (*e*/$Mc1r^e$) at the *extension* locus and A^y at the *agouti* locus. The *extension* (*E/Mc1r*) locus increases EM in hair follicle melanocytes when dominant, but blocks EM synthesis, extending the range of PM when recessive (Silvers 1979, Fig. 2.5e, Table 2.1). EM content in cultured B10-$Mc1r^e/Mc1r^e$ melanocytes was higher than in *Mc1r/Mc1r* melanocytes. In contrast, PM contents in cultured $Mc1r^e/Mc1r^e$ melanocytes was lower than in *Mc1r/Mc1r* melanocytes. EM contents in the epidermis and dermis of $Mc1r^e/Mc1r^e$ mice were much lower than those of *Mc1r/Mc1r* mice, whereas PM contents in the epidermis and dermis of $Mc1r^e/Mc1r^e$ mice was much greater than those of *Mc1r/Mc1r* mice (Hirobe et al. 2007a). EM and PM contents in the dorsal hair of female B10-$Mc1r^e/Mc1r^e$ mice were greater than those in male mice, suggesting that expression of the *recessive yellow* allele is regulated in a sex-dependent manner (Hirobe et al. 2007a). We have suggested that estrogen is a major factor that determines the higher content of EM and PM in the hair of female $Mc1r^e/Mc1r^e$ mice (Hirobe et al. 2010).

A^y represents the top dominant of the *agouti* locus. Phenotypically A^y/-mice mostly produce pheomelanic hairs (Table 2.1). A^y/-mice produce large amounts of Asip that suppress the action of α-MSH toward Mc1r, consequently producing PM only during normal hair growth (Miller et al. 1993). The EM and PM contents in hairs from B6-A^y/a and B6-$Mc1r^e/Mc1r^e$ mice are at a similar level (Ozeki et al. 1995), suggesting that melanin synthesized in hair bulb melanocytes does not differ between A^y/a and $Mc1r^e/Mc1r^e$ mice.

The *pink-eyed dilution* mutation was discovered in the mouse fancy and is known to reduce the pigmentation of both the coat and the eyes (Fig. 2.5f). The eyes of *pink-eyed dilution* mice resemble those of *albino* mice, possessing a pink tint. However, in contrast to the eyes of *albino* mice, the eyes of *pink-eyed dilution* mice are not entirely free of pigmentation (Silvers 1979). *P* (*Oca2*), the wild-type allele at the *pink-eyed dilution* locus, produces intense pigmentation of both EM and PM in the skin and eyes, whereas *p* (*Oca2*^{p}), the recessive allele, greatly reduces pigmentation in both the coat and the eyes (Silvers 1979). The *Oca2*^{p} allele greatly inhibits the synthesis of EM but not of PM (Hirobe et al. 2011b, Table 2.1).

Slaty (*slt/Dct*^{slt}) is a recessive autosomal mutation that occurs in a heterogeneous stock of mice carrying limb-deformity (*ld*^{j}) and *mg* (Fig. 2.5g). The *slaty* locus encodes Tyrp/Dct and thus wild-type animals produce DHICA-rich EM. The *slaty* mutation greatly reduced the PTCA value with a mild reduction in TM (Table 2.1). Therefore, the PTCA/TM ratio was reduced four- to sixfold, suggesting that DHICA-poor EM is produced in *Dct*^{slt}/*Dct*^{slt} melanocytes. The *slaty* mutation occurs due to a change from an arginine to glutamine in the first copper-binding domain of Dct, which converts DC to DHICA in the EM synthesis pathway (Korner and Pawelek 1980; Jackson et al. 1992; Tsukamoto et al. 1992). Dct is produced in both wild-type and *slaty* mutant mice, but the protein level of Dct in *slaty* mutant mice is greatly reduced (Kroumpouzos et al. 1994). The EM and PM contents in the epidermis and dermis of *Dct*^{slt}/*Dct*^{slt} mice were lower than those of *Dct/Dct* mice (Hirobe et al. 2006). The EM content in *Dct*^{slt}/*Dct*^{slt} hairs (5-week-old mice) was lower than that in *Dct/Dct* hairs, whereas the PM content in *Dct*^{slt}/*Dct*^{slt} hairs did not differ from that in *Dct/Dct* hairs (Hirobe et al. 2006, Table 2.1).

In 2006, a spontaneous autosomal recessive mutant with brown coat color and ruby eyes occurred in B10 mice in Hirobe's Laboratory (Hirobe et al. 2011b). The phenotype of this mutant is similar to that of ruby-eye (*ru/Hps6*^{ru}) and ruby-eye 2 (*ru2/Hps5*^{ru2}). Hermansky–Pudlak syndrome (HPS) is a recessive inherited disease that affects several organs such as the skin (hypopigmentation), eyes (low visual acuity), blood cells (prolonged bleeding), and lungs (interstitial pulmonary fibrosis) (Wei 2006). Tyr activity, the expression of Tyr, Tyrp1, Dct, and EM synthesis were markedly decreased in *Hps5*$^{ru2\text{-}d}$/*Hps5*$^{ru2\text{-}d}$ melanocytes (Hirobe 2011). The EM content in cultured *Hps5*$^{ru2\text{-}d}$/*Hps5*$^{ru2\text{-}d}$ melanocytes was lower than that in *Hps5/Hps5* melanocytes (Hirobe 2011, Table 2.1). In contrast, the PM content in cultured *Hps5*$^{ru2\text{-}d}$/*Hps5*$^{ru2\text{-}d}$ melanocytes as well as in culture media did not differ from *Hps5/Hps5* melanocytes (Hirobe 2011, Table 2.1). The EM content in the epidermis, dermis, and hairs (5-week-old) of *Hps5*$^{ru2\text{-}d}$/*Hps5*$^{ru2\text{-}d}$ mice was much lower than those of *Hps5/Hps5* mice. However, a two- to threefold increase in PM content was observed in the hairs of 5-week-old *Hps5*$^{ru2\text{-}d}$/*Hps5*$^{ru2\text{-}d}$ mice (Hirobe 2011, Table 2.1). These results suggest that the synthesis of PM in *Hps5*$^{ru2\text{-}d}$/*Hps5*$^{ru2\text{-}d}$ mice was increased in the hair bulbs and that EM and PM synthesis in melanocytes is regulated by many coat color genes in a complicated manner.

Premelanosome protein (PMEL) is an amyloidogenic protein that appears to be exclusively expressed in pigment cells and forms intralumenal fibrils within early stage melanosomes upon which EM deposits in later stages. PMEL is well conserved among vertebrates, and allelic variants in several species are associated with reduced levels of EM in epidermal tissues. Inactivation of *Pmel* has only a mild effect on the coat color phenotype in four different genetic backgrounds, with the clearest effect in mice also carrying the *brown/Tyrp1* mutation. This phenotype is similar to that observed with the spontaneous *silver* mutation in mice mutations. Despite a mild effect on visible pigmentation, inactivation of *Pmel* led to a substantial reduction in

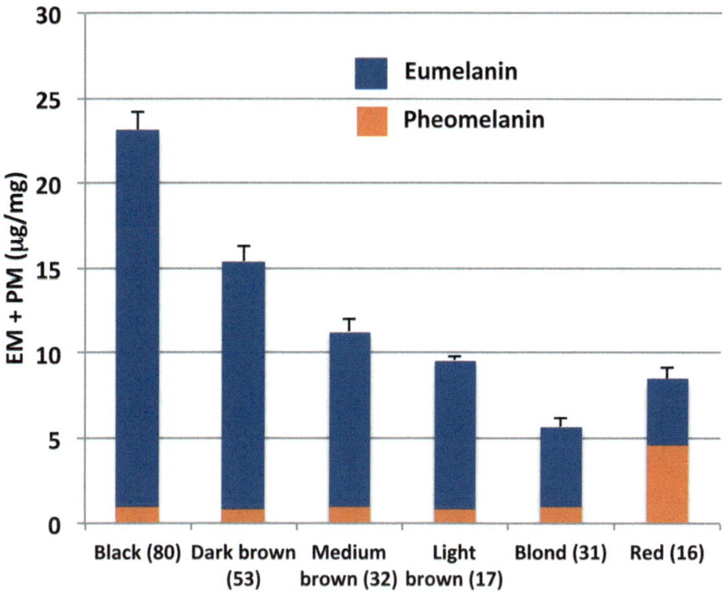

Fig. 2.6 Contents of EM and PM (the "chemical" phenotype) in human hair samples ($n = 228$) of various colors (the "visual" phenotype). PTCA and TTCA were used to calculate the contents of EM and PM, respectively. Note that if we use 4-AHP as a marker for PM, its content in eumelanic hairs (black to blond) becomes several folds smaller than those shown in this graph, indicating that PM in human hairs consists mainly of BZ moieties. (Adapted from Ito and Wakamatsu 2011b)

EM content in hair, which demonstrates that PMEL has a critical role for maintaining efficient epidermal pigmentation (Hellström et al. 2011).

Variations in human pigmentation are also involved in homologs of most mouse coat color genes. Therefore, a comparison of the pigmentary effects of human hair color genes to those of mouse coat color genes should be informative. The effects of interactions between two (even three) coat color genes can be precisely analyzed in mice but are not so straightforward in humans. Hair color in humans is one of the most conspicuous phenotypes, ranging from black, brown, and blond to red. This diversity arises mostly from the quality, quantity, and ratio of the black-dark brown EM and the reddish-brown PM. To study the diversity of hair color, we have developed several chemical methods as described above to quantify these pigments. EM and PM contents (chemical phenotype) assessed by chemical analysis correlate well with hair color (visual phenotype) (Ito et al. 2011c). The genetic basis for the diversity of human hair color has been extensively studied (for reviews, see Ito and Wakamatsu 2011a; Sturm and Duffy 2012). Based on the analysis of melanin content in human hair samples ($n = 228$) of various colors, the EM content decreases in human hair of black, dark brown, brown, light brown, blond, and red color, in that order, while the PM content remains low but constant, except for red hair (Fig. 2.6) (Valenzuela et al. 2010; Ito et al. 2011c). This result shows that these hairs are eumelanic, despite their great diversity in color. The fact that PM content remains low but at a constant level fits well with the casing model of mixed melanogenesis (Ito 2006; Ito and Wakamatsu 2008; Simon et al. 2009), which was originally suggested by Agrup et al. (1982) based on biochemical findings. The casing model of mixed melanogenesis is that PM is always produced first, and then EM is deposited on that PM (Ito and Wakamatsu 2008; Simon and Peles 2010; Simon et al. 2009). A study by Bush et al. (2006) provided biophysical evidence supporting this model. It is thought that melanosomes, which are organelles within melanocytes involved in the synthesis of melanin and their eventual transfer to keratinocytes in the epidermis, are assembled through a casing process in which an initially formed PM core is encapsulated within an EM coating. Evidence for this architecture was provided by studies of iridal melanosomes and NM (Peles et al. 2009; Bush et al. 2006).

Only red hair contained comparable levels of EM and PM. Red hairs can be unambiguously differentiated from light brown to blond hair, based on their contents of PTCA (a EM marker) and TTCA (a BZ-derived PM marker), although color intensities are rather similar between red hair and light brown to blond hair. Thus, it is important to analyze melanin pigments in EM and PM by measuring PTCA and TTCA after H_2O_2 oxidation and 4-AHP (a BT-derived PM) after hydroiodic acid hydrolysis.

Hair melanin content assessed in patients with various hypopigmentary disorders, including HPS (Kono et al. 2012; Okamura et al. 2018), oculocutaneous albinism (OCA) (Yoshizawa et al. 2014), and Menkes disease (Tomita et al. 1992), helped to evaluate the precise effects of each disease on pigmentation. HPS, first described in 1959, is a rare autosomal recessive disorder characterized by OCA, a bleeding tendency due to the absence of platelet dense granules and the accumulation of ceroid lipofuscin in lysosomes (Hermansky and Pudlak 1959). Kono et al. (2012) reported the contents of EM and PM in the ivory white hair of a patient with OCA1MP. The PM content of that patient was almost the same as that obtained from normally pigmented Japanese volunteers and was twice that of patients with OCA1A (tyrosinase negative OCA1). However, the EM content was far lower than that of normal volunteers and was almost the same as that of patients with OCA1A. Okamura et al. (2018) reported the melanin content in one patient each with HPS4, HPS6, and HPS9, which showed a decrease in total melanin (50–90% of normal subjects). The ratios of BZ-PM/EM and BT-PM/EM increased in all HPS patients examined. These results suggest that the tendency toward pheomelanic pigment observed in HPS patients is explained by the increases in the ratios of BZ-PM/EM and BT-PM/EM, compared with the control samples. It has been known that OCA resulted in reduced melanin synthesis, skin hypopigmentation, increased risk of UV-induced malignancy, and developmental eye abnormalities affecting vision. No treatments exist at present. Adams et al. (2018) have recently shown that oral nitisinone increases ocular and fur pigmentation in a mouse model of one form of albinism, OCA-1B, due to hypomorphic mutations in the *tyrosinase* gene. They showed that nitisinone did not result in an increase in iris melanin content but may increase hair and skin pigmentation in patients with OCA-1B.

PM has been shown to contribute to melanomagenesis through both ultraviolet (UV)-dependent and independent pathways (Mitra et al. 2012). In particular, the UVA photoexcitation of BT chromophores is associated with melanomagenesis through the generation of ROS (Napolitano et al. 2014), indicating that the ratio of BZ-PM/BT-PM is important for carcinogenesis. Patients with various types of OCA, including HPS, are known to be at increased risk of melanoma as well as for other skin cancers (Iwata et al. 2017; Ozaki et al. 2017). Although the correlation between hair and skin pigmentation remains unknown (Thody et al. 1991; Del Bino et al. 2015), it is conceivable that the ratio of BZ-PM/BT-PM in hair samples might serve as a marker to predict the risk of developing skin cancer.

Various changes appear in hair by aging. Hair graying is the most remarkable one and is well studied, nevertheless, little was known about the aging of pigmented hairs before graying. Itou et al. (2019) have studied the age changes in hair color, melanin compositions, and melanosomes. They found a significant change in the morphology of hair melanosomes with age. This suggested that the enlargement of melanosome seems to be a cause of the age-related color change of pigmented hairs from brown to black. A significant positive correlation with age was found in TM and the mol% of DHI units, while there was no correlation in PM mol%. They found that the increase in DHI mol% contributes to the darkening of hair color by aging. In addition, the DHI mol% showed a significant negative correlation with the aspect ratio of melanosomes, suggesting a contribution of DHI melanin to the change in melanosome morphology by aging. Regarding this report, another review that summarizes the characteristics and properties of Asian hair has recently appeared (Leerunyakul and Suchonwanit 2020).

Evolution is usually viewed as a slow process, with changes in traits emerging over thousands of generations only. Over the recent years, however, research has indicated that adaptation in specific traits can occur more quickly. However, very few studies outside microorganisms were able to demonstrate empirically how quickly natural selection shapes the whole genome. The research team led by Hoekstra has now provided the evidence for rapid evolution within a single generation (Barrett et al. 2019). Barrett et al. used wild populations of deer mice in different habitats to identify both ecological and molecular biological mechanisms that drive the trait adaptation. Investigating the relationship between the color of the coat and the genome that encodes it for survival when the wild mice are kept in the wild in the sandy area and the green areas, they examined what kind of mice could survive on each sand, and found that mice with pale back color survived on brown sand and black rats on black soil survived. The nucleotide sequences before and after the Agouti gene, which controls the brown color, were examined in detail, and the diversification occurred in the mice that survived in each sand. They identified seven types of SNPs that were naturally selected from each SNP and formed three types of blocks linked by these seven types, one of which was the coding of Agouti. Thus, it was confirmed that the mutation was a deletion of serine in the region. Focusing on the serine-deficient mutation (ΔSer) of the amino acid in Agouti, they found that ΔSer mice had significantly lower amounts of PM than the hair of wild-type mice (Fig. 2.7). The adaptation to novel environments can occur rapidly, with evolution in ecologically relevant phenotypes arising within a few generations. Such rapid

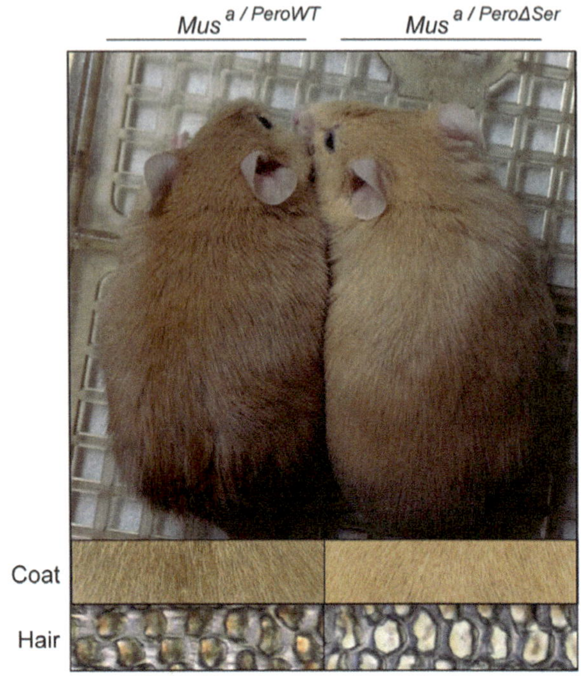

Fig. 2.7 Matched lines of transgenic Mus (in C57BL/6, an Agouti knockout strain) expressing the wild-type (WT; dark) or the ΔSer (light) Peromyscus Agouti allele. Close-up pictures show the intensity of pheomelanin in dorsal coats and individual dorsal hairs from transgenic mice. (Adapted for Barrett et al. 2019)

phenotypic evolution has sometimes been linked to changes in allele frequencies at underlying genetic loci. Laurentino et al. (2020) have now provided evidence for rapid evolution within a single generation, using three spine stickleback fish as model organism.

2.5 Human Skin

The constitutive pigmentation of human skin shows extensive variations. Differences in skin color are attributed to differences in the type, number, and distribution of pigment granules within keratinocytes. The melanin content produced by melanocytes was determined by the same method as used for the hair. To the best of our knowledge, only one study has been conducted to determine the correlation between skin pigmentation and hair pigmentation in the same individuals (Thody et al. 1991), wherein it was reported for the first time that PM is present in human skin. A fairly good correlation in the ratio of EM to PM between skin and hair pigmentation was found in that study. However, a recent study showed that skin and hair melanocytes in mice are regulated quite independently (Van Raamsdonk et al. 2009). In humans, pigmentation in follicular melanocytes from red hair individuals may be an exception in that it is dominated by the production of PM through loss-of-function

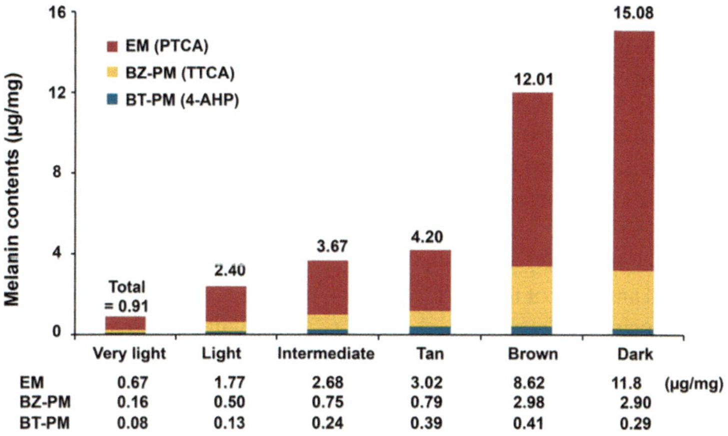

Fig. 2.8 Contribution of EM (PTCA × 38), BZ-PM (TTCA × 34), and BT-PM (4-AHP × 9) in each skin color group. For the conversion of PTCA, TTCA and 4-AHP to EM, BZ-PM and BT-PM, previously defined conversion factors of 38, 34, and 9 were used, respectively. (Adapted from Del Bino et al. 2015)

mutations in *MC1R* (Ito and Wakamatsu 2008, 2011a). The concentration of melanins in cultured melanocytes derived from individuals with *MC1R* mutations was measured (Wakamatsu et al. 2006) and the results showed no striking differences in the levels of PM or EM or their ratio in mutant *MC1R* compared with wild-type *MC1R*+melanocytes. Thus, the correlation between skin and hair pigmentation remains an open issue in the field of melanogenesis, awaiting further studies.

EM and PM contents were measured in various skin color groups by using H_2O_2 oxidation and HI hydrolysis, followed by the analysis of the degradation products using HPLC (Del Bino et al. 2015). The EM content was rather constant (72–79%) regardless of the degree of pigmentation. As a result, the PM content was also rather constant (21–28%). Thus, the human epidermis is comprised of approximately 74% EM, 20% BZ-PM, and 6% BT-PM, regardless of the degree of pigmentation (Fig. 2.8). The ratio of EM to PM (2.5–2.9) was also rather constant, independent of the degree of pigmentation, except for dark skin, which has a ratio of 3.7. This study showed for the first time that epidermal melanin contains more PM than expected with a constant level of about 26% regardless of the degree of pigmentation.

Regarding the sensitivity of melanins toward UV, it is known that EM is photoprotective and limits the extent of UV penetration within the epidermis as well as scavenging ROS radicals, while PM is not only weakly photoprotective against UV but also highly phototoxic and can enhance the UV-induced production of ROS and further damage cells (Hill et al. 1997; Hill and Hill 2000; Kadekaro et al. 2003; Napolitano et al. 2014). Very light to tan skin had lower levels of photoprotective EM; thus, these skin types have been clearly shown to be affected by UV exposure, which induces DNA damage in the keratinocytes of the basal layer as well as in melanocytes and in fibroblasts of the upper dermis. In contrast, brown and dark skin types with higher EM content seem to be protected from UV exposure (Del Bino et al. 2006; Del Bino and Bernerd 2013). This may explain the higher susceptibility of lighter skin types to skin cancers, including melanoma (Elder 1995; Halder and Ara 2003), premature aging (Chung 2003; Nouveau-Richard et al. 2005; Hourblin et al. 2014), and pigmentary disorders (Ho and Chan 2009).

One major risk factor for melanoma is the red hair phenotype, characterized by a high ratio of red PM to black EM in the melanin pigment that serves as a natural sunscreen. Melanomagenesis, caused by UVA, requires melanin and is associated with increased amounts of the DNA oxidation product 8-oxodG (8-oxo-7,8-dihydro-2′-deoxyguanosine) (Noonan et al. 2012). Mice with the red hair phenotype containing a cancer-initiating "driver" mutation have also been found to progress to melanoma even in the absence of UV, whereas in black and albino mice, it occurs gradually (Mitra et al. 2012). A higher level of oxidative DNA damage in the red hair mice suggested that PM biosynthesis might either produce ROS or oxidize radical scavenging thiols. Premi et al. (2015) demonstrated a new pathway leading to cyclobutane pyrimidine dimers (CPDs) that involve melanin and were determined by carefully monitoring the concentration of CPDs after UV irradiation. In doing so, they discovered that the levels of CPDs continued to increase for 2–3 h after irradiation of normal melanocytes with UVA and/or UVB before decreasing due to DNA repair, whereas the levels of CPDs decreased immediately in albino melanocytes. The increase in CPDs was even greater in melanocytes with the red hair phenotype, underscoring the importance of PM in this process. The CPDs formed during the dark reaction (dark CPD) also had a higher fraction of C-containing CPDs than those produced by UVA alone, which suggests that this pathway is more mutagenic. The increase in CPDs could not be explained by a direct photochemical or indirect photosensitization mechanism, both of which take place in a fraction of a second upon absorption of light. Such an unusual behavior can only be explained by a chemosensitization mechanism, in which the triplet electronic excited state of a sensitizer is produced slowly from the thermal decomposition of a high-energy precursor molecule (chemiexcitation). The excited molecule then transfers its triplet energy to DNA in a nonradiative manner to produce CPDs. Premi et al. (2015) elucidated a possible pathway involving the reaction of superoxides, formed by photoexcitation of PM or by UV-induced NOX (reduced nicotinamide adenine dinucleotide phosphate (NADPH) oxidase), with nitric oxide generated enzymatically by inducible nitric oxide synthase (iNOS) to form peroxynitrite. Peroxynitrite was shown to react with melanin and melanin precursors to chemosensitize C-containing CPD formation, presumably via dioxetane intermediates (Fig. 2.9). Melanin also plays a role in DNA damage induced by visible light, which results in DNA strand breaks and oxidative pyrimidine modification in melanocytes in vitro but not in the melanin deficient (albino) equivalent (Chiarelli-Neto et al. 2014). Although melanin acts as an excellent photoprotectant (Fajuyigbe et al. 2018) (particularly in darker skin types), there is evidence of its photosensitizing potential through the photogeneration of ROS (Beltrán-García et al. 2014; Ito et al. 2018b; Szewczyk et al. 2016). Melanin photosensitization could explain the results of

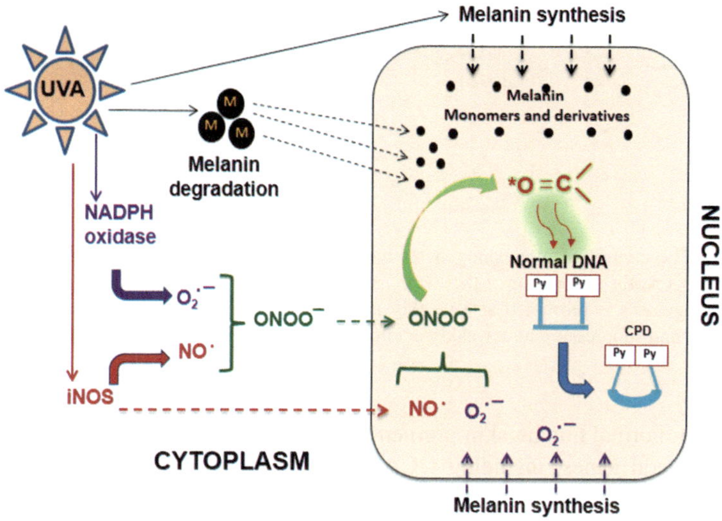

Fig. 2.9 A mechanistic model for the generation of dark CPDs in melanocytes by chemiexcitation, with melanin as an active participant. (Adapted from Premi and Brash 2015)

Tewari et al. (2012), who reported increased UVA1 (340–400 nm)-induced CPDs with epidermal depth in human skin in vivo (in direct contrast with the results observed with UVB (300 nm)), given that melanin concentration also increases with epidermal depth (Tewari et al. 2012; Fajuyigbe et al. 2018). Lawrence et al. (2018) demonstrated that UV/visible boundary wavelengths (385–405 nm) cause significant biologically relevant damage in vitro and in vivo, including dark CPD formation. This damage is most likely caused by oxidative stress generated by chromophores in the skin, such as protoporphyrin IX, β-carotene, and melanins, which absorb strongly in this region, although this requires further investigation.

We recently examined the photobleaching of PM with BT and BZ moieties. The results showed that synthetically prepared BZ-PM exhibited a higher efficiency to photogenerate singlet oxygen than the synthetic BT-PM, and that 7-(2-amino-2-carboxyethyl)-5-hydroxy-dihydro-1,4-benzothiazine-3-carboxylic acid (DHBTCA) and 6-(2-amino-2-carboxyethyl)-4-hydroxybenzothiazole (BZ-AA) as BT and BZ monomers produced superoxide anion, singlet oxygen, and H_2O_2. It was also shown that redox reactions may occur via singlet oxygen and H_2O_2 in BZ-AA and via superoxide anions in DHBTCA. These results showed that UVA enhanced the pro-oxidant activity of PM, in particular BZ-PM (Zadlo et al. 2019; Tanaka et al. 2018).

The effect of pH in the control of mixed melanogenesis is now receiving much attention because it was found that melanosomes in melanocytes derived from white/fair skin are acidic, while those derived from black/dark skin are near neutral (Smith et al. 2004). Also, tyrosinases produced by white or black melanocytes have the same optimum pH 7.4 with greatly reduced activities at acidic pHs, i.e., activities at pH 5.8 were ~20% those at pH 6.8 (Fuller et al. 2001). Furthermore, the great

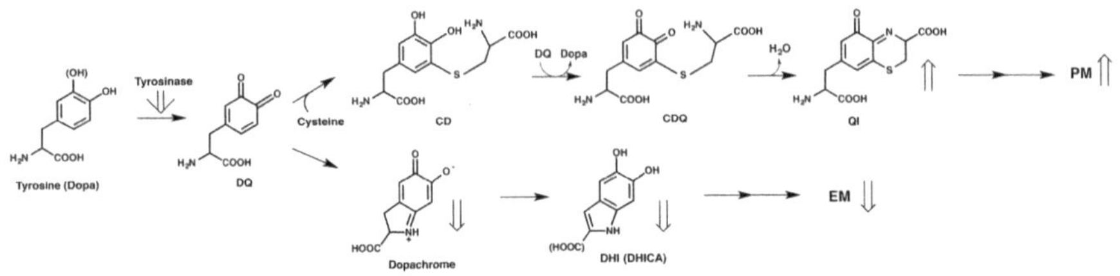

Fig. 2.10 The control of eumelanogenesis and pheomelanogenesis at a weakly acidic pH. An acidic pH greatly suppresses the late stages of eumelanogenesis beyond DC, and pheomelanogenesis is promoted at a weakly acidic pH. The up-arrow and down-arrow indicate acceleration and suppression by a weakly acidic pH, respectively. (Adapted from Wakamatsu et al. 2017)

diversity in normal human skin pigmentation appears to stem from polymorphisms in only several genes, including OCA2, membrane-associated transporter protein (MATP), and SLC24A5 (Lamason et al. 2005; Lao et al. 2007; Norton et al. 2007). Available evidence suggests that polymorphisms in those genes may result in the acidification of melanosomes. However, the significance of pH in controlling mixed melanogenesis has only been addressed directly in a study by Ancans et al. (2001). They showed by neutralizing intramelanosomal pH in human melanocytes and in melanoma cells that cells with greatly increased tyrosinase activity after neutralization had a preferential increase in the production of EM as evidenced by an increase in the EM to PM ratio. The activation of the cAMP pathway by α-MSH or forskolin leads to an alkalization (neutralization) of melanosomes and a four-fold increase in melanin content (Cheli et al. 2009). Although that study links the pH of melanosomes to the cAMP pathway for the first time, the issue of mixed melanogenesis was not examined. Ito et al. (2013a) reported that an acidic pH greatly suppresses the late stages of eumelanogenesis and that Cu^{2+} ions accelerate the conversion of DC to DHICA and its subsequent oxidation. Wakamatsu et al. (2017) reported that tyrosinase-catalyzed production of PM either from DOPA or tyrosine in the presence of CYS was greatest at pH values of 5.8–6.3, while the production of EM was suppressed at pH 5.8. This suggests that mixed melanogenesis is chemically shifted to more pheomelanogenesis at a weakly acidic pH (Fig. 2.10).

Rhododendrol (4-(4-hydroxyphenyl)-2-butanol, RD) known as a skin-whitening ingredient was added to cosmetics by a cosmetic company in Japan. In July 2013, cosmetics containing RD were recalled because a considerable number of consumers developed leukoderma on their face, neck, and hands (Nishigori et al. 2015). RD was shown to exert melanocyte toxicity via a tyrosinase-dependent mechanism (Sasaki et al. 2014). Ito et al. reported that the oxidation of RD by mushroom tyrosinase produced RD-quinone, which is quickly converted to RD-cyclic quinone through an intramolecular addition of the hydroxy function (Ito et al. 2014a), and that human tyrosinase is able to oxidize both enantiomers of RD (Ito et al. 2014b) (Fig. 2.11). The in vitro binding of RD-quinone to cellular thiol proteins as well as to non-protein thiols, GSH and CYS, was confirmed in B16F1 mouse melanoma cells exposed to RD. Further, Ito et al. have shown that the RD-oligomer derived from RD-quinone

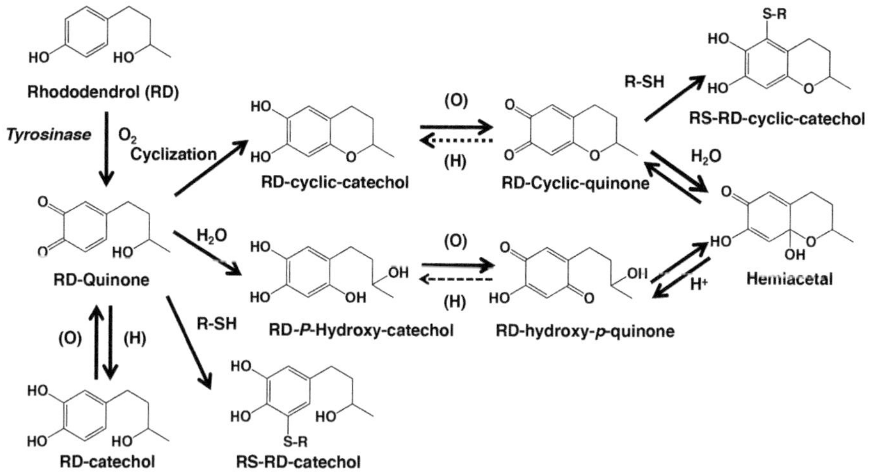

Fig. 2.11 Scheme for the oxidation of rhododendrol (RD) by tyrosinase. Oxidation of RD gives RD-quinone as an immediate product, which gives rise to RD-cyclic quinone and RD-hydroxy-*p*-quinone. These quinones are produced by redox exchange of the corresponding catechols, RD-cyclic catechol and RD-hydroxy-catechol, which are produced through an intramolecular cyclization of RD-quinone and addition of a water molecule to 2, respectively. These redox exchanges reduce RD-quinone to RD-catechol. RD-cyclic quinone produces RD-hydroxy-*p*-quinone instantaneously upon acidification via the hemiacetal intermediate. This conversion may take place slowly at neutral pH. These quinones decay gradually, probably through a coupling reaction between them. Solid arrows indicate pathways that are confirmed. Dotted arrows indicate pathways that are not confirmed. (Adapted from Ito et al. 2014a, b)

exerts a potent pro-oxidant activity by oxidizing GSH and other cellular antioxidants and by concomitantly producing H_2O_2 and that this pro-oxidant activity is enhanced by UVA radiation (Ito and Wakamatsu 2018; Ito et al. 2017).

trans-Resveratrol (3,5,4′-trihydroxy-*trans*-stilbene, RES) is a naturally occurring polyphenol and is well known for its antioxidant, antiplatelet, anti-inflammatory, anti-aging, anti-cancerous, anti-diabetic, cardioprotective, cancer chemopreventive properties, and neuroprotective properties. However, results from the available human clinical trials were controversial concerning the protective effects of RES against diseases and their sequelae. While there is a considerable body of literature on the protective effects of RES against diseases, there are relatively few reports investigating its possible toxicity. RES is generally considered as a good inhibitor of tyrosinase rather than a substrate (Na et al. 2019). For example, Park et al. (2014) showed that RES strongly inhibits mushroom tyrosinase. Since RES has a *p*-substituted phenol structure like RD, RES is expected to be a substrate for tyrosinase and to produce a toxic *o*-quinone metabolite. In fact, RES was found to be a good substrate for tyrosinase and was oxidized to produce a highly reactive *o*-quinone form. This RES-quinone decayed rapidly to produce an oligomer, which exhibited a pro-oxidant activity (Ito et al. 2019). Ito et al. showed that the cosmetic use of RES should be considered with caution. Indeed, toxicity and adverse effects were reported following consumption of RES; therefore, extensive future studies on the long-term effects, as well as the in vivo adverse effects, of RES supplementation in humans are needed (Shaito et al. 2020).

It is known that the production of melanin promotes skin pigmentation and reduces the risk of skin cancer. As mentioned above, the production of melanin is dependent on the pH of the melanosomes, with lighter skin tones being more acidic than those with darker skin. Zhou et al. (2018) have shown that inhibition of soluble adenylyl cyclase (sAC) increases melanosome pH and regulates pigmentation, both in cells and in vivo. Distinct from the canonical melanocortin 1 receptor (MC1R)-dependent cAMP pathway that controls pigmentation by altering gene expression, they found that inhibition of sAC increased pigmentation by increasing the activity of tyrosinase, the rate-limiting enzyme in melanin synthesis, which is more active at basic pH. They also demonstrated that the effect of sAC activity on pH and melanin production in human melanocytes depended on the skin color of the donor. Thus, Zhou et al. identified sAC inhibitors as a new class of drugs that increase melanosome pH and pigmentation in vivo, suggesting that pharmacologic inhibition of this pathway may affect skin cancer risk or pigmentation conditions.

2.6 Bird Feathers

Birds are among the most diverse animals with regard to melanin-based coloration, especially in their plumage. In contrast to other vertebrates such as mammals, but common with humans, birds mainly rely on visual cues for communication (Negro et al. 2016). The pigments responsible for this diversity are deposited in feathers and bare areas such as the bills and legs. These pigments include carotenoids, flavins, melanins, porphyrins, psittacofulvins, pterins, purines, and turacin (McGraw 2006). However, in this chapter, we focus on melanin pigments. Melanins and carotenoids are widespread among birds, but melanins are usually more abundant than carotenoids, and in some species, such as barn swallows, the levels of melanin are orders of magnitude greater than carotenoids (McGraw et al. 2004).

The most reliable methods for melanin identification and quantification in feathers are chemical degradation methods followed by HPLC analysis, as mentioned above. To develop direct methods to study the melanin structures in an unaltered state, Liu et al. (2014) used synchrotron-based photoionization mass spectrometry to determine the composition of feather melanins. They found that brown plumage is related to a greater content of BZ moieties, while black plumage to dihydroxyindole oligomers, and gray to BT moieties. EPR and X-ray photoelectron spectroscopies have also been applied to explore the properties of EM in silky fowl tissues (Chen et al. 2008). The use of Raman spectrometry as a simple, noninvasive technique to identify and quantify melanins in feathers and in hairs has been recently proposed (Galván and Jorge 2015; Galván and Solano 2015). Raman spectroscopy has different spectra for EM and PM without the need to extract the pigments and is potentially applicable to any biological tissues.

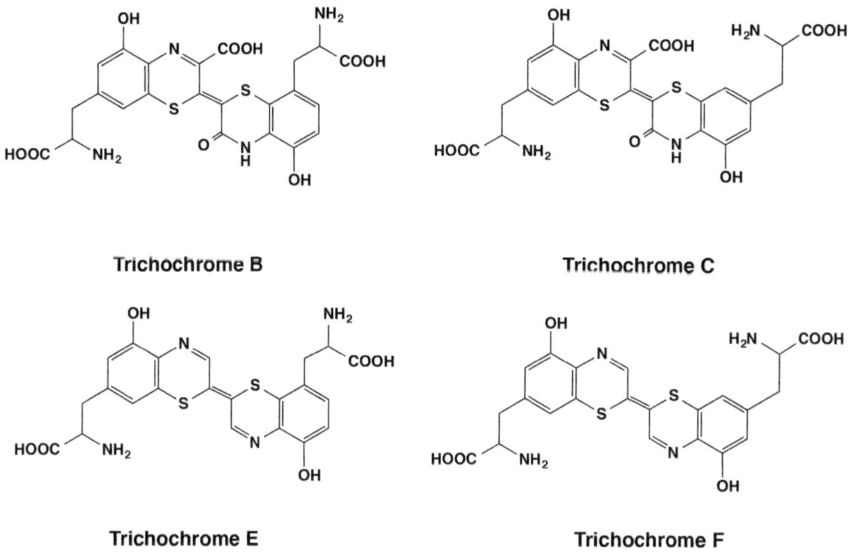

Fig. 2.12 The chemical structures of trichochromes B, C, E, and F

Melanogenesis in birds proceeds via the same pathway as that in mammals. Both EM and PM are present in birds; PM is more widely distributed and may be more chemically diverse among birds. In addition to EM and PM, there are two minor types of avian melanins, termed trichochromes and erythromelanins (Hudon 2005). Trichochromes are dimers of BT, and therefore they can be considered as a particular type of soluble PMs. Two most abundant trichochromes isolated from red chicken feathers are trichochromes B and C (Rorsman et al. 1979) (Fig. 2.12). Trichochrome B is formed by the conjugation of 5SCD with 2SCD, whereas trichochrome C is formed by the conjugation of two units of 5SCD. Similar natural trichochromes E and F, which are devoid of carboxyl and keto groups in the BT moiety, were also isolated from red feathers (Rhode Island hens), although in lower amounts than trichochromes B and C. It has been hypothesized that erythromelanins justify the wide range of reddish color found in some bird species; however, they have never been chemically characterized, and it is doubtful whether they are melanin granules composed of PM or mixtures of PM and EM (Prum 2006).

The ubiquitous nature of black melanin (i.e., EM) in living organisms is probably related to their capacity to protect cells from damage by harmful UV radiation, thus avoiding pro-oxidant substrates generated by the highly energetic UV radiation. The variation in melanization levels in birds has mainly been studied with regard to plumage. However, skin melanization, which might be the most relevant for the protection of the organism against UV radiation, remains unexplored in birds (Bortolotti 2006).

In the wide absorbance spectrum of melanins in the UV-visible range, melanin pigments protect cells from the damaging effects of energetic UV radiation, but this also implies that melanized biological structures usually increase their temperature more easily than in non-melanized, lighter structures. Birds, similar to mammals, maintain a relatively constant physiological temperature, although the thermal function of melanins remains unclear (Bortolotti 2006). However, maintaining a constant body temperature may incur a high physiological cost for birds when exposed to low ambient temperatures (Bech and Præsteng 2004), and under such circumstances, the ability of black melanized plumage to rapidly absorb radiant energy may be adaptive. The differential thermoregulatory properties of different chemical forms of melanins have not been explicitly evaluated in birds, but Koskenpato et al. (2016) recently reported larger and denser plumulaceous parts of dorsal feathers in the gray morph (mainly EM) than in the brown morph (mainly PM) in tawny owls, which suggests a superior insulative capacity in the former. However, it may be related to the cost represented by the consumption of CYS/GSH during the synthesis of PM under thermal stress, which generates oxidative stress and reduces GSH levels (Gümüşlü et al. 2002). PM-based color traits will be particularly costly to produce under high stress levels, when CYS/GSH is more necessary to combat oxidative stress and the synthesis of PM represents consumption of this antioxidant resource (Galván and Alonso-Alvarez 2009; Galván and Solano 2009). Thus, it is expected that high stress levels promote the evolution of PM-based signals when they can only be produced by high-quality individuals. In contrast, low stress levels may promote the evolution of EM-based color signals, as producing EM under such circumstances may require the elimination of CYS/GSH. Arai et al. (2017) demonstrated that high-quality individuals produce high concentrations of PM in their plumage without experiencing increased oxidative stress, despite PM production, which is triggered by ASIP, potentially actively consuming the sulfhydryl group from GSH. Arai et al. (2014) also reported that PM-based coloration may have evolved via sexual selection for PM pigmentation in barn swallows. In this regard, studies on several species of birds have shown that PM production can be beneficial and can increase survival if there is no evidence of particularly high levels of environmental oxidative stress (Galván and Møller 2013), but physiological trade-offs occur, as demonstrated by the negative relationships between the extent of plumage colored by PM and brain size (Galván and Møller 2011), the capacity to resist the effects of ionizing radiation (Galván et al. 2014), the physiological stress caused by corticosterone (Almasi et al. 2008), and survival during adverse environmental conditions (Karell et al. 2011). Previous studies have shown a conversion of the BT moiety in PM to the BZ moiety under heat or energetic radiation exposure (Wakamatsu et al. 2012b; Galván et al. 2014; Tanaka et al. 2018), but it is unknown whether endogenous conditions can also produce this structural change. Rodríguez-Martínez et al. (2020) induced an endogenous free radical production by exposing male zebra finches *Taeniopygia guttata* experimentally to diquat dibromide in drinking water during the development of PM-pigmented flank feathers. The BZ-to-BT ratio of feather PM increased more and was significantly lower in diquat-treated birds than in controls. Therefore, it was shown that endogenous free radicals, and not only environmental radiation, can also promote structural changes in PM.

The domesticated rock pigeon (*Columba livia*) has been bred for hundreds of years to display an immense variety of ornamental attributes such as feather color and color patterns. Color is influenced by multiple loci that impact the type and amount of melanin deposited on the feathers. Pigeons homozygous for the "recessive red" mutation, which causes downregulation of *Sry-Box 10* (*Sox10*), display brilliant red feathers instead of blue/black feathers. *Sox10* encodes a transcription factor important for melanocyte differentiation and function, but the genes that mediate its promotion of black versus red pigment are unknown. Domyan et al. (2019) presented a transcriptomic comparison of regenerating feathers from wild-type and recessive red pigeons to identify candidate SOX10 targets. They identified both known and novel targets, including many genes not previously implicated in pigmentation, which highlighted the value of using novel, emerging model organisms to gain insight into the genetic basis of pigment variation.

Melanin-based colorations in birds constitute a paradigm for the study of the molecular basis of phenotypic variation. Variation in *MC1R* gene, a key regulator of melanin synthesis in feather melanocytes, can lead to changes in the production of melanin and hence in feather color. Avilés et al. (2020) investigated the proximate mechanisms behind color plumage polymorphism in the Eurasian Scops-owl *Otus scops*, a species showing pronounced variation in the degree of redness. Although EM pigment was three times more abundant than PM pigment, the degree of plumage redness was more strongly associated with the amount of PM than EM pigments. They detected only one synonymous substitution and one non-synonymous substitution in MC1R which were, however, not associated with variation in plumage coloration. Redness variation in Eurasian Scops-Owls is primarily due to variation in PM, and to genes or regulatory elements other than *MC1R*.

2.7 Fish, Reptiles, and Amphibians

Although the nature of melanins synthesized in higher vertebrates has been extensively studied, the melanins produced by aquatic animals, including aquacultured fishes, are not well understood. Red seabream, one of the most commercially valuable species of fish in Japan, is usually found at a 20 m depth of seawater, and its skin is a bright scarlet. Their skin turns dark when they are bred in net cages, and fish farmers call these "suntanned" seabream, which causes severe loss in their market value (Fig. 2.13). The cause of this suntanning is thought to be due to the production of melanin. Adachi et al. (2005) demonstrated the first detailed chemical

Fig. 2.13 The fish described in this review and their histology. (**a**) Wild red seabream. (**b**) Suntanned red seabream bred in net cages without curtains. (**c**) Shaded red seabream bred in net cages with black curtains. (Adapted from Adachi et al. 2005)

quantification of melanin in fish. As shown in Fig. 2.13, 2.3–2.8μg PTCA per cm^2 skin was detected in suntanned red seabream, while only 0.3–0.7μg PTCA per cm^2 tissue was detected in the corresponding areas of shaded and wild red seabream. This result showed that the skin of suntanned fish accumulated five times higher levels of EM than the corresponding areas of shaded and wild red seabream. Regarding PM, the content of 4-AHP was below the detection limits in fish. It should be noted that aquacultured fish were bred using a formulated diet with a high level of unsaturated fatty acids. It has been clearly shown that unsaturated fatty acids, such as oleic acid, linoleic acid, and α-linolenic acid, decrease melanin synthesis and tyrosinase activity, while saturated fatty acids increase it (Shono and Toda 1981; Ando et al. 1998). This indicates that farmed fish are reared under negative conditions for melanogenesis in the view of the nutritional facts. Nevertheless, a higher accumulation of EM was observed in suntanned and shaded fish compared with wild red seabream, indicating that the UV overwhelmed the effects of unsaturated fatty acids on tyrosinase in the fish. In fish farms, the heads of male red seabream turn black during the spawning period, while females retain their color throughout the year. Chemical analysis of EM indicated that female red seabream deposit EM in June (the spawning period), which disappears in September. This indicates that the discoloration is induced by melanin accumulation as well as by suntanning. This blackening is induced by the oral administration of androgen (methyl testosterone) in a dose-dependent manner, indicating that the discoloration is induced by hormonal control (Adachi et al. 2010). In addition to these cases, the blackening of red seabream is observed, which is induced by low temperature, background adaptation, feeding density, stress, and so on. However, no detailed investigation has been published except for a preliminary study on the effects of low temperature (Adachi et al. 2006). To suppress the melanogenesis in aquacultured red seabream, the net cages are shielded by black curtains to block UV, which inhibits 80% of EM accumulation (Adachi et al. 2005). Thus, shielding is an effective method against the melanogenesis in suntanning; however, no countermeasures have been established to date for other types of discolorations described above because of the lack of information.

In reptiles, body coloration is often structural or attributable to carotenoid and pteridine pigments (Olsson et al. 2007). Reptiles were thought to produce only EM, but not PM. However, Roulin et al. (2013) reported that in the Eastern Hermann tortoise (*Eurotestudo boettgeri*), the EM/PM ratio increased along with the change from yellowish to the black coloration of the shell, and the yellow part of the shell mainly contained PM; however, the occurrence of PM is still very rare in lower organisms. In reptiles, the melanosomes and subcellular organelles that synthesize and accumulate melanin are under hormonal and neural control, and they may be rapidly and reversibly transported into specific skin regions because of the melanin-concentrating hormone (MCH), which is much more active in these animals than in mammals (Baker 1993). Hormonal control of melanization by MSH is also important in these animals, as well as in birds and mammals; however, this aspect is beyond the scope of this review. Melanins in these animals are generally EM, except for the presence of PM in tortoises (Roulin et al. 2013). Melanin in Kupffer cells of amphibian could also be formed and the melanin produced in the liver was found to be DHICA-rich EM similar to Sepia melanin (Gallone et al. 2007).

2.8 Internal Melanin in Vertebrates

In vertebrates, melanin is present in various organs and tissues, including the spleen, liver, kidney, brain, pineal gland, inner ear, eye, lung, connective tissues, testes, choroid, heart, peritoneum, muscles, venom glands, and tumors (Dubey and Roulin 2014) (Fig. 2.14). Several studies have suggested that the variation in externally visible melanin-based color is correlated with the amount of internal melanin (Crespo and Pizarr 2006; Larsen et al. 2012). This demonstrates that the mechanism involved in melanin production in the skin and in internal tissues is similar. Although melanin production in the internal organs is considered to be important, the amount of melanin produced in these organs is not well understood. In humans, correlations between aging and both internal and external melanins are also present. For example, the level of NM localized in the pineal gland increases with aging (Koshy and Vettivel 2001), and in older people, the skin is usually more pigmented than in younger individuals, owing to a greater functional activity in older melanocytes.

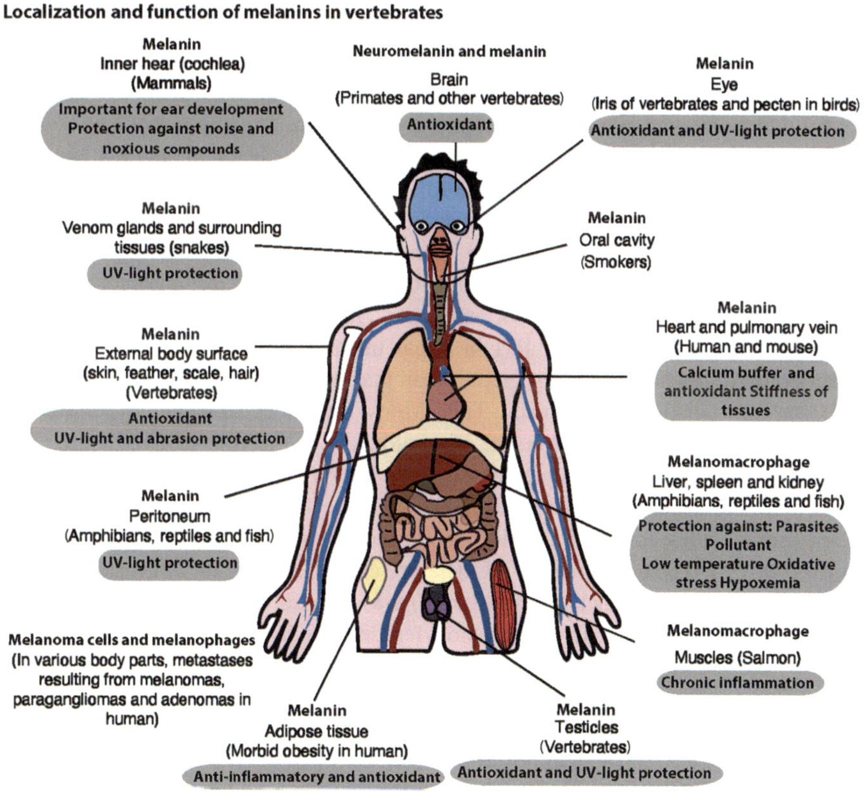

Fig. 2.14 Localization and functions of melanin in vertebrates. (Adapted from Dubey and Roulin 2014)

This occurs despite a decrease in melanocyte density with aging (Costin and Hearing 2007). Associations between internal melanin and coloration have also been reported between subspecies, breeds of domestic animals, and between laboratory mutants (Dubey and Roulin 2014). These studies have shown that darker individuals produce more internal melanin (Faraco et al. 2001). A correlation between the coloration of the skin and internal organs and tissues could occur if the expression of melanogenic genes located in various parts of the body has common regulatory processes and/or similar biosynthetic pathways. However, such covariations could also be purely correlational with independent factors leading to similar variations in internal and external melanin production. The internal and external melanin located throughout most parts of the body have a common embryonic origin; however, internal and external melanocytes are partly distinct despite their common embryonic origin (Fedorow et al. 2005). Although melanin-based coloration of the integument and internal organs is correlated, it remains unclear how such a covariation occurs. For example, whether some hormones simultaneously regulate the expression levels of melanogenic genes involved in the production of internal and external melanin.

In humans, PM and EM are present in various parts of the eye, such as in the iris and choroid-retinal pigment epithelium (Sturm and Frudakis 2004) (Fig. 2.15a), where it mainly plays photoprotective and antioxidant roles (Weiter et al. 1986; Biesemeier et al. 2011; Rózanowska 2011; Peles and Simon 2012). The variation in iris color is related to the incidence of several important ocular diseases, including uveal melanoma and age-related macular degeneration. The amount of EM and PM in cultured human uveal melanocytes was measured by chemical degradation using HPLC methods (Wakamatsu et al. 2008) (Fig. 2.15b). Uveal melanocytes derived from eyes with dark-colored irides (dark brown and brown irides) contained a significantly greater quantity of EM than uveal melanocytes derived from eyes with light-colored irides (blue, yellow-brown, green and hazel-colored irides). A similar, but less pronounced difference was observed between dark brown and brown irides. The amount of PM in uveal melanocytes derived from eyes with light-colored irides was slightly greater than that from dark-colored irides, although this was not statistically significant. The EM/PM ratio in uveal melanocytes was also related to iris color: the darker the iris color, the higher the EM/PM ratio. This indicates that darker melanocytes produce not only a higher quantity of melanin but also more eumelanic pigment than lighter melanocytes. Melanocytes derived from dark brown irides contain 93% EM, while those derived from blue irides contain only 44% EM. The total quantity of melanin (either EM+PM or TM measured with HPLC or spectrophotometry, respectively) was also correlated with iris color. EM +PM and TM values in uveal melanocytes derived from eyes with dark-colored irides were greater than those in uveal melanocytes derived from eyes with light-colored irides. This suggests that uveal melanin granules in eyes with dark-colored irides are eumelanic at the surface and act as antioxidants, while those in eyes with light-colored irides have an exposed PM core and behave as a pro-oxidant.

Melanin has been described in the cochlear labyrinth and has been suggested to protect the cochlea from various types of trauma. The quantity of melanin has been shown to change with aging in several organs, but to the best of our knowledge,

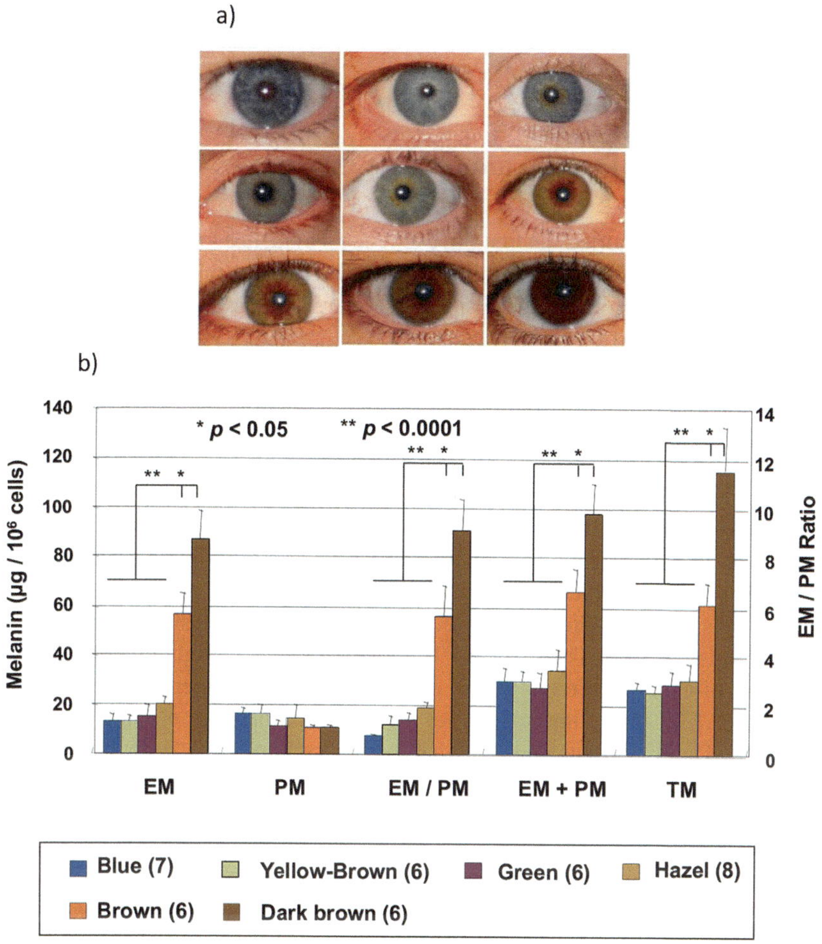

Fig. 2.15 (**a**) Representative eye colors ranging from blue, gray, green, hazel, light brown to dark brown. (Adapted from Sturm and Frudakis 2004). (**b**) Quantity and ratio of melanin in uveal melanocytes derived from eyes with various colored irides. Data are presented as means ± SEM. **P* and ***P* in growing melanocytes were compared between dark brown and brown and between brown and light-colored irides (blue, yellow-brown, green, and hazel combined, as indicated by the horizontal bars), respectively. (Adapted from Wakamatsu et al. 2008)

aging changes in the cochlea have not been documented. Therefore, we chemically quantified cochlear EM and PM contents and compared them with young and old C57BL/6 mice (Hayashi et al. 2007). The quantities of cochlear EM and PM were 421 and 480 ng per cochlea in young mice, and 2060 and 765 ng per cochlea in old mice, respectively. Transmission electron microscopy observations showed that the number of pigment granules were greater in older mice than in younger mice, especially in marginal cells. This is the first quantitative evidence of an age-related overexpression of cochlear melanin and an alteration in the proportion of EM and

PM with aging, suggesting a possible otoprotective function of EM against age-related cochlear deterioration.

2.9 Melanin in Fossil Animals

Glass et al. (2012) reported the first direct chemical evidence to confirm the presence and distribution of EM in >160 million years (Myr) Jurassic cephalopod ink sacs. The chemical degradation of fossil specimen using alkaline H_2O_2 produced PTCA, PDCA, isoPTCA, and PTeCA. The ratio of PTeCA/PTCA was greater than that of modern *S. Officinalis*. As the increase in this ratio occurs when synthetic EM are exposed to elevated temperatures for prolonged periods, they attributed the increased concentrations of PTeCA marker in the fossil samples to cross-linking in EM subunits as a consequence of diagenesis and thermal maturation during sedimentary burial (Ito et al. 2013b).

The evolution of vision in vertebrates is an important theme in the history of animal life. We have reported on the tissues of the eye of the acanthodii fish *Acanthodes bridgei*, which provides the first record of rods and cones in a fossil and indicates that fish likely possessed color vision for at least 300 Myr (Tanaka et al. 2014; Fig. 2.16a). The PTeCA/PTCA ratio of 2.7 is extremely high compared with those of extant fish retina (0.10) and bovine retinal pigment epithelium melanosomes (0.13), indicating extensive thermal modification of EM (Ito et al. 2013b, c). These findings indicate that such receptors have been conserved in vertebrate eyes for at least 300 Myr. The existence of rods, cones, and melanin pigments in the eyes of *A. bridgei* suggests that retinomotor activity (two modes of

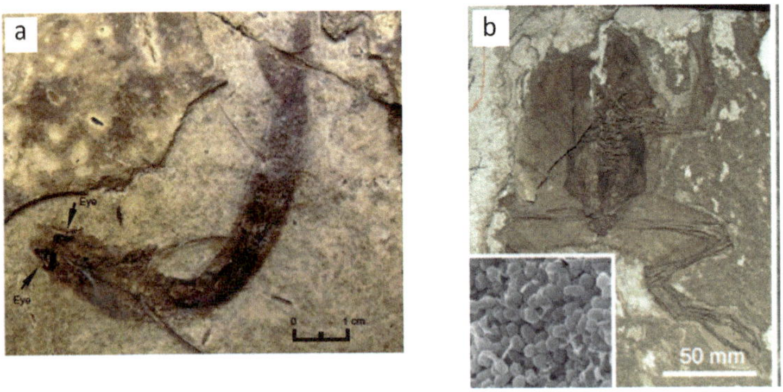

Fig. 2.16 (**a**) Optical photograph of Carboniferous fish A. bridge. Complete dorsoventrally compressed specimen (NSM PV22244); the head, including a pair of black/dark brown eyes, is at the lower left. (Taken from Tanaka et al. 2014). (**b**) Fossil amphibian soft tissue. Frog from Libros (Miocene, Spain). NMP 39449, National Museum Prague. Inset in b shows melanosomes from the torso. (Adapted from McNamara et al. 2018)

vision: daylight vision by cones and crepuscular vision by the more sensitive rods) probably already existed 300 Myr ago.

Many recent studies on fossil color have assumed that fossilized melanin granules—melanosomes—come from the skin. However, McNamara et al. (2018) showed that other tissues, such as the liver, lungs, and spleen, can also contain melanosomes, suggesting that fossil melanosomes may not always provide information on fossil color (Fig. 2.16b). They studied internal tissues in modern frogs with powerful microscopes and chemical techniques and showed that internal melanosomes are highly abundant, indicating that these internal melanosomes could comprise the majority of melanosomes preserved in some fossils. Melanosomes in the internal organs and the skin can easily be distinguished owing to their distinct sizes and shapes, thereby facilitating more accurate reconstructions of the original colors of ancient vertebrates. In this connection, Rossi et al. (2019) have found a novel way to reconstruct a whole-body picture of an extinct vertebrate by analyzing the chemical composition of fossilized visceral melanosomes. Previous studies of fossil melanin were limited to the appearance color of skin and feathers. Using the recent advanced synchrotron X-ray fluorescence analysis and detailed chemical analyses, this study revealed that it is possible to detect trace metals in melanosomes, and melanin is abundant in modern amphibians, reptiles, birds, mammals and soft tissues from fossil (Fig. 2.17). This discovery is groundbreaking in that it opens new avenues for reconstructing paleontological structures. Fossil specimens could identify skin, lungs, liver, digestive organs, heart, and even connective tissue. Furthermore, it has been suggested that melanin was involved in the regulation of metal ions as far back as 10 or 100 Myr ago. The chemical data from alkaline H_2O_2 analyses of tissues in the extant frog confirm the presence of PTCA, PDCA, and PTeCA in the liver, lung, and spleen. 4-AHP was also detected after hydroiodic acid hydrolysis (Rossi et al. 2019). The amounts of 4-AHP are much lower than those of PTCA and PTeCA, indicating that the melanin produced in these internal organs is eumelanic. Chemical analyses of the fossil soft tissues confirmed the presence of PTCA, PDCA, and PTeCA. The PTeCA/PTCA ratio in the fossil tissue was two- to fourfold higher than that in the extant tissues.

The research group of Lindgren et al. (2018) investigated the composition of skin tissue in fossil specimens of the well-preserved ichthyosaur *Stenopterygius*, which is said to be about 180 Myr old. As a result, it became clear that the original smooth skin remnants of *Stenopterygius* are still flexible and consist of a distinct inner layer (dermis layer) and outer layer (epidermal layer), with a fat layer underneath. The fat layer is a characteristic of modern marine mammals and has a heat insulating property that protects the body from low temperatures, increases buoyancy, and stores fat. This is the first time a fat layer has been identified in a fossil ichthyosaur, confirming that the ichthyosaur was a warm-blooded animal. In addition, they suggested a pigmented pattern on the skin of ichthyosaurs, which suggests countershading of reptiles (brighter shaded ventral side, darker dorsal side). They identified endogenous proteinaceous and lipid constituents, together with keratinocytes and branched melanophores that contain EM pigment characterized by a high PTeCA/PTCA ratio. It was shown that this physiological adjustment of color, found in many

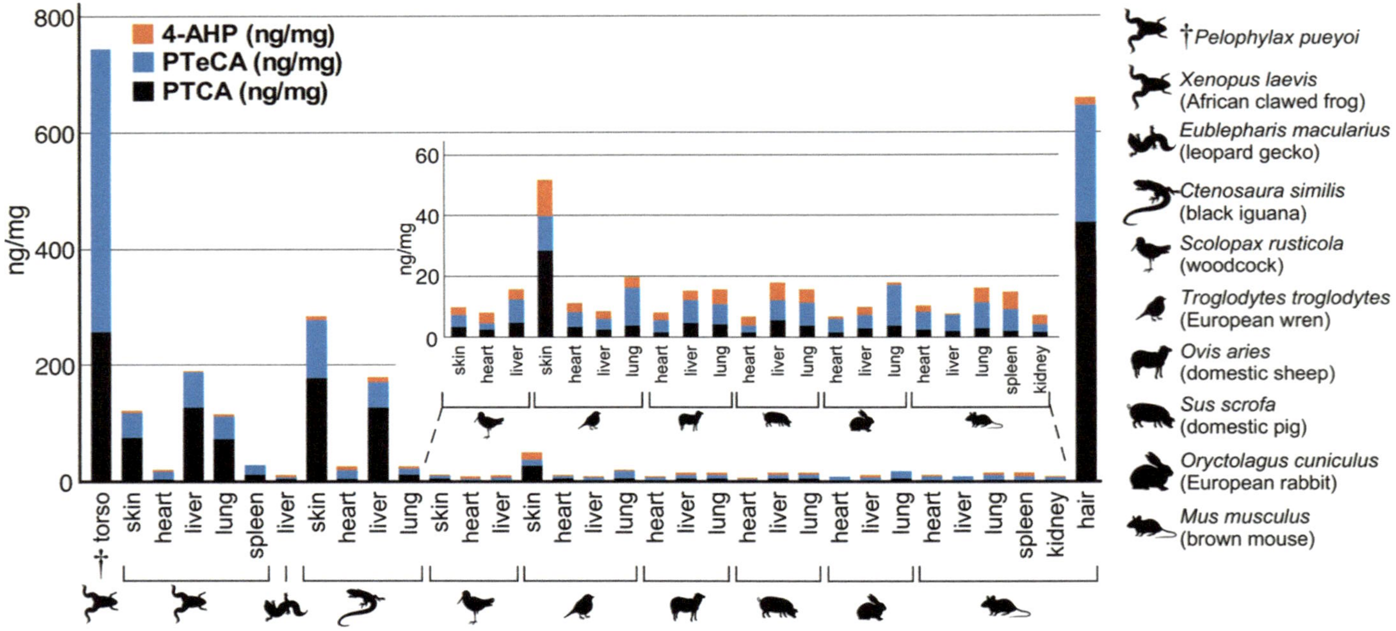

Fig. 2.17 Alkaline hydrogen peroxide oxidation (AHPO) and HI hydrolysis of extant and fossil tissues. All extant tissues analyzed contain PTCA and 4-AHP. The fossil sample (†torso, P. pueyoi) contains PTCA and abundant PTeCA, but no 4-AHP. (Adapted from Rossi et al. 2019)

Fig. 2.18 Stenopterygius specimen MH 432. Photographic representation of the fossil in oblique ventral view. (Adapted form Lindgren et al. 2018)

modern marine mammals, acts as a camouflage, photoprotection, and thermoregulation (Fig. 2.18).

Lindgren et al. (2019) also obtained the first record of melanic screening pigments in arthropods from a 54 Myr old fossil crane-flies. Evidence of secondary calcifications of lens cuticle was found in the fossilized insect eye, and these structures may have been chitinous during life. They also showed that eumelanin is present in the facet walls of living crane-flies, in which it forms the outermost ommatidial pigment shield in compound eyes incorporating a chitinous cornea. To the best of our knowledge, this is the first record of melanic screening pigments in arthropods and reveals a fossilization mode in insect eyes that involves a decay-resistant biochrome coupled with early diagenetic mineralization of the ommatidial lenses. It is commonly accepted that the trilobite eyes were rich in calcite, but this study suggests that living trilobite eyes contained far less calcite.

2.10 Conclusions

In this chapter, we focused on external and internal melanins produced by pigment cells in vertebrates: mammals, birds, fish, reptiles, and amphibians.

The diversity of the color in the external and internal tissues in vertebrates is mainly attributed to the quantity and ratio of EM and PM. Thus, it is thus important to analyze these two types of melanin pigments in studies of the biochemical and genetic bases of pigmentation in vertebrates. This can be achieved by analyzing PTCA, PDCA, and PTeCA (EM markers), as well as TTCA and TDCA (PM markers) after alkaline H_2O_2 oxidation. PM can also be analyzed as 4-AHP after HI hydrolysis. By using these methods for the evaluation of the "chemical phenotype," we estimated the mutual triangular relationship among "chemical phenotype," "visual phenotype," and "genotype." The alkaline H_2O_2 oxidation and HI hydrolysis methods are simple and reproducible and are thus suitable for increased use among researchers who do not have expertise in analytical chemistry.

It is expected that melanin analysis will continue to be a powerful tool for measuring melanin not only in vertebrates but also in invertebrates or pigment-producing organisms.

Acknowledgements The authors thank useful discussions with Dr. Tomohisa Hirobe and for his permission for allowing us to use the figure and table. And also thanks Professor Alexandre Roulin, Dr. Sylvain Dubey, and Assistant Professor Sanjay Premi for the permission to use the figure.

References

Adachi K, Kato K, Wakamatsu K (2005) The histological analysis, colorimetric evaluation, and chemical quantification of melanin content in suntanned fish. Pigment Cell Res 18:465–468

Adachi K, Kato K, Wakamatsu K et al (2006) Low temperature induced discoloration of juvenile red sea bream, Pagrus major. Aquacult Sci 54:31–35

Adachi K, Wakamatsu K, Ito S et al (2010) A close relationship between androgen levels and eumelanogenesis in the teleost red seabream (Pagrus major): quantitative analysis of its seasonal variation and effects of oral treatment with methyl testosterone. Comp Biochem Physiol A Mol Integr Physiol 156:184–189

Adams DR, Menezes S, Jauregui R et al (2018) One-year pilot study on the effects of nitisinone on melanin in patients with OCA-1B. J Clin Invest 4(2):e124387

Adelmann CH, Traunbauer AK, Chen B et al (2020) MFSD12 mediates the import of cysteine into melanosomes and lysosomes. Nature 588: 699–704

Adhikari K, Mendoza-Revilla KJ, Sohail A et al (2019) A GWAS in Latin Americans highlights the convergent evolution of lighter skin pigmentation in Eurasia. Nat Commun 10:358

Affenzeller S, Frauendorf H, Licha T et al (2019) Quantitation of eumelanin and pheomelanin markers in diverse biological samples by HPLC-UV-MS following solid-phase extraction. PLoS One 14(10):e0223552

Agrup G, Hansson C, Rorsman H et al (1982) The effect of cysteine on oxidation of tyrosine, dopa, and cysteinyldopas. Arch Dermatol Res 272:103–115

Almasi B, Roulin A, Jenni-Eiermann S et al (2008) Parental investment and its sensitivity to corticosterone is linked to melanin-based coloration in barn owls. Horm Behav 54:217–223

Ancans J, Tobin DJ, Hoogduijn MJ et al (2001) Melanosomal pH controls rate of melanogenesis, eumelanin / phaeomelanin ratio and melanosome maturation in melanocytes and melanoma cells. Exp Cell Res 268:26–35

Ando H, Ryu A, Hashimoto A et al (1998) Linoleic acid and alpha-linolenic acid lightens ultraviolet-induced hyperpigmentation of the skin. Arch Dermatol Res 290:375–381

Arai E, Hasegawa M, Nakamura M et al (2014) Male pheomelanin pigmentation and breeding onset in barn swallows Hirundo rustica gutturalis. J Ornithol 156:419–427

Arai E, Hasegawa M, Makino T et al (2017) Physiological conditions and genetic controls of phaeomelanin pigmentation in nestling barn swallows. Behav Ecol 28:706–716

Avilés JM, Cruz-Miralles Á, Ducrest A-L et al (2020) Redness variation in the Eurasian scops-owl *Otus Scops* is due to pheomelanin but is not associated with variation in the melanocortin-1 receptor gene (*MC1R*). Ardeola 67:3–13

Baker BI (1993) The role of melanin-concentrating hormone in color change. Ann N Y Acad Sci 680:279–289

Barrett RDH, Laurent S, Mallarino R et al (2019) Linking a mutation to survival in wild mice. Science 363:499–504

Barsh GS (1996) The genetics of pigmentation: from fancy genes to complex traits. Trends Genet 12:299–305

Baxter LL, Watkins-Chow DE, Pavan WJ et al (2019) A curated gene list for expanding the horizons of pigmentation biology. Pigment Cell Melanoma Res 32:348–358
Bech C, Præsteng KE (2004) Thermoregulatory use of heat increment of feeding in the tawny owl (*Strix. aluco*). J Therm Biol 29:649–654
Beltrán-García MJ, Prado FM, Oliveira MS et al (2014) Singlet Molecular oxygen generation by light-activated DHN-melanin of the fungal pathogen *Mycosphaerella fijiensis* in Black Sigatoka disease of bananas. PLoS One 9:e91616
Bennett DC, Lamoreux ML (2003) The color loci of mice – a genetic century. Pigment Cell Res 16:333–344
Biesemeier A, Schraermeyer U, Eibl O (2011) Chemical composition of melanosomes, lipofuscin and melanolipofuscin granules of human RPE tissues. Exp Eye Res 93:29–39
Boissy RE, Sakai C, Zhao H et al (1998) Human tyrosinase related protein-1 (TRP-1) does not function as a DHICA oxidase activity in contrast to murine TRP-1. Exp Dermatol 7:198–204
Bortolotti GR (2006) Natural selection and coloration: protection, concealment, advertisement, or deception? In: Hill GE, McGraw KJ (eds) Bird coloration, vol 2. Harvard University Press, Cambridge, pp 3–35
Bush WD, Garguilo J, Zucca FA et al (2006) The surface oxidation potential of human neuromelanin reveals a spherical architecture with a pheomelanin core and a eumelanin surface. Proc Natl Acad Sci U S A 103:14785–14789
Carballo-Carbajal I, Laguna A, Romero-Giménez J et al (2019) Brain tyrosinase overexpression implicates age-dependent neuromelanin production in Parkinson's disease pathogenesis. Nat Commun 10:973
Cassidy CM, Zucca FA, Girgis RR et al (2019) Neuromelanin-sensitive MRI as a noninvasive proxy measure of dopamine function in the human brain. Proc Natl Acad Sci U S A 116:5108–5117
Cheli Y, Luciani F, Khaled M et al (2009) αMSH and cyclic AMP elevating agents control melanosome pH through a protein kinase A-independent mechanism. J Biol Chem 284:18699–18706
Chen SR, Jiang B, Zheng JX et al (2008) Isolation and characterization of natural melanin derived from silky fowl *(Gallus gallus domesticus Brisson*). Food Chem 111:745–749
Chiarelli-Neto O, Ferreira AS, Martins WK et al (2014) Melanin photosensitization and the effect of visible light on epithelial cells. PLoS One 9:e113266
Chintala S, Li W, Lamoreux ML et al (2005) Slc7a11 gene controls production of pheomelanin pigment and proliferation of cultured cell. Proc Natl Acad Sci U S A 102:10964–10969
Chung JH (2003) Photoaging in Asians. Photodermatol Photoimmunol Photomed 19:109–121
Costin GE, Hearing VJ (2007) Human skin pigmentation: melanocytes modulate skin color in response to stress. FASEB J 21:976–994
Crawford NG, Kelly DE, Hansen MEB et al (2017) Loci associated with skin pigmentation identified in African populations. Science 358:eaan8433
Crespo R, Pizarr M (2006) Skin and abdominal fascia melanization in Broiler Chickens. Avian Dis 50:309–311
d'Ischia M, Wakamatsu K, Napolitano A et al (2013) Melanins and melanogenesis: methods, standards, protocols. Pigment Cell Melanoma Res 26:616–633
Del Bino S, Bernerd F (2013) Variations in skin colour and the biological consequences of ultraviolet radiation exposure. Br J Dermatol 169(Suppl 3):33–40
Del Bino S, Sok J, Bessac E et al (2006) Relationship between skin response to ultraviolet exposure and skin color type. Pigment Cell Res 19:606–614
Del Bino S, Ito S, Sok J et al (2015) Chemical analysis of constitutive pigmentation of human epidermis reveals constant eumelanin to pheomelanin ratio. Pigment Cell Melanoma Res 28:707–717
del Marmol V, Ito S, Bouchard B et al (1996) Cysteine deprivation promotes eumelanogenesis in human melanoma cells. J Invest Dermatol 107:698–702

Domyan ET, Hardy J, Wright T et al (2019) SOX10 regulates multiple genes to direct eumelanin versus pheomelanin production in domestic rock pigeon. Pigment Cell Melanoma Res 32:634–642
Dubey S, Roulin A (2014) Evolutionary and biomedical consequences of internal melanins. Pigment Cell Melanoma Res 27:327–338
Eisenhofer G, Kopin IJ, Goldstein DS (2004) Catecholamine metabolism: a contemporary view with implications for physiology and medicine. Pharmacol Rev 56:331–349
Elder DE (1995) Skin cancer. Melanoma and other specific nonmelanoma skin cancers. Cancer 75:245–256
Engelen M, Vanna R, Bellei C et al (2012) Neuromelanins of human brain have soluble and insoluble components with dolichols attached to the melanic structure. PLoS One 7:e48490
Enshell-Seijiffers D, Lindon C, Wu E et al (2010) β-catenin activity in the dermal papilla of the hair follicle regulates pigment-type switching. Proc Natl Acad Sci U S A 107:21564–21569
Fajuyigbe D, Lwin SM, Diffey BL et al (2018) Melanin distribution in human epidermis affords localized protection against DNA photodamage and concurs with skin cancer incidence difference in extreme phototypes. FASEB J 32:3700–3706
Faraco CD, Vaz SAS, Pástor MAD et al (2001) Hyperpigmentation in the silkie fowl correlates with abnormal migration of fate-restricted melanoblasts and loss of environmental barrier molecules. Dev Dyn 220:212–225
Fedorow H, Tribl F, Halliday G et al (2005) Neuromelanin in human dopamine neurons: comparison with peripheral melanins and relevance to Parkinson's disease. Prog Neurobiol 75:109–124
Foster M (1965) Mammalian pigment genetics. Adv Genet 13:311–339
Fuller BB, Spaulding DT, Smith DR (2001) Regulation of the catalytic activity of preexisting tyrosinase in black and Caucasian human melanocyte cell cultures. Exp Cell Res 262:197–208
Gahl WA, Bashan N, Tietze F et al (1982) Cystine transport is defective in isolated leukocyte lysosomes from patients with cystinosis. Science 217:1263–1265
Gallone A, Sagliano A, Guida G et al (2007) The melanogenic system of the liver pigmented macrophages of Rana esculenta L. –tyrosinase activity. Histol Histopathol 22:1065–1075
Galván I, Alonso-Alvarez C (2009) The expression of melanin-based plumage is separately modulated by exogenous oxidative stress and a melanocortin. Proc R Soc Lond B Biol Sci 276:3089–3097
Galván I, Jorge A (2015) Dispersive Raman spectroscopy allows the identification and quantification of melanin types. Ecol Evol 5:1425–1431
Galván I, Møller AP (2011) Brain size and the expression of pheomelanin-based colour in birds. J Evol Biol 24:999–1006
Galván I, Møller AP (2013) Pheomelanin-based plumage coloration predicts survival rates in birds. Physiol Biochem Zool 86:184–192
Galván I, Solano F (2009) The evolution of eu- and pheomelanic traits may respond to an economy of pigments related to environmental oxidative stress. Pigment Cell Melanoma Res 22:339–342
Galván I, Solano F (2015) Melanin chemistry and the ecology of stress. Physiol Biochem Zool 88:352–355
Galván I, Bonisoli-Alquati A, Jenkinson S et al (2014) Chronic exposure to low-dose radiation at Chernobyl favours adaptation to oxidative stress in birds. Funct Ecol 28:1387–1403
Gaudel C, Soysouvanh F, Leclerc J et al (2020) Regulation of melanogenesis by the amino acid transporter SLC7A5. I Invest Dermatol 140:2253–2259
Glass K, Ito S, Wilby PR et al (2012) Direct chemical evidence for eumelanin pigment from the Jurassic period. Proc Natl Acad Sci U S A 109:10218–10223
Goedert M (2001) Alpha-synuclein and neurodegenerative diseases. Nat Rev Neurosci 2:492–501
Gümüşlü S, Sarikçioĝlu SB, Sahin E et al (2002) Influences of different stress models on the antioxidant status and lipid peroxidation in rat erythrocytes. Free Radic Res 36:1277–1282
Gunn TM, Inui T, Kitada K et al (2001) Molecular and phenotypic analysis of attractin mutant mice. Genetics 158:1683–1695
Halder RM, Ara CJ (2003) Skin cancer and photoaging in ethnic skin. Dermatol Clin 21:725–732

Hayashi H, Sone M, Schachern PA et al (2007) Comparison of the quantity of cochlear melanin in young and old C57BL/6 mice. Arch Otolaryngol Head Neck Surg 133:151–154

Hellström AR, Watt B, Fard SS et al (2011) Inactivation of *Pmel* alters melanosome shape but has only a subtle effect on visible pigmentation. PLoS Genet 7:e1002285

Hermansky F, Pudlak P (1959) Albinism associated with hemorrhagic diathesis and unusual pigmented reticular cells in the bone marrow: Report of two cases with histochemical studies. Blood 14:162–169

Hill HZ, Hill GJ (2000) UVA, pheomelanin and the carcinogenesis of melanoma. Pigment Cell Res 13(Suppl 8):140–144

Hill HZ, Li W, Xin P et al (1997) Melanin: a two edged sword? Pigment Cell Res 10:158–161

Hirobe T (1984) Effects of genic substitution at the brown locus on the differentiation of epidermal melanocytes in newborn mouse skin. Anat Rec 209:425–432

Hirobe T (1992) Control of melanocyte proliferation and differentiation in the mouse epidermis. Pigment Cell Res 5:1–11

Hirobe T (2011) How are proliferation and differentiation of melanocytes regulated? Pigment Cell Melanoma Res 24:462–478

Hirobe T (2012) The coat color genes regulate eumelanin and pheomelanin synthesis in melanocytes. In: Ma X-P, Sun X-X (eds) Melanin: biosynthesis, functions and health effects. Nova Science Publishers, Inc., New York, pp 109–137

Hirobe T, Abe H (1999) Genetic and epigenetic control of the proliferation and differentiation of mouse epidermal melanocytes in culture. Pigment Cell Res 12:147–163

Hirobe T, Wakamatsu K, Ito S (1998) Effects of genic substitution at the agouti, brown, albino, dilute, and pink-eyed dilution loci on the proliferation and differentiation of mouse epidermal melanocytes in serum-free culture. Eur J Cell Biol 75:184–191

Hirobe T, Wakamatsu K, Ito S (2003) Changes in the proliferation and differentiation of neonatal mouse pink-eyed dilution melanocytes in the presence of excess tyrosine. Pigment Cell Res 16:619–628

Hirobe T, Takeuchi S, Hotta E et al (2004) Pheomelanin production in the epidermis from newborn agouti mice is induced by the expression of the agouti gene in the dermis. Pigment Cell Res 17:506–514

Hirobe T, Wakamatsu K, Ito S et al (2006) The slaty mutation affects eumelanin and pheomelanin synthesis in mouse melanocytes. Eur J Cell Biol 85:537–549

Hirobe T, Wakamatsu K, Ito S (2007a) The eumelanin and pheomelanin contents in dorsal hairs of female recessive yellow mice are greater than in male. J Dermatol Sci 45:55–62

Hirobe T, Abe H, Wakamatsu K et al (2007b) Excess tyrosine rescues the reduced activity of proliferation and differentiation of cultured recessive yellow melanocytes derived from neonatal mouse epidermis. Eur J Cell Biol 86:315–330

Hirobe T, Kiuchi M, Wakamatsu K et al (2010) Estrogen increases hair pigmentation in female recessive yellow mice. Zool Sci 27:470–476

Hirobe T, Ito S, Wakamatsu K (2011a) The mouse pink-eyed dilution allele of the P-gene greatly inhibits eumelanin, but not pheomelanin synthesis. Pigment Cell Melanoma Res 24:241–246

Hirobe T, Yoshihara C, Takeuchi S et al (2011b) A novel deletion mutation of mouse ruby-eye 2 named ru2d/Hps5ru2-d inhibits melanocyte differentiation and its impaired differentiation is rescued by L-tyrosine. Zool Sci 28:790–801

Ho SGY, Chan HHL (2009) The Asian dermatologic patient: review of common pigmentary disorders and cutaneous diseases. Am J Clin Dermatol 10:153–168

Hourblin V, Nouveau S, Roy N et al (2014) Skin complexion and pigmentary disorders in facial skin of 1204 women in 4 Indian cities. Indian J Dermatol Venereol Leprol 80:395–401

Hudon J (2005) Considerations in the conservation of feathers and hair, particularly their pigments. In: Proc. CAC/ACCR 31st Annual Conf., Jasper, AB, Canada, May, pp. 127–147

Ito S (2006) Encapsulation of a reactive core in neuromelanin. Proc Natl Acad Sci U S A 103:14647–14648

Ito S, Fujita K (1985) Microanalysis of eumelanin and pheomelanin in hair and melanomas by chemical degradation and liquid chromatography. Anal Biochem 144:527–536
Ito S, Prota GA (1977) A facile one-step synthesis of cysteinyldopas using mushroom tyrosinase. Experientia 33:1118–1119
Ito S, Wakamatsu K (2003) Quantitative analysis of eumelanin and pheomelanin in humans, mice, and other animals: a comparative review. Pigment Cell Res 16:523–531
Ito S, Wakamatsu K (2008) Chemistry of mixed melanogenesis – pivotal roles of dopaquinone. Photochem Photobiol 84:582–592
Ito S, Wakamatsu K (2011a) Human hair melanins: what we have learned and have not learned from mouse coat color pigmentation. Pigment Cell Melanoma Res 24:63–74
Ito S, Wakamatsu K (2011b) Diversity of human hair pigmentation as studied by chemical analysis of eumelanin and pheomelanin. J Eur Acad Dermatol Venereol 25:1369–1380
Ito S, Wakamatsu K (2018) Biochemical mechanism of rhododendrol-induced leukoderma. Int J Mol Sci 19:e552
Ito S, Nakanishi Y, Valenzuela RK et al (2011c) Usefulness of alkaline hydrogen peroxide oxidation to analyze eumelanin and pheomelanin in various tissue samples: application to chemical analysis of human hair melanins. Pigment Cell Melanoma Res 24:605–613
Ito S, Wakamatsu K, d'Ischia M et al (2011d) Structure of melanins: melanins and melanosomes: biosynthesis, biogenesis, physiological, and pathological functions. Wiley-Blackwell, Amsterdam, pp 167–185
Ito S, Suzuki N, Takebayashi S et al (2013a) Neutral pH and copper ions promote eumelanogenesis after the dopachrome stage. Pigment Cell Melanoma Res 26:817–825
Ito S, Wakamatsu K, Glass K et al (2013b) High-performance liquid chromatography estimation of cross-linking of dihydroxyindole moiety in eumelanin. Anal Biochem 434:221–225
Ito S, Pilat A, Gerwat W et al (2013c) Photoaging of human retinal pigment epithelium is accompanied by oxidative modifications of its eumelanin. Pigment Cell Melanoma Res 26:357–366
Ito S, Ojika M, Yamashita T et al (2014a) Tyrosinase-catalyzed oxidation of rhododendrol produces 2-methylchromane-6,7-dione, the putative ultimate toxic metabolite: implication for melanocyte toxicity. Pigment Cell Melanoma Res 27:744–753
Ito S, Gerwat W, Kolbe L et al (2014b) Human tyrosinase is able to oxidize both enantiomers of rhododendrol. Pigment Cell Melanoma Res 27:1149–1153
Ito S, Okura M, Wakamatsu K et al (2017) The potent pro-oxidant activity of rhododendrol-eumelanin induces cysteine depletion in B16 melanoma cells. Pigment Cell Melanoma Res 30:63–67
Ito S, Wakamatsu K, Sarna T (2018b) Photodegradation of eumelanin and pheomelanin and its pathophysiological implications. Photochem Photobiol 94:409–420
Ito S, Miyake S, Maruyama S et al (2018c) Acid hydrolysis reveals a low but constant level of pheomelanin in human black to brown hair. Pigment Cell Melanoma Res 31:393–403
Ito S, Fujiki Y, Matsui N et al (2019) Tyrosinase-catalyzed oxidation of resveratrol produces a highly reactive ortho-quinone: implications for melanocyte toxicity. Pigment Cell Melanoma Res 32:766–776
Ito S, Sugumaran M, Wakamatsu K (2020a) Chemical reactivities of *ortho*-quinones produced in living organisms: fate of quinonoid products formed by tyrosinase and phenoloxidase action on phenols and catechols. Int J Mol Sci 21:6080
Ito S, Del Bino S, Hirobe T et al (2020b) Improved HPLC conditions to determine eumelanin and pheomelanin contents in biological samples using an ion pair reagent. Int J Mol Sci 21:5134
Itou T, Ito S, Wakamatsu K (2019) Effects of aging on hair color, melanosome morphology, and melanin composition in Japanese females. Int J Mol Sci 20:3739
Iwata Y, Kobayashi T, Arima M et al (2017) Case of Japanese Hermansky-Pudlak syndrome patient with deeply invasive squamous cell carcinoma and multiple lesions of actinic keratosis on the face and neck. J Dermatol 44:219–220

Jackson IJ, Chambers D, Rinchik EM et al (1990) Characterization of TRP-1 mRNA levels in dominant and recessive mutations at the mouse brown (b) locus. Genetics 126:451–459
Jackson IJ, Chambers DM, Tsukamoto K et al (1992) A second tyrosinase-related protein, TRP-2, maps to and is mutated at the mouse slaty locus. EMBO J 11:527–535
Jiménez-Cervantes C, Solano F, Kobayashi T et al (1994) A new enzymatic function in the melanogenic pathway. The 5,6-dihydroxyindole-2-carboxylic acid oxidase activity of tyrosinase-related protein-1 (TRP1). J Biol Chem 269:17993–18000
Jonas AJ, Smith ML, Schneider JA (1982) ATP-dependent lysosomal cystine efflux is defective in cystinosis. J Biol Chem 257:13185–13188
Kadekaro AL, Kavanagh RJ, Wakamatsu K et al (2003) Cutaneous photobiology. The melanocyte versus the sun: who will win the final round? Pigment Cell Res 16:434–447
Karell P, Ahola K, Karstinen T et al (2011) Climate change drives microevolution in a wild bird. Nat Commun 2:208
Kobayashi T, Urabe K, Winder AJ et al (1994) Tyrosinase related protein 1 (TRP1) functions as a DHICA oxidase in melanin biosynthesis. EMBO J 13:5818–5825
Kono M, Kondo T, Ito S et al (2012) Genotype analysis in a patient with oculocutaneous albinism 1 minimal pigment type. Br J Dermatol 166:896–898
Körner AM, Pawelek J (1980) DOPAchrome conversion: a possible control point in melanin biosynthesis. J Invest Dermatol 75:192–195
Koshy S, Vettivel S (2001) Melanin pigments in human pineal gland. J Anat Soc India 50:122–126
Koskenpato K, Ahola K, Karstinen T et al (2016) Is the denser contour feather structure in pale grey than in pheomelanic brown tawny owls *Strix aluco* an adaptation to cold environments? J Avian Biol 47:1–6
Kroumpouzos G, Urabe K, Kobayashi T et al (1994) Functional analysis of the slaty gene product (TRP2) as dopachrome tautomerase and the effect of a point mutation on its catalytic function. Biochem Biophys Res Commun 202:1060–1068
Lamason RL, Mohideen M-APK, Mest IR et al (2005) SLC24A5, a putative cation exchanger, affects pigmentation in zebrafish and humans. Science 310:1782–1786
Lamoreux ML, Wakamatsu K, Ito S (2001) Interaction of major coat color gene functions in mice as studied by chemical analysis of eumelanin and pheomelanin. Pigment Cell Res 14:23–31
Land EJ, Riley PA (2000) Spontaneous redox reactions of dopaquinone and the balance between the eumelanic and phaeomelanic pathways. Pigment Cell Res 13:273–277
Land EJ, Ito S, Wakamatsu K et al (2003) Rate constants for the first two chemical steps of eumelanogenesis. Pigment Cell Res 16:487–493
Lao O, de Gruijter JM, van Duijin K et al (2007) Signatures of positive selection in genes associated with human skin pigmentation as revealed from analyses of single nucleotide polymorphisms. Ann Hum Genet 71:354–369
Larsen HAS, Austbø L, Mørkøre T et al (2012) Pigment-producing granulomatous myopathy in Atlantic salmon: a novel inflammatory response. Fish Shellfish Immunol 33:277–285
Laurentino TG, Moser D, Roesti M et al (2020) Genomic release-recapture experiment in the wild reveals within-generation polygenic selection in stickleback fish. Nat Commun 11:1928
Lawrence KP, Douki T, Sarkany RPE et al (2018) The UV/visible radiation boundary region (385–405 nm) damages skin cells and induces "dark" cyclobutane pyrimidine dimers in human skin *in vivo*. Sci Rep 8:12722
Leerunyakul K, Suchonwanit P (2020) Asian hair: a review of structures, properties, and distinctive disorders. Clin Cosmet Investig Dermatol 13:309–318
Lindgren J, Sjövall P, Thiel V et al (2018) Soft-tissue evidence for homeothermy and crypsis in a Jurassic ichthyosaur. Nature 564:359–365
Lindgren J, Nilsson D-E, Sjövall P et al (2019) Fossil insect eyes shed light on trilobite optics and the arthropod pigment screen. Nature 573:122–125
Liu SY, Shawkey MD, Parkinson D et al (2014) Elucidation of the chemical composition of avian melanin. RSC Adv 4:40396–40399

Mann DMA, Yates PO (1983) Possible role of neuromelanin in the pathogenesis of Parkinson's disease. Mech Aging Dev 21:193–203
Marsden CD (1961) Pigmentation in the nucleus substantiae nigrae of mammals. J Anat 95:256–261
Marsden CD (1983) Neuromelanin and Parkinson's disease. J Neural Transm 19:121–141
McGraw KJ (2006) Mechanics of uncommon colors: Pterins, porphyrins, and psittacofulvins. In: Hill GH, McGraw KJ (eds) Bird coloration, vol 1. Harvard University Press, Harvard, pp 354–398
McGraw KJ, Safra RJ, Evans MR et al (2004) European barn swallows use melanin pigments to color their feathers brown. Behav Ecol 15:889–891
McNamara ME, Kaye JS, Benton MJ et al (2018) Non-integumentary melanosomes can bias reconstructions of the colours of fossil vertebrates. Nat Commun 9:2878
Miller MW, Duhl DM, Vrieling H et al (1993) Cloning of the mouse agouti gene predicts a secreted protein ubiquitously expressed in mice carrying the lethal yellow mutation. Genes Dev 7:454–467
Mitra D, Luo X, Morgan A et al (2012) An ultraviolet-radiation-independent pathway to melanoma carcinogenesis in the red hair/fair skin background. Nature 491:449–453
Na J-I, Shin J-W, Choi H-R et al (2019) Resveratrol as a multifunctional topical hypopigmenting agent. Int J Mol Sci 20:E956
Napolitano A, Panzella L, Monfrecola G et al (2014) Pheomelanin-induced oxidative stress: bright and dark chemistry bridging red hair phenotype and melanoma. Pigment Cell Melanoma Res 27:721–733
Negro JJ, Blasco R, Rosell J et al (2016) Potential exploitation of avian resources by fossil hominins: An overview from ethnographic and historical data. Quat Int 421:6–11
Nicolaus RA (1969) Melanins. Hermann, Paris
Nishigori C, Aoyama Y, Ito A et al (2015) Guide for medical professionals (i.e. dermatologists) for the management of Rhododendrol-induced leukoderma. J Dermatol 42:113–128
Noonan FP, Zaidi MR, Wolnicka-Glubisz A et al (2012) Melanoma induction by ultraviolet A but not ultraviolet B radiation requires melanin pigment. Nat Commun 3:e884
Norton HL, Kittles RA, Parra E et al (2007) Genetic evidence for the convergent evolution of light skin in Europeans and East Asians. Mol Biol Evol 24:710–722
Nouveau-Richard S, Yang Z, Mac-Mary S et al (2005) Skin ageing: a comparison between Chinese and European populations. A pilot study. J Dermatol Sci 40:187–193
Okamura K, Abe Y, Araki Y et al (2018) Characterization of melanosomes and melanin in Japanese patients with Hermansky-Pudlak Syndrome Type 1, 4, 6 and 9. Pigment Cell Melanoma Res 31:267–276
Olivares C, Jiménez-Cervantes J, Lozano A et al (2001) The 5,6-dihydroxyindole-2-carboxylic acid (DHICA) oxidase activity of human tyrosinase. Biochem J 354:131–139
Olsson M, Healey M, Wapstra E et al (2007) Mating system variation and morph fluctuations in a polymorphic lizard. Mol Ecol 16:5307–5315
Ozaki S, Funasaka Y, Otsuka Y et al (2017) Melanotic malignant melanoma in oculocutaneous albinism type 4. Acta Derm Venereol 97:287–288
Ozeki H, Ito S, Wakamatsu K et al (1995) Chemical characterization of hair melanins in various coat-color mutants of mice. J Invest Dermatol 105:361–366
Ozeki H, Ito S, Wakamatsu K et al (1996) Spectrophotometric characterization of eumelanin and pheomelanin in hair. Pigment Cell Res 9:265–270
Ozeki H, Wakamatsu K, Ito S et al (1997) Chemical characterization of eumelanins with special emphasis on 5,6-dihydroxyindole-2-carboxylic acid content and molecular size. Anal Biochem 248:149–157
Pan T, Zhu J, Hwu W-J et al (2012) The role of alpha-synuclein in melanin synthesis in melanoma and dopaminergic neuronal cells. PLoS One 7:e45183
Park J, Park JH, Suh H-J et al (2014) Effects of resveratrol, oxyresveratrol, and their acetylated derivatives on cellular melanogenesis. Arch Dermatol Res 306:475–487

Pawelek JM, Körner AM, Bergstrom A et al (1980) New regulators of melanin biosynthesis and the autodestruction of melanoma cells. Nature 286:617–619
Peles DN, Simon JD (2012) The UV-absorption spectrum of human iridal melanosomes: a new perspective on the relative absorption of eumelanin and pheomelanin and its consequences. Photochem Photobiol 88:1378–1384
Peles DN, Hong L, Hu DN et al (2009) Human iridal stroma melanosomes of varying pheomelanin contents possess a common eumelanic outer surface. J Phys Chem B 113:11346–11351
Pisoni RL, Acker TL, Lisowski KM et al (1990) A cysteine-specific lysosomal transport system provides a major route for the delivery of thiol to human fibroblast lysosomes: possible role in supporting lysosomal proteolysis. J Cell Biol 110:327–335
Potterf SB, Virador V, Wakamatsu K et al (1999) Cysteine transport in melanosomes from murine melanocytes. Pigment Cell Res 12:4–12
Premi S, Wallisch S, Mano CM et al (2015) Chemiexcitation of melanin derivatives induces DNA photoproducts long after UV exposure. Science 347:842–847
Premi S, Brash DE (2015) Unanticipated role of melanin in causing carcinogenic cyclobutane pyrimidine dimers. Mol Cell Oncol 3 (1):e1033588
Prota G (1995) The chemistry of melanins and melanogenesis. Fortschr Chem Org Naturst 64:93–148
Prum RO (2006) Anatomy, physics, and evolution of avian structural colors. In: Hill GE, McGraw KJ (eds) Bird coloration, vol 1. Harvard University Press, Cambridge, pp 295–353
Rioux B, Rouanet J, Akil H et al (2019) Determination of eumelanin and pheomelanin in melanomas using solid-phase extraction and high performance liquid chromatography – diode array detection (HPLC-DAD) analysis. J Chromatogr B 1113:60–68
Rodríguez-Martínez S, Wakamatsu K, Galván I (2020) Increase of the benzothiazole moiety content of pheomelanin pigment after endogenous free radical inducement. Dyes Pigments 180:108516
Rorsman H, Agrup G, Hansson C et al (1979) Detection of Phaeomelanins. Pigment Cell 4:244–252
Rossi V, McNamara ME, Webb SM et al (2019) Tissue-specific geometry and chemistry of modern and fossilized melanosomes reveal internal anatomy of extinct vertebrates. Proc Natl Acad Sci U S A 116:17880–17889
Roulin A, Mafli A, Wakamatsu K (2013) Reptiles produce pheomelanin: evidence in the Eastern Hermann's Tortoise (*Eurotestudo boettgeri*). J Herpetol 47:258–261
Rózanowska M (2011) Properties and functions of ocular melanins and melanosomes. In: Borovansky J, Riley PA (eds) Melanins and melanosomes: biosynthesis, biogenesis, physiological, and pathological functions. Wiley, Weinheim, pp 187–224
Sakurai T, Ochiai H, Takeuchi T (1975) Ultrastructural change of melanosomes associated with agouti pattern formation in mouse hair. Dev Biol 47:466–471
Sasaki M, Shibata E, Tohyama K et al (2006) Neuromelanin magnetic resonance imaging of locus ceruleus and substantia nigra in Parkinson's disease. Neuroreport 17:1215–1218
Sasaki M, Kondo M, Sato K et al (2014) Rhododendrol, a depigmentation-inducing phenolic compound, exerts melanocyte cytotoxicity via a tyrosinase-dependent mechanism. Pigment Cell Melanoma Res 27:754–763
Sealy RC, Hyde JS, Felix CC et al (1982) Novel free radicals in synthetic and natural pheomelanins: distinction between dopa melanins and cysteinyldopa melanins by ESR spectroscopy. Proc Natl Acad Sci U S A 79:2885–2889
Shaito A, Posadino AM, Younes N et al (2020) Potential adverse effects of resveratrol: a literature review. Int J Mol Sci 21:2084
Shono S, Toda K (1981) Phenotypic expression in pigment cells. In: Seiji M (ed) Pigment cell. University of Tokyo Press, Tokyo, pp 263–268
Silvers WK (1979) The coat colors of mice: a model for mammalian gene action and interaction. Springer, Basel
Simon JD, Peles DN (2010) The red and the black. Acc Chem Res 43:1452–1460

Simon JD, Peles D, Wakamatsu K et al (2009) Current challenges in understanding melanogenesis: bridging chemistry, biological control, morphology, and function. Pigment Cell Melanoma Res 22:563–579

Smith DR, Spaulding DT, Glenn HM et al (2004) The relationship between Na^+/H^+ exchanger expression and tyrosinase activity in human melanocytes. Exp Cell Res 298:521–534

Sturm RA (2009) Molecular genetics of human pigmentation diversity. Hum Mol Genet 18:R9–R17

Sturm RA, Duffy DL (2012) Human pigmentation genes under environmental selection. Genome Biol 13:248

Sturm RA, Frudakis TN (2004) Eye colour: portals into pigmentation genes and ancestry. Trends Genet 20:327–332

Sulzer D, Mosharov E, Talloczy Z et al (2008) Neuronal pigmented autophagic vacuoles: lipofuscin, neuromelanin, and ceroid as macroautophagic responses during aging and disease. J Neurochem 106:24–36

Sulzer D, Cassidy C, Horga G et al (2018) Neuromelanin detection by magnetic resonance imaging (MRI) and its promise as a biomarker for Parkinson's disease. NPJ Parkinson's Dis 4:11

Szewczyk G, Zadlo A, Sarna M et al (2016) Aerobic photoreactivity of synthetic eumelanins and pheomelanins: generation of singlet oxygen and superoxide anion. Pigment Cell Melanoma Res 29:669–678

Tamate HB, Hirobe T, Wakamatsu K et al (1989) Levels of tyrosinase and its mRNA in coat-color mutants of C57BL/10J congenic mice: effects of genic substitution at the agouti, brown, albino, dilute, and pink-eyed dilution loci. J Exp Zool 250:304–311

Tanaka S, Yamamoto H, Takeuchi S et al (1990) Melanization in albino mice transformed by introducing cloned mouse tyrosinase gene. Development 108:223–227

Tanaka G, Parker AR, Hasegawa Y et al (2014) Mineralized rods and cones suggest colour vision in a 300 Myr-old fossil fish. Nat Commun 5:5920

Tanaka H, Yamashita Y, Umezawa K et al (2018) The Pro-Oxidant activity of pheomelanin is significantly enhanced by UVA irradiation: benzothiazole moieties are more reactive than benzothiazine moieties. Int J Mol Sci 19:2889

Tewari A, Sarkany RP, Young AR (2012) UVA1 induces cyclobutane pyrimidine dimers but not 6-4 photoproducts in human skin in vivo. J Invest Dermatol 132:394–400

Thody AJ, Higgins EM, Wakamatsu K et al (1991) Pheomelanin as well as eumelanin is present in human epidermis. J Invest Dermatol 97:340–344

Tomita Y, Kondo Y, Ito S et al (1992) Menkes' disease: report of a case and determination of eumelanin and pheomelanin in hypopigmented hair. Dermatology 185:66–68

Trujillo P, Summers PE, Ferrari E et al (2017) Contrast mechanisms associated with neuromelanin-MRI. Magn Reson Med 78:1790–1800

Tsukamoto K, Jackson IJ, Urabe K et al (1992) A second tyrosinase-related protein, TRP-2, is a melanogenic enzyme termed DOPAchrome tautomerase. EMBO J 11:519–526

Valenzuela RK, Henderson MS, Walsh MH et al (2010) Predicting phenotype from genotype: normal pigmentation. J Forensic Sci 55:315–322

Van Raamsdonk CD, Barsh GS, Wakamatsu K et al (2009) Independent regulation of hair and skin color by two G-protein-coupled pathways. Pigment Cell Melanoma Res 22:816–826

Vsevolodov ED, Ito S, Wakamatsu K et al (1991) Comparative analysis of hair melanins by chemical and electron spin resonance methods. Pigment Cell Res 4:30–34

Wakamatsu K, Ito S, Rees JL (2002) The usefulness of 4-amino-3-hydroxyphenylalanine as a specific marker of pheomelanin. Pigment Cell Res 15:225–232

Wakamatsu K, Fujikawa K, Zucca FA et al (2003) The structure of neuromelanin as studied by chemical degradative methods. J Neurochem 86:1015–1023

Wakamatsu K, Kavanagh R, Kadekaro AL et al (2006) Diversity of pigmentation in cultured human melanocytes is due to differences in the type as well as quantity of melanin. Pigment Cell Res 19:154–162

Wakamatsu K, Hu D-N, McCormick SA et al (2008) Characterization of melanin in human iridal and choroidal melanocytes from eyes with various colored irides. Pigment Cell Melanoma Res 21:97–105

Wakamatsu K, Ohtara K, Ito S (2009) Chemical analysis of late stages of pheomelanogenesis: conversion of dihydrobenzothiazine to a benzothiazole structure. Pigment Cell Melanoma Res 22:474–486

Wakamatsu K, Murase T, Zucca FA et al (2012a) Biosynthetic pathway to neuromelanin and its aging process. Pigment Cell Melanoma Res 25:792–803

Wakamatsu K, Nakanishi Y, Miyazaki N et al (2012b) UVA-induced oxidative degradation of melanins: fission of indole moiety in eumelanin and conversion to benzothiazole moiety in pheomelanin. Pigment Cell Melanoma Res 25:434–445

Wakamatsu K, Tanaka H, Tabuchi K et al (2014) Reduction of the nitro group to amine by hydroiodic acid to synthesize o-aminophenol derivatives as putative degradative markers of neuromelanin. Molecules 19:8039–8050

Wakamatsu K, Tabuchi K, Ojika M et al (2015) Norephinephrine and its metabolites are involved in the synthesis of neuromelanin derived from the locus coeruleus. J Neurochem 135:768–776

Wakamatsu K, Nagao A, Watanabe M et al (2017) Pheomelanogenesis is promoted at a weakly acidic pH. Pigment Cell Melanoma Res 30:372–377

Walker WP, Gunn TM (2010) Shades of meaning: the pigment-type switching system as a tool for discovery. Pigment Cell Melanoma Res 23:485–495

Wei ML (2006) Hermansky-Pudlak syndrome: a disease of protein trafficking and organelle function. Pigment Cell Res 19:19–42

Weiter JJ, Delori FC, Wing GL et al (1986) Retinal pigment epithelial lipofuscin and melanin and choroidal melanin in human eyes. Invest Ophthalmol Vis Sci 27:145–152

Yamamoto H, Takeuchi S, Kudo T et al (1989) Melanin production in cultured albino melanocytes transfected with mouse tyrosinase cDNA. Jpn J Genet 64:121–135

Yoshizawa J, Abe Y, Oiso N et al (2014) Variants in melanogenesis-related genes associate with skin cancer risk among Japanese populations. J Dermatol 41:296–302

Zadlo A, Szewczyk G, Sarna M et al (2019) Photobleaching of pheomelanin increases its phototoxic potential; physicochemical studies of synthetic pheomelanin subjected to aerobic photolysis. Pigment Cell Melanoma Res 32:359–372

Zecca L, Stroppolo A, Gatti A et al (2004) The role of iron and copper molecules in the neuronal vulnerability of locus coeruleus and substantia nigra during aging. Proc Natl Acad Sci U S A 101:9843–9848

Zecca L, Zucca FA, Albertini A et al (2006) A proposed dual role of neuromelanin in the pathogenesis of Parkinson's disease. Neurology 67(7 Suppl 2):S8–S11

Zecca L, Bellei C, Costi P et al (2008) New melanic pigments in the human brain that accumulate in aging and block environmental toxic metals. Proc Natl Acad Sci U S A 105:17567–17572

Zhou D, Ota K, Nardin C et al (2018) Mammalian pigmentation is regulated by a distinct cAMP-dependent mechanism that controls melanosome pH. Sci Signal 11(555):eaau7987

Zucca FA, Basso E, Cupaioli FA et al (2014) Neuromelanin of the human substantia nigra: an update. Neurotox Res 25:13–23

Zucca FA, Segura-Aguilar J, Ferrari E et al (2017) Interactions of iron, dopamine and neuromelanin pathways in brain aging and Parkinson's disease. Prog Neurobiol 155:96–119

Zucca FA, Vanna R, Cupaioli FA et al (2018) Neuromelanin organelles are specialized autolysosomes that accumulate undegraded proteins and lipids in aging human brain and are likely involved in Parkinson's disease. NPJ Parkinson's Dis 4:17

Chapter 3
Body Color Expression in Birds

Toyoko Akiyama and Keiji Kinoshita

Abstract Birds display various body colors and patterns; some of them exhibit very beautiful colors and complex patterns. The main part in birds that shows body color is feathers. There are three known systems for the expression of plumage colors: (1) eumelanin and pheomelanin produced by integumental melanocytes, (2) accumulation of dietary pigments such as carotenoids, pterin, and psittacofulvins, from foods; and (3) structural colors as a result of fine structure of feathers and without special pigments. The pigment cell type in the trunk region of birds is the only melanocyte that produces melanin (eumelanin and pheomelanin), which is the same as in mammals. Since birds have additional two mechanisms (2 and 3) as noted above, they exhibit colorful body colors that are not shown in mammals. Based on the contents and ratio of eumelanin and pheomelanin, the feathers exhibit white, orange, yellow, brown, gray, or black colors. Other colors such as bright yellow, red, or glossy metallic colors of red, blue, green, or purple, are derived by the second and third mechanisms. The body and feather color patterns and color combinations are determined in the feather root, during the feather developmental process. Body colors and patterns in birds often exhibit differences in body parts, age, sex, and season. Although during evolution, birds lost the colorful erythrophore, xanthophore, and iridophores found in lower vertebrates, they obtain dietary pigments from foods and show structural color due to the fine structure of feathers. With these colorations, feathers in birds often show amazing complex patterns. The diversity of body colors and patterns in birds are very important for their survival strategy.

Keywords Bird · Feather color · Melanocyte · Melanin · Dietary pigments · Structural colors

T. Akiyama (✉)
Department of Biology, Keio University, Yokohama, Japan
e-mail: akiyama@a3.keio.jp

K. Kinoshita
Reproduction and Developmental Lab., Yunnan Agricultural University, Yunnan, China

H. Hashimoto et al. (eds.), *Pigments, Pigment Cells and Pigment Patterns*,
https://doi.org/10.1007/978-981-16-1490-3_3

3.1 Introduction

Avian species display various body colors; some exhibit very attractive body colors (Crawford 1990; Delhey et al. 2017) (Fig. 3.1). Each body part, such as plumage, skin, eyes, comb, ear lobe, feet, often has a different color, implying the presence of different mechanisms underlying the expression of a different color in each body part (Fig. 3.2). However, the main effector organ of body color in birds is feathers. Therefore, the mechanisms underlying expression of plumage color are described in this chapter. Plumage colors often differ from chick to adult, male to female, and between body parts in many species (Crawford 1990). In a species of Grouse, *Lagopus muta* shows seasonal change in feather color from brown in summer to white in winter. In addition, each feather sometimes exhibits extremely complicated patterns. Despite the fact that birds show various plumage colors, the "pigment cell" in the trunk region is known to be only type of melanocytes. This is different compared to lower vertebrates that possess multiple pigment cells (Chapters 7, 8, 10, and 13). The special plumage colors and patterns in birds are expressed as a result of three different control mechanisms: (1) production of eumelanin or

Fig. 3.1 Various body colors expressed in birds. Examples of the typical plumage colors in Red jungle fowl (*Gallus gallus*) (**a**), Black Silky (BS) chicken (*Gallus gallus domesticus*) (**b**), White Silky (WS) chicken (*Gallus gallus domesticus*) (**c**), Nagoya (*Gallus gallus domesticus*) (**d**), Europe Flamingo (*Phoenicopterus roseus*) (**e**), Scarlet ibis (*Eudocimus ruber*) (**f**), Budgerigar (*Melopsittacus undulatus*) (**g**, **h**), scarlet macaw (*Ara macao*) (**i**), Kingfisher (*Alcedo atthis*) (**j**), Indian peafowl (wild type) (*Pavo cristatus*) (**k**), White mutant of Indian peafowl (**l**). (**l**: Photo by Nogeyama Zoo in Japan. Other photos are provided by the authors.)

Fig. 3.2 Various body colors expressed in parts of birds. Face, combs, ear lobes, and beaks from a variant of Shokoku synthetic strain (*Mo*$^+$/*mo*^{w}, see Sect. 3.2.1 (4)) (**a**), Ayam Cemani (**b**), and WS (**c**). Skin and plumages from the BMC (**d**) and WS (**e**). Legs of a hybrid between *Fm*/*Fm* (see Sect. 3.2.1 (8)) and *mo*^{w}/*mo*^{w} (**f**), Shokoku synthetic strain (*Mo*$^+$/*mo*^{w}) (**g**), Ayam Cemani (**h**), and another hybrid between *Fm*/*Fm* and *mo*^{w}/*mo*^{w} (**i**), Nazca Booby (*Sula leucogaster*) (**j**), Blue-footed Booby (*Sula nebouxii*) (**k**), and Red-footed Booby (*Sula sula*) (**l**). (**j**)–(**l**): photos provided with permission from Groupe Pico, Ltd, Japan

pheomelanin by the integumental melanocytes, (2) accumulation and metabolic conversion of pigments incorporated from foods, and (3) structural colors due to fine structure of feathers and without special pigments.

Melanocytes in birds are derived from the neural crest, similar to that in mammals (Fig. 3.3) (Erickson 1993). In general, melanoblasts/melanocytes migrate through the dorsolateral route from the neural crest and are located in whole-body integument. In addition, some birds contain considerable melanocytes in the inner organs or connective tissues (Fig. 3.4). The developmental process has been disclosed by the immunofluorescence assay using the melanoblast-specific antibody (Kitamura et al. 1992; Erickson and Goins 1995). Dermal melanization suggests that the melanoblasts migrate through the dorsoventral route from the neural crest (Faraco et al. 2001) in addition to the dorsolateral route. The content of inner dermal melanocytes seems to be inversely proportional to the outer feather color

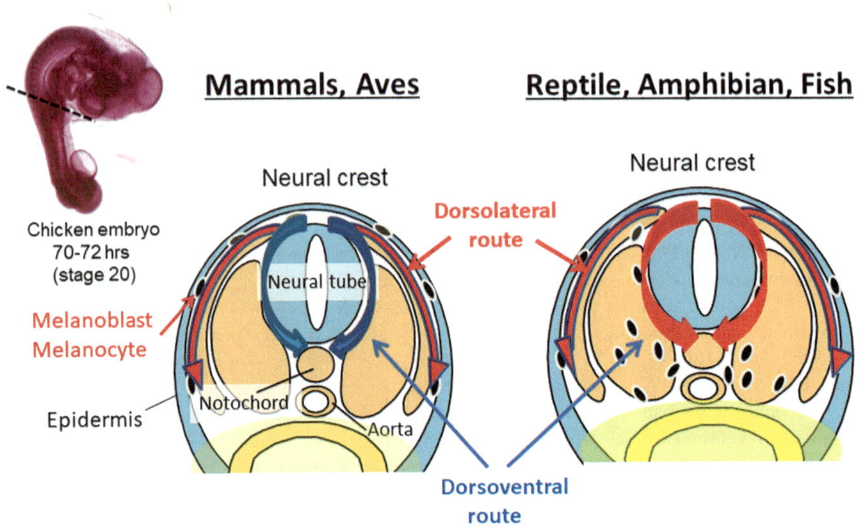

Fig. 3.3 Migration and differentiation of melanocyte precursors from neural crest in developmental process. In mammals and avian species, melanoblasts migrate out from the neural crest at an early developmental stage and mainly pass through the dorsolateral route. On the other hand, in reptiles, amphibians, and fish, many melanoblasts migrate out from the neural crest and pass through the dorsoventral route

(Fig. 3.4). The number of original melanoblasts in feather roots may be relatively constant.

In feather development, pigment production proceeds via a very complicated process, as shown in Figs. 3.5, 3.6, 3.7, and 3.8 (Crawford 1990). Birds have two types of feathers: down of juvenile plumage in freshly hatched birds and contour feather of adult plumage in young and adult birds. Young chicks produce projections at the roots in the skin. Melanocytes located at the inner edge of the projections produce melanin, and the melanin is packed into melanosomes and transferred to keratinocytes that constitute the down barbs (Figs. 3.5 and 3.6). On the other hand, the adult plumage is constructed as a ring structure at the feather root in the skin (Figs. 3.7 and 3.8). Melanocytes are located at the inner edge of the ring and transfer melanosomes to the keratinocytes surrounding the ring shape. Then, keratinocytes produce many barbs and barbules at the periphery of the ring (Figs. 3.7 and 3.8). The plumage colors and patterns are determined in keratinocytes by the content and quality of melanin received from melanocytes (Boissy 2003; Boissy and Hornyak 2006). As the feather shaft elongates, barbs circularly arranged in the periphery of the ring of feather root move up. Finally, the feather shaft allocates with the whole barb on both sides and the adult feather completes the structure (Figs. 3.7 and 3.8).

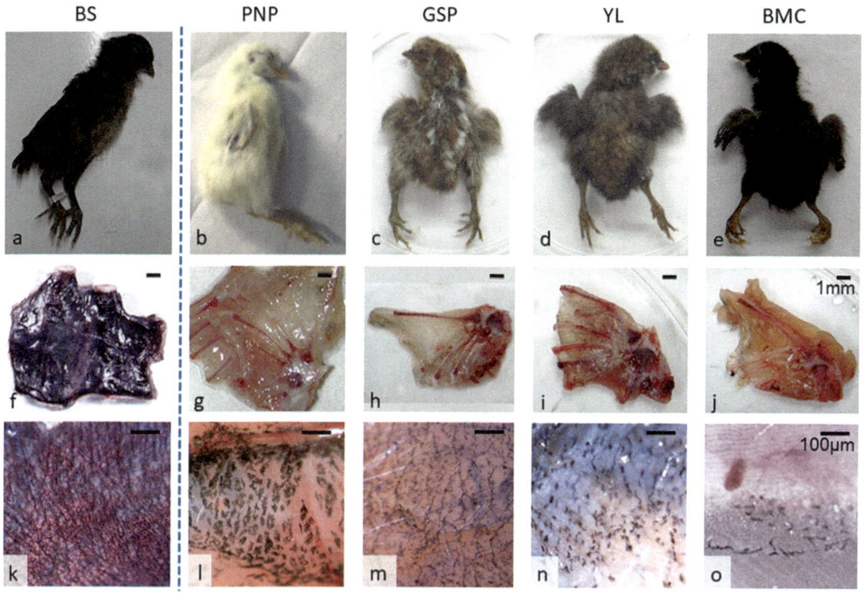

Fig. 3.4 Examples showing the pigmentation in feather and inner body. Plumages of 2-weeks-old chicks of Black Silky (BS) (**a**), Fayoumi (PNP) (**b**), Fayoumi (GSP) (**c**), Fayoumi (YL) (**d**), Black Minorca (BMC) (**e**). Tissue colors in the peritoneum of the chicks (**f–j**). High-power images of the peritoneum of **f–j**, respectively (**k–o**). Silky chicken shows extreme pigmentation in the inner body, but it is an exception with a mutation in *Fm*. Generally, chickens with darker feathers have lighter pigmentation in their inner bodies. Bars in **f–j**; 1 mm, in **k–o**; 100μm. Reproduced from "Skin pigmentation" P187, Fig. 8 by Nova Science Publishers

Eumelanin and pheomelanin are produced by melanocytes in the feather roots, based on the genetic background for melanization. In some cases, not only the degree of melanization is determined, but the switching on and off to produce eumelanin or pheomelanin also occurs in feather roots (Fig. 3.8). As the barb elongates, keratinocytes die, and the color pattern is almost fixed until molting. Although the new feather after molting commonly grows with the same color and the pattern of the old feather, sometimes the color or pattern is different due to age or season in some species (see Chap. 11).

The three above-mentioned mechanisms are involved independently or cooperatively in the expression of body colors (Figs. 3.1 and 3.2). In addition to the whole-body colors, special feathers for display or to attract females in some species reveal very beautiful and/or complex patterns. Birds are valuable for research on pigmentation mechanisms since they have two pigmentation systems in addition to the melanin system that is similar to that in mammals. In this chapter, we introduce the mechanisms underlying the expression of plumage colors and patterns in birds, and discuss the significance of body color in their survival strategy.

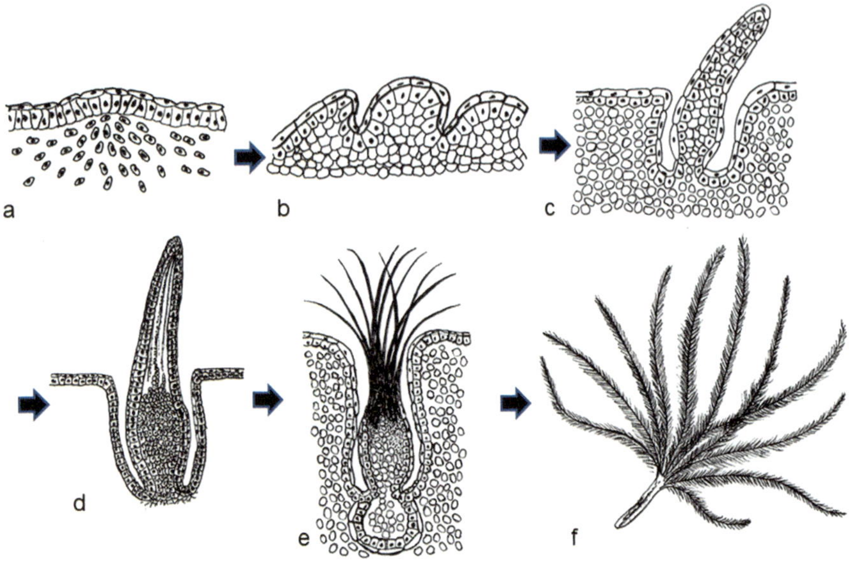

Fig. 3.5 Developmental process of the down feather. Dermal cells are concentrated under epidermal cells and form feather rudiments at 7–8 d embryos (**a**). The skin at the point becomes thick and melanoblasts gather to the point, and the feather bud becomes a pile with multiple layers (**b**). The epidermis surrounding a protrusion gets down and protrusion called the feather follicle is formed (**c**). Keratinocytes produce the fibers inside the feather follicle and melanocytes transfer the melanosomes to the fibers (**d**); the upper cap of the feather follicle is removed, and fibers extend (**e**). Down feather that appears at the first feather in the embryo is completed (**f**). Illustrations are drawn with reference to Carlson (1988)

3.2 Mechanism of Feather Color Expression

3.2.1 Regulation of the Melanization Disclosed from Genetic Traits

In vertebrates, integumental melanocytes in the trunk region originate from the neural crest as members of pluripotent cells during the early developmental process (Fig. 3.3). In mammals and birds, melanoblasts, precursors of melanocytes, have been known to migrate from the neural crest through the dorsolateral pathway and settle in entire body's integument (Erickson and Goins 1995). On the other hand, melanocytes in fish and amphibians are known to migrate through the dorsoventral route in addition to the dorsolateral route. Although in birds, the dorsolateral route is considered the main migration route of melanoblasts, similar to what is seen in mammals, some birds have a large number of melanocytes in inner organs, suggesting migration through the dorsoventral route (Fig. 3.3); for example, silky chicken shows heavy melanization in inner organs and connective tissue, and the

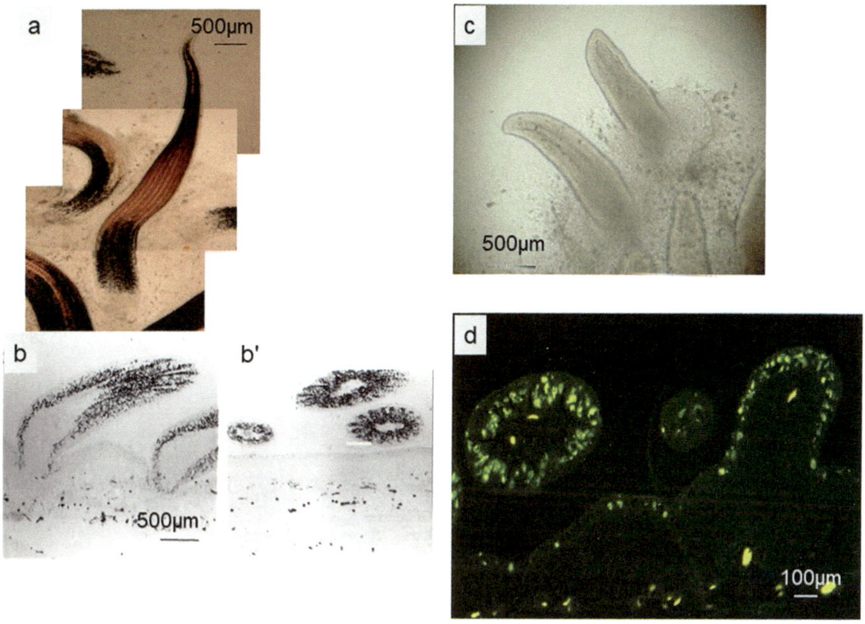

Fig. 3.6 Pigmentation of down feather in WS and BS chickens. At the feather root in the BS chicken with normal tyrosinase (C^{+}), melanocytes provide melanosomes (**a**, **b**, and **b′**); this does not happen in WS (*c/c*) (**c**). Down feather in the skin of 14-d embryos in BS under binocular microscope (**a**), thin sectioned image of the down feather at the same stage under transmission microscopy (**b**, **b′**). Longitudinal section (**b**) and cross section (**b′**). Down feathers in the skin of 14-d embryos from WS, observed under the binocular microscope (**c**). A thin sectioned immunofluorescence image of the same sample of **c** under transmission microscopy, labeled by MEBL-1 antibody (**d**). Note the locations of the pigmented area in **b**, **b′**, and **d**

genetic mechanism responsible for this mutation has been reported (Dorshorst et al. 2011; Shinomiya et al. 2012) as described in detail in the following section.

Although many birds have beautiful colorful feathers, such birds are often difficult to use in developmental biology or genetic expression research. In many cases, these birds are registered as endangered species and their survival is heavily protected. Therefore, domestic poultry such as chicken or quail are generally used as representatives for birds in research. Red jungle fowl (Fig. 3.1a), known as the genetic origin of domestic chickens, has colored plumage, but no special patterns in feathers. Therefore, chickens with white plumage or spot-, stripe-, and eyespot-patterns have been thought to provide insights on some mutation(s) related to pigmentation. Since differences of the colors and patterns are easy to detect, birds are recognized as useful animals to analysis on the genes related to pigmentation or to consider the history of evolution in each species (Hutt 1949; Crawford 1990; Gunnarsson et al. 2007). Over 200 loci have been suggested to be related to body

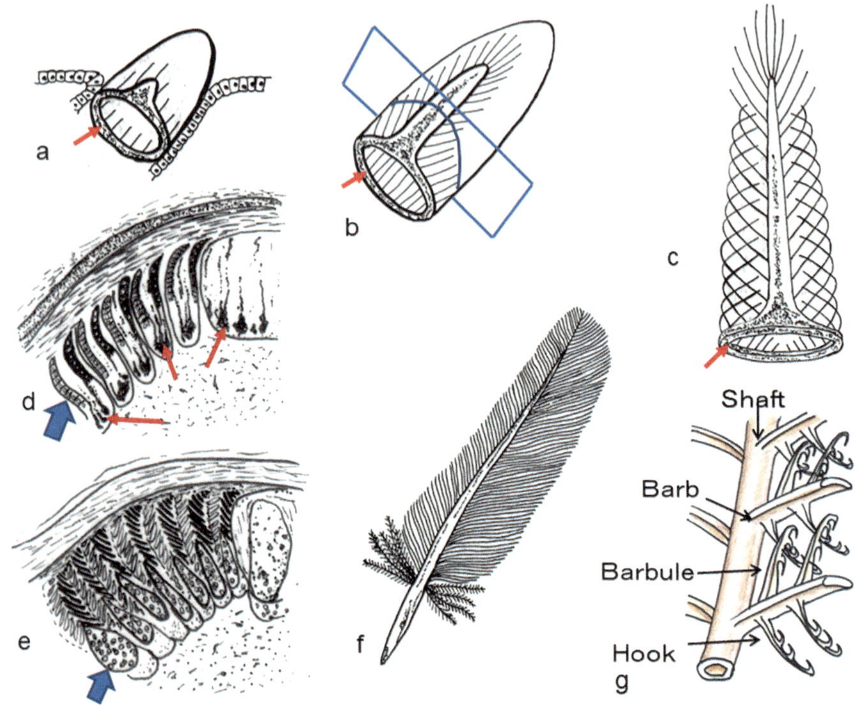

Fig. 3.7 Illustrations of developmental process of the contour feather. At the first stage of contour feather development, a cap-like structure is formed by the dermis and epidermis on the skin (**a**). Elongation phases of cap structure (**b**, **c**). Transverse sectioned profiles in A and B (**d**, **e**). A contour feather completed (**f**), and the names of each part in feather (**g**). Illustrations are drawn with reference to Carlson (1988). Melanocytes locate at the periphery of the ring structure (red arrows) and keratinocytes (blue arrows) form the cap structure and whole barbule

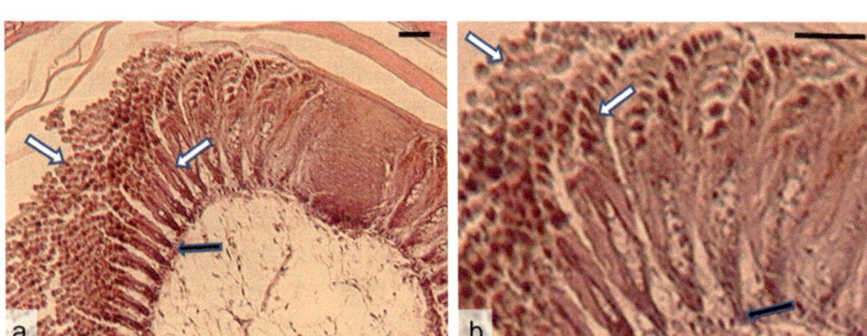

Fig. 3.8 Histochemical profiles of feather roots in developmental process. Feather root areas in BMC (**a**, **b**). Melanocytes (black arrows) are located at the barb ridges in the feather root and transfer the melanosomes into the columns of barbule keratinocytes (melanized; white arrows). Bar = 300 μm

Fig. 3.9 Pigmentation in the cultured melanocytes in various chicken lines. Melanocytes from Barred Plymouth rock (**a**), White Plymouth rock (*c/c*) (**b**), Albino (c^a/c^a) (**c**), White Leghorn (*I/I*) (**d**), Black Silky (BS) (*Fm/Fm*) (**e**), and Nagoya (e^y/e^y) (**f**). Bar = 50 μm

colors or patterns in mice and humans (Nordlund et al. 2006). The genes responsible for mutations in birds have been determined as mice or human homologs (Nordlund et al. 2006; Lamoreux et al. 2010). In addition, the whole genome sequence of red jungle fowl has been determined (International Chicken Genome Sequencing Consortium 2004) and subsequent comparative studies on genes or chromosomal structure in avian species have proceeded rapidly (Osman et al. 2005; Griffin et al. 2008).

To determine the melanin content and condition of melanocytes, we isolated melanoblasts from the neural crest cells of 3-day embryos (stage 20 by Hamburger and Hamilton 1951) and cultured them with our medium supplied with endothelin (EDN) 3 (Akiyama and Shinomiya 2013, Fig. 3.9). The proliferation and differentiation of melanoblasts into melanocytes, and melanin production depend on the genetic control system for pigmentation (Fig. 3.9). Main enzyme for melanization is tyrosinase (*C*) and an example of the three-dimensional structure of tyrosinase is shown in Fig. 3.10 (Matoba et al. 2006). From the chicken line with wild type of tyrosinase, well developed and melanized melanocytes are obtained in the culture system (Fig. 3.9a). Several recessive mutants of tyrosinase with hypopigmentation (*c/c*) (Chang et al. 2006) (Fig. 3.9b) or with complete white plumage and red eyes

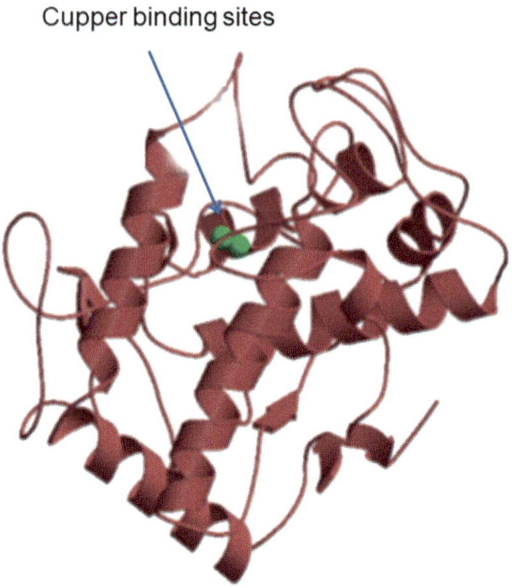

Fig. 3.10 Three-dimensioned structure model and the copper binding sites of Tyrosinase. The illustration shows tyrosinase from *Streptomyces*. It is provided by Prof. Masanori Sugiyama at Hiroshima University (Matoba et al. 2006)

(c^a/c^a) (Brumbaugh et al. 1983) (Fig. 3.9c) produce less pigmented or complete amelanotic melanocytes, respectively. Other gene mutations, such as dominant white (white leghorn; *Pmel17*) (Kerje et al. 2004) (Fig. 3.9d), mottling (Shokoku, etc.; endothelin receptor (EDNR) B2 mutant; mo^w) (Kinoshita et al. 2014), autoimmune vitiligo (Smith line; the responsible gene is unknown) (Smyth Jr. 1990; Erf et al. 1995; Jang et al. 2014), etc., are reported. These chicken lines represent "loss of the function" by the mutation of the responsible genes. On the other hand, the unique Silky chicken is a rare example of "gain of the function" mutation that reveals dermal hyperpigmentation in internal organs and connective tissues; this is caused by excess proliferation and irregular migration of melanoblasts during early developmental processes (Fig. 3.9e). The mechanism of ectopic pigmentation in this line is introduced in detail below. From Nagoya breed, only brownish melanocytes containing pheomelanin are appeared (Fig. 3.9f).

The characteristics of the phenotype in the following examples showing white plumage were examined, and the genes responsible for some of these mutations were disclosed.

1. Albino; Albino chicken was reported by Brumbaugh et al. (1983) as an autosomal recessive mutant of the tyrosinase (Fig. 3.10) gene, *C*. This line was isolated from the white Wyandotte group and called c^a/c^a. Since these chickens have complete white plumage and "red eyes," they are called "albino" (Fig. 3.9c). This is the only albino chicken line maintained in the world at present. The melanocytes produce many melanosomes that are filled with non-melanized matrix (amelanotic melanosomes) (Fig. 3.11). We transfected the normal mouse

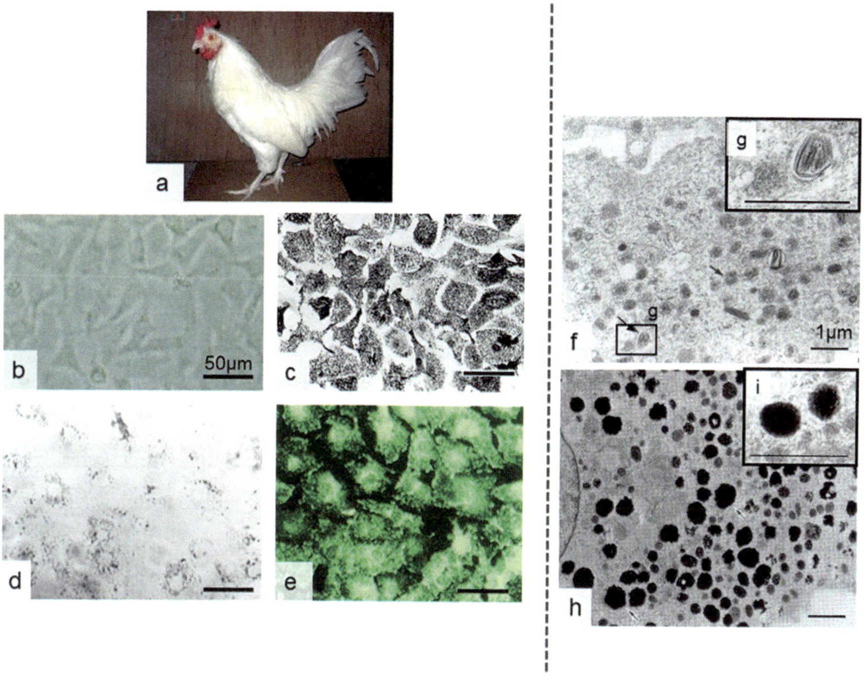

Fig. 3.11 Effects of transfection of mouse normal tyrosinase gene to the melanocytes from albino chicken. Albino chicken (c^a/c^a) with red eyes and white plumage (**a**). Cultured melanocytes from c^a/c^a chicken with defective pigmentation (**b**). After gene transfection, the melanocytes from the c^a/c^a chicken showed heavy melanization. The specimen was fixed with ethanol without staining (**c**). Cultured melanoblast populations from the transfected c^a/c^a (**d**) and stained with the specific mouse melanoblast antibody (**e**). Bar in (**b**)–(**e**); 50μm. Electron microscopic images of the melanocyte from the c^a/c^a chicken (**f–g**) and the transfected melanocytes (**h**, **i**). Arrows in (**f**) show amelanotic melanosomes. (**g**) and (**i**) show higher magnified images of melanosomes. (**g**) amelanotic melanosomes in a square in (**f**), and (**i**) typical melanosomes in the same cell of (**h**). Bar in (**f**)–(**i**); 1 μm

tyrosinase gene construct with quail promoter in retroviral vector into cultured albino melanocytes and then confirmed the occurrence of clear melanization in cells (Akiyama et al. 1994) (Fig. 3.11c–e, h). In addition, we determined the gene responsible for the mutation of the tyrosinase gene with six base losses at one of the two copper binding sites (Tobita-Teramoto et al. 2000) (Fig. 3.12). We determined that the albino line also has the *c/c* mutation described below. It is convincing that the original line, white Wyandotte, already had *c/c* mutation and revealed white plumage.

2. Recessive white lines (*c/c*; insertion of a virus sequence) (Chang et al. 2006); these chickens have white plumage and black eyes, and all melanocytes produce less melanin, as shown in Fig. 3.9b. However, melanization in melanosomes is suggested to be abnormal, and such melanosomes might not be transferred to the

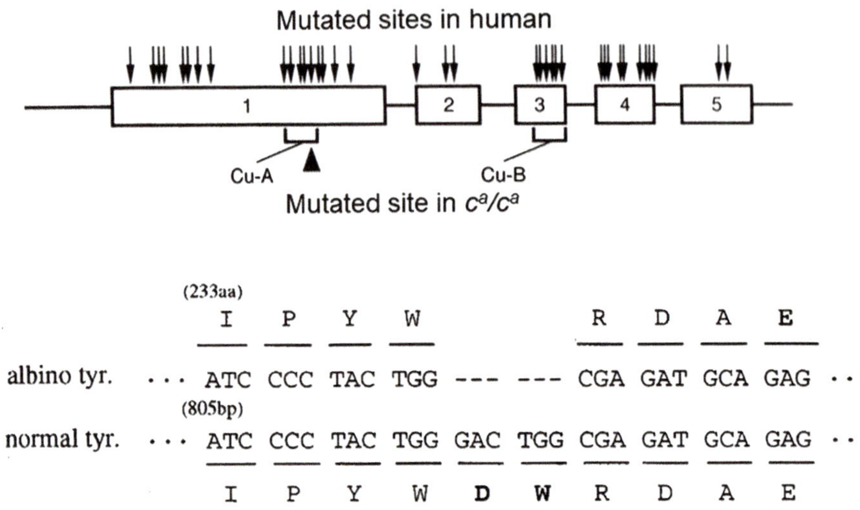

Fig. 3.12 A mutated site at tyrosinase gene in c^a/c^a chicken. Six-base losses at the copper binding site of tyrosinase (Tobita-Teramoto et al. 2000)

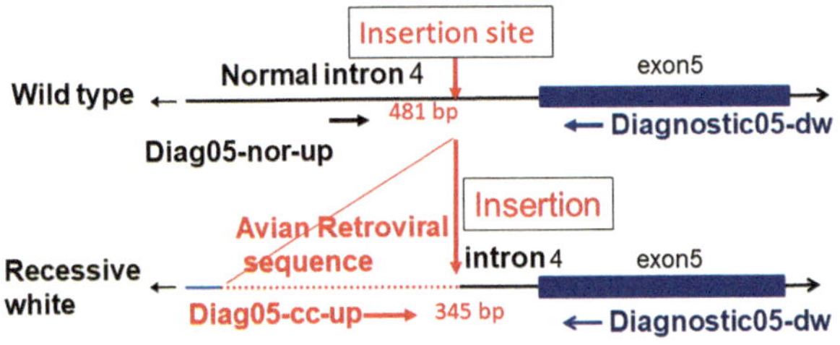

Fig. 3.13 *c/c* mutation in the chicken tyrosinase gene. A recessive white mutation in chicken is caused by the ALV insertion in intron 4 of the tyrosinase gene. To detect mutations, two digestive processes are effective. Two fragments are obtained: 481 bp between Diag05-nor-up and Diagnostic05-dw for normal C^+ allele, and 345 bp between Diag05-cc-up and Diagnostic05-dw for the *c* allele (Chang et al. 2006)

keratinocytes in feather roots. This *c/c* mutation is caused by the insertion of a full sequence of a virus into the tyrosinase gene (Chang et al. 2006; Sato et al. 2007) (Figs. 3.13 and 3.14). This white mutation was found sporadically in many genetically different lines such as white Plymouth Rock, white Tosa-Jidori, Albino (c^a/c^a), white Silky (WS), among others (Figs. 3.15 and 3.16). Although the tyrosinase in this line is functional to some extent, the mutated site of tyrosinase is located in a cell membrane binding site; the mutated tyrosinase is supposed to be fixed unstably with the cell membrane of melanocytes. Hence, unstable and abnormal tyrosinase may produce insufficient melanin (Figs. 3.15 and 3.16) and abnormal melanosomes that may not be transferred to keratinocytes

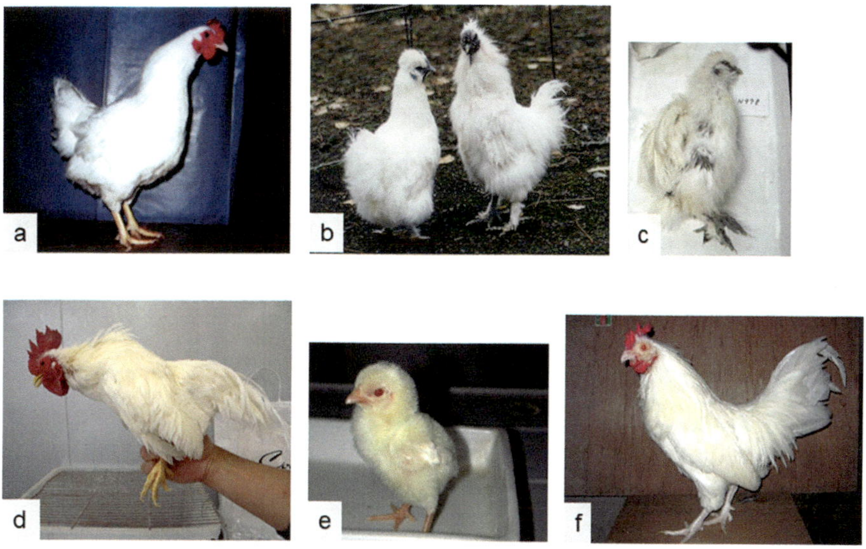

Fig. 3.14 Chicken lines with the recessive white mutation; *c/c*. White Plymouth Rock (**a**), WS (**b**), a white variant derived from BS (**c**), Tosa-Jidori white (**d**), chick (**e**), and adult (**f**) of albino c^a/c^a. Albino line has double mutation of *c/c* and c^a/c^a

in the feather root. In conclusion, this mutation is thought to occur accidentally by viral infection in various lines (Fig. 3.14).

3. Dominant white line (White leghorn; *Pmel 17* mutant) (Kerje et al. 2004); white leghorn has white plumage and black eyes (Fig. 3.9d), and is a line of the most popular domestic fowls because it is easy to handle, sturdy, and shows clean images of light-colored meat. So far, many different lines have been repeatedly crossed with this line to make a useful domestic fowl. The genetic background became very complicated, but the cause of the white plumage was identified as the dominant mutation of *Pmel 17* by microsatellite analysis (Kerje et al. 2004). This gene encodes a matrix protein of the melanosome and functions in generating normal melanosomes. The mutation provokes a part\ial defect in the process, and abnormal melanization occurs in melanosomes. Due to abnormal melanization, radical intermediate substances in the melanin metabolism may be produced and dispersed in the melanoblasts/melanocytes. Finally, almost all melanoblasts/melanocytes die and finally disappear from the trunk skin. Melanoblasts/melanocytes are very difficult to maintain even in culture systems (Fig. 3.9d). Because of the dominant mutation, white leghorn may include other recessive mutations that cause white plumage. Despite the white plumage with such a dominant mutation in melanization, white leghorn still maintains good health.

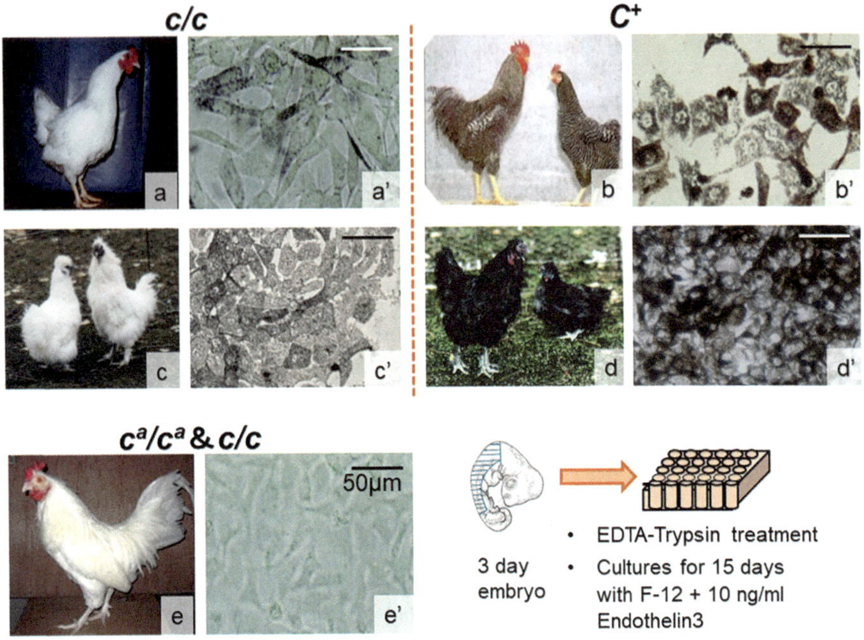

Fig. 3.15 *c/c* chickens produce less amount of melanin in their melanocytes. Melanocytes from White Plymouth Rock (**a**) produce less melanin (**a′**) than Barred Plymouth Rock (**b**) in the culture system (**b′**). In addition, melanocytes from WS (**c**) showed significant pigmentation (**c′**) but slightly light gray color compared with BS (**d, d′**). The albino chicken (**e**) showed no melanization (**e′**) with two mutations by c^a/c^a and *c/c*. Bar = 50 μm. Our routine culture system is illustrated

4. Endothelin receptor B2 mutant (mo^w; base substitutions) (Kinoshita et al. 2014); Endothelin receptors (EDNRs) are receptors of endothelins (EDNs) (Hirata 1996; Lecoin et al. 1998) and their genes are known to be responsible for the white coat color (Sakurai et al. 1990, 1992; Arai et al. 1990; Baynash et al. 1994; Hosoda et al. 1994). White variants with this mutation were isolated from "panda" and "dotted white" quail, named from their white body color and a few black spots (Mizutani et al. 1974; Tsudzuki et al. 1993; Miwa et al. 2006, 2007), and "Minohiki/Shokoku" chicken line (Tadano et al. 2007; Kinoshita et al. 2014). One of the variants in the chicken line shows complete white plumage and another variant shows white spotting plumage (Fig. 3.17). We determined the gene responsible for the EDNRB2 (Kinoshita et al. 2014), named mo^w for the complete white mutant and *mo* for the white spotting (mottling) (Figs. 3.17 and 3.18a). Among these long tail chickens, Shokoku has been known as the ancestor line of the long tail fowls such as Onagadori (Fig. 3.17d) that is Special National Treasure in Japan and Minohiki. Therefore, we concluded that Shokoku introduced a white plumage mutation into these long tail chicken lines. Hence,

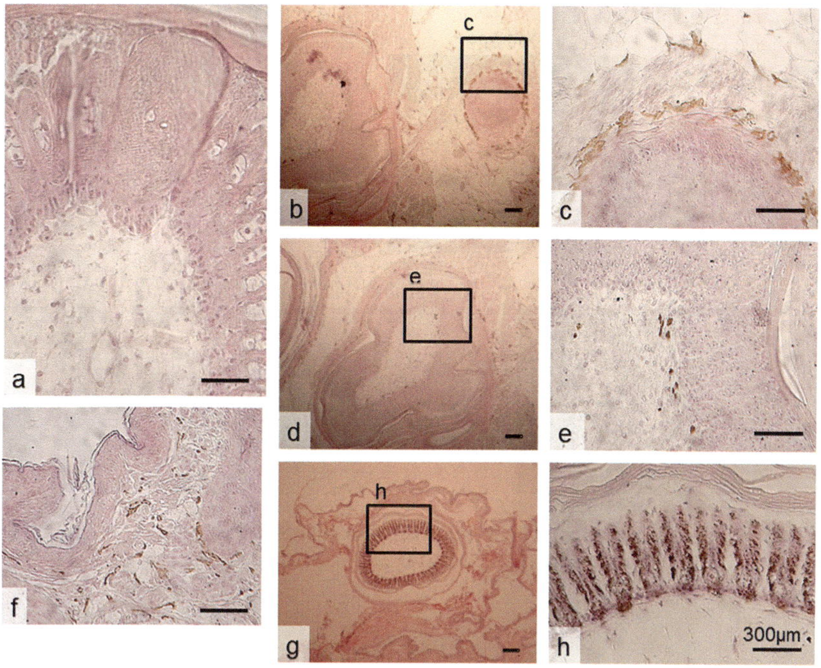

Fig. 3.16 Pigmentation of *c/c* mutant, WS chicken. Skin and feather buds in 28 d WS chicks (**a**–**f**). There is no pigment in the feather bud (**a**, **d**, **e**), but some brown pigmentation in surrounding feather roots (**b**, **c**) and dermal skin tissue (**f**) in WS because WS has *Fm* revealing hyperpigmentation in the internal organs and dermis. Feather root in 28 d chick of C^{+} (BMC) as a control (**g**, **h**). Bar = 300 μm

EDNRB2 is a valuable tool to study the phylogenic relationship in these chickens. The EDNRB2 receptor is a seven-pass transmembrane protein (Fig. 3.18b). mo^{w} has a mutation in the extracellular loop between the 4th and 5th intramembrane regions and *mo* has a mutation in the sixth intramembrane region. These lines have lower EDNRB2 synthesis and are defective in receiving signal from EDN3. The melanoblasts/melanocytes in this mutant did not proliferate well even in the culture system (Fig. 3.19) with a high dose of EDN3. Especially, when melanocytes from mo^{w} were isolated from neural crest and cultured, only 15.9% of melanocytes compared with that of a wild line, Black Minorca (BMC). Although EDN3-EDNRB homologous mutants in humans and mice have severe defects such as being lethal or causing the Hirschsprung disease (Hosoda et al. 1994; Pavan and Tilghman 1994; Metallinos et al. 1998; McCallion and Chakravarti 2001), the EDN3-EDNRB2 mutants in birds exhibit only hypomelanization and no other health problems. Since mammals have EDNRB but not EDNRB2, mammals probably did not obtain or lost EDNRB2 during evolutionary history (Fig. 3.20) (Braasch et al. 2009).

Fig. 3.17 Chicken lines with *mo* and mo^w mutation. A male Shokoku (*Mo/Mo*) (**a**), Mottled Ehime-jidori (*mo/mo*) (**b**), White Shokoku (mo^w/mo^w) (**c**), and White long tail chicken, Onagadori; (mo^w/mo^w) (**d**)

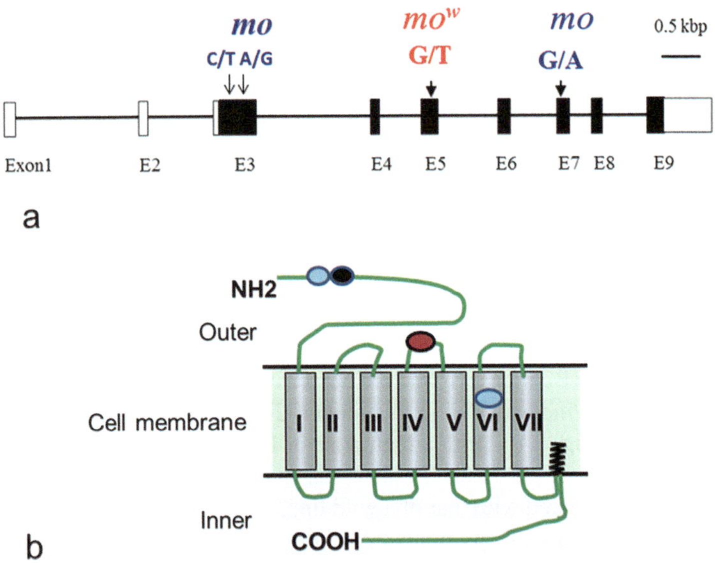

Fig. 3.18 Mutation map of *EDNRB2*. Mutated sites of the two alleles of Mottled (*mottling; mo*) and an allele of white plumage (mo^w) (**a**). Illustration showing EDNRB2 protein structure and the mutated sites; *mo* (two light blue spots) and mo^w (red spot) (**b**)

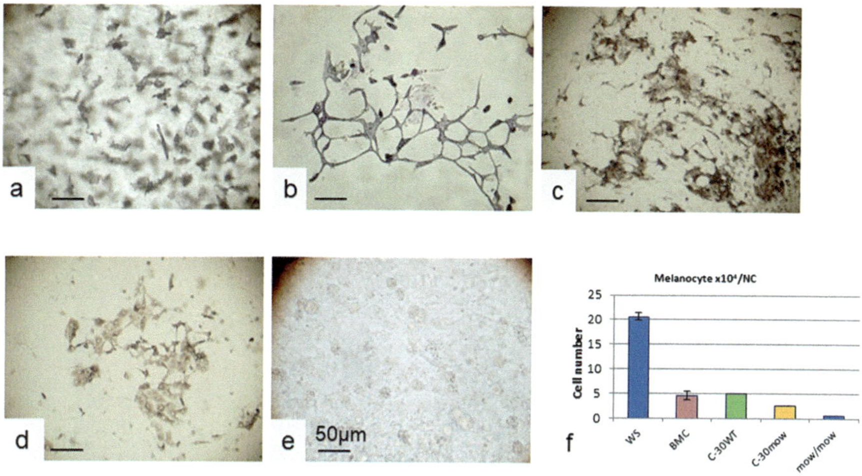

Fig. 3.19 Effects of mo^w mutation in melanocyte culture from three-day embryos. Melanocyte cultures from WS (*Fm/Fm, Mo/Mo*) (**a**), BMC (*fm/fm, Mo/Mo*) (**b**), *Fm* hetero (*Fm/fm, Mo/Mo*) (C-30 WT) (**c**), *Fm* and mo^w hetero ($Fm/fm, Mo/mo^w$)(C-30 hetero) (**d**), and *fm* and mo^w homo ($fm/fm, mo^w/mo^w$) (**e**). Bar = 50 μm. Melanocyte numbers after two-week culture from chickens in the above figures (**f**). The cell number of melanocytes from mo^w/mo^w is 15.9% of BMC (a wild type of melanization)

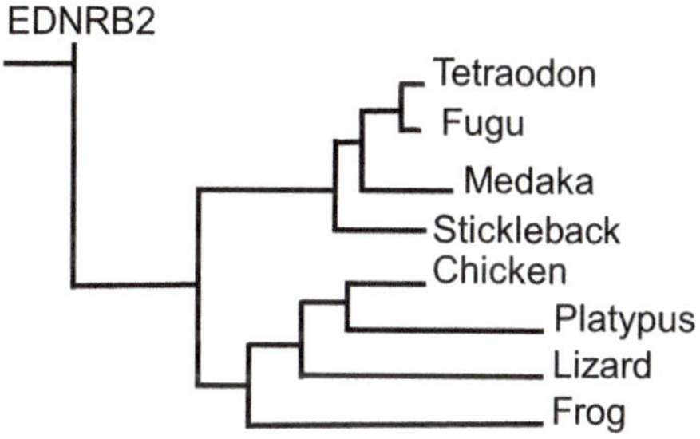

Fig. 3.20 Phylogenetic map of the EDNRB2 gene. In mammals, only the platypus is located in this phylogenetic tree. Reproduction of the figure by Braasch et al. (2009)

From the importance of body color for the survival in lower vertebrates, the EDN3-EDNRB signal transduction system should share the functions of body color and inner organ morphogenesis (Kawasaki-Nishihara et al. 2011). This differentiation of the function is suggested to work for a fail-safe system for survival. Therefore, EDNRB2 is a valuable tool to study the phylogenic relationship in these chickens (Fig. 3.20) (Braasch et al. 2009).

Fig. 3.21 Profiles of YL line chick, hens, and feathers. YL chicks in a week after hatching (**a**), YL young female in the early whitening stage (**b**) and more processed stage (**c**), females of complete whitening YL (left) and wild type (right) (**d**). Whitening plumage in process (**e**). White feathers gradually increase with growth in this mutant. Finally, almost all the long feathers become white and a part of the feather shaft sometimes shows abnormal curling (arrow)

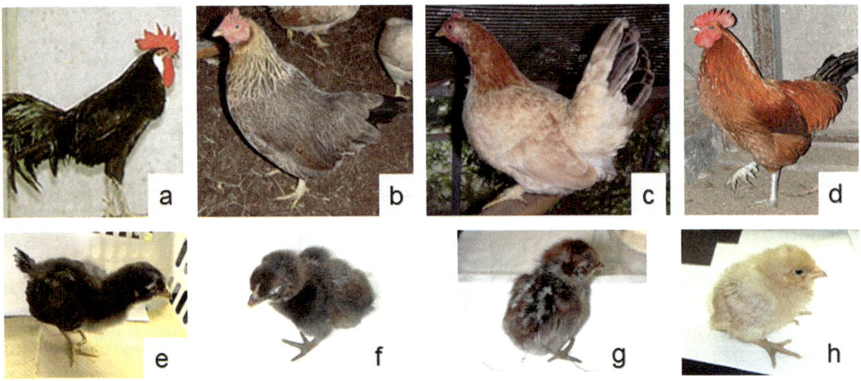

Fig. 3.22 Function of MC1R (*E*) and variant chickens of the gene. Various plumage colors of chickens and chicks with different phenotypes of the *E* locus. BMC male with *E* mutation (**a**), Ehime-jidori (e^{+}/e^{+}) (**b**), Ehime-jidori (e^{y}/e^{y}) (**c**), Nagoya male (e^{y}/e^{y}, *Co*/*Co*) (**d**), Chicks of BMC (**e**), YL (**f**), GSP (**g**), and Nagoya (**h**)

5. Smyth line (autoimmune vitiligo); Delayed amelanotic (DAM) lines similar to human vitiligo are reported as Smyth lines (Boissy et al. 1986; Erf et al. 1995; Smyth 1990; Smyth and McNeil 1999). The chicken gradually shows white plumage and via dotted white, finally becomes completely white (Fig. 3.21). The mechanism has not yet been completely elucidated, but the cause seems to be an autoimmune attack on some factor(s) related to melanization (Jang et al. 2014).

Fig. 3.23 A Mitf mutant in chicken. A *Bl* chicken with MITF mutation is covered with a single lacing plumage. This mutant chicken of Mitf did not show severe symptoms of vitiligo or microphthalmia

The Smyth line and similar YL line have been maintained at Arkansas University and Nagoya University, respectively. The phenomenon of hypomelanization is thought to be quite similar to human autoimmune vitiligo. This Smith line is a useful model for studying human vitiligo.

6. Melanocortin 1 receptor (MC1R) mutant; The intracellular signaling system via MC1R is one of the major pathways in the regulation mechanism of melanin production in avian species, similar to mammals (Nordlund et al. 2006). MC1R is a receptor of alpha-melanocyte stimulating hormone (α-MSH) and adrenocorticotropic hormone (ACTH) (Takeuchi et al. 1996; Theron et al. 2001) and has been determined as the product of the *E* locus in birds. By binding with α-MSH, MC1R is activated to induce eumelanin production. By binding with agouti signaling protein (ASIP), MC1R activity is suppressed and the produced pigment type switches from eumelanin to pheomelanin (Fig. 3.22). As a result, MC1R signaling affects plumage and skin color according to the degree of melanization (Valverde et al. 1995; Mundy 2005; Gluckman and Mundy 2017). With the multiple effects of this gene, birds continue to produce various amounts of melanin in the melanocytes of feather roots during the feather development process and continue to transfer to keratinocytes. As a result, the gradation of melanin or the pattern by the combination with eumelanin and pheomelanin appears in feathers (Fig. 3.22). A dominant mutation of *MC1R, E,* induces the unbinding of ASIP and causes black plumage throughout the body, even in

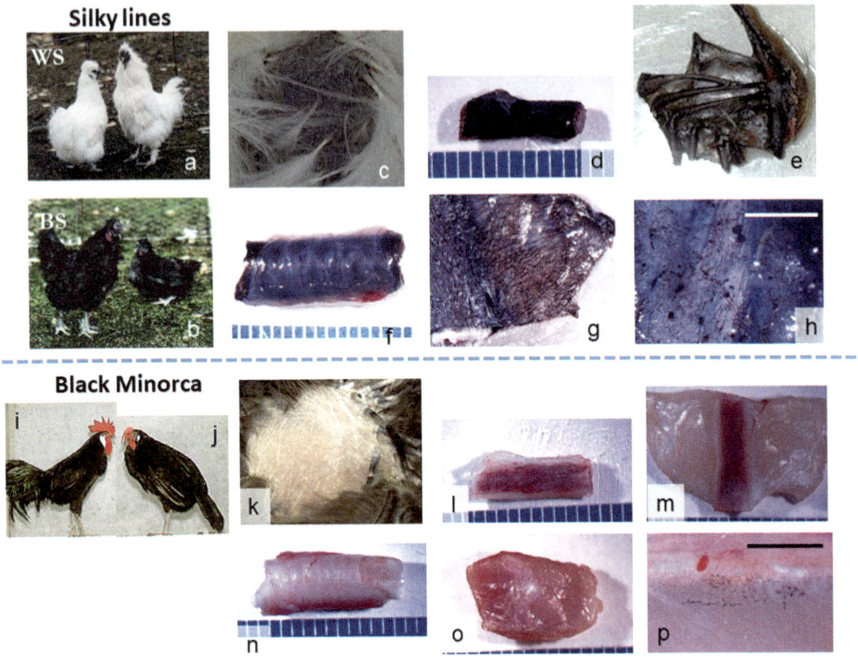

Fig. 3.24 Heavy pigmentation of inner organs in Silky chicken. Both WS (**a**) and BS (**b**) show hyperpigmentation (*Fm*) in the skin (**c**), bone (**d**), peritoneum (**e**), trachea (**f**), muscle (**g**), and the high-power magnified profile (**h**). Outlook and parts of BMC (**i, j**) are shown as a control (wild type; *fm/fm*) that reveals no pigmentation in inner organs (**k–p**). **k–p** are profiles of the same organs in as **c–h**, respectively. Scale = 1 mm

ventral areas, like in BMC chicken (Fig. 3.22a, e) or crow. This autosomal mutation in *MC1R* causing heavy melanization overcomes pheomelanin production or hypomelanization effects regulated by other MC1R-related genes. This MC1R signaling pathway is also thought to involve strongly in the pattern formation of entire body. In the juvenile quail, the characteristic periodic pattern is revealed in the dorsal side but the regulation mechanisms to express the pattern have been remained unknown. Recently, using natural variants of the juvenile galliform birds, it is clarified that the one member gene in the MC1R signaling pathway, *agouti* works to form the periodic coloration through a prepattern of somite origin (Haupaix et al. 2018). *agouti* is the responsible gene for agouti protein that induces pheomelanin production by binding to MC1R and inhibiting its activity. Haupaix et al. report that the autonomous signaling from the somite sets stripe prepattern marked by the expression profile of *agouti* and then *agouti* regulate stripe width through dose-dependent control of local pigment production. This evidence will surely become a breakthrough to understand the mechanism of the pattern formation in entire body of birds.

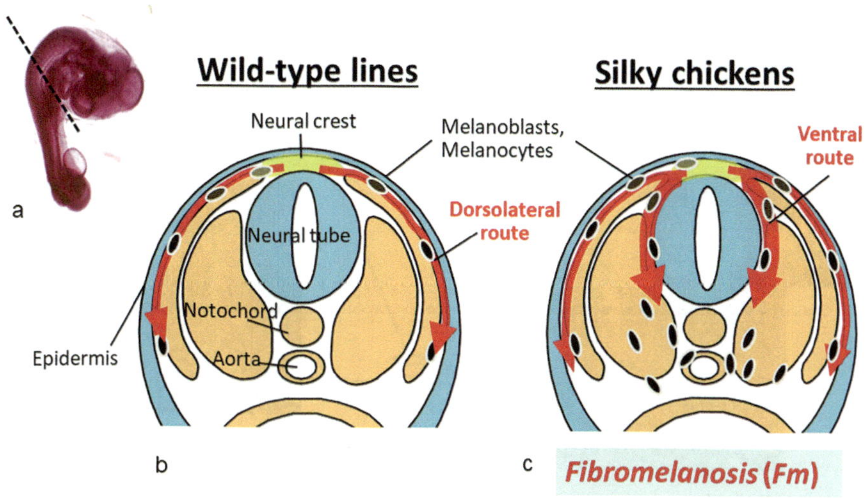

Fig. 3.25 Migration route and pigmentation in inner organs in Silky. At 3-day embryo in Silky, melanoblasts/melanocytes migrate in both dorsolateral and dorsoventral routes during developmental processes. At 10 days, almost all inner organs reveal heavy pigmentation. Three-day embryo and cutting plane (**a**) shown in (**b**, **c**). The dorsolateral route in wild-type lines (**b**) and the additional dorsoventral route in silky chicken (**c**)

7. Mitf (a master gene for melanization); the gene responsible for microphthalmia-associated transcription factor (*MITF*) is found in humans and mice (Tachibana et al. 1994) and codes a transcription factor for tyrosinase expression (Tachibana et al. 1996). In birds, *Mitf* also plays an important role as the master gene and the mutants display hypopigmented plumage in chicken and quail (Minvielle et al. 2010). We found a chicken line having the mutation that show hypopigmentation but not the symptom like small eyes found in microphthalmia (Fig. 3.23).
8. Fibromelanosis (*Fm*) (hyperpigmentation of inner organs, duplication of EDN3 in silky);

Silky chicken reveals heavy melanization in the inner body organs such as black bone, muscle, blood vessel, and connective tissue (Fig. 3.24) (Bateson and Punnett 1911; Dunn and Jull 1927; Ortolani-Machado et al. 2007). The large number of melanoblasts in the chicken migrates through the dorsoventral route from the neural crest (Figs. 3.25 and 3.26) and enters almost all inner organs or connective tissue (Figs. 3.27 and 3.28). Irregularly shaped melanosomes in WS are observed in melanocytes (Fig. 3.28). This unique phenotype is called *fibromelanosis* (*Fm*), and its responsible gene was identified by microsatellite analysis as the duplication region including EDN3 (Fig. 3.29) (Shinomiya et al. 2012). Also, Dorshorst et al. (2011) disclosed the complex structure of the duplicated region. EDN3 has been isolated by Yanagisawa et al. (1988) and is known as a strong mitogen for melanoblasts (Lahav et al. 1996, 1998; Opdecamp et al. 1998); it is thought to be the gene

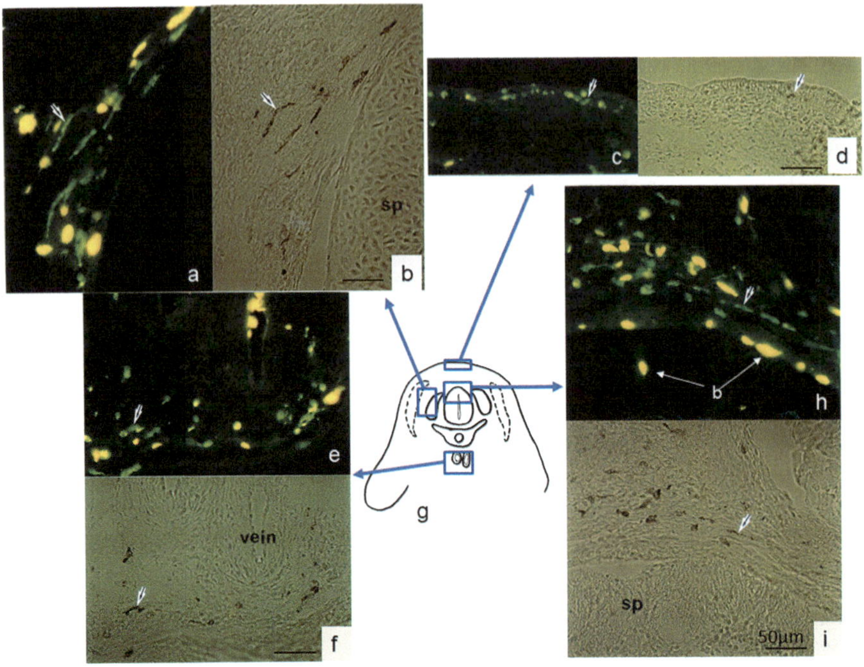

Fig. 3.26 Immunofluorescent profiles in Silky embryo. A transverse section of the embryo was stained with MEBL-1 (melanoblast-specific antibody; Kitamura et al. 1992). Melanoblasts and melanocytes (arrows) are located in both the dorsoventral route (**a**, **b**, **e**, **f**, **h**, **i**) and dorsolateral route (**c**, **d**). **a**, **c**, **e**, and **h** are immunofluorescence profiles of the same fields of **b**, **d**, **f**, and **i**, respectively. **g** is an illustration of transverse section of embryo at stage 36. *sp* spiral cord, *b* blood cell, *vein* blood vessel. Bar = 50 μm

responsible for *Fm*. Furthermore, we reported that the origin of the mutation was determined approximately 6600–9100 years ago, as a mutation in one of the jungle fowls, the origins of domestic chickens. The phenotype has been taken over by domestic chickens by strong artificial selection, including the Silky line (Dharmayanthi et al. 2017). We discovered that this phenotype has been retained for generations in other lines like Ayam Cemani (Fig. 3.30a, b), which has the same origin as Silky, but obtained the mutation after Silky. As described above, this phenotype is regulated by a single gene duplication of EDN3; it can be transferred by artificial crossing. By such crossing, a black bantam line has the same *Fm* phenotype and duplication of the EDN3 region (Fig. 3.30c) (personal communication). On the other hand, crow has whole black plumage, black beak, legs, and nail, but not heavy melanization in inner organs. Crow is suggested to have only *E* (*MC1R*) but not *Fm* (*EDN3* duplication) mutation (Fig. 3.31).

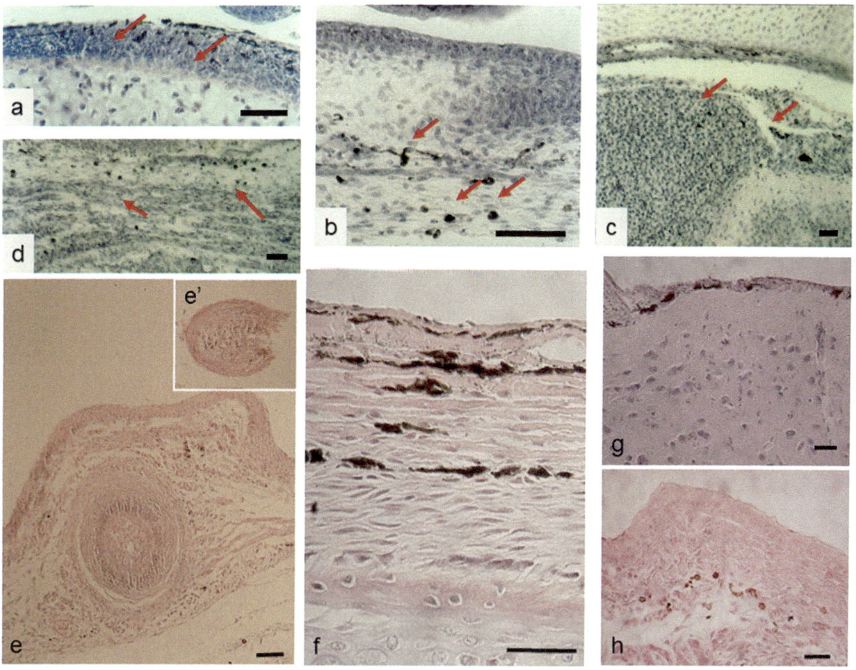

Fig. 3.27 Profiles of the melanocytes in internal organs of BS and WS by histochemical analysis. BS has black feathers and pigmentation in cartilage (**a**), epidermis and dermis (**b**), brain (notochord) (**c**), and muscle (**d**) in the paraffin section (stage 36). Feather root (**e**), a cross sectioned feather (**e'**), epidermis and dermis (**f**), brain (**g**), and muscle (**h**) in WS. WS has white feathers without melanization (**e**) but dermal pigmentation (**f–h**), similar to BS. Paraffin section (15 d and 28 d embryos). Bar = 100 μm

As previously described, pluripotent cells migrate out from the neural crest from the early developmental stage (Le Douarin and Kalcheim 2009). The precursors of melanocytes, melanoblasts, are derived from pluripotent cell group and mainly migrate through the dorsolateral route. But a few melanoblasts enter the dorsoventral route and survive depending upon the factors secreted in the environment. The inner contents of the melanoblasts may be different from outer plumage colors because the total number of melanoblasts in the neural crest is supposed to be determined as a certain number (Fig. 3.4). EDN3 induces the proliferation of the melanoblasts in the neural crest and also probably the expression of some adhesive factor(s) combining melanoblasts with surrounding cells such as fibroblasts. Melanoblasts/melanocytes migrate together with fibroblasts through the dorsoventral route, and melanoblasts differentiate into melanocytes in dermal connective tissues. Even in many lines of chicken or wild-type quail, we found a considerable number of melanocytes in the inner organs (Fig. 3.4), although not as much as in Silky chicken. Therefore,

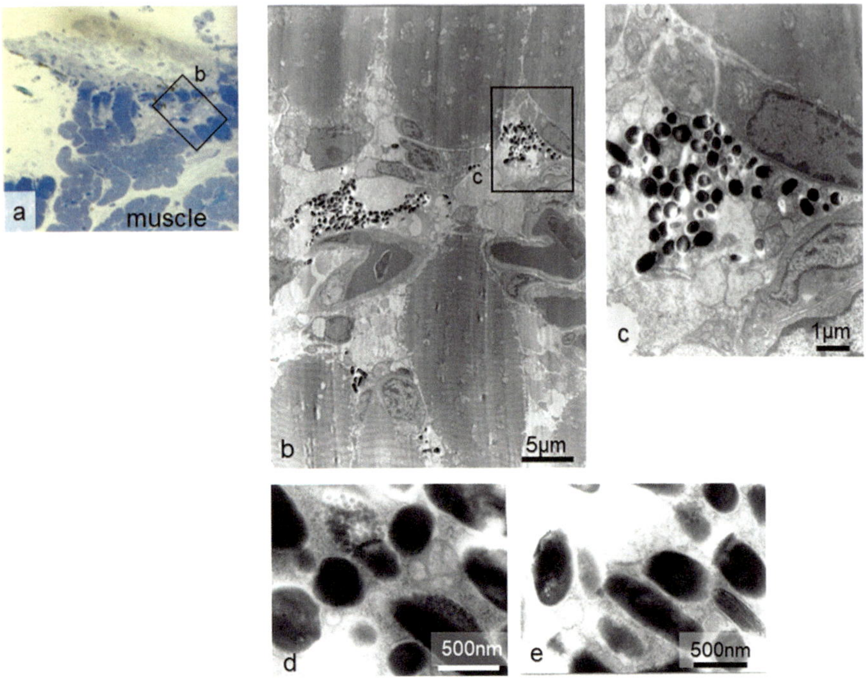

Fig. 3.28 Fine structures of dermal melanocytes in internal organs of BS. Toluidine blue staining of a thin section of the internal organ of BS (**a**) Melanocytes with both heavy melanized and immature melanosomes are located in muscle tissue (**b**, **c**). Irregularly shaped melanosomes in WS are observed in melanocytes (**d**, **e**)

melanoblasts are thought to migrate through not only the dorsolateral route but also the dorsoventral route. However, the major part of the melanoblasts entering the dorsoventral route are supposed to disappear in incompatible circumstances.

In lower vertebrates, many melanoblasts migrate through the dorsoventral route from the neural crest, enter the dermal connective tissue, and survive in the inner organs. Some of these melanoblasts and melanocytes are large in size and appear around the digestive tube earlier than in the integument during the developmental process. These pigment cells may be useful to protect the inner organs, especially digestive organs. The cell type commitment of pluripotent cells derived from the neural crest and the selectivity of the migration route have not been completely clarified yet in either mammals or birds.

EDN3 is clearly involved in melanoblast proliferation and differentiation after the migration phase from the neural crest (Rubanyli and Polokoff 1994; Reid et al. 1996) but the reason why large numbers of melanoblasts migrate through the dorsoventral route in Silky is not yet disclosed. When melanoblasts isolated from the neural crest in Silky are cultured, differentiated melanocytes show strong contact with each other

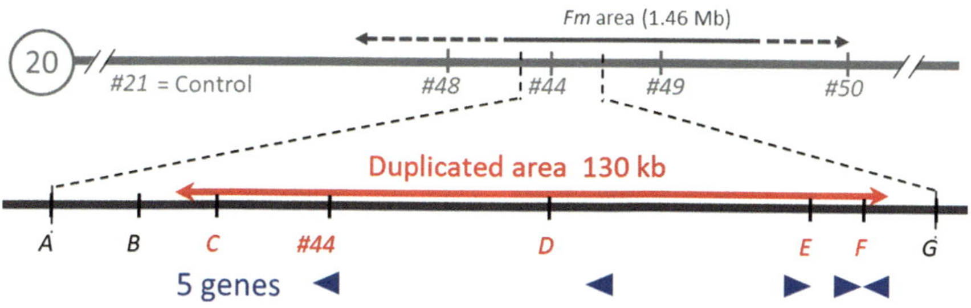

Symbol	O	Description
LOC768509	-	similar to endothelin 3
LOC419309	-	similar to dJ492J12.1 (novel protein similar to zinc finger protein human immunodeficiency virus type I enhancer-binding protein 1 (HIVEP1))
LOC419310	+	similar to Protein C20orf45 (CGI-107)
LOC768660	+	similar to H+ transporting F1 ATP synthase epsilon subunit
TUBB3	-	tubulin, beta 3

Fig. 3.29 Genes in the duplicated region linked to *Fm*. This region contains five genes: endothelin 3, human immunodeficiency virus type I enhancer-binding protein 1 (HIVEP1), similar to protein C20orf45 (CGI-107), similar to H^+ transporting F1 ATP synthase epsilon subunit, and tubulin beta 3 (Shinomiya et al. 2012)

and form flat cell sheets. When the dissected neural crest is cultured, melanoblasts migrate out from the tissue associated with fibroblasts, revealing in a parallel rod-shaped fashion (Fig. 3.32a). These shapes imply the strong migration activity of melanocytes or the attraction of the surrounding fibroblasts. On the other hand, melanocytes from a wild type (*fm/fm*) Tomaru chicken with black plumage migrated with a flat quadrangle or trapezium shape (Fig. 3.32b) at the same migrating phase. This phenomenon implies that twice the dose of EDN3 induces the production of some cell adhesion factor(s) in addition to the stimulation of proliferation and differentiation of melanocytes. At the current moment, the adhesion of melanoblasts to fibroblasts is strongly suggested to be one of the candidates leading to the dorsoventral route.

3.2.2 Accumulation of Pigments Other Than Melanin

Some species of birds, such as flamingo (Fig. 3.33a), scarlet ibis (Fig. 3.33b), and ruddy kingfisher (Fig. 3.33c), exhibit matte-pink, -red, or -orange colors. In addition, a group of Budgcrigar rcvcals various plumagc colors of mattc-ycllow, -orangc, -rcd, -green, -blue, and -purple (Fig. 3.33d–g). Almost all such cases accumulate less melanin than carotenoids (McGraw et al. 2004; Lopes et al. 2016; Weaver et al.

Fig. 3.30 Other lines showing the *Fm* phenotype. *Ayam Cemani* provides the same *Fm* phenotype and responsible gene (**a**, **b**), but the origin of the *Fm* mutation is supposed to have occurred approximately 6600–9100 years ago, after that of Silky (Dharmayanthi et al. 2017). A black chabo (Bantam) also has the *Fm* phenotype by repeated artificial crossing with Silky and selection (**c**). This photo was provided by Sachio Kono. These *Fm* lines prove that the phenotype is expressed by Mendelian law and has been maintained with the power to survive for a long time

2018), pteridines (Thomas et al. 2013), or psittacofulvins (Stradi et al. 2001; McGraw and Nogare 2005; Mundy 2018) in their feathers (Figs. 3.33a–g and 3.34). In the case of peacocks, the feather color mainly appears by the structural color, but with the support of melanin and other pigments (Fig. 3.34h). When these feathers of budgerigars were observed under a microscope, it was disclosed that the plumage colors were emitted from the upper part of barbs, probably carotenoids for yellow and psittacofulvins for blue or green with structural color (Fig. 3.35a–i). Chemical structures of these pigments are shown in Fig. 3.36. Colorful feathers with high contents of psittacofulvin are reported to resist feather-degrading *Bacillus licheniformis* better than white ones (Burtt et al. 2010). Carotenoids are well known to be widely included in yellow-orange colored plants, example, ß-carotene in carrots. However, major part of the animals generally cannot synthesize carotenoids directly by themselves in the body and accumulate the chemicals as it is after incorporation from foods, or convert to other forms by the metabolic mechanism, or a combination of both forms (McGraw et al. 2004; McGraw and Nogare 2005;

Fig. 3.31 Crow with whole black outlook. Crow has whole black body color (**a**) with black plumage, beak, leg, toe, and nail (**b**) but not pigmentation in inner organs (**c**, **d**). Crow was provided by Dr. Naoki Tsukahara

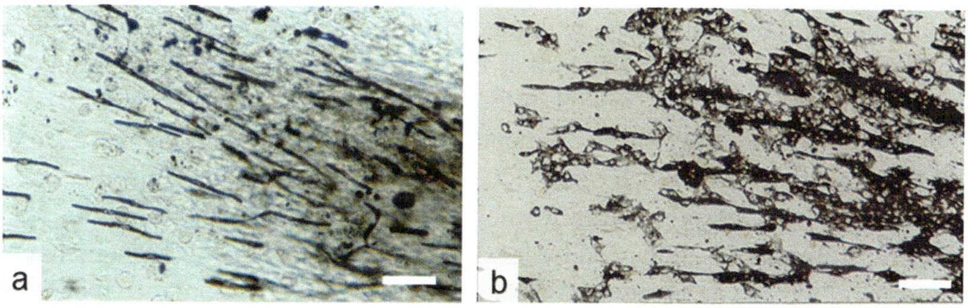

Fig. 3.32 Organ culture of the neural crest from BS. After 7 days of culture, melanocytes in Silky (*Fm/Fm*) migrated out with a straight rod shape (**a**). Melanocytes from Tomaru with *fm/fm* migrated out with flat polygonal shapes. It is shown as a reference for wild type (**b**). Bar = 100 μm

Lopes et al. 2016; Twyman et al. 2018). These accumulated chemicals in feathers are also responsible for the body color in other vertebrates, such as fish, amphibians, and reptiles. In Aves, Europe Flamingo (*Phoenicopterus roseus*) (Fig. 3.37) or Japanese crested ibis (Toki; *Nipponia Nippon*) reveal pink color and are supposed to accumulate red substances from aquatic insects that were previously reported as canthaxanthin (Thommen and Wackernagel 1963; Esatbeyoglu and Rimbach 2017). The pink feathers in flamingoes are almost transparent, but are observed as pale pink or

Fig. 3.33 Variation of plumage color expression by the accumulation of pigments from foods. Europe flamingo (*Phoenicopterus roseus*) (**a**), Scarlet ibis (*Eudocimus ruber*) (**b**), Ruddy kingfisher (*Halcyon coromanda)* (**c**), Budgerigar (*Melopsittacus undulatus*) (**d**, **e**), and variants of Psittacidae (**f**, **g**)

red in color under transmission microscopy (Fig. 3.37). This proves that such plumage color is not derived from structural color but is supposed to be derived from the pigment particles or the cytosol accumulating pigments inside the feathers. In addition, the observation by scanning electric microscopy on the case of the dietary pigment coloration clarifies that these birds have intracellular sponge-like structure, but melanin is not packed in the structure (see Chap. 11, Fig. 11.9).

On the other hand, the yellow pigment in canary is a carotenoid; canary xanthophyll B is incorporated from foods (McGraw et al. 2004). Scarlet ibis (*Eudocimus ruber*) accumulates red pigment in feathers (Mendes-Pinto et al. 2012) (Fig. 3.33b) and the red coloration is enhanced by foods such as capsaicin, which is a main color substance of red cayenne pepper. Such birds possess a special mechanism for packing the red pigment into some particles or binding these substances with structural proteins in the cytosol. The mechanism to incorporate these substances into the feathers is not yet completely understood. The birds in Fig. 3.33d–g showing variable feather colors (Fig. 3.34b–g); they belong to budgerigar (good cockatoo; *Melopsittacus undulatus*) and are reported to contain psittacofulvins (Tinbergen et al. 2013; Mundy 2018) and probably other pigments also. These pigments and sponge-like nanoscale structures of barbs or barbules in feathers are responsible for these various plumage colors (Fig. 3.35). In these feathers, some genes and a regulation system would work to pack pteridines, carotenoids, or psittacofulvin, etc., instead of melanin into the particles in amelanotic melanocytes, and the particles would be transferred to keratinocytes in feathers. CRYP2J19 is reported to be one of the genes controlling the feather color that depends upon carotenoids (Lopes et al. 2016).

Fig. 3.34 Variation of colored feathers. Europeflamingo (*Phoenicopterus roseus*) (**a**), Psittacidae species (**b–g**), and Indian peafowl (*Pavo cristatus*) (**h**)

The birds utilizing these dietary pigments have less melanin at first and accumulate these dietary pigments in the amelanotic feathers. Colored cells containing pigments from foods are proposed to be called the new type of pigment cells. The carotenoid-modified coloration is thought to affect health and body toughness in addition to the special plumage color. As described before, the expression of body color is an important tool for the survival of birds (Weaver et al. 2018). Psittacofulvins that are responsible for the bright-red, orange, and yellow colors in parrots (McGraw and Nogare 2005; Delhey et al. 2017; Mundy 2018) are reported to be better-resistant against a feather-degrading *Bacillus licheniformis* than white feather (Burtt et al. 2010). These chemicals belong to one type of linear polyene and can be extracted using an organic solvent. Since these polyenes were not found

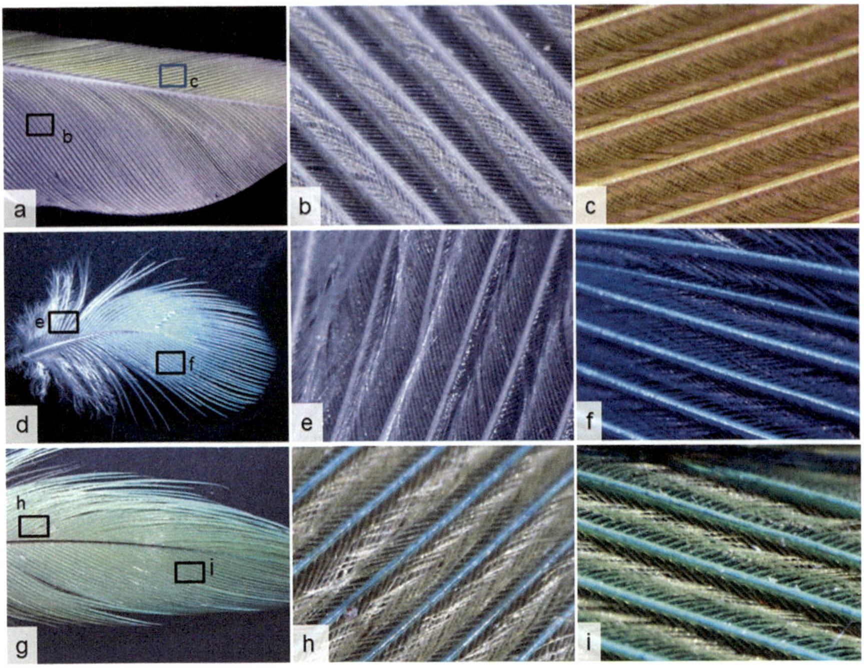

Fig. 3.35 Structure color in the feathers of Psittacidae (*Melopsittacus undulatus*). Yellow feather (**a**), white part of A (**b**), yellow part of A (**c**), blue feather (**d**), white part of D (**e**), blue part of D (**f**), green and blue feather (**g**), blue part of G (**h**), green part of G (**i**)

in the blood during molting, these pigments were thought to be obtained not from diet but from metabolic synthesis in the inner body during the feather production process (McGraw and Nogare 2005).

As shown in Fig. 3.2, the leg color is also very variable, showing white, yellow, light green, brown, blue, red, gray, black, etc. The colors are determined by the melanin content in the epidermis, dermis, and scale (keratinocyte). For the special colors of light green, yellow, red, blue, and gray (Crawford 1990) in legs, the structural colors by keratin fine structure in scale or accumulation of pigments from foods are strongly suggested to be related to the color expression.

3.3 The Propositions for the Concept of Pigment Cells

From pigment research in avian species and lower vertebrates, the following points are suggested to reconstruct the concept of "pigment cell":

1. Although lower vertebrates have multiple pigment cell types, avian species only have melanocytes, similar to mammals. However, birds show erythrophore–xanthophore-like color by pigments from foods and iridophore-like color due to

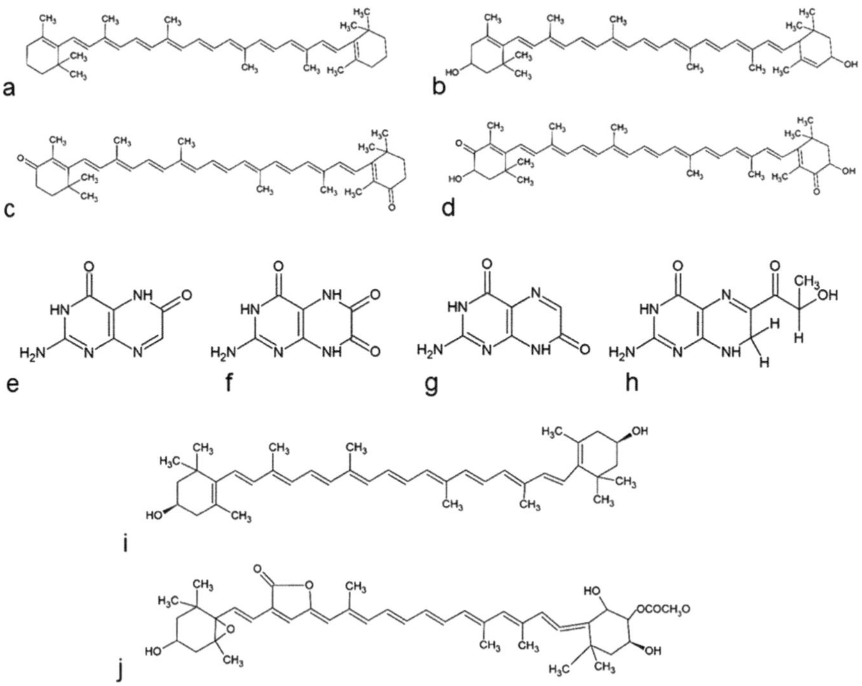

Fig. 3.36 Chemical structures of main carotenoids, pteridines, and psittacofulvin accumulated in the feathers. ß-carotein (**a**), lutein (**b**), anthaxanthin (**c**), astaxanthin (**d**), xanthopterin (**e**), leucopterin (**f**) iso-xanthopterin (**g**), sepiapterin (**h**), zeaxanthin (**i**), psittacofulvin (**j**)

the structural color caused by the fine structure of keratin in feathers. Regarding the former colors, the special cells first incorporate carotenoids, pteridine, and psittacofulvins, etc., from foods, and then transfer these pigments to keratinocytes. Even though these cells incorporate and accumulate pigments from foods, these cells might also be called "pigment cells," in addition to melanocytes.

2. In lower vertebrates, melanoblasts generally migrate not only through the dorsolateral route but also through the dorsoventral route from the neural crest. The priority to select the route may be different for each species and depend on the environmental conditions of the inner organs. After or during movement through the dorsoventral route, the melanoblasts/melanocytes would either survive in the inner organs in suitable environments, or die in an unsuitable environment. When they continue to live in the internal body, they keep the melanosomes inside the cells because they do not have host cells such as keratinocytes to pass the melanosomes. Therefore, as the basic concept for pigment cell development in vertebrates, it is suggested that pigment cells are able to migrate through the dorsoventral route also and not only through the dorsolateral route, depending on the environmental factors.

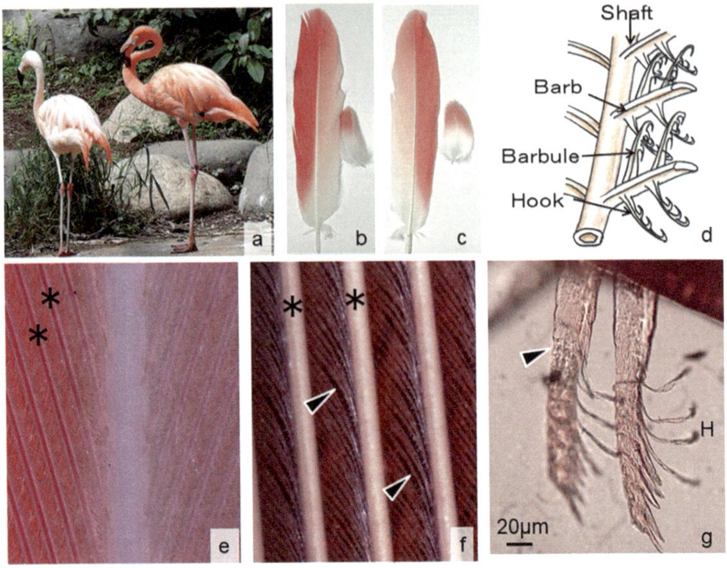

Fig. 3.37 Plumage color expression by accumulation of carotenoid. Europe flamingo (*Phoenicopterus roseus*) (**a**) accumulates ß-carotene or canthaxanthin in the feathers (**b**, **c**). The structure of a feather is illustrated (**d**). Barbs (**e**, **f**, *) and barbules (**f** and **g**, arrowheads) reveal pinky colors. Barbules provide hooks (H) shown in (**d**)

Acknowledgements The authors deeply thank Dr. John Brumbaugh, Dr. Makoto Mizutani, Dr. Ai Shinomiya, Dr. Yoko Satta, Dr. Kunio Kitamura, Mrs. Mizuho Nakamura, Dr. Takayuki Tobita-Teramoto, Dr. Atsushi Kurabayashi, Dr. Yasunori Kayashima and Dr. Tomoko Adachi for their valuable help with the research. We also appreciate Dr. Katsutoshi Kino, Dr. Naoki Tsukahara, Dr. Masanori Sugiyama, Dr. Gen Morimoto, Dr. Harumi Kusano, Mr. Sachio Kono, Mr. Tsutomu Matsuda, Mr. Hiroto Hayashi, Ms. Tamaki Shimosaka, Groupe Pico, Ltd, Tokyo Tama Zoo, Fukuoka City Zoo, Ishikawa Zoo, and Nogeyama Zoo in Japan for the donation or supply of birds, illustrations, photos, and feathers. We are also grateful to the Avian Bioscience Research Center in Nagoya University for supplying chickens and valuable information. This research was supported in part by grants from the Keio University.

References

Akiyama T, Shinomiya A (2013) Overview on the melanocyte precursor migration from the neural crest. In: Smith JB, Haworth MB (eds) Skin pigmentation. Nova Science Publishers, Inc., New York, pp 175–196

Akiyama T, Whitaker B, Federspiel M, Hughes S, Yamamoto H, Takeuchi T, Brumbaugh JA (1994) Tissue-specific expression of mouse tyrosinase gene in cultured chicken cells. Exp Cell Res 214:154–162

Arai H, Hori S, Aramori I, Ohkubo H, Nakanishi S (1990) Cloning and expression of cDNA encoding an endothelin receptor. Nature 348:730–732

Bateson W, Punnett R (1911) The inheritance of the peculiar pigmentation of the silky fowl. J Genet 1:185–203
Baynash AG, Hosoda A, Giaid JA, Richardson N, Emoto RE, Hammer M, Yanagisawa M (1994) Interaction of endothelin-3 with endothelin-B receptor is essential for development of epidermal melanocytes and enteric neurons. Cell 79:1277–1285
Boissy RE (2003) Melanosome transfer to and translocation in the keratinocyte. Exp Dermatol 12:5–12
Boissy RE, Hornyak TJ (2006) Extracutaneous melanocytes. In: Nordlund JJ, Boissy RE, Hearing VJ, King RA, Oetting WS, Ortonne JP (eds) The pigmentary system: physiology and pathophysiology, 2nd edn. Oxford University Press, New York, pp 91–107
Boissy RE, Moellmann G, Trainer AT, Smyth JR Jr, Lerner AB (1986) Delayed-amelanotic (DAM or Smyth) chicken: melanocyte dysfunction in vivo and in vitro. J Invest Dermatol 86 (2):149–156
Braasch I, Volff JN, Schartl M (2009) The endothelin system: evolution of vertebrate-specific ligand-receptor interactions by three rounds of genome duplication. Mol Biol Evol 26:783–799
Brumbaugh JA, Barger TW, Oetting WS (1983) A "new" allele at the *C* pigment locus in the fowl. J Hered 74:331–336
Burtt EH, Schroeder MR, Smith LA, Sroka JE, McGraw KJ (2010) Colourful parrot feathers resist bacterial degradation. Biol Lett. https://doi.org/10.1098/rsbl.2010.0716
Carlson BM (1988) The skin and its derivatives. In: Pattern's foundations of embryology, 5th edn. McGraw-Hill Publishing Company, New York, p 363
Chang CM, Coville JL, Coquerelle G, Gourichon D, Oulmouden A, Tixier-Boichard M (2006) Complete association between a retroviral insertion in the tyrosinase gene and the recessive white mutation in chickens. BMC Genomics 7:19–34
Crawford RD (1990) Poultry breeding and genetics. Elsevier Science Publishers, Amsterdam
Delhey K, Szecsenyi B, Nakagawa S, Peters A (2017) Conspicuous plumage colours are highly variable. Proc Biol Sci 284(1847):20162593. https://doi.org/10.1098/rspb.2016.2593
Dharmayanthi AB, Terai Y, Sulandari S, Zein MS, Akiyama T, Satta Y (2017) The origin and evolution of fibromelanosis in domesticated chickens: genomic comparison of Indonesian Cemani and Chinese Silkie breeds. PLoS One 12(4):e0173147. https://doi.org/10.1371/journal.pone.0173147
Dorshorst B, Molin AM, Rubin CJ, Johansson AM, Strömstedt L, Pham MH, Chen Hallböök F, Ashwell C, Andersson L (2011) A complex genomic rearrangement involving the endothelin 3 locus causes dermal hyperpigmentation in the chicken. PLoS Genet 7(12):e1002412
Dunn L, Jull M (1927) On the inheritance of some characters on the silky fowl. J Genet 19:27–63
Erf GF, Trejo-Skalli AV, Smyth JR (1995) T cells in regenerating feathers of Smyth line chickens with vitiligo. Clin Immunol Immunopathol 76(2):120–126
Erickson CA (1993) From the crest to the periphery: control of pigment cell migration and lineage segregation. Pigment Cell Res 6:336–347
Erickson CA, Goins TL (1995) Avian neural crest cells can migrate in the dorsolateral path only if they are specified as melanocytes. Development 121(3):915–924
Esatbeyoglu T, Rimbach G (2017) Canthaxanthin: from molecule to function. Mol Nutr Food Res 61(6). https://doi.org/10.1002/mnfr.201600469
Faraco CD, Vaz SA, Pastor MV, Erickson CA (2001) Hyperpigmentation in the Silkie fowl correlates with abnormal migration of fate-restricted melanoblasts and loss of environmental barrier molecules. Dev Dyn 220:212–225
Gluckman T-L, Mundy NI (2017) The differential expression of MC1R regulators in dorsal and ventral quail plumages during embryogenesis: implications for plumage pattern formation. PLoS ONE 12(3):e0174714
Griffin DK, Robertson LB, Tempest HG, Vignal A, Fillon AV, Crooijmans RP, Groenen MA, Deryusheva S, Gaginskaya E, Carré W, Waddington D, Talbot R, Völker M, Masabanda JS, Burt DW (2008) Whole genome comparative studies between chicken and turkey and their

implications for avian genome evolution. BMC Genomics 9:168. https://doi.org/10.1186/1471-2164-9-168
Gunnarsson U, Hellström AR, Tixier-Boichard M, Minvielle F, Bed'hom B, Ito S, Jensen P, Rattink A, Vereijken A, Andersson L (2007) Mutations in SLC45A2 cause plumage color variation in chicken and Japanese quail. Genetics 175(2):867–877
Hamburger V, Hamilton HL (1951) A series of normal stages in the development of the chick embryo. J Morphol 88:49–92
Haupaix N, Curantz C, Bailleul R, Beck S, Robic A, Manceau M (2018) The periodic coloration in birds forms through a prepattern of somite origin. Science 361(6408):eaar4777. https://doi.org/10.1126/science.aar.4777
Hirata Y (1996) Endothelin peptides. Curr Opin Nephrol Hypertens 5:12–15
Hosoda K, Hammer RE, Richardson JA, Baynash AG, Cheung JC, Giaid A, Yanagisawa M (1994) Targeted and natural (Piebald-Lethal) mutations of endothelin-B receptor gene produce megacolon associated with spotted coat color in mice. Cell 79:1267–1276
Hutt FB (1949) Genetics of the fowl. McGraw-Hill, New York
International Chicken Genome Sequencing Consortium (2004) Nature 432(7018):695–718
Jang H, Erf GF, Rowland KC, Kong BW (2014) Genome resequencing and bioinformatic analysis of SNP containing candidate genes in the autoimmune vitiligo Smyth line chicken model. BMC Genomics 15:707–728. https://doi.org/10.1186/1471-2164-15-707
Kawasaki-Nishihara A, Nishihara D, Nakamura H, Yamamoto H (2011) ET3/Ednrb2 signaling is critically involved in regulating melanophore migration in *Xenopus*. Dev Dyn 240:1454–1466
Kerje S, Sharma P, Gunnarsson U, Kim H, Bagchi S, Fredriksson R, Schützb K, Jensen P, Heijne GV, Okimoto R, Andersson L (2004) The *Dominant white*, *Dun* and *Smoky* color variants in chicken are associated with insertion/deletion polymorphisms in the *PMEL17* gene. Genetics 168:1507–1518
Kinoshita K, Akiyama T, Mizutani M, Shinomiya A, Ishikawa A, Hassan Younis HH, Tsudzuki M, Namikawa T, Matsuda Y (2014) *Endothelin receptor B2* (*EDNRB2*) is responsible for the tyrosinase-independent recessive white (mo^w) and mottled (*mo*) plumage phenotypes in the chicken. PLoS One 9:1–14
Kitamura K, Takiguchi-Hayashi K, Sezaki M, Yamamoto H, Takeuchi T (1992) Avian neural crest cells express a melanogenic trait during early migration from the neural tube; observations with the new monoclonal antibody, 'MEBL-1'. Development 114:367–378
Lahav R, Ziller C, Dupin E, Le Douarin NM (1996) Endothelin 3 promotes neural crest cell proliferation and mediates a vast increase in melanocyte number in culture. Proc Natl Acad Sci U S A 93:3892–3897
Lahav R, Dupin E, Lecoin L, Glavieux C, Champeval D, Ziller C, Le Douarin NM (1998) Endothelin 3 selectively promotes survival and proliferation of neural crest-derived glial and melanocytic precursors in vitro. Proc Natl Acad Sci U S A 95:14214–14219
Lamoreux ML, Delmas V, Larue L, Bennett DC (2010) The colors of mice. A model genetic network. Wiley-Blackwell, Chichester
Le Douarin NM, Kalcheim C (2009) The neural crest. Cambridge University Press, Cambridge
Lecoin L, Sakurai T, Ngo MN, Abe Y, Yanagisawa M, Le Douarin NM (1998) Cloning and characterization of a novel endothelin receptor subtype in the avian class. Proc Natl Acad Sci U S A 95:3024–3029
Lopes RJ, Johnson JD, Toomey MB, Ferreira MS, Araujo PM, Melo-Ferreira J, Andersson L, Hill GE (2016) Genetic basis for red coloration in birds. Curr Biol 26(11):1427–1434. https://doi.org/10.1016/j.cub.2016.03.076
Matoba Y, Kumagai T, Yamamoto A, Yoshitsu H, Sugiyama M (2006) Crystallographic evidence that the dinuclear copper center of tyrosinase is flexible during catalysis. J Biol Chem 281:8981–8990
McCallion AS, Chakravarti A (2001) EDNRB/EDN3 and Hirschsprung disease type II. Pigment Cell Res 14:161–169

McGraw KJ, Nogare MC (2005) Distribution of unique red feather pigments in parrots. Biol Lett 1 (1):38–43

McGraw KJ, Hill GE, Navara KJ, Parker RS (2004) Differential accumulation and pigmenting ability of dietary carotenoids in colorful finches. Physiol Biochem Zool 77(3):484–491

Mendes-Pinto MM, LaFountain AM, Stoddard MC, Prum RO, Frank HA, Robert B (2012) Variation in carotenoid-protein interaction in bird feathers produces novel plumage coloration. J R Soc Interface 9(77):3338–3350. https://doi.org/10.1098/rsif.2012.0471

Metallinos DL, Bowling AT, Rine J (1998) A missense mutation in the endothelin-B receptor gene is associated with lethal white foal syndrome: an equine version of Hirschsprung disease. Mamm Genome 9:426–431

Minvielle F, Bed'hom B, Coville J-L, Ito S, Inoue-Murayama M, Gourichon D (2010) The "silver" Japanese quail and the *MITF* gene: causal mutation, associated traits and homology with the "blue" chicken plumage. BMC Genet 11:15. https://doi.org/10.1186/1471-2156-11-15

Miwa M, Inoue-Murayama M, Kobayashi N, Kayang BB, Mizutani M, Takahashi H, Ito S (2006) Mapping of panda plumage color locus on the microsatellite linkage map of the Japanese quail. BMC Genet 7:2–6

Miwa M, Inoue-Murayama M, Aoki H, Kunisada T, Hiragaki T, Mizutani M, Ito S (2007) *Endothelin receptor B2* (*EDNRB2*) is associated with the *panda* plumage color mutation in Japanese quail. Anim Genet 38:103–108

Mizutani M, Chiho K, Umezawa H, Kuramasu S (1974) Genetic analysis of a new plumage—panda in Japanese quail (in Japanese with English summary). Exp Anim 23:59–61

Mundy NI (2005) A window on the genetics of evolution: MC1R and plumage coloration in birds. Proc R Soc B 272:1633–1640. https://doi.org/10.1098/rspb.2005.3107

Mundy NI (2018) Colouration genetics: pretty polymorphic parrots. Curr Biol 28(3):R113–R114. https://doi.org/10.1016/j.cub.2017.12.045

Nordlund JJ, Boissy RE, Hearing VJ, King CY, Oetting WS, Ortonne JP (2006) The pigmentary system: physiology and pathophysiology. Blackwell Publishing Ltd., Malden, p 1229

Opdecamp K, Kos L, Arnheiter H, Pavan WJ (1998) Endothelin signaling in the development of neural crest-derived melanocytes. Biochem Cell Biol 76:1093–1099

Ortolani-Machado C, De Freitas P, Borges ME, Faraco C (2007) Special features of dermal melanocytes in WS chicken embryos. Anat Rec 291:55–64

Osman SAM, Sekino M, Nishibori M, Yamamoto Y, Tsudzuki M (2005) Genetic variability and relationships of native Japanese chickens assessed by microsatellite DNA profiling—Focusing on the breeds established in Kochi Prefecture, Japan. Asian-Australas J Anim Sci 18:755–761

Pavan WJ, Tilghman SM (1994) Piebald lethal (s^l) acts early to disrupt the development of neural crest-derived melanocytes. Proc Natl Acad Sci U S A 91:7159–7163

Reid K, Turnley AM, Maxwell GD, Kurihara Y, Kurihara H, Bartlett PF, Murphy M (1996) Multiple roles for endothelin in melanocyte development: regulation of progenitor number and stimulation of differentiation. Development 122:3911–3919

Rubanyli GM, Polokoff MA (1994) Endothelins: molecular biology, biochemistry, pharmacology, physiology, and pathophysiology. Pharmacol Rev 46:325–415

Sakurai T, Yanagisawa M, Takuwa Y, Miyazaki H, Kimura S, Goto K, Masaki T (1990) Cloning of a cDNA encoding a non-isopeptide-selective subtype of the endothelin receptor. Nature 348:732–735

Sakurai T, Yanagisawa M, Masaki T (1992) Molecular characterization of endothelin receptors. Trends Pharmacol Sci 13:103–108

Sato S, Otake T, Suzuki C, Saburi J, Kobayashi E (2007) Mapping of the recessive white locus and analysis of the tyrosinase gene in chickens. Poult Sci 86:2126–2133

Shinomiya A, Kayashima Y, Kinoshita K, Mizutani M, Namikawa T, Matsuda Y, Akiyama T (2012) Duplication of the *endothelin 3* gene is closely correlated with *Fibromelanosis* (*Fm*), the hypermelanization of the internal organs of Silky chickens. Genetics 190(2):627–638. https://doi.org/10.1534/genetics.111.136705

Smyth JR Jr, McNeil M (1999) Alopecia areata and universalis in the Smyth chicken model for spontaneous autoimmune vitiligo. J Investig Dermatol Symp Proc 4:211–215. https://doi.org/10.1038/sj.jidsp.5640213

Smyth JR Jr (1990) Genetics of plumage, skin and eye pigmentation in chickens. In: Crawford RD (ed) Poultry breeding and genetics. Elsevier, Amsterdam, pp 109–167

Stradi R, Pini E, Celentano G (2001) The chemical structure of the pigments in Ara macao plumage. Comp Biochem Physiol Part B 130:57–63

Tachibana M, Perez-Jurado LA, Naknama A, Hodgkinson CA, Li X, Schneider M, Miki T, Fex J, Francke U, Arnheiter H (1994) Cloning of MITF, the human homolog of the mouse *microphthalmia* gene and assignment to chromosome 3p14.1-p12.3. Hum Mol Genet 3:553–557

Tachibana M, Takeda K, Nobukuni Y, Urabe K, Long JE, Meyers KA, Aaronson SA, Miki T (1996) Ectopic expression of MITF, a gene for Waardenburg syndrome type 2, converts fibroblasts to cells with melanocyte characteristics. Nat Genet 14:50–54

Tadano R, Sekino M, Nishibori M, Tsudzuki M (2007) Microsatellite marker analysis for the genetic relationships among Japanese long-tailed chicken breeds. Poult Sci 86:460–469

Takeuchi S, Suzuki H, Yabuuchi M, Takahashi S (1996) A possible involvement of melanocortin 1-receptor in regulating feather color pigmentation in the chicken. Biochim Biophys Acta 1308:164–168

Theron E, Hawkins K, Bermingham E, Ricklefs RE, Mundy NI (2001) The molecular basis of an avian plumage polymorphism in the wild: a melanocortin-1-receptor point mutation is perfectly associated with the melanic plumage morph of the bananaquit, Coereba flaveola. Curr Biol 11 (8):550–557. https://doi.org/10.1016/s0960-9822(01)00158-0

Thomas DB, McGoverin CM, McGraw KJ, James HF, Madden O (2013) Vibrational spectroscopic analyses of unique yellow feather pigments (spheniscins) in penguins. J R Soc Interface 10 (83):20121065. https://doi.org/10.1098/rsif.2012.1065

Thommen H, Wackernagel H (1963) Isolation and identification of canthaxanthin in the lesser flamingo (Phoenicolnaias minor). Biochim Biophys Acta 69:387–396

Tinbergen J, Wilts BD, Stavenga DG (2013) Spectral tuning of Amazon parrot feather coloration by psittacofulvin pigments and spongy structures. J Exp Biol 216:4358–4364. https://doi.org/10.1242/jeb.091561

Tobita-Teramoto T, Jang GY, Kino K, Salter DW, Brumbaugh JA, Akiyama T (2000) Autosomal albino chicken mutation (c^a/c^a) deletes hexanucleotide (-△GACTGG817) at a copper-binding site of the tyrosinase gene. Poult Sci 79:46–50

Tsudzuki M, Nakane Y, Wakasugi N, Mizutani M (1993) Allelism of panda and dotted white plumage genes in Japanese quail. J Hered 84:225–229

Twyman H, Prager M, Mundy NI, Andersson S (2018) Expression of a carotenoid-modifying gene and evolution of red coloration in weaverbirds (Ploceidae). Mol Ecol 27(2):449–458

Valverde P, Healy E, Jackson I, Rees JL, Thody AJ (1995) Variants of the melanocyte-stimulating hormone receptor gene are associated with red hair and fair skin in humans. Nat Genet 11 (3):328–330

Weaver RJ, Santos ESA, Tucker AM, Wilson AE, Hill GE (2018) Carotenoid metabolism strengthens the link between feather coloration and individual quality. Nat Commun 9(1):73–81

Yanagisawa M, Kurihara H, Kimura S, Tomobe Y, Kobayashi M, Mitsui Y, Yazaki Y, Goto K, Masaki T (1988) A novel potent vasoconstrictor peptide produced by vascular endothelial cells. Nature 332:411–415

Chapter 4
Pigments in Teleosts and their Biosynthesis

Tetsuaki Kimura

Abstract Although chromatophores are mainly derived from neural crest cells, the properties of pigment production vary greatly between these cells. During differentiation, pigment cells form pigment-containing organelles; melanosomes in melanophores, pterinosomes in xanthophores and erythrophores, reflecting platelets in iridophores, and leucosome in leucophores. Cyanophores produce a blue pigment; however, the biochemicals involved in the production of this pigment are unknown. Our understanding of fish pigments and their biosynthesis has been largely derived from medaka and zebrafish mutant analysis. Thus, this chapter describes the different types of pigments in medaka and zebrafish and their biosynthesis.

Keywords Melanin · Guanine · Pteridine · Carotenoid · Uric acid

4.1 De Novo Purine Synthesis Is Indispensable for Embryonic Pigmentation

Medaka and zebrafish have four kinds of chromatophores, namely melanophores, xanthophores, iridophores, and leucophores (Schartl et al. 2016).

The main pigments in pigment cells are eumelanin (melanophores), pteridines and carotenoids (xanthophores), guanine (iridophores), and uric acid (leucophores). Except for carotenoids, which originate from food, these are all synthesized in a fish. The pteridine, guanine, and uric acid are made from purines, and eumelanin is made from tyrosine. Purines are generated by de novo synthesis or recycled through the salvage pathway. In embryonic development, the de novo synthesis pathway is an important part of pigmentation. Indeed, mutations in enzymes involved in de novo purine synthesis result in pigmentation defects (Ng et al. 2009). Conversely, pigmentation in de novo pyrimidine synthesis mutants is normal (Willer et al. 2005).

T. Kimura (✉)
National Institute of Genetics, Mishima, Japan
e-mail: tekimura@nig.ac.jp

H. Hashimoto et al. (eds.), *Pigments, Pigment Cells and Pigment Patterns*,
https://doi.org/10.1007/978-981-16-1490-3_4

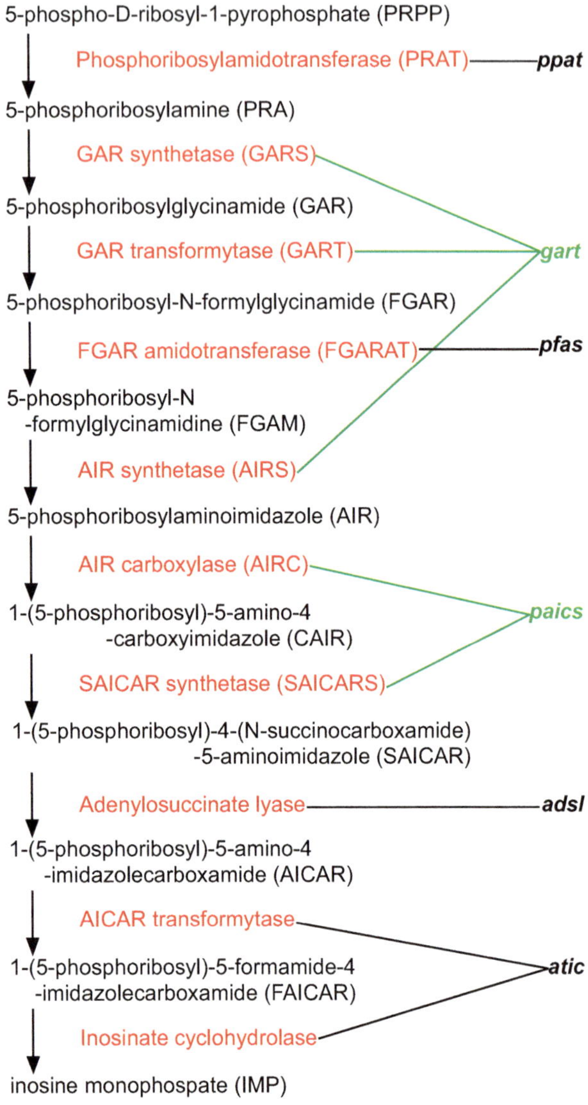

Fig. 4.1 The de novo purine synthesis pathway. IMP is synthesized from PRPP via multiple enzymatic pathways, in which *gart* and *paics* proteins (green) catalyze multiple steps (from Ng et al. 2009)

The de novo synthesis of purine starts from phosphoribosyl diphosphate and produces purine nucleotides, adenosine monophosphate (AMP), and guanosine monophosphate (GMP). Inosine monophosphate (IMP) is an intermediate at a branch point during de novo purine synthesis (Figs. 4.1 and 4.2). From IMP, de novo purine synthesis branches to the AMP and GMP pathways. *Gart* and *paics* are indispensable in the de novo synthesis of IMP, the precursor for AMP or GMP. The *gart* genes encode phosphoribosylglycinamide formyltransferase, phosphoribosylglycinamide synthetase, and phosphoribosylaminoimidazole synthetase, while the *paics* gene encodes phosphoribosylaminoimidazole carboxylase and

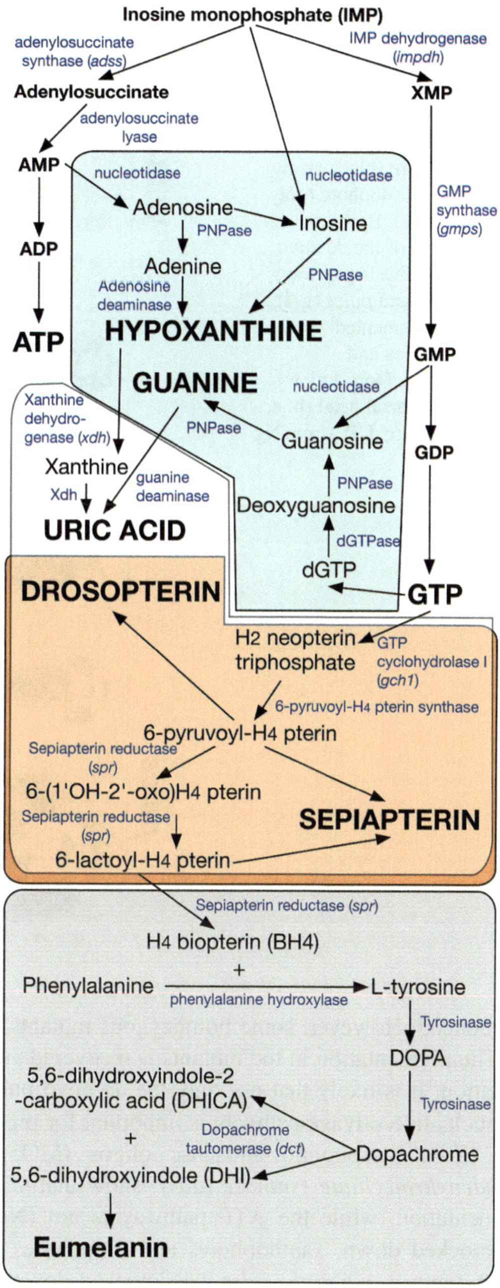

Fig. 4.2 Pigment synthesis pathway in medaka embryo. ATP and GTP are made from IMP. The process of producing pigment from purines is shared between pigment cells. Enzymes are shown in red. The gray background indicates iridophore, white indicates leucophore, yellow indicates xanthophore, and black indicates melanophore. In embryonic stage, the medaka leucophore has xanthophore character, so that leucophore contains white and yellow pigments (modified from Ng et al. 2009)

phosphoribosylaminoimidazole succinocarboxamide synthetase. *Gart* and *paics* mutants result in pigmentation defects and microphthalmia (Fig. 4.3). As a result, xanthophores and iridophores become invisible, and melanophore pigmentation is

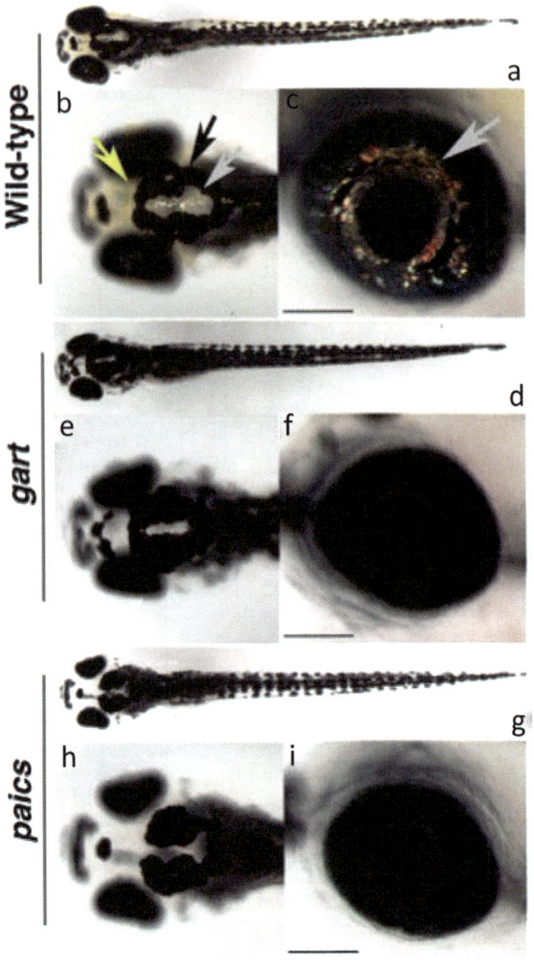

Fig. 4.3 Mutant phenotype of *gart* and *paics*. *gart* and *paics* show pigmentation defects in early larval stage. Wild-type has pigmented xanthophore (yellow arrow), melanophore (black arrow), and silver iridophore (gray arrow) (**a**–**c**). Both mutants of enzyme of the de novo purine synthesis pathway, *gart* (**d**–**f**) and *paics* (**g**–**i**), have no pigmented xanthophores and iridophores. Dorsal view (**a**, **d**, **g**), dorsal head (**b**, **e**, **h**), and eye (**c**, **f**, **i**). (from Ng et al. 2009)

reduced. However, some homozygous mutants of *gart* and *paics* reach adulthood. The pigmentation in the mutants is recovered and resembles the wild-type pigmentation. It is likely that not only the de novo purine synthesis pathway but also the nucleotide salvage pathway is important for pigmentation in adults.

Two morpholino antisense oligos (MO) for *GMP synthase* (*gmps*) and *adenylosuccinate synthase* (*adss*) show that GTP pathway is dispensable for pigmentation, while the ATP pathway is not (Ng et al. 2009). When *gamps* were knocked down, xanthophores and iridophores pigmentation defects were visible. In contrast, when *the adss* was knocked down, only eye defects were visible, and pigmentation defects were not observed. Additionally, *IMP dehydrogenase 1a* (*impdh1a*) MO induces the loss of nearly all iridophore and xanthophore pigmentation (Li et al. 2015). Thus, GTP derived from IMP is the main precursor for the synthesis of pigment molecules of all types of pigment cells in the embryonic stages.

4.2 Melanophore Pigment and Mutants

The black pigment melanophore is a melanin that is highly conserved in animals. Melanin is synthesized in the melanosomes of melanophores in vertebrates. Mammals and birds produce two types of melanin, the black eumelanin and the lighter pheomelanin. However, fishes produce only eumelanin.

Eumelanin is synthesized from tyrosine by the tyrosinase gene family, *tyrosinase* (*try*), *tyrosinase-related protein 1* (*tyrp1*), and *dopachrome tautomerase* (*dct*) (Fig. 4.4) in vertebrates. Tyrosinase catalyzes the first two rate-limiting steps of melanin synthesis from tyrosine to DOPA and from DOPA to DOPA quinone. Dct catalyzes the production of DHICA from DOPA chrome, while Tyrp1 is involved in the stabilization of tyrosinase and in the formation of indole-5,6-quinone carboxylic acid. Indole-5,6-quinone carboxylic acid is converted to eumelanin by premelanosome protein a (Pmela) or Pmelb.

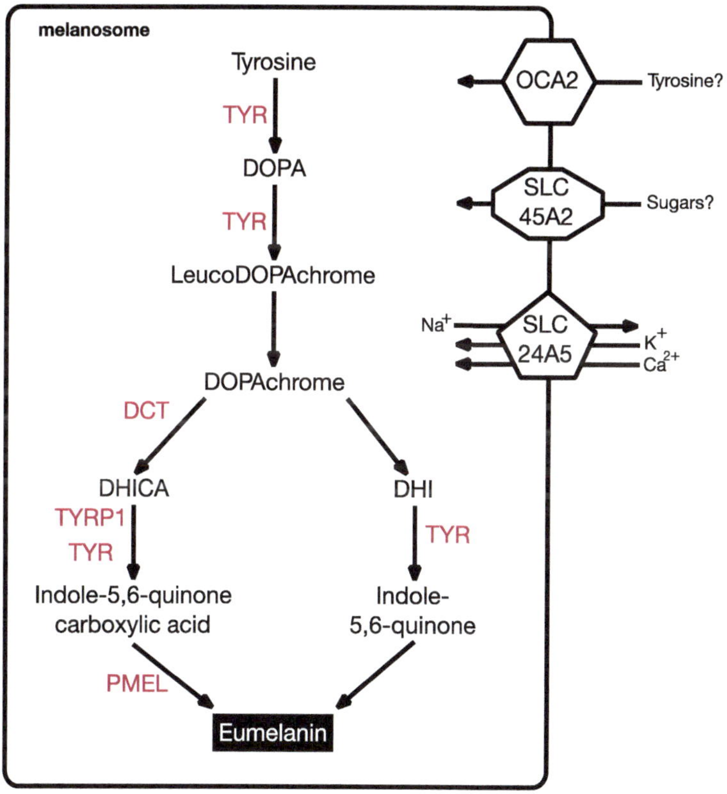

Fig. 4.4 Eumelanin synthesis pathway in fish. Eumelanin is made from tyrosine by an enzymatic process in the melanosome. OCA2 and SLC45A2 transport the substrates of eumelanin. SLC24A5 is an anion transporter and is dispensable for tyrosinase activity. Enzymes are shown in red (modified from Braasch et al. 2007)

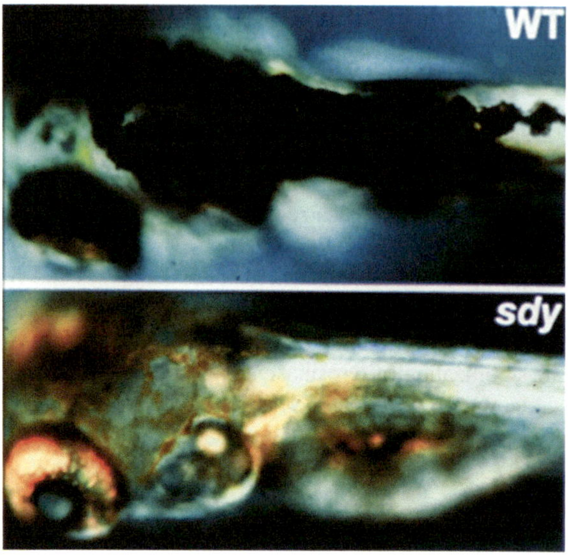

Fig. 4.5 Zebrafish tyrosinase mutant 6 days phenotype of wild-type and tyrosinase mutant (previously known as sdy). Zebrafish tyrosinase mutant has gaps in xanthphore pattern. (from Kelsh et al. 1996)

The zebrafish mutant collection is comprised of tyrosinase gene family mutants. The *tyr* mutant has unpigmented melanophores and an albino phenotype (Fig. 4.5) (Kelsh et al. 1996), while the *tyrp1a* mutant has a melanophore-loss phenotype caused by melanophore death (Fig. 4.6) (Krauss et al. 2014). *The pmela* mutant, known as *fading vision*, has reduced melanin pigmentation and small eyes caused by degenerating retina (Haffter et al. 1996) and the *pmelb* mutant is mainly expressed in the retinal pigmented epithelium. *The pmelb* morphant shows irregular melanin deposition in the melanosome and an uneven melanosome morphology (Burgoyne et al. 2015).

Melanosomal transporter mutants also show an albino phenotype. *oca2* (*oculocutaneous albinism II*) encodes protein P, which is a putative tyrosine transporter. The *oca2* mutant has reduced melanin pigmentation in zebrafish and medaka (Figs. 4.7 and 4.8) (Fukamachi et al. 2004; Beirl et al. 2014). *Solute carrier family 45 member 2* (*slc45a2*) encodes a transporter protein, but its exact function is unknown. In medaka, *slc45a2* is known as the B gene, which is a causal gene of the orange-red variant, hi-medaka (Fukamachi et al. 2001). In zebrafish, the *slc45a2* mutant is known as *albino* (Dooley et al. 2013). Both mutants show reduced melanin pigmentation, since *slc45a2* is involved in melanin production. Slc24a5 is a potassium-dependent sodium/calcium exchanger. The zebrafish mutant of *slc24a5* is known as *golden* (Lamason et al. 2005) and shows reduced melanin pigmentation. These transporters most likely transport the substrates of melanin production or control tyrosinase gene family activities. However, the relationship between these transporters and enzymes remains unclear.

Melanosomes are one of the lysosome-related organelles. BLOC-1 (biogenesis of lysosome-related organelles complex (1), BLOC-2, BLOC-3, and HOPS (homotypic fusion and protein sorting) are required for normal biogenesis of

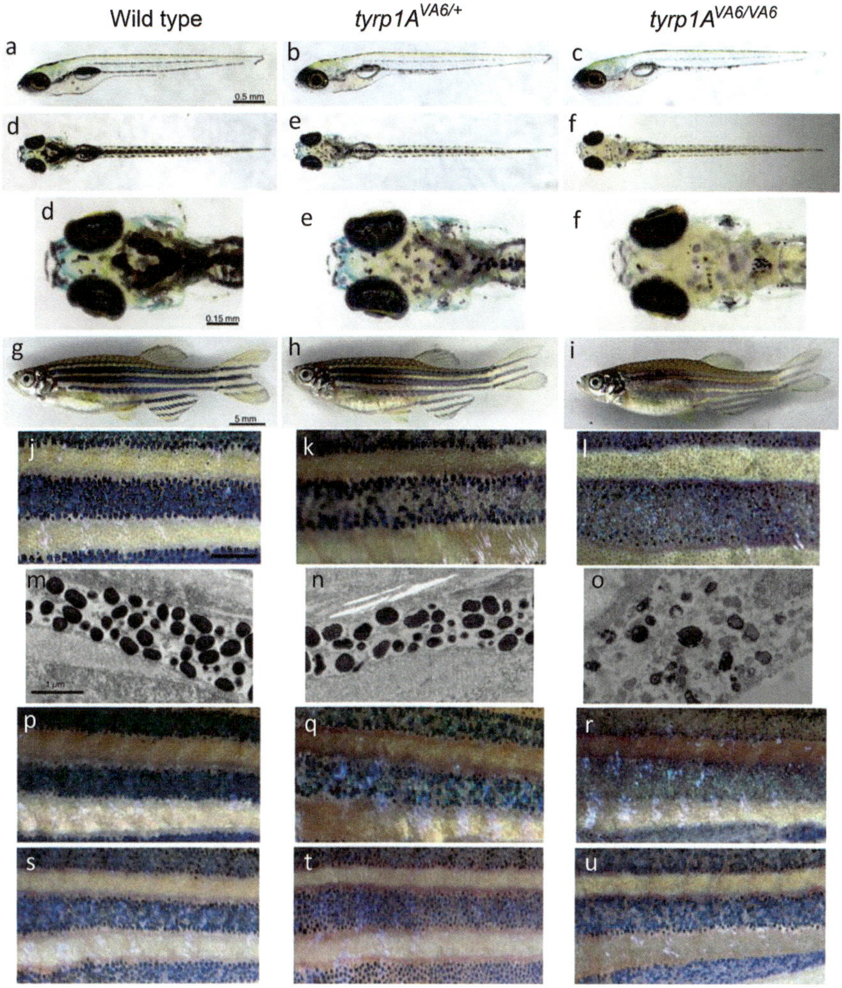

Fig. 4.6 Zebrafish *tyrp1a* mutant. *tyrp1a*^{dtva6} allele causes melanophore death. This mutation has a semi-dominant effect. Not only homozygous, but also heterozygous show a reduction of melanophores in 5 dpf larvae (**b**, **c**). Mutants have faint blue strips in 10-week adult fish (**g**–**i**). Electron micrographs (**m**–**o**) show irregular melanosomes with less electron dense pigment in the homozygous mutants (**o**). Heterozygous mutants have a normal pigmentation (**m**, **n**). Melanophore density and stripe pigmentation is recovered in later stages (14 weeks: P-R, and 18 weeks: **s**-**u**) (from Krauss et al. 2014)

lysosome-related organelles. In human, mutations of BLOC-1 lead to Hermansky–Pudlak syndrome (HPS), which shows oculocutaneous albinism. Mutation in the Xenopus homolog of the human *HSP6* gene also shows pigmentation deficiency in melanophores and iridophores (Nakayama et al. 2017). Zebrafish mutants of genes encoding subunits of BLOC-1, BLOC-2, and HOPS showed pigmentation defects in

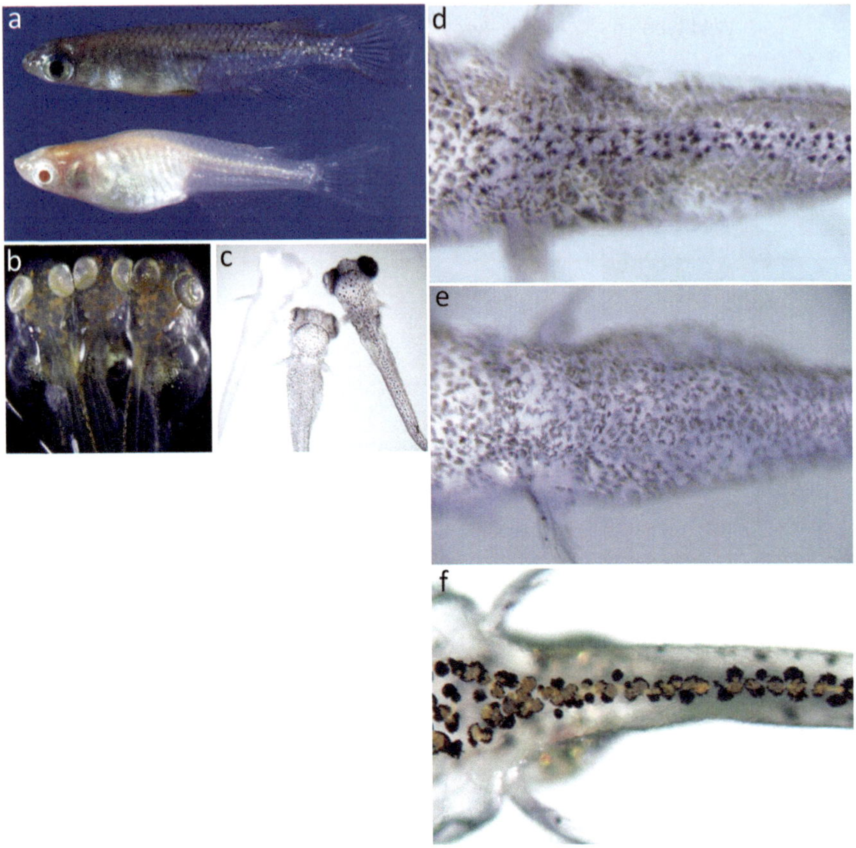

Fig. 4.7 Medaka *oca2* mutant shows the albino phenotype Medaka *oca2* mutant has no pigmented melanophores (**b**). (**b**) *tyr* mutant (left), *oca2* mutant (middle), and *slc45a2* mutant (right). The *oca2* mutant is indistinguishable from the *tyr* mutant. The *slc45a2* mutant has slightly pigmented eyes. (**c**) *tyr* mutant (left), *oca2* mutant (middle), and *slc45a2* mutant (right). *tyr* mutant shows no tyrosinase activity, but *oca2* and *slc45a2* mutants show tyrosinase activity. (**d**, **e**) Higher magnification of the dorsal parts of b^{g8} (**d**) and *oca2* (**e**) after the tyrosinase reaction. (**f**) denotes wild-type fry (Photos from Fukamachi)

melanophore and iridophore (Bowman et al. 2019). These complexes are involving to transport the proteins required for melanin synthesis to the melanosome (Fig. 4.9).

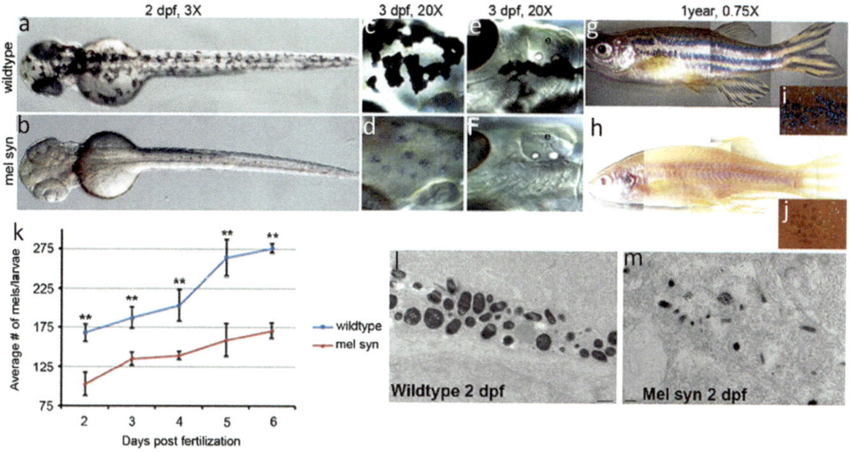

Fig. 4.8 Zebrafish *oca2* mutant. The *oca2* mutant shows the albino phenotype, similar to medaka *oca2* mutant (**a–h**). Melanin synthesis is low in this mutant (**k**). In the mutant, the electron density of pigments is low (**m**) (from Beirl et al. 2014)

4.3 Iridophore Pigments and Mutants

Iridophores have platelets derived from guanine. Platelets reflect light and generate iridescent colors. The structural differences of platelets (e.g. orientation, size, and density) determine the nature of the color observed. Thus, guanine is the most important pigment in iridophores.

There are two guanine synthesis pathways, one starting from IMP and the other starting from GTP. In the first pathway, IMP is dehydrogenated to XMP by IMP dehydrogenase, XMP is converted to GMP by GMP synthase, GMP is dephosphorylated to guanosine by 5′-nucleotidase, and guanosine is converted to guanine by PNPase (Polynucleotide phosphorylase) (Fig. 4.2). In the second pathway, GTP is reduced to dGTP by ribonucleoside-triphosphate reductase, dGTP is hydrolyzed to deoxyguanosine by dGTPase, and deoxyguanosine is converted to guanosine by PNPase. Therefore, PNPase plays a major role in the guanine synthesis pathway.

Medaka and zebrafish have three families of PNPase genes: *pnp4*, *pnp5*, and *pnp6*. Among them, *pnp4a* plays the main role in guanine synthesis and is known as a differentiated marker of iridophores in both species (Curran et al. 2010; Kimura et al. 2017). The *pnp4a* mutant of medaka, known as *gu*, shows severe defects in iridophore pigmentation (Kimura et al. 2017) (Fig. 4.10).

mpv17 (*mitochondrial inner membrane protein MPV17*) is another important gene involved in iridophore pigmentation (Krauss et al. 2013). Zebrafish mutants of *mpv17*, known as *roy orbison* or *transparent*, lack reflective iridophores (Fig. 4.11) (D'Agati et al. 2017). However, the mechanism for *mpv17* works in iridophore pigmentation is not known.

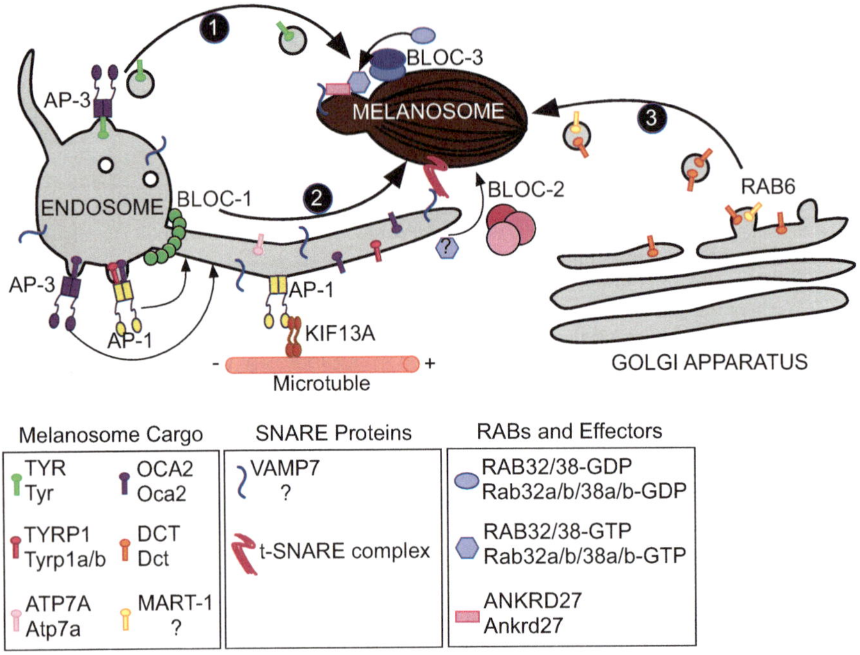

Fig. 4.9 HPS complexes and mechanisms of cargo delivery to melanosomes in human. The enzymes required for the synthesis of melanin are delivered to melanosomes as melanosome-destined transmembrane protein cargoes. Tyrosinase is mainly transported via the AP-3 essential vesicular pathway (black circle 1). TYRP1 (zebrafish orthologues are Tyrp1a and b), OCA2 (zebrafish orthologue is Oca2), and ATP7A (zebrafish orthologue is Atp7a) pass through endosome-derived membrane tubes on their way to melanosomes (black circle 2). BLOC-1 is required to generate the tubules. BLOC-2 is required to direct the tubular transport carriers to melanosome membranes. DCT (zebrafish orthologue is Dct) and MART-1 are transported from the Golgi apparatus to the melanosome in a separate vesicle transport pathway that requires RAB6 (black circle 3) (modified from Bowman et al. 2019)

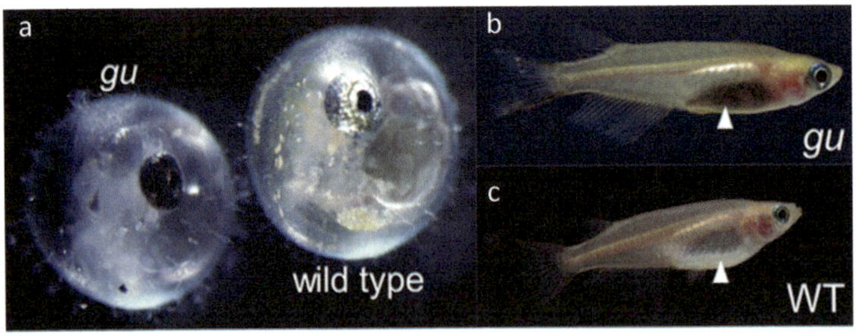

Fig. 4.10 *pnp4a* medaka mutant shows pigmentation defects in iridophores. In medaka, the *pnp4a* mutant shows remarkable iridophore pigmentation defect throughout the life cycle. (**a**) The *pnp4a* mutant (left) has black eyes due to the lack of iridescent in the iridophores. (**b**, **c**) The *pnp4a* mutant has a black abdominal region compared to the wild-type (from Kimura et al. 2017)

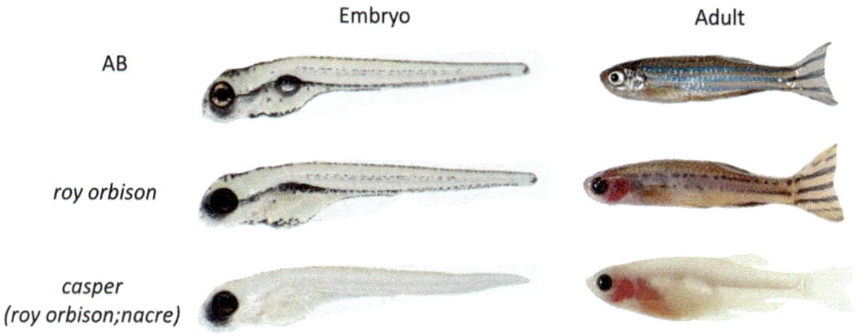

Fig. 4.11 *mpv17* mutant zebrafish lack reflective iridophore. AB is wild type. The *mpv17* mutant, known as *roy orbison*, shows severe iridophore pigmentation defects. *Casper* is double mutant of *mpv17* and *mitfa* (*nacre*). (from D'Agati et al. 2017)

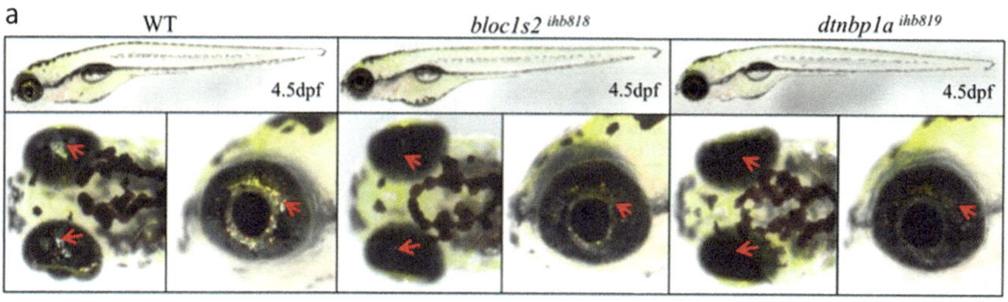

Fig. 4.12 *bloc1s2* and *dtnbp1a* mutant zebrafish show reduced reflective iridophores. Both mutants also show reduced iridophore reflection (arrows) (from Chen et al. 2018)

Zebrafish mutants of genes encoding subunits of BLOC-1, *bloc1s1*, *bloc1s2*, and *dtnbp1a* (*dystrobrevin binding protein 1a*), show a deficiency of melanophores and iridophores (Chen et al. 2018) (Fig. 4.12). Probably, many membrane proteins involving guanine synthesis are transported by BLOC-1 activity like a melanosome.

4.4 Xanthophore/Erythrophores Pigments and Mutants

Xanthophores and erythrophores include pteridines and carotenoids as yellow, orange, and red pigments. The ratio of yellow, orange, and red pigments determines the hue of the chromatophore. Therefore, it is very difficult to classify both chromatophores. In the following, only xanthophores are discussed because many genetic studies have investigated them in zebrafish and medaka mutants. Those two fish have only xanthophores.

4.4.1 Pteridine Synthesis

Of the two pigments, pteridines are synthesized from GTP (Fig. 4.13) and stored in the pteridine-containing organelle, pterinosome. Pteridine synthesis includes three

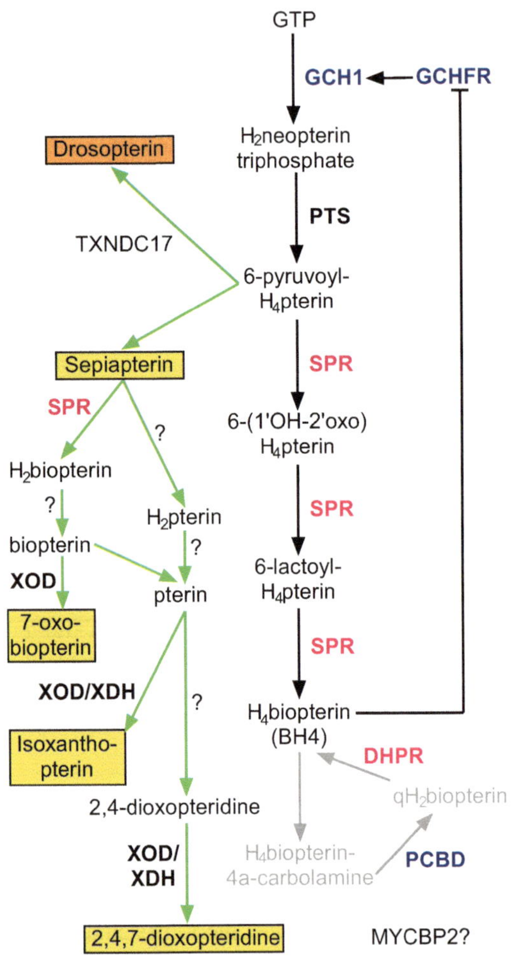

Fig. 4.13 Pteridine pigments synthesis pathway. The pathway is constructed by three processes: H_4biopterin de novo synthesis (black), H_4biopterin (BH4) reproduction (gray), and pteridine synthesis (green) (modified from Ziegler 2003 and Braasch et al. 2007). TXNDC17 (previously known as clot) probably is involved in the synthesis of drosoptein. MYCBP2 is probably working on this pathway, but the details are unknown. Bold types indicate enzymes. *GCH1* GTP cyclohydrolase 1, *GCHFR* GTP cyclohydrolase 1 feedback regulatory protein, *PTS* 6-pyruvoyltetrahydropterin/6-carboxytetrahydropterin synthase, *SRP* sepiapterin reductase, *DHPR* dihydropteridine reductase, *PCBD* 4a-hydroxytetrahydrobiopterin dehydratase, *XOD* xanthine oxidase, *XDH* xanthine dehydrogenase

component pathways: H_4biopterin (BH4) de novo synthesis from GTP, BH4 reproduction, and yellow pteridine pigment synthesis.

During BH4 de novo synthesis, GTP cyclohydrolase 1 (Gch1) catalyzes the first rate-limiting step of pteridine synthesis. Gch1 activity is regulated by the H_4biopterin-dependent GTP cyclohydrolase 1 feedback regulatory protein (Gchfr). Subsequent pteridine synthesis steps are catalyzed by 6-pyrovoyltetrahydroptein synthase and sepiapterin reductase (Spr). Spr catalyzes the three subsequent steps of BH4 synthesis.

Pcbd (pterin-4 alpha-carbinolamine dehydratase/dimerization cofactor of hepatocyte nuclear factor 1 alpha (TCF1)) and Dhpr (dihydropteridine reductase) catalyze the reaction during BH4 reproduction (Fig. 4.13). Pcbd dehydrates H4biopterin-4a-carbolamine to qH_2biptrerin, and then Dhper reduces qH_2biptrerin to BH4.

Pteridine pigment synthesis is catalyzed by Spr and Xod/Xdh (xanthine oxidase/xanthine dehydrogenase) (Fig. 4.13). In zebrafish, the Xod/Xdh family genes are *aldehyde oxidase 5* (*aox5*), *aox6*, and *xdh*.

The biosynthetic pathway of drosopterin has yet to be elucidated in vertebrates. However, two steps in the biosynthesis of drosopterin in Drosophila are known to be conserved in the pteridine pigment synthesis pathway of vertebrates, which are catalyzed by Gch1 and Pts. Drosopterin in vertebrates is most likely synthesized from 6-pyruvoyl-H4pterin. The *thioredoxin domain containing 17* (*txndc17*) is a vertebrate ortholog of Drosophila clot, which is an enzyme involved in drosopterin synthesis. Thus, it is likely that *txndc17* is involved in drosopterin synthesis.

MYC binding protein 2 (*mycbp2*) is necessary to produce high levels of sepiapterin in xanthophores (Le Guyader et al. 2005). The zebrafish *mycbp2* mutant, *esrom*, has reduced yellow pigmentation caused by a deficiency of sepiapterin synthesis. Mycbp2 most likely regulates pteridine synthesis by mediating signal transduction.

Because *slc22a7a* (*solute carrier family 22 member 7a*) is expressed in zebrafish xanthophores (Saunders et al. 2019) and human *SLC22A7* is known as a facilitative transporter of cGMP and other guanine nucleotides (Cropp et al. 2008), *slc22a7a* is likely to be involved in xanthophore pigmentation. *slc2a11b* medaka and zebrafish mutants show xanthophore pigmentation deficiency at the embryonic stage (Kimura et al. 2014). In medaka, the *slc2a15b* mutant shows not only leucophore pigmentation deficiency but also xanthophore pigmentation deficiency in the embryonic stage (Kimura et al. 2014). Thus, these Slc proteins most likely transport the substrates of pteridines into the xanthophores.

4.4.2 Carotenoids

Unlike pteridines, carotenoids are not synthesized in vertebrates. Therefore, fish acquire carotenoids from the diet. Carotenoids are yellow pigments in adult zebrafish xanthophores (Fig. 4.14). Carotenoid pigments are normally localized to lipid

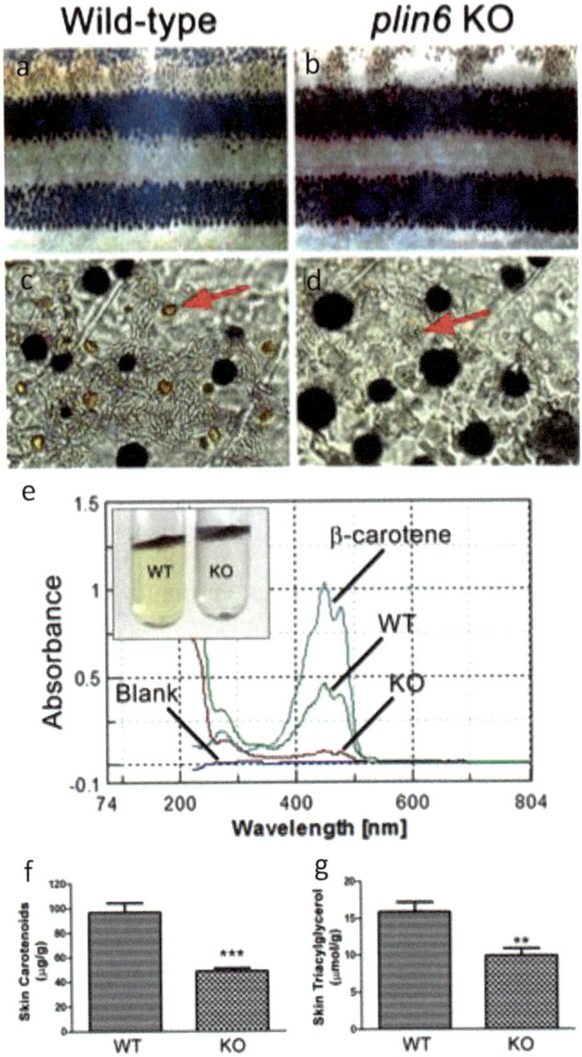

Fig. 4.14 *plin6* knockout in zebrafish decreases the concentration of carotenoids. (**a–d**) Brightfield images showing that *plin6* mutant has reduced yellow carotenoid pigmentation (arrows in **c** and **d**). (**e**) Extraction of carotenoid pigment from wild-type and *plin6* mutant integument (inset) and the corresponding absorbances of carotenoids from a b-carotene standard, wild-type skin (WT), and *plin6* mutant skin (KO). (**f**) Quantification of skin carotenoid levels from wild-type (WT) and *plin6* mutant skin (KO). N = 4; *** $p < 0.001$. (**g**) Quantification of skin triacylglycerol levels from wild-type (WT) and *plin6* mutant skin carotenoid droplets (KO). N = 3; ** $p < 0.02$. (from Granneman et al. 2017)

droplets in xanthophores (Fig. 4.15). Therefore, once carotenoids are acquired from food, they are transported, deposited, and processed.

Since carotenoids are highly hydrophobic, they are transported by lipoproteins through the plasma to xanthophores. Scavenger receptors recognize these lipoproteins and facilitate the movement of carotenoids into xanthophores (Toews et al. 2017). Zebrafish *scarb1* (*scavenger receptor class B, member 1*) mutant loses the yellow color of xanthophores (Saunders et al. 2019).

plin6 (*perilipin 6*) is another gene involved in carotenoid transport and deposition (Granneman et al. 2017). *plin6* transports and concentrates carotenoids into lipid

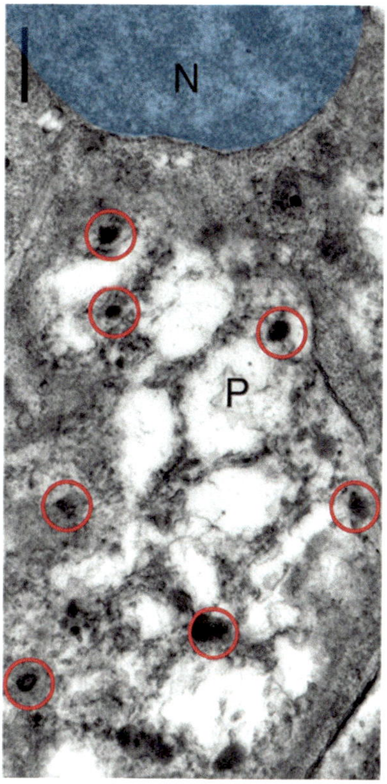

Fig. 4.15 Carotenoids containing lipid droplets in xanthophore. Transmission electron microscopy image showing carotenoids and lipids as electron dense carotenoid vesicles (red circles). *N* nucleus. *P* pterinosome (from Saunders et al. 2019)

droplets in xanthophores. In fact, the zebrafish *plin6* knockout showed reduced carotenoid concentration (Fig. 4.14).

In vertebrates, β-carotene oxygenases are involved in the cleavage of carotenoids. *bco1* (*beta-carotene oxygenase 1*), also known as β-carotene oxygenase, plays a role in the visual system. In sheep, BCO2 mutations affect the accumulation of carotenoids in adipose tissue (Våge and Boman 2010). *bco1* and *bco2b*, zebrafish orthologs, are expressed in xanthophores (Saunders et al. 2019). These results suggest that these genes are related to the production of yellow pigments from dietary β-carotene in xanthophores.

4.4.3 Pigment Shift from Pteridines to Carotenoids

Interestingly, some medaka and zebrafish mutants have an embryonic/early larval xanthophore pigmentation phenotype, but have normal pigmented adult xanthophores (Kimura et al. 2014; Kelsh et al. 2004; Odenthal et al. 1996). The pteridines synthesis enzyme *gch2* (*GTP cyclohydrolase 2*) mutants show a similar phenotype (Lister 2019). This suggests that the embryonic/early larval xanthophore pigment is mainly pteridines.

In contrast, the homozygous allele *r* of the *r* locus of medaka results in color-less xanthophores in adults, but has no pigmentation phenotype at the embryonic stage. Similarly, *scarb1* mutants show pigmentation defects only in the adult stage, but not in the embryonic/early larval stage (Saunders et al. 2019). Therefore, the adult xanthophore pigment is mainly carotenoids in both species. Since fish are unable to process food during the embryonic stages, pteridines are suitable for xanthophore pigment. By contrast, adult fish are able to easily uptake carotenoids from their food. Thus, carotenoids are suitable for xanthophore pigments in the adult stage. Thyroid hormones regulate the pigment shift from pteridine at the embryonic/early larval stages to carotenoids in adults (Saunders et al. 2019).

4.5 Leucophore Pigments and Mutant

In vertebrate, melanophore, iridophore, and xanthophore/erythrophore are the common pigment cells. However, medaka has other pigment cell, leucophore. Leucophores have white granules laden with uric acid (Hama 1975). In the larval stage, the leucophores contain orange pigments. Orange pigments are in drosopterinosomes that contain red pigments, drosopterin, isodrosopterin, and neodrosopterin (Hama 1975), similar to xanthophores. These leucophores lose their orange pigments during larval development and are transformed into white in adults. One of the leucophore pigmentation medaka mutants, *wl* (*white leucophore*), the causal gene of which is *slc2a11b*, has white leucophores and no yellow xanthophores in the embryo to the early larval stage (Kimura et al. 2014) (Fig. 4.16). However, the *wl* mutant has orange-pigmented xanthophores in adults. Thus, in leucophores, there is probably no pigment shift from pteridines to carotenoids, simply the loss of pteridines. Since leucophores and xanthophores have common precursor cells in medaka (Kimura et al. 2014; Nagao et al. 2014), larval leucophores have both features in the early stage.

The white hue of leucophores is given by uric acid in medaka. The uric acid in leucophores was first identified by paper chromatographic and spectroscopic analysis (Hama 1975). Uric acid is produced by xanthine dehydrogenase (XDH), a functional enzyme that catalyzes hypoxanthine to xanthine and xanthine to uric acid (Fig. 4.2). Xanthine is also produced from guanine by guanine deaminase. Melamine is known to inhibit uric acid transport and accumulation. When medaka is immersed in a saturated melamine solution, the leucophores disappear within 24 h.

leucophore free (*lf*) is a mutant that has a pigmentation defect in the white pigment of leucophores (Fig. 4.16). The *lf* mutant lacks leucophores throughout life. The causal gene of *lf* is *slc2a15b*, which is a putative sugar transporter (Kimura et al. 2014). As Slc2a15b is most similar to Slc2a9, which is a urate transporter, it is likely that is transports uric acid to leucosome.

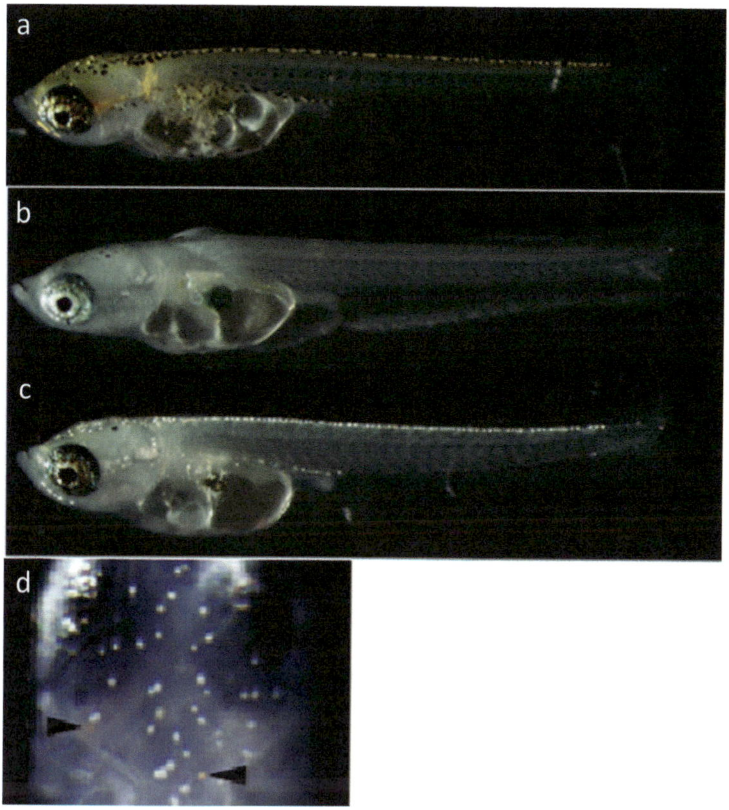

Fig. 4.16 Medaka leucophores and mutants. Medaka *lf/slc2a15b* mutant has no pigmented leucophores or xanthophores (**a**, **b**). *wl/slc2a11b* mutant has white leucophores and no pigmented xanthophores (**a**, **c**). (**d**) BAC (which contains *slc2a11b*)-injected *wl/slc2a11b* mutant. Some leucophores were rescued and recovered their orange pigment (black arrowheads). White arrowheads denote leucophores (from Kimura et al. 2014)

Lewis and colleagues showed that zebrafish and related species have two types of leucophores in the adult stage: xantholeucophores and melanoleucophores (Lewis et al. 2019) (Fig. 4.17). Xantholeucophores arise from xanthophore-like cells, while mleanoleucophores arise directly from melanophores. Both leucophores are located in adult unpaired fins.

Xantholeucophores contain carotenoids and pteridines as orange pigments, such as xanthophores. Additionally, xantholeucophores have pterinosome-like organelles and carotenoid vesicles (Fig. 4.18d). Although these resemble medaka leucophores, uric acid accumulation was not detected by mass spectrometry analysis. The guanine content was found to increase; however, the crystalline forms of guanines were not detected in xantholeucophores (Fig. 4.18f). Thus, the source of the white pigment of xantholeucophores remains unclear.

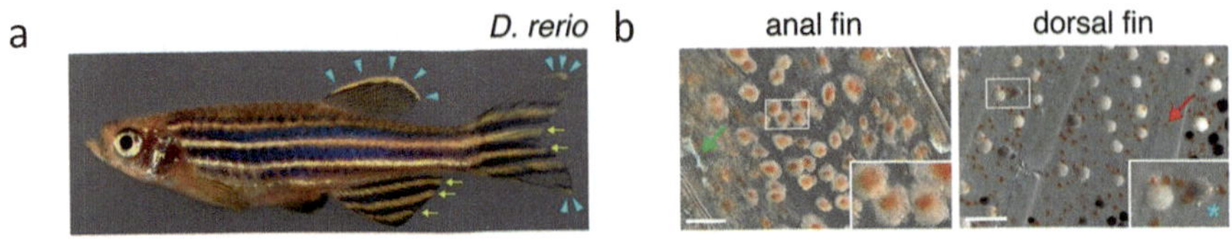

Fig. 4.17 Zebrafish leucophores. (**a**) Zebrafish has two types of leucophores: xantholeucophores (yellow arrowheads) and melanoleucophores (blue arrowheads). (**b**) Brightfield images of xantholeucophores (left) and melanoleucophores (right). Xantholeucophores contain white and orange pigments. Some melanoleucophores contain white and black pigments (asterisk). Green arrows indicate iridophore and red arrows indicate xanthophore (from Lewis et al. 2019)

As in xantholeucophores, the white pigment in melanoleucophores is not uric acid. Raman spectroscopy revealed that the white pigment is crystalline β-guanine, typical of iridophores (Fig. 4.18f). Melanoleucophores are similar to iridophores, although they lack reflecting platelets, which are characteristic structures and confer iridescence to iridophores (Fig. 4.18a, c). In total, melanoleucophores are completely different from the white pigment cells of medaka leucophores.

4.6 Genetic Ablation of Pigment Cells from the Body Wall in Medaka and Zebrafish

Some of the pigment mutants mentioned above are recessive and fertile. These can be crossed to produce fish that are transparent throughout life. These fish allow for non-invasive observation of internal organs. Such transparent fishes are known as see-through (Wakamatsu et al. 2001) in medaka and *casper* (White et al. 2008) in zebrafish.

See-through medaka was established by combining recessive mutations at four loci: *i-3* (albino mutation, *pinked-eye dilution/OCA2*) to remove melaninization in melanophores and to reduce yellow pigmentation in xanthophores (Fukamachi et al. 2004); *guanineless/pnp4a* to remove light-reflecting iridophores in the peritoneum and the iris (Kimura et al. 2017); *leucophore free/slc2a15b* to remove white pigmentation in leucophores and to reduce their interference in the transmission light through the body (Kimura et al. 2014); *il-1* (unidentified locus) to reduce the degree of reflectance of the skin and the operculum.

The zebrafish line *casper* was generated by crossing two recessive mutations *nacre/mitfa* (see Sect. 7.5.3) and *roy/mpv17*: the *nacre* locus removes melanophores and reduces xanthophores, while *roy* contributes to ablation of reflective iridophores.

Both medaka and zebrafish transparent adult fish are strikingly transparent, as most of the internal organs are visible from the outside of the body in an intact state. Since those are genetically stable, they are amendable to genetic manipulation, e.g. generating GFP transgenic lines and serve as an ideal model system for in vivo imaging of the dynamics of a single cell. Note that, in medaka, strong

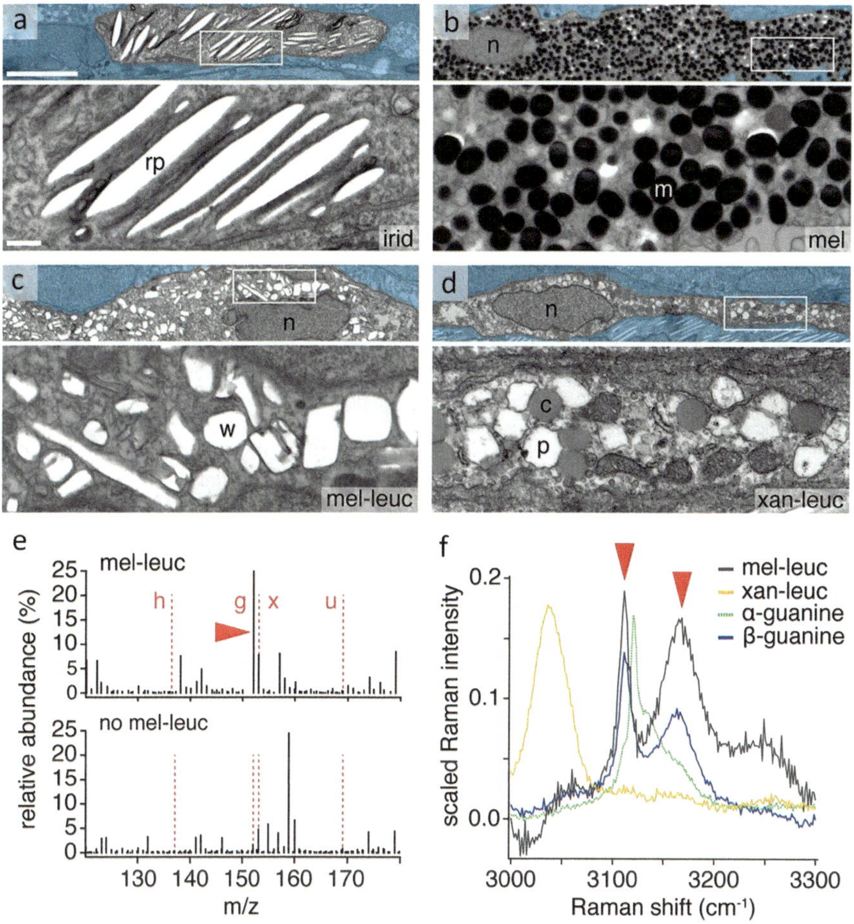

Fig. 4.18 Ultrastructural and chemical characteristics of melanoleucophores and xantholeucophores. (**a–d**) Fin iridophores (**a**) exhibit reflecting platelets (rp), while melanophores had typical melanosomes (m). Melanoleucophores (**c**) contained irregularly shaped and arranged organelles (w), presumptively containing white material, whereas xantholeucophores (**d**) presumptively contained pterinosomes (p) and carotenoid vesicles (c), without other organelles likely to harbor white pigment. (Top) Low magnification with adjacent cells masked. (Bottom) Boxed regions. n, nucleus. (**e**) Mass spectrometry of fin tissue containing melanoleucophores (Upper) revealed more abundant guanine (arrowhead) compared with fin tissue without melanoleucophores (Bottom) (*h* hypoxanthine, *g* guanine, *x* xanthine, *u* uric acid). (**f**) Representative Raman spectra of melanoleucophores (gray; n = 19 cells) and xantholeucophores (orange; n = 14) compared with alpha-guanine and beta-guanine. High-energy peak pattern (arrowheads) indicates that β-guanine is present in melanoleucophores but is not detectable in xantholeucophores. Scale bar: 5 μm (top), 500 nm (bottom) (from Lewis et al. 2019)

autofluorescence of leucophore interrupts the imaging of fluorescent protein-labeled cells, e.g. neurons in the central nervous system under the leucophores, but can be avoided by taking advantages of the pigmentation mutants, e.g. see-through and *leucophore free*.

References

Beirl AJ, Linbo TH, Cobb MJ, Cooper CD (2014) oca2 regulation of chromatophore differentiation and number is cell type specific in zebrafish. Pigment Cell Melanoma Res 27(2):178–189

Bowman SL, Bi-Karchin J, Le L, Marks MS (2019) The road to lysosome-related organelles: insights from Hermansky-Pudlak syndrome and other rare diseases. Traffic 20(6):404–435

Braasch I, Schartl M, Volff JN (2007) Evolution of pigment synthesis pathways by gene and genome duplication in fish. BMC Evol Biol 7:74

Burgoyne T, O'Connor MN, Seabra MC, Cutler DF, Futter CE (2015) Regulation of melanosome number, shape and movement in the zebrafish retinal pigment epithelium by OA1 and PMEL. J Cell Sci 128(7):1400–1407

Chen T, Song G, Yang H, Mao L, Cui Z, Huang K (2018) Development of the Swimbladder surfactant system and biogenesis of lysosome-related organelles is regulated by BLOS1 in zebrafish. Genetics 208(3):1131–1146

Cropp CD, Komori T, Shima JE, Urban TJ, Yee SW, More SS, Giacomini KM (2008) Organic anion transporter 2 (SLC22A7) is a facilitative transporter of cGMP. Mol Pharmacol 73 (4):1151–1158

Curran K, Lister JA, Kunkel GR, Prendergast A, Parichy DM, Raible DW (2010) Interplay between Foxd3 and Mitf regulates cell fate plasticity in the zebrafish neural crest. Dev Biol 344 (1):107–118

D'Agati G, Beltre R, Sessa A, Burger A, Zhou Y, Mosimann C, White RM (2017) A defect in the mitochondrial protein Mpv17 underlies the transparent Casper zebrafish. Dev Biol 430 (1):11–17

Dooley CM, Schwarz H, Mueller KP, Mongera A, Konantz M, Neuhauss SC (2013) Nüsslein-2Volhard C, Geisler R. Slc45a2 and V-ATPase are regulators of melanosomal pH homeostasis in zebrafish, providing a mechanism for human pigment evolution and disease. Pigment Cell Melanoma Res 26(2):205–217

Fukamachi S, Asakawa S, Wakamatsu Y, Shimizu N, Mitani H, Shima A (2004) Conserved function of medaka pink-eyed dilution in melanin synthesis and its divergent transcriptional regulation in gonads among vertebrates. Genetics 168(3):1519–1527

Fukamachi S, Shimada A, Shima A (2001) Mutations in the gene encoding B, a novel transporter protein, reduce melanin content in medaka. Nat Genet 28(4):381–385

Granneman JG, Kimler VA, Zhang H, Ye X, Luo X, Postlethwait JH, Thummel R (2017) Lipid droplet biology and evolution illuminated by the characterization of a novel perilipin in teleost fish. Elife 28(6):e21771

Haffter P, Odenthal J, Mullins MC, Lin S, Farrell MJ, Vogelsang E, Haas F, Brand M, van Eeden FJ, Furutani-Seiki M, Granato M, Hammerschmidt M, Heisenberg CP, Jiang YJ, Kane DA, Kelsh RN, Hopkins N, Nüsslein-Volhard C (1996) Mutations affecting pigmentation and shape of the adult zebrafish. Dev Genes Evol 206(4):260–276

Hama T (1975) Chromatophores and iridocytes. In: Yamamoto T (ed) Medaka (killifish) biology and strains. Keigaku Publishing, Tokyo, Japan, pp 138–153

Kelsh RN, Brand M, Jiang YJ, Heisenberg CP, Lin S, Haffter P, Odenthal J, Mullins MC, van Eeden FJ, Furutani-Seiki M, Granato M, Hammerschmidt M, Kane DA, Warga RM, Beuchle D, Vogelsang L, Nüsslein-Volhard C (1996) Zebrafish pigmentation mutations and the processes of neural crest development. Development 123:369–389

Kelsh RN, Inoue C, Momoi A, Kondoh H, Furutani-Seiki M, Ozato K, Wakamatsu Y (2004) The Tomita collection of medaka pigmentation mutants as a resource for understanding neural crest cell development. Mech Dev 121(7–8):841–859

Kimura T, Nagao Y, Hashimoto H, Yamamoto-Shiraishi Y, Yamamoto S, Yabe T, Takada S, Kinoshita M, Kuroiwa A, Naruse K (2014) Leucophores are similar to xanthophores in their specification and differentiation processes in medaka. Proc Natl Acad Sci U S A 111 (20):7343–7348

Kimura T, Takehana Y, Naruse K (2017) pnp4a Is the Causal Gene of the Medaka Iridophore Mutant guaninelessr. G3 (Bethesda) 7(4):1357–1363

Krauss J, Astrinidis P, Astrinides P, Frohnhöfer HG, Walderich B, Nüsslein-Volhard C (2013) Transparent, a gene affecting stripe formation in zebrafish, encodes the mitochondrial protein Mpv17 that is required for iridophore survival. Biol Open 2(7):703–710

Krauss J, Geiger-Rudolph S, Koch I, Nüsslein-Volhard C, Irion U (2014) A dominant mutation in tyrp1A leads to melanophore death in zebrafish. Pigment Cell Melanoma Res 27(5):827–830

Lamason RL, Mohideen MA, Mest JR, Wong AC, Norton HL, Aros MC, Jurynec MJ, Mao X, Humphreville VR, Humbert JE, Sinha S, Moore JL, Jagadeeswaran P, Zhao W, Ning G, Makalowska I, McKeigue PM, O'donnell D, Kittles R, Parra EJ, Mangini NJ, Grunwald DJ, Shriver MD, Canfield VA, Cheng KC (2005) SLC24A5, a putative cation exchanger, affects pigmentation in zebrafish and humans. Science 310(5755):1782–1786

Le Guyader S, Maier J, Jesuthasan S (2005) Esrom, an ortholog of PAM (protein associated with c-myc), regulates pteridine synthesis in the zebrafish. Dev Biol 277(2):378–386

Lewis VM, Saunders LM, Larson TA, Bain EJ, Sturiale SL, Gur D, Chowdhury S, Flynn JD, Allen MC, Deheyn DD, Lee JC, Simon JA, Lippincott-Schwartz J, Raible DW, Parichy DM (2019) Fate plasticity and reprogramming in genetically distinct populations of Danio leucophores. Proc Natl Acad Sci U S A 116(24):11806–11811

Li Y, Li G, Görling B, Luy B, Du J, Yan J (2015) Integrative analysis of circadian transcriptome and metabolic network reveals the role of de novo purine synthesis in circadian control of cell cycle. PLoS Comput Biol 11(2):e1004086

Lister JA (2019) Larval but not adult xanthophore pigmentation in zebrafish requires GTP cyclohydrolase 2 (gch2) function. Pigment Cell Melanoma Res 32(5):724–727

Nagao Y, Suzuki T, Shimizu A, Kimura T, Seki R, Adachi T, Inoue C, Omae Y, Kamei Y, Hara I, Taniguchi Y, Naruse K, Wakamatsu Y, Kelsh RN, Hibi M, Hashimoto H (2014) Sox 5 functions as a fate switch in medaka pigment cell development. PLoS Genet 10(4):e1004246

Nakayama T, Nakajima K, Cox A, Fisher M, Howell M, Fish MB, Yaoita Y, Grainger RM (2017) No privacy, a Xenopus tropicalis mutant, is a model of human Hermansky-Pudlak syndrome and allows visualization of internal organogenesis during tadpole development. Dev Biol 426 (2):472–486

Ng A, Uribe RA, Yieh L, Nuckels R, Gross JM (2009) Zebrafish mutations in gart and paics identify crucial roles for de novo purine synthesis in vertebrate pigmentation and ocular development. Development 136(15):2601–2611

Odenthal J, Rossnagel K, Haffter P, Kelsh RN, Vogelsang E, Brand M, van Eeden FJ, Furutani-Seiki M, Granato M, Hammerschmidt M, Heisenberg CP, Jiang YJ, Kane DA, Mullins MC, Nüsslein-Volhard C (1996) Mutations affecting xanthophore pigmentation in the zebrafish. Danio rerio Dev 123:391–398

Saunders LM, Mishra AK, Aman AJ, Lewis VM, Toomey MB, Packer JS, Qiu X, McFaline-Figueroa JL, Corbo JC, Trapnell C, Parichy DM (2019) Thyroid hormone regulates distinct paths to maturation in pigment cell lineages. Elife 29(8):e45181

Schartl M, Larue L, Goda M, Bosenberg MW, Hashimoto H, Kelsh RN (2016) What is a vertebrate pigment cell? Pigment Cell Melanoma Res 29(1):8–14

Toews DPL, Hofmeister NR, Taylor SA (2017) The evolution and genetics of carotenoid processing in animals. Trends Genet 33(3):171–182

Våge DI, Boman IA (2010) A nonsense mutation in the beta-carotene oxygenase 2 (BCO2) gene is tightly associated with accumulation of carotenoids in adipose tissue in sheep (Ovis aries). BMC Genet 11:10

Wakamatsu Y, Pristyazhnyuk S, Kinoshita M, Tanaka M, Ozato K (2001) The see-through medaka: a fish model that is transparent throughout life. Proc Natl Acad Sci U S A 98:10046–10050

White RM, Sessa A, Burke C, Bowman T, Leblanc J, Ceol C, Bourque C, Dovey M, Goessling W, Burns CE, Zon LI (2008) Transparent adult zebrafish as a tool for in vivo transplantation analysis. Cell Stem Cell 2:183–189

Willer GB, Lee VM, Gregg RG, Link BA (2005) Analysis of the zebrafish perplexed mutation reveals tissue-specific roles for de novo pyrimidine synthesis during development. Genetics 170:1827–1837

Ziegler I (2003) The pteridine pathway in zebrafish: regulation and specification during the determination of neural crest cell-fate. Pigment Cell Res 16(3):172–182

Chapter 5
Bioluminescence and Pigments

José Paitio and Yuichi Oba

Abstract Bioluminescence is present in organisms across the tree of life and inhabiting diverse environments on the planet. Light emissions are used for specific communication purposes such as camouflage, attracting preys, and mating. The light produced in a chemical reaction is often modulated by pigmented tissues in the light organs. Although the chemical compounds vary, pigments can be categorized according to functions of light shields, control of light intensity, and spectral modification. In this chapter, we discuss the diversity of pigments present in different tissues of the light organs and their role in the alteration of the light emitted from bioluminescent organisms.

Keywords Bioluminescence · Pigment · Communication · Photophore · Reflection · Fluorescence

5.1 Introduction

Diurnal animals manipulate sunlight and utilize pigments to display colorful patterns in visual communication, e.g., butterflies and tropical fishes. Animals that live in dark environments have to produce their own light to present such displays. Bioluminescence is the production of light by organisms through an oxidative chemical reaction, which generally consists of three players: a substrate, "luciferin"; an enzyme, "luciferase"; and a cofactor, oxygen (Shimomura 2006). Chemical structures of the luciferins have been determined from various marine and terrestrial luminous organisms (Fig. 5.1); however, the structures of several luciferins remain a mystery in some invertebrates, teleosts, and sharks.

Photocytes are cells responsible for light production (Morin 1983; Thompson and Rees 1995; Haddock et al. 2010). Simple bioluminescent structures consist of solely

J. Paitio · Y. Oba (✉)
Department of Environmental Biology, Chubu University, Kasugai, Japan
e-mail: yoba@isc.chubu.ac.jp

H. Hashimoto et al. (eds.), *Pigments, Pigment Cells and Pigment Patterns*,
https://doi.org/10.1007/978-981-16-1490-3_5

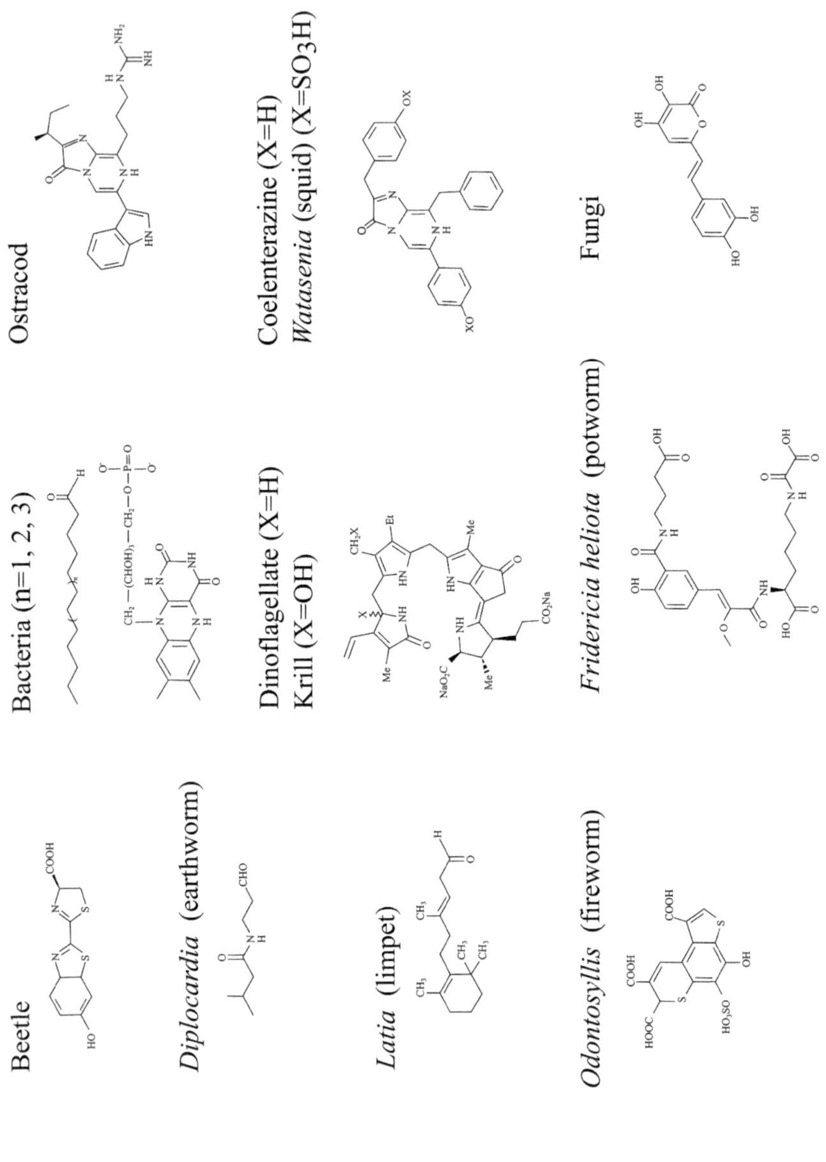

Fig. 5.1 Examples of luciferins from diverse taxa of terrestrial and marine animals and their respective molecular structures

these cells, such as in ctenophores (Harvey 1952), cnidarians (Harvey 1952; Morin 1974; Morin and Reynolds 1974), scale worms (Harvey 1952; Bassot 1966b), springtails (Sano et al. 2019), amphipods (Herring 1981a), and brittle stars (Deheyn et al. 2000); however, other animals utilize complex organs (Morin 1983; Kotlobay et al. 2019; Tsarkova et al. 2016). The structures of these organs are greatly diverse among various taxa; however, they exhibit certain common characteristics: photocytes in a matrix, which is covered by an inner reflector and a pigment layer, and an outer lens (Clarke 1963; Denton et al. 1985). The reflector ensures minimal loss of the light produced via redirecting photons to outside of the organ, while the pigmented layer prevents the light from penetrating into the tissues surrounding the organ. Some species possess colored reflectors, filters, or lenses to adjust the spectra of light emission. Because the terminology used for lenses and filters varies among authors (Haneda 1949, 1951, 1966; Bassot 1966a; Lawry 1973; Denton et al. 1985; Cavallaro et al. 2004), we adopted the following definitions for lenses and filters (Haneda 1949, 1951, 1952): Filters are internal tissues with light-absorbing pigments for spectral selection, while lenses are outer tissues that perform the primary dioptric function (Denton et al. 1970; Lawry 1973), i.e., refract light on a ventral angle. Bioluminescent organs are called light organs or photophores (Morin 1983; Thompson and Rees 1995; Haddock et al. 2010). Light organs are symbiotic and open to outside, while photophores, including photocytes, are closed organs, except in blackchin fishes of the genus *Neoscopelus* for unknown reasons (Herring and Morin 1978; Karplus 2014; Paitio et al. 2016).

Most luminous organisms produce their own light—intrinsic bioluminescence; however, some squids and fishes maintain cultures of glowing bacteria in their light organs—symbiotic bioluminescence (Morin 1983; Karplus 2014). These symbionts can be either facultative or obligatory; however, the biotic relationship provides them nutrition and growing conditions in exchange for light emission. Rare exceptions are certain anglerfishes of the *Centrophryne* and *Linophryne* genera that, in addition to the symbiotic escal light, possess intrinsic luminous barbels (Karplus 2014).

Luminous organisms are spread throughout the web of life from bacterial unicellular life forms to vertebrates (Fig. 5.2). The apparently random dispersion of light-emitting taxa has been debated for decades. It is difficult to determine the number of times that bioluminescence has evolved independently since its rise likely 400 million years ago; however, estimations indicate that this number is most likely more than 50 times (Haddock et al. 2010). This estimation is even more challenging in the case of hosts of symbiotic bacteria, such as squids or fishes, considering the evolution of the hosts and the symbionts.

Our planet is inhabited by light-emitting organisms either living in terrestrial or aquatic habitats. Several luminescent species live on land in the most diverse habitats, such as the flying fireflies, fungus, snails, and millipedes on the vegetation and soil and the earthworms below the soil. In freshwater, apart from the firefly larval forms and certain brackish water bacteria, the only bioluminescent species known so far is the limpet *Latia* (Harvey 1956). In contrast to freshwater, seawater exhibits the highest number of bioluminescent species, which inhabit the seas from shallow

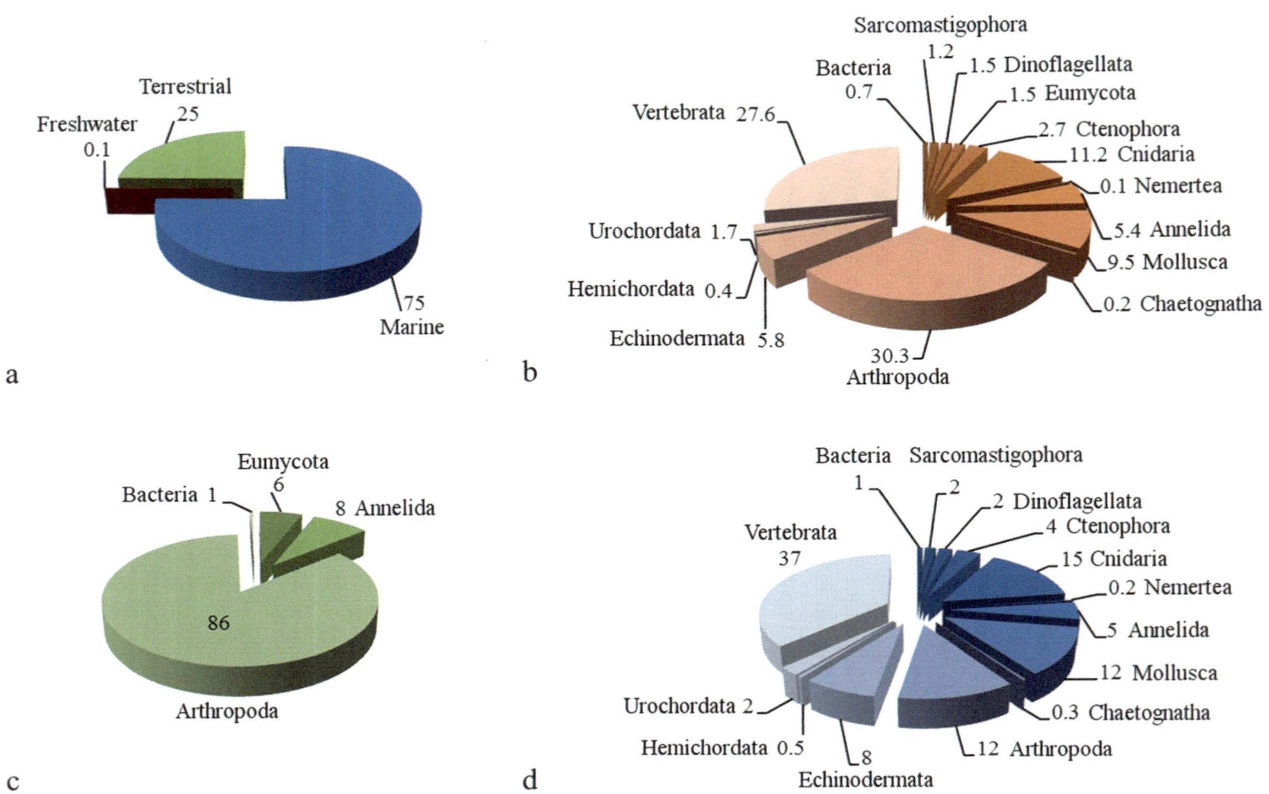

Fig. 5.2 Biodiversity of bioluminescent genera in diverse habitats (%). (**a**) according to habitat; (**b**) globally; (**c**) among terrestrial organisms, per phyla; (**d**) and among marine organisms, per phyla. Adapted from Oba (2019), Haddock et al. (2010), Oba and Schultz (2014), and Paitio et al. (2016)

coasts to deep-sea floors. Approximately 80% of bioluminescent organisms are estimated to be marine species (Shimomura 2006; Widder 2010), living in the upper 1000 m depth (Widder 2001). Some researchers (Young 1983) claim that the mesopelagic zone (200–1000 m depths) is the primary domain of bioluminescence in the planet, in terms of species diversity and abundance. Additionally, the most complex bioluminescence systems exist at these depths (Young 1983). The high richness of bioluminescent creatures is derived from the light parameters of this environment (Young 1983; Widder 2001). This region, which is also known as twilight zone, contains no visual obstacles, and only low-intensity light penetrates it, allowing luminous signals to be easily seen at large distances (Young 1983; Widder 2001). Most light is absorbed by the shallow waters, and only a narrow wavelength of blue-green light reaches the mesopelagic zone (Warrant and Locket 2004; Johnsen 2014) in a highly directional manner at a vertical angle (Widder 2001; Warrant and Locket 2004; Johnsen 2014), which explains the development of bioluminescent camouflage in several species in this zone (Widder 2001; Johnsen 2014).

Bioluminescence is primarily an ecological "tool" for interspecific communication. The primordial purpose of bioluminescence is believed to be defense against predators (Morin 1983). The ecological roles of light emission are highly diverse, as well as multifunctional in some animals that exhibit different light signals (Haddock et al. 2010) and/or different light organs or photophores (Young 1983). The ecological functions of bioluminescence can be classified as interspecific functions, for attracting prey and protecting against predators; intraspecific functions, for reproduction or recognition; and functions for illumination of the surroundings (Table 5.1).

5.2 Pigments in Bioluminescent Tissues

Light organs and photophores are similar with respect to their function, irrespective of their organic differences, as indicated by the similarity in the basic structure of the light organ in animals from different phyla. Convergent evolution of tissues has clearly provided a vast variety of light organs and photophores that work in astonishingly similar manners. This is only possible through the organization of tissues with the same functions, despite the difference in the organic composition of these tissues. In contrast, luminous structures are dependent on their functions and the complexity of the organism (Fig. 5.3). Single-cell life forms such as bacteria and dinoflagellates possess specific organelles as light-emitting structures (Morin 1983). Animals, which represent most luminous organisms, have developed photogenic tissues in complex light organs and photophores (Morin 1983). This complexity is characterized by organs containing differentiated tissues for specific functions. Some

Table 5.1 Ecological functions of bioluminescence among diverse luminous organisms (Morin 1983; Haddock et al. 2010; Dunlap and Urbanczyk 2013; Oba and Schultz 2014; Paitio et al. 2016)

Ecological role			Taxonomic group
Interspecific	Defense	Aposematism	Fungi, Cnidaria, Arthropoda, Annelida, Echinodermata,[a] Chordata
		Illumination	Chordata
		Camouflage	Arthropoda, Mollusca, Chordata
		Startle	Dinoflagellata, Arthropoda, Annelida, Mollusca, Chordata
		Smokescreen	Ctenophora, Cnidaria, Arthropoda, Annelida, Mollusca, Chaetognatha, Chordata
		Distraction	Cnidaria, Annelida, Mollusca, Echinodermata
		Burglar alarm	Dinoflagellata, Cnidaria
		Decoy	Cnidaria, Arthropoda, Annelida, Mollusca, Echinodermata
	Offense	Attraction (hosts and feeders)	Proteobacteria
		Attraction (prey)	Fungi,[a] Cnidaria, Arthropoda, Mollusca, Chordata
		Stun	Mollusca, Chordata
		Illumination (prey)	Chordata
Intraspecific		Recognition	Arthropoda, Annelida, Mollusca, Chordata
		Mating	Arthropoda, Mollusca, Chordata
		Schooling	Chordata
Illumination of surroundings			Chordata

[a]Uncertain

tissues are responsible for the modulation of the primordial light produced in the photocytes, namely the pigmented layer, reflector,[1] and filter.

For this purpose, several chemical compounds are used (Table 5.2), such as melanin for light shielding (Nicol 1957, 1964; Karplus 2014), guanine for reflection (Bassot 1966a; Best and Bone 1976; Karplus 2014), and carotenoids for spectral absorption (Herring 1972; Herring and Locket 1978; Denton et al. 1985), beyond taxonomical groups (Fig. 5.4). Pigments limited by function and animal taxon are observed as well, e.g., uric acid in the reflectors of fireflies (Goh et al. 2013), porphyrins in the filters of fishes (Denton et al. 1985), and lumazine in a blue-luminescent strain of bacteria (Koka and Lee 1979; Visser and Lee 1980; Vervoort et al. 1982).

[1]According to Bagnara (1966), Bagnara and Hadley (1974), and Kelsh (2004), pigmentation in animals is achieved by all chromatophore types, including melanophores and iridophores. Additionally, the same authors state that iridophores are pigment reflective cells. On the basis of these descriptions, we include reflectors with iridophores as pigmented tissues.

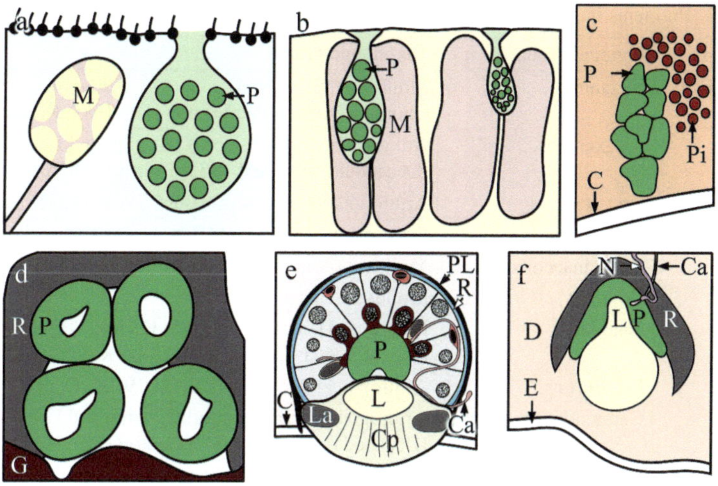

Fig. 5.3 Schematic illustrations of the diversity of bioluminescent structures. (**a**) Cnidaria *Pelagia*. (**b**) Tubeworm *Chaetopterus*. (**c**) Amphipod *Parapronoe*. (**d**) Larvae of fungus gnat *Arachnocampa*. (**e**) General Euphausiidae. (**f**) Midshipman fish *Porichthys*. *M* mucous cell, *P* photocyte, *Pi* pigment cells, *C* cuticle, *R* reflector, *G* gut, *PL* pigment layer, *L* lens, *La* laminar ring; *Cp* cap, *Ca* capilar, *N* nerve, *D* dermis, *E* epidermis. Adapted from Dahlgren (1916), Harvey (1952), Nicol (1957), Herring and Locket (1978), Herring (1981a), and Rigby and Merritt (2011)

5.2.1 Light Shields

Bioluminescence is undoubtedly an ecological advantage; however, without the right structure, light organs and photophores can become a disadvantage for the user. A light signal has the potential to attract preys and predators as well; thereby, if an organism glows constantly it is putting itself in danger. Thus, it is not surprising that the light organs and photophores of amphipods, mysids, euphausiids, decapods, cephalopods, and certain fishes are internally covered by a pigmented layer that acts as a "light wall" (Haneda 1952; Nicol 1964; Herring and Locket 1978; Herring and Morin 1978; Herring 1981c). This layer is primarily composed of chromatophores that absorb the produced light that is not emitted outside the organ; thus, it prevents the light from reaching the inner body tissues and exposing the animal. The amphipods *Parapronoe*, *Megalanceola*, *Chevreuxiella*, and *Dannaela* have been reported to possess photocytes covered with brownish pigment (Herring 1981a). Secretory bioluminescence of the mysid *Gnathophausia* is attributed to pigmented photophores on the maxilla (Herring 1985; Meland and Aas 2013). The pigmentation coating the light organs and photophores is usually black or dark-brown and composed of melanophores, such as in fishes (Fig. 5.5) (Nicol 1957; Denton et al. 1985; Karplus 2014) and squids (Nicol 1964). The outer pigment layer in the photophores of certain squids can be reddish-brown, such as in *Pterygioteuthis* (Arnold et al. 1974), *Pyroteuthis* (Butcher et al. 1982), and *Histioteuthis* (Dilly and Herring 1981). Euphausiids possess photophores that are covered with

Table 5.2 Pigments in light organs, classified according to function, chemistry, and organisms

Functions	Pigment group	Taxonomic group
Light shield	Melanin	Mollusca (Nicol 1964) Chordata (Nicol 1957; Denton et al. 1985; Karplus 2014)
	Astaxanthin	Arthropoda (Euphausiacea) (Herring and Locket 1978), (Decapoda)[a] (Dennell 1955; Herring 1981b)
Reflection	Uric acid	Arthropoda (Coleoptera) (Buck 1948; Goh et al. 2013)
	Lipid[a]	Arthropoda (Decapoda) (Herring 1981b)
	"Proteinaceous"	Mollusca (Arnold et al. 1974; Herring 2000)
	Collagen	Mollusca (Herring 2000)
	Guanine	Mollusca (Denton et al. 1985; Herring 2000), Chordata (Bassot 1966a; Denton et al. 1985; Karplus 2014)
Spectra	Lipochrome[a]	Arthropoda (Decapoda) (Kemp 1910b, Dennell 1955)
	Carotenoprotein[a]	Arthropoda (Euphausiacea) (Herring and Locket 1978), (Decapoda) (Herring 1972; Denton et al. 1985)
	Protoporphyrin[a]	Mollusca (Dilly and Herring 1981; Denton et al. 1985)
	Porphyrin[a]	Chordata (Stomiidae; *Malacosteus*, *Chaulodius*, *Stomias*) (Denton et al. 1985)
	Cytochrome *c*[a]	Chordata (Sternoptychidae, *Argyropelecus*) (Denton et al. 1985)
	Dicarboxylic porphyrin	Chordata (Sternoptychidae, *Valenciennellus*) (Denton et al. 1985)
Fluorescence	FMN	Proteobacteria (Daubner et al. 1987; Macheroux et al. 1987)
	6,7-Dimethyl-8-(1-D-ribityl)lumazine	Proteobacteria (Koka and Lee 1979; Visser and Lee 1980; Vervoort et al. 1982)
	Porphyrin[a]	Cnidaria (Haddock et al. 2005)
	7,8-Dihydropterin-6-carboxylic acid	Arthropoda (Kuse et al. 2001)
	Riboflavin	Annelida (*Odontosyllis*) (Deheyn et al. 2013; Branchini et al. 2014; Rawat and Deheyn 2016)
	Phycobiliprotein[a]	Chordata (Stomiidae) (Campbell and Herring 1987)

[a]Uncertain

chromatophores containing astaxanthin (Fig. 5.6a, b) (Herring and Locket 1978). As this pigment is present in many decapods (Herring 1972), the red carotenoid pigmented outer layer observed on the photophores (Fig. 5.6d, e, g) (Nowel et al. 1998) is possibly astaxanthin as well (Dennell 1955; Herring 1981b). In a previous report, although it is not stated by the author, the description of the photophores of the fungus gnat *Orfelia* as "black bodies" might indicate cover pigmentation (Fulton 1941). In cnidarians, neither chromatophores nor reflectors have been reported to be associated with photocytes (Harvey 1952; Morin 1974; Morin and Reynolds 1974).

Intrinsic bioluminescence is under neural or hormonal control, and it cannot be achieved in symbiotic luminous animals because of the impossibility of directly regulating the constant light produced by the bacterial symbionts (Nicol 1959). To control the intensity of the light emitted by the bacteria cultures, hosts have

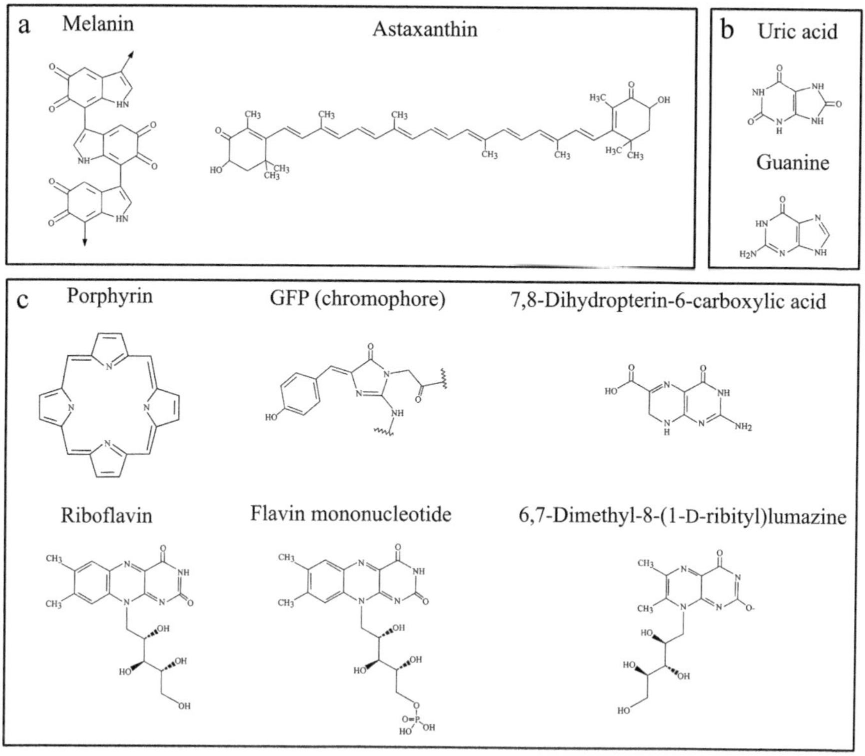

Fig. 5.4 Examples of pigments present on bioluminescent organs, with their respective molecular structures, by function: (**a**) light shields; (**b**) reflection; (**c**) colored filters and fluorescence. Arrows indicate the position where the polymer continues. Adapted from Koka and Lee (1979), Delgado-Vargas et al. (2000), Kuse et al. (2001), Lubczak et al. (2002), Nuevo et al. (2009), Schweitzer-Stenner (2014), and Shinoda et al. (2018)

developed accessory structures for light organs—pigmented shutters. These structures are similar to Venetian blinds, which are used for controlling the sunlight illuminating living rooms; however, they act on outgoing light instead of incoming light. Similar to blinds and curtains, the shutters for light organs occur as diverse structures to achieve the same purpose.

Squids can control emissions from symbiotic light organs via tissues densely pigmented with melanophores, shutters, and ink sacs. In certain species, colored chromatophores on the skin absorb visible light ranging from purple to green and may act as screens for spectral selection to the underlying photophores (Nicol 1959). Chromatophores present on the organs can act as movable screens in *Spirula*, *Vampyroteuthis* (Schmidt 1922; Pickford 1949), and *Histioteuthis* (Dilly and Herring 1981). Additionally, researchers have observed that light output is controlled via contraction and expansion of the brown chromatophores covering the photophores in *Watasenia* (Fig. 5.7) (Berry 1920) and that of purple red chromatophores in *Bathothauma* (Dilly and Herring 1974). In *Uroteuthis*, the rectum is flanked by

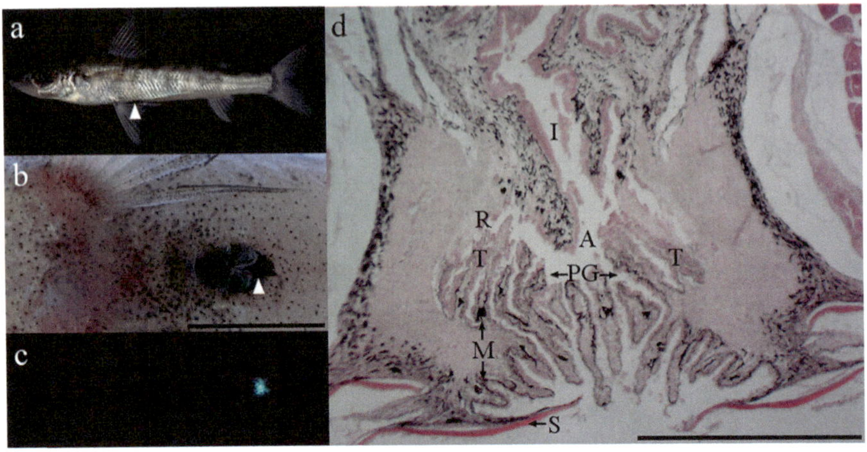

Fig. 5.5 Melanophores on the circumanal light organ of the green-eye fish *Chlorophthalmus*. (**a**) Lateral view of the full body of the fish. (**b**) Ventral view of the light organ under white light. (**c**) Ventral view of bioluminescence from the light organ in the dark. Arrowheads indicate the position of the light organ. (**d**) Histological section of the light organ. *I* intestine, *R* reflector, *A* anus, *PG* perianal groove, *T* tubules with luminous bacteria, *M* melanophore layers, *S* scale. Scale bars: (**b**) 1 cm; (**d**) 500 μm. Photography: whole fish, courtesy of Hiromitsu Endo; the light organ and bioluminescence, by Yuichi Oba; histological section, by José Paitio

Fig. 5.6 Red chromatophores on photophores of the euphausiid *Euphausia* and decapod *Lucensosergia*. (**a**) Lateral and (**b**) ventral views of the full body of *Euphausia* under white light and (**c**) ventral view of bioluminescent light emissions in the dark. (**d**) Lateral and (**e**) ventral views of the full body of *Lucensosergia* under white light, (**f**) ventral view of bioluminescent light emissions in the dark, and (**g**) close view of red-pigmented chromatophore layers on the skin and around the photophore. *PL* chromatophores in photophores pigment layer. *Sk* Chromatophores in skin. White arrowheads indicate the position of photophores in *Euphausia*. Black arrowheads indicate the position of the photophore of (**g**) on *Lucensosergia*. Scale bars: (**a–f**) 1 cm; (**g**) 500 μm. Photography by Yuichi Oba and José Paitio

two light organs, which are covered by a black membrane that acts as a diaphragm and regulates the light emission enabling the light to fade away (Haneda 1963). Pigmented muscular flaps can rapidly block the photophores of *Leachia* (Young 1975). The bobtail squid *Euprymna* is a special case; it controls light emissions through movements of the ink sac and reflector (McFall-Ngai and Montgomery 1990).

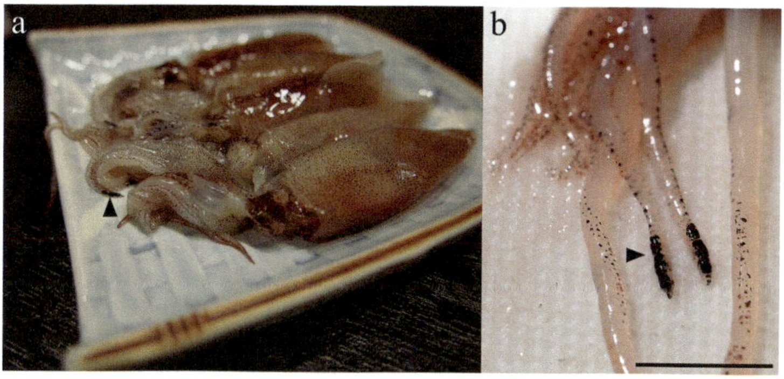

Fig. 5.7 Arm-tip photophores on the squid *Watasenia*. (**a**) Full body showing darkened area on arm-tip. (**b**) High concentration of chromatophores on arm-tip photophores. Arrowheads indicate the photophore position. Scale bar = 1 cm. Photography by Yuichi Oba

Although several squids possess photophores positioned near the ink sac, till date, control over the light emission by this organ has been proven only in the bobtail squid.

Symbiotic luminous fishes with internal luminescence share the feature of melanophores in the skin surrounding light organs, which leads to the theory that these fishes regulate light output through aggregation and dispersion of melanocytes on the outer tissue layer. Chromatophores on the skin of non-luminous fishes are under neural, paracrine, and hormonal control (Fujii 2000; Sköld et al. 2016). A similar mechanism is assumed to be utilized in luminous organs even though all existing reports are based on observations and suppositions. Regulation of light emission via chromatophores, on the skin or internally surrounding the light organs or photophores, have been proposed for the families Acropomatidae (Fig. 5.8), Pempheridae, Trachichthyidae, Macrouridae (Fig. 5.9), Apogonidae, Leiognathidae, Monocentridae, Chlorophthalmidae, Evermannellidae, Merluciidae, and Moridae, (Haneda 1949, 1951; Haneda and Johnson 1962; Cohen 1964; Herring 1977; Herring and Morin 1978; Tebo et al. 1979; Somiya 1981; McFall-Ngai and Dunlap 1983). Chromatophore patterns around the light organ in *Lumiconger* (Congridae) is similar to that in Gadiformes (Castle and Paxton 1984); although the function is not clearly stated, it should achieve a similar purpose. In Opisthoproctidae species, the chromatophores are located on the ventral–lateral body scales in a species-specific pattern, and they may be additionally used for intraspecific recognition (Bertelsen 1958; Poulsen et al. 2016).

Accessory structures are common on light organs, and shutters in symbiotic bioluminescent fishes are no exception. In contrast to the chromatophores in the skin that are adapted to a specific function, shutters must have developed because of the necessity to control bioluminescent emissions. Evolution facilitated the development of these structures, allowing a faster and precise regulation of light emissions and possibly introducing time-lapse signals for communication. Thus, shutters opened the doors to the eco-ethological roles of luminescence in fishes. Besides camouflage, owing to the versatility of light signals using shutters, new roles and multipurpose luminescence may have been introduced at that time for intra- and interspecific communication. It is difficult not to provide such a hypothesis,

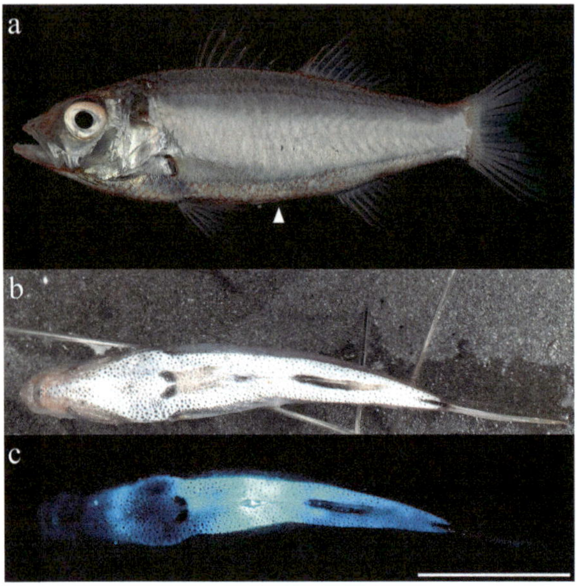

Fig. 5.8 Internal bioluminescence in *Acropoma*. (**a**) Lateral view of fish, chromatophores visible on ventral skin (arrowhead). (**b**) Ventral view exhibiting chromatophore pattern on all luminous areas, under white light. (**c**) Ventral view of bioluminescent emission in the dark. Scale bar = 1 cm. Photography: lateral view, courtesy of Hiromitsu Endo; others, by Yuichi Oba

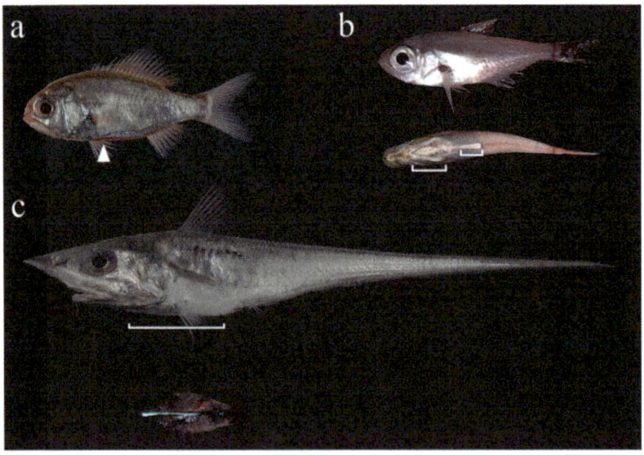

Fig. 5.9 Light organs in roughy (Trachichthyidae), sweeper (Pempheridae), and grenadier (Macrouridae) fishes. (**a**) Lateral view of the roughy *Aulotrachichthys*. (**b**) Lateral view of the sweeper *Parapriacanthus* under white light (upper) and bioluminescence in the dark (lower). (**c**) Lateral view of the grenadier *Coelorinchus* under white light (upper) and bioluminescence in the dark (lower). White arrowheads or bracket lines indicate the position and extension of the bioluminescent organs. Photography: *Aulotrachichthys*, courtesy of Hiromitsu Endo; *Parapriacanthus*, courtesy of Manabu Bessho-Uehara; *Coelorinchus*, by Yuichi Oba

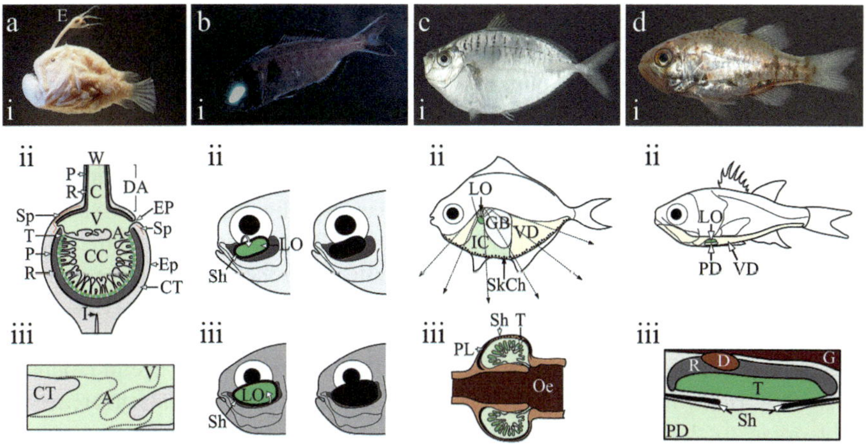

Fig. 5.10 Shutters on symbiotic bioluminescent organs of fishes. (**a**) The anglerfish *Himantolophus* (i), diagrams of generic structure of esca (ii), and shutter mechanism (iii). (**b**) The flashlight fish *Anomalops* (i), scheme of shutter mechanisms in *Anomalops* (ii), and *Photoblepharon* (iii). (**c**) The ponyfish *Secutor* (i), illustrations of the bioluminescence system (ii), and light organ (iii) of *Gazza*. (**d**) The cardinalfish *Siphamia* (i), schematics of the light emission system (ii), and light organ (iii). *E* esca, *P* pigment layer, *R* reflector, *Sp* sphincter, *T* tubules with luminous bacteria, *W* window, *C* light guide core, *V* vestibule, *A* central escal aperture, *CC* central escal cavity, *I* illicium, *DA* distal appendage, *EP* escal pore, *Ep* epiderm, *LO* light organ, *Sh* shutter, *GB* gas bladder, *VD* ventral diffuser (transparent muscle), *SkCh* skin chromatophores, *Oe* esophagus, *PD* primary diffuser, *D* duct (link light organ to gut), *G* gut. Dashed line on Aiii represents aperture under contraction by the sphincter. Dashed arrows on Cii indicates the direction of light emission. Photography: *Himantolophus*, by Yuichi Oba; *Anomalops*, courtesy of Takehito Miyatake; *Secutor* and *Siphamia*, courtesy of Hiromitsu Endo. Schemes adapted from Haneda (1949), McCosker (1977), Munk (1999), Sparks et al. (2005), and Dunlap and Nakamura (2011)

considering the best known cases of ponyfishes and flashlight fishes. Besides these two families of shallow-water fishes, only deep-sea anglerfishes present shutters on their light organs. The structures of the light organs are slightly different among these fishes primarily because anglerfishes are deep-sea creatures, and between the shallow-water fish groups, ponyfishes exhibit internal luminescence, while flashlight fishes exhibit external light organs (Karplus 2014). Despite their differences, all three groups of fishes use shutters in the same manner: smooth muscles control a dark-pigmented tissue to cover or uncover the emission window of the light organ (Karplus 2014).

Anglerfishes are one of the strangest creatures in the world, and their luminous species share the characteristic of the light organ—the esca—that looks like a fishing rod (Fig. 5.10a), generally hanging from their heads (Munk 1999). The luminous structures are species-specific, and their emissions illuminate the dark deep-sea and act as torches to attract preys. On the frontal top, the light organs possess an aperture for light emission, which can be completely shut through contracting a sphincter-like muscle that works similar to a curtain (Herring and Munk 1994; Munk 1998, 1999). In addition to the shutter, oxygen depletion on the light organ has also been proposed to regulate light flashes; however, this theory remains to be confirmed (Karplus 2014).

Flashlight fishes appear as "blinking schools" on moonless nights around shallow coral reefs (Fig. 5.10b) (Morin et al. 1975). These small animals use light emissions to avoid predators, attract and search for preys, illuminate the surroundings, schooling, and mating. Distinct structures, shutter mechanism, and even bacterial species differentiate the two defined types of sub-ocular large light organs for *Anomalops* and *Photoblepharon* (McCosker 1977; Hendry et al. 2016). The ancestral light organ had been smaller and shut off by a primitive system of rotation, which was derived first and refined by *Anomalops* to fully rotate downwards into a pouch (Rosenblatt and Johnson 1991). Later, the primitive rotatable small organ developed a shutter-like structure that upwardly encloses the light organ, similar to an eyelid (Johnson et al. 1988; Baldwin et al. 1997). The evolutionary stages from the ancient archetype light organ to the light organ in *Photoblepharon* are represented in the genera *Protoblepharon*, *Parmops*, *Phthanophaneron*, and *Kryptophanaron*, in the order of evolution. Finally, at the edge of this family phylogenetic tree is *Photoblepharon*, which completely lost the ancestral rotation character and sophisticated the eyelid membrane. Presently, flashlight fishes are possibly the best living case study on bioluminescence evolution in shallow-water fishes.

Ponyfishes are possibly the zenith of multifunctional bioluminescence systems among fishes (Fig. 5.10c). The meticulous regulation of light signals involves the skin, shutters, and swim bladder; all these tissues and the light organ itself possess chromatophores for light intensity control (Herring and Morin 1978; Sparks et al. 2005). The circumesophageal light organ exhibits translucent windows covered by muscular shutters that are filled with melanophores, which together with the skin chromatophores regulate bioluminescent signals (Herring and Morin 1978; McFall-Ngai and Dunlap 1983). The light organ morphology and pattern of melanophores is species-specific and even sexually dimorphic, which appears to correlate with variations in bioluminescent behavior (McFall-Ngai and Dunlap 1984). The light organ in *Gazza* and *Leiognathus* are heavily covered with chromatophores, which allow a gradual control of light intensity compared to that in other species (McFall-Ngai and Morin 1991). The luminous tissue and overlap shutter are restricted to the dorsal area of the esophagus; however, some species possess two additional lateroventral bacterial pouches and independent shutters (McFall-Ngai and Dunlap 1983; Sparks et al. 2005) *Photoplagios* has a non-identified dark-blue pigmented patch along the median lateral area of the body that may regulate and absorb the blue-green luminescence to prevent lateral light emissions when necessary (Sparks et al. 2005).

In addition to ponyfishes, flashlight fishes, and anglerfishes, another example of fishes with shutters is Apogonidae, which is one of the most puzzling and astonishing cases of evolution in bioluminescent fishes; in this family, some genera are symbiotic, while others produce their own light (Thacker and Roje 2009). The symbiotic genus *Siphamia* possess an eyelid shutter covering the ventral face of the internal light organ (Fig. 5.10d) (Dunlap and Nakamura 2011). The tissue consists of two halves intercepting on the midline of the anterior–posterior axis of

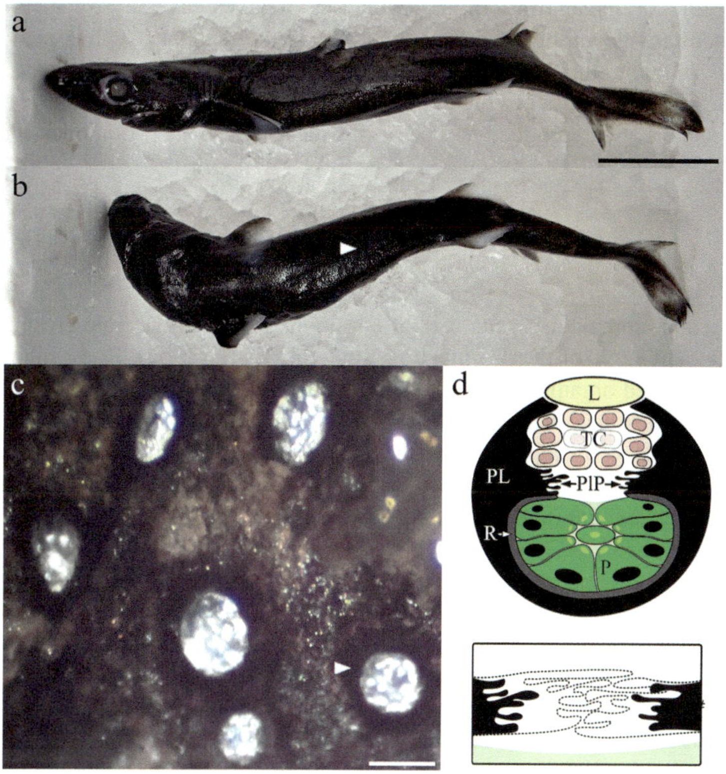

Fig. 5.11 Photophores on the lanternshark *Etmopterus*. (**a**) Lateral view of full body. (**b**) Ventral view of full body presenting the photophore patterns (arrowhead indicates photophores area of **c**). (**c**) Photophores surrounded by a pigmented layer (arrowhead). (**d**) Schemes representing photophore structure (upper) and "pseudopodia-like" projections when the iris-like ring is opened (filled) or closed (dashed outline). *L* lens, *TC* transition cells, *PL* pigment layer, *PlP* "parapodia-like" projections, *R* reflector, *P* photocyte. Scale bars: (**a**) 5 cm; (**c**) 100 μm. Photography by José Paitio. Schemes adapted from Claes and Mallefet (2010) and Renwart et al. (2014)

the body of the fish. Mediated by adrenalin neuromuscular control, both halves of the shutter retract or contract laterally to expose or cover the light emission from the symbiotic bacteria.

Luminous sharks developed a specific mechanism to control the light intensity (Fig. 5.11). The photocytes of Etmopteridae are surrounded by a dark iris-like ring tissue in which lie chromatophores with "pseudopodia-like projections" (Ohshima 1911). The pigmented projections spread across the organ, covering the luminous tissue below. Under hormonal control, the iris-like structure uncovers the photocytes through contraction of the chromatophore projections, enhancing the light output (Claes and Mallefet 2010). Additionally, two types of cell layers between the photocytes and the lens have been suggested to be involved in the light control; however, this theory has not been confirmed till date (Renwart et al. 2014).

The role of melanin for light regulation involves diverse structures spread throughout a wide variety of fishes, either in terms of habitats or phylogeny. In the authors' opinion, this can be an example of evolutionary convergence, which illustrate per se not only the crucial importance of bioluminescence signal regulation for the survival and reproductive success of fishes but also the evolutionary pressure.

5.2.2 Colored Filters

Mesopelagic animals use ventral light emissions to match the blue-green spectra, intensity, and angle of the downwelling sunlight (Clarke 1963). Body silhouette of the camouflaged organism is concealed from the eyes of the predators lying below. This bioluminescence camouflage is called counterillumination and used by euphausiids, decapods, cephalopods, and fishes (Herring 1983; Haddock et al. 2010). The spectra of light produced in the photogenic tissues do not perfectly match the downwelling sunlight (under 250 m depth, approximately 475 nm); therefore, filters and reflectors are used to optimize camouflage (Denton et al. 1985).

Marine invertebrates possess light organs and photophores with colored filters. The photophores of the euphausiids *Nematobrachion*, *Thysanopoda*, and *Euphausia* contain purple-blue pigmentation on the chitin overlying the photophores or in its center, which has been characterized as carotenoproteins (Kemp 1910b; Herring and Locket 1978). Decapods include different photophores, the organs of Pesta of sergestiids and general dermal photophores (Kemp 1910b; Dennell 1955). Filters on both types present violet to blue color (Kemp 1910a; Herring 1972, 1981b; Nowel et al. 1998) The deep-violet-blue pigment in *Systellaspis* is likely to be a lipochrome (Kemp 1910b; Dennell 1955), while those in *Sergestes*[2] and *Oplophorus* appear to be a carotenoprotein (Denton et al. 1985). The blue filters in *Sergestes* have been reported to cut out long wavelengths to match the downwelling sunlight (Denton et al. 1985). The filters appear to fulfill the same purpose in *Oplophorus* and *Systellaspis*, as the light spectra of the photophores differ with respect to the excreted fluid (Herring 1983; Latz et al. 1988). *Histioteuthis* is the only reported cephalopod to have developed pigmented purple-reddish filters, which appear to be protoporphyrins (Dilly and Herring 1981). The absence of filters on the arm-tip photophores (Dilly and Herring 1981) might be explained by their role other than counterillumination (Denton et al. 1985).

Deep-sea fishes commonly use pigmented filters to alter the spectra of emitted light. Researchers have demonstrated the achievement of camouflage through filters in the photophores of these fishes; removing the filters broadens the spectra of

[2]The analyzed specimens of *Sergestes* (Herring 1981b; Denton et al. 1985) were identified up to the genus taxon. Since this publication in 1985, according to the genera alteration of *Sergestes* species (Vereshchaka et al. 2014), the species cited as *Sergestes* (Herring 1981b; Denton et al. 1985) may now belong to a different genus.

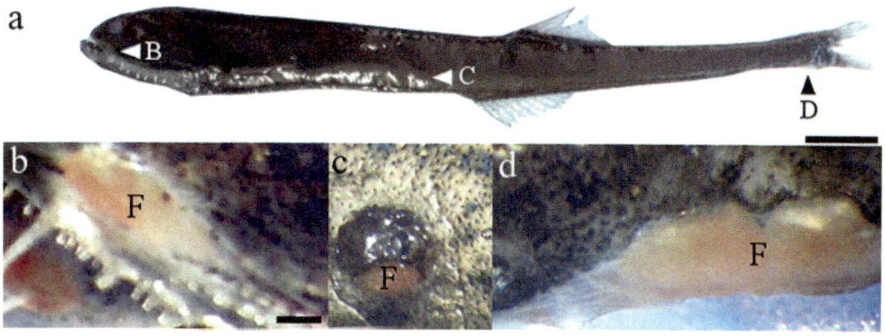

Fig. 5.12 Red-pigmented filter in photophores of the bristlemouth fish *Sigmops*. (**a**) Full body. (**b**) Sub-ocular photophore. (**c**) Ventral–lateral body photophore. (**d**) Caudal photophore. *F* pigmented filter. Arrowheads indicate the photophores in photographs **b–d**. Scale bars: (**a**) 1 cm; (**b–d**) 1 mm. Photography by José Paitio

emitted light (Widder et al. 1983; Denton et al. 1985). Pigmented orange-red filters (Fig. 5.12, pers. obs.) on *Sigmops* may be used for spectral alteration (Copeland 1991). Green filters have been reported in the photophores of the lightfish *Photichthys* (Haneda 1952). Red filters have been observed in other lightfishes, *Polymetme* (Fig. 5.13a), *Yarrella*, and *Ichthyococcus* (Haneda 1952, 1985; Herring 1981c; Denton et al. 1985), and the bristlemouths *Vinciguerria*, *Margrethia*, and *Bonapartia* (Denton et al. 1985; Haneda 1985). Reddish-purple filters have been observed in the hatchetfishes *Maurolicus* (Fig. 5.13b), *Sternoptyx*, *Argyropelecus*, and *Polyipnus*, and red filters in *Valenciennellus* (Haneda 1952, 1985; Herring 1981c; Denton et al. 1985). Red filters in the lightfishes *Ichthyococcus* and *Polymetme* and the hatchetfish *Valencienellus* have been reported to be similar to porphyrins (Herring 1981c; Denton et al. 1985). The reddish-purple pigment on the hatchetfish *Argyropelecus* appears to be similar to cytochrome *c* (Denton et al. 1985), which contains a covalently bound heme group (Schweitzer-Stenner 2014). All fishes in the family Stomiidae exhibit lilac-red-pigmented ventral photophores and lilac, blue, or yellow barbels (Herring 1981c; Denton et al. 1985; Haneda 1985). *Chauliodus* and *Stomias* dragonfishes possess lilac filters in body photophores that appear similar to porphyrin (Herring 1981c; Denton et al. 1985). As adult stomiids live in the bathypelagic zone, where light does not penetrate, counterillumination would have no significance; therefore, it has been suggested (Herring 1981c; Denton et al. 1985) that the pigmentation in the photophores has remained since juvenile forms migrated from the surface, where the larvae used counterillumination to avoid predators. The stomiid fishes *Pachystomias*, *Malacosteus*, and *Aristostomias* possess red-light-emitting sub-ocular photophores. In contrast to most deep-sea animals, the red-light-emitting dragonfishes *Aristostomias*, *Malacosteus*, and *Pachystomias* are visually sensitive to red, which indicates the possibility that this "private bandwidth" is used for intraspecific communication and illumination of preys (Widder et al. 1984; Denton et al. 1985). The dark-red suborbital photophores in *Malacosteus* emit far-red light with a red-brown filter that absorbs all wavelengths below 650 nm; they appear to contain a porphyrin pigment (Widder et al. 1984; Denton et al. 1985).

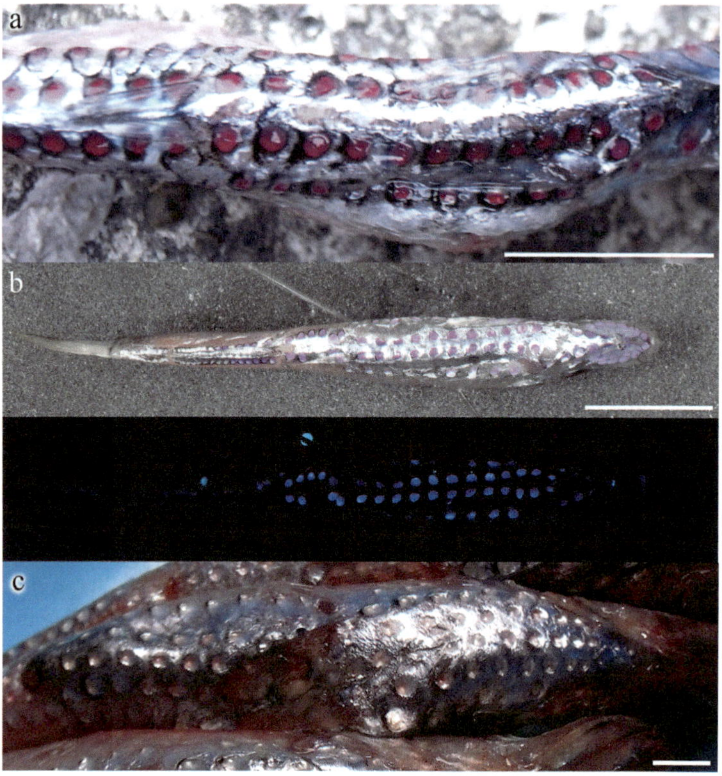

Fig. 5.13 Ventral view of pigmented filters in ventral–lateral photophores in lightfish, hatchetfish, and blackchin fish. (**a**) Red-pigmented filters of the lightfish *Polymetme*. (**b**) Purple filters on the photophores of the hatchetfish *Maurolicus* under white light (upper) and bioluminescent emission in the dark (lower). (**b**) Orange-pigmented filters in the photophores of the blackchin *Neoscopelus*. Scale bars = 1 cm. Photography by Yuichi Oba

Light absorption by filters appears to be lacking in red photophores in other genera of dragonfishes (Denton et al. 1985; Campbell and Herring 1987; Herring and Cope 2005). The slickhead fishes *Xenodermychthys* and *Photostylus* possess small photophores with reddish-violet filters, which can be used to match ambient light (Best and Bone 1976). Orange-colored filters have been reported in another slickhead fish, *Bathytroctes* (Haneda 1952); these filters may be used for counterillumination as proposed for other species of Alepocephalidae. Blackchin *Neoscopelus* (Fig. 5.13c) possess photophores with internal filters exhibiting faint orange coloration (Kuwabara 1953), similar to the filters in Stomiiformes species.

Filters are present in symbiotic luminous fishes as well. The barreleyes *Opisthoproctus* possesses a violet filter between the luminous bacteria and the ventral sole, which ensures that the bandwidth of all light produced is narrowed (Denton et al. 1985). The pinecone fish *Cleidopus* (Fig. 5.14) possesses a pigmented filter on the bacterial light organ that alters the spectra of light produced by the

Fig. 5.14 School of the pinecone fish *Monocentris*. Arrowhead indicates the position of the lower-lip light organ. Photography by Yuichi Oba

symbionts to longer wavelengths, i.e., from blue to blue-green color, after the light passes through the orange-reddish filter (Haneda 1966; Widder et al. 1983).

5.2.3 *Reflection*

Internal reflectors are a widespread characteristic of light organs and photophores; they ensure that the light produced by the photocytes is emitted effectively to the environment (Herring 2000). Additional reflective tissues peripheral to the inner reflector act as light guides, conducting the light through a precise path inside the organ. For reflecting the entire light spectrum produced by the photocytes, the coloration of the reflector is generally silver or white. Specular structural colored reflectors are also used to alter the spectra of the luminous output. An angle-dependent constructive interference is produced by alternating layers of materials with different refractive indexes—high for the reflective substance and low for the cytoplasm, in which the reflective layers are perfectly positioned and spaced, resulting in colored reflectors (Herring 2000).

On firefly photophores, immediately behind the photogenic tissue, one can easily discriminate the thick white reflective layer (Buck 1948). The reflector is composed of uric acid spherulites (Goh et al. 2013), which function as a diffusive mirror (Herring 2000). The presence of a reflective layer is not restricted to the adult phase; it is observed in the firefly larvae as well (Fig. 5.15) (Okada 1935). The reflector is observed on the photophores of another bioluminescent larval form on land; the swollen distal tips of the Malpighian tubules form the light organ of the larvae of the fungus gnat *Arachnocampa*, and the ventral air-filled tracheal layer supplies oxygen and functions as a reflector as well (Green 1979; Rigby and Merritt 2011).

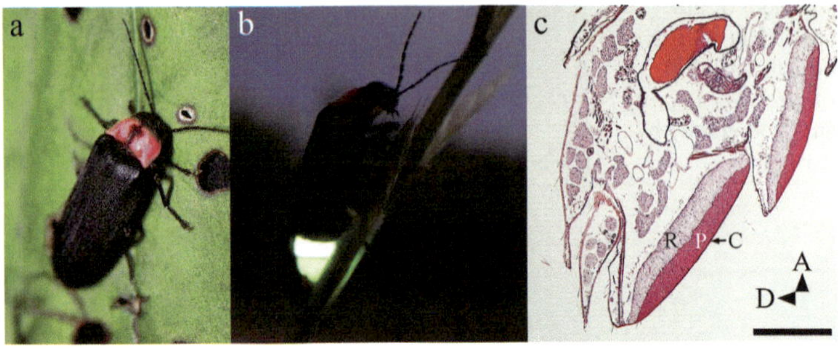

Fig. 5.15 Photophores of the firefly *Luciola*. (**a**) Full body dorsal view under white light. (**b**) Lateral view of full body in twilight, showing light emission from photophores. (**c**) Lateral view of histological section of light organs. *R* reflector, *P* photocytes, *C* cuticle, *A* anterior, *D* dorsal. Scale bar = 500 μm. Photography of full body, by Ken-ichi Onodera and Yuichi Oba; histological section, courtesy of Keisuke Kawano

The mesopelagic zone is a "bioluminescent hotspot," where most luminous organisms on this planet live (Young 1983); therefore, it is not surprising that the light organs and photophores of the organisms in this habitat exhibit high complexity. In mesopelagic crustaceans, the photophores are highly complex and widespread throughout the taxa in euphausiids than in decapods (Kemp 1910b; Yaldwyn 1957). The photophores of some euphausiids exhibit blue-green multilayered reflectors, which act as light interference (Herring and Locket 1978). Colored reflectors have been observed in the body photophores of *Thysanopoda*, *Nematobrachion*, and *Euphausia*, as well in the eye-stalk photophores of *Thysanopoda*. In contrast, *Meganyctiphanes* has a red inner reflector that might enable spectral modification through light interference (Bassot 1966a). In addition to the inner reflectors, euphausiids possess an iridescent lamellar ring on the proximal side of the photophores that narrow the angle of blue light emission and limit lateral emissions of light (Bassot 1966a; Herring and Locket 1978). Although several researchers have observed the structure of colored reflectors in euphausiids, there is no report on the chemical composition.

Granular diffuse reflectors on decapods have been observed in a wide range of species; however, the reflective material remains unknown (Herring 1981b; Nowel et al. 1998). Nevertheless, the reflectors in the organs of Pesta of *Sergestes* appear to be lipidic spheres and non-lipidic in the hepatic organs of *Plesionika*,[3] *Thalassocaris*, and *Chlorotocoides* (Herring 1981b). A crustacean sea-firefly *Vargula hilgendorfii*, known to produce a brilliant secretion of luminous cloud from the upper lip, is a coastal species; however, it possesses reflector structure posterior to the anus. (Abe et al. 2000) suggested that the function might be related to their bioluminescence.

[3] *Parapandalus richardi* in the original reference (Herring 1981b) is accepted as *Plesionika richardi* (WoRMS 2019).

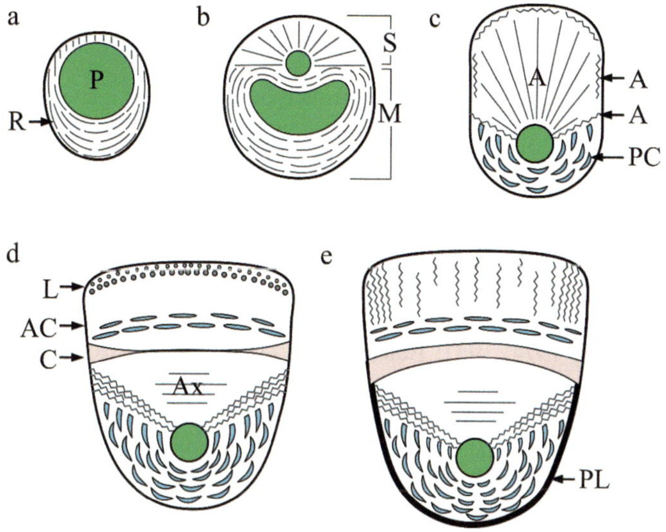

Fig. 5.16 Schematic representation of the variability of the reflective structure of the photophores in the squid *Pyroteuthis*. The silver reflectors are represented in silver and light interference colored reflectors in blue. (**a**) Individual photophores of tentacles. (**b**) Double photophores on tentacles. (**c**) Mantle photophores. (**d**) Ocular photophores. (**e**) Anal photophores. *P* photocytes, *R* reflector; *S* small outer photophore of double photophore, *M* main photophore of double photophore, *A* accessory reflector, *PC* posterior cup, *L* lens, *AC* anterior cap, *C* connective tissue, *Ax* axial reflector, *PL* pigmented layer. Adapted from Butcher et al. (1982)

Reflectors in the light organs and photophores of cephalopods are generally composed of collagen fibers, such as in *Abralia* (Young and Arnold 1982), *Spirula* (Herring et al. 1981), and *Pyroteuthis* (Butcher et al. 1982). Other specific materials can be found, such as the endoplasmic reticulum in *Sepiola* and *Selenoteuthis* (Herring et al. 1981; Herring 2000), the proteinaceous platelets in Sepiolidae and Octopoteuthidae (Dilly and Herring 1981; Herring et al. 1992), and guanine in *Histioteuthis* (Denton et al. 1985). The structure of the photophores and their reflectors vary according to different body parts in Octopoteuthidae (Herring et al. 1992), Sepiolidae (Dilly and Herring 1978, 1981), and Pyroteuthidae (Arnold et al. 1974). The squid *Pyroteuthis* possesses collagenous reflective systems that include photophores along the tentacle and mantle and ocular and anal photophores (Fig. 5.16) (Butcher et al. 1982). The tentacles contain individual and double photophores, the latter consisting of a major elongated organ and an outer smaller one (Butcher et al. 1982). The tentacle and mantle photophores contain green-colored main reflectors and accessory silver reflective tissues as light diffusers (Butcher et al. 1982). The violet to green hue of the ocular and anal photophores of *Pyroteuthis* (Butcher et al. 1982) is imparted by one of the most complex reflective structures in all photophores known to date. The inner reflector—the posterior cup—has a central aperture where lies the axial reflector, covered by another reflective layer—the anterior cap—and the outer lens, which also integrates reflective fibers (Butcher et al. 1982). While the posterior cup and anterior cap are colored tissues for spectral selection, the axial reflector and lens act as light guides (Butcher et al. 1982).

The direction of light emission is controlled by iridophores, generally for dispersion purposes, acting as diffusers (Herring et al. 1981, 1992, 2002; Herring 2000). The reflector can be used for light collimation (*Pterygioteuthis*) (Arnold et al. 1974). Additionally, the symbiotic photophores attached to the ink sac of *Heteroteuthis* have a blue distal cap with additional lateral collar-like reflectors and a proximal main reflector that may be multifunctional, both for diffusion and interference reflection (Dilly and Herring 1978, 1981).

In addition to the spectral selective function of the reflectors (Arnold et al. 1974; Dilly and Herring 1974, 1981; Butcher et al. 1982; Herring 2000; Herring et al. 2002), some species went one step further, not only matching the bioluminescence color to the ambient light but also controlling the emitted spectra. Spectral manipulation via changing the platelet spacing or muscular contraction has been observed in *Abralia* (Young and Arnold 1982), *Pyroteuthis* (Butcher et al. 1982; Latz et al. 1988), and *Leachia* (Latz et al. 1988). Bioluminescence color variation in *Abraliopsis* and *Abralia* is triggered by the water temperature, corresponding to the light phenomena encountered during vertical migration (Young and Mencher 1980).

The major material in the photophore's reflectors is assumed to be guanine in sharks and teleosts (Bassot 1966a; Herring 2000); however, few studies have confirmed this assumption, similar to that for flashlight fishes (Watson et al. 1978), midshipman (Nicol 1957), and pearlsides (Barraud et al. 1959). However, slickhead fishes appear to be an exception. According to a previous report (Best and Bone 1976), blue-green specular reflectors are not composed of guanine platelets but closely and regularly spaced flattened cells.

Guanine is commonly present in light organs and photophores to guide light emissions for different purposes. The flashlight fish *Anomalops* has a thick internal diffusive guanine reflector lined behind the photocytes and a thin external ventral oblique reflector to enhance the dorsal output and reduce the downward emission (Watson et al. 1978). The internal reflector of the esca of anglerfishes covers the entire core of the bacterial space, except for an aperture where it extends distal from the organ toward the tubular appendages to output the bacterial light through one or multiple windows (Munk 1998, 1999).

Only comparable to that in squids, the complex light guidance systems in luminous fishes ensure adequate light emission angle for camouflage. Simple reflectors lined along the sole tube of barreleyes allow specular reflection of light from the posterior light organ through the entire ventral side of the body (Bertelsen 1958; Poulsen et al. 2016).

The bristlemouth *Sigmops* possesses double glandular photophores with reflectors that are thicker on the ventral-median region but lacking on its lower surface and auxiliary reflectors that concentrate and guide the light to the lens (Copeland 1991). The pearlside fish *Maurolicus* possesses ventral photophores with a specular inner reflector and a lamellar ring composed of flat cells with guanine, which act as light guides to limit the wide angle of lateral emission (Barraud et al. 1959; Bassot 1966a). A similar inner reflector has been observed in *Yarrella* and *Polyipnus* (Haneda 1952), and (Denton et al. 1969) described the structure and function in the hatchetfish *Argyropelecus*. Each side of the ventral surface contains a tubular photogenic chamber aligned on the anterior–posterior axis of the body (Denton et al. 1969). The chamber is internally covered with guanine crystal, excluding the bottom apertures, where the light is directed to each individual wedge-shaped photophore (Denton et al. 1969). Elongated crystals aligned vertically in each tubular photophore, in a randomly positioned short axis, direct light from the photogenic chamber and spread it anterior–posteriorly (Denton et al. 1969). Additionally, the front surface of each organ is half-silvered, containing broad crystals parallel to its surface with long axes obliquely positioned for higher reflection at an oblique incidence, allowing the ventral angular light emitted to match the downwelling sunlight (Denton et al. 1969).

The body photophores of myctophids are a rare exception among fishes, lacking pigmented filters but incorporating a colored inner reflector (Fig. 5.17) (Denton et al. 1985). The outer portion of the photocytes is covered by a thick silver reflector, which ensures that the light produced by these cells reaches the inner reflector before being emitted to the environment (Lawry 1973). The inner blue-green reflector of *Diaphus*, which is composed of a regular hexagonal arrangement of iridophores, exhibits a mathematically parabolic shape that allows all light to be simultaneously reflected ventrally from the photophore while minimizing light loss (Paitio et al. 2020). This reflective system ensures that light is emitted at the same angle as that of the ambient light below 200 m (Warrant and Locket 2004; Paitio et al. 2020). Each iridophore is composed of stacked guanine crystals that induce colored light interference (Paitio et al. 2020). The light produced from the photocytes is modulated by the reflector to longer wavelengths and directed outside the organ. The reflected light output matches the downwelling light spectra at mesopelagic depths for successful camouflage (Denton et al. 1985; Johnsen 2014; Paitio et al. 2020). The ability of these fishes to regulate color, along with the variation of the reflection spectra (Young 1983), suggests that the iridophores can be modulated by the fishes to adapt a camouflage color during diel migrations (Paitio et al. 2020).

In addition to photophores, accessory reflective tissues have been reported in ponyfishes and sabertooths. Iridophores between the esophagus and the light organ direct light through the light organ windows of ponyfishes, which is reflected by the swimbladder to the outer tissues (Herring 2000). The swimbladder exhibits a species-specific spatial orientation of guanine, allowing different patterns of light to be reflected through the skin patches, which contains guanine as well (Herring 2000). The reflective accessory system of concentric multilayered guanine-like platelets is present exceptionally on the photophores of the isthmus on the sabertooth fish *Coccorella* (Herring 1977). The reflector surrounds all internal organs except for the ventral aperture directing the light to the translucent muscle below, which itself contains additional dorsolateral reflective layers.

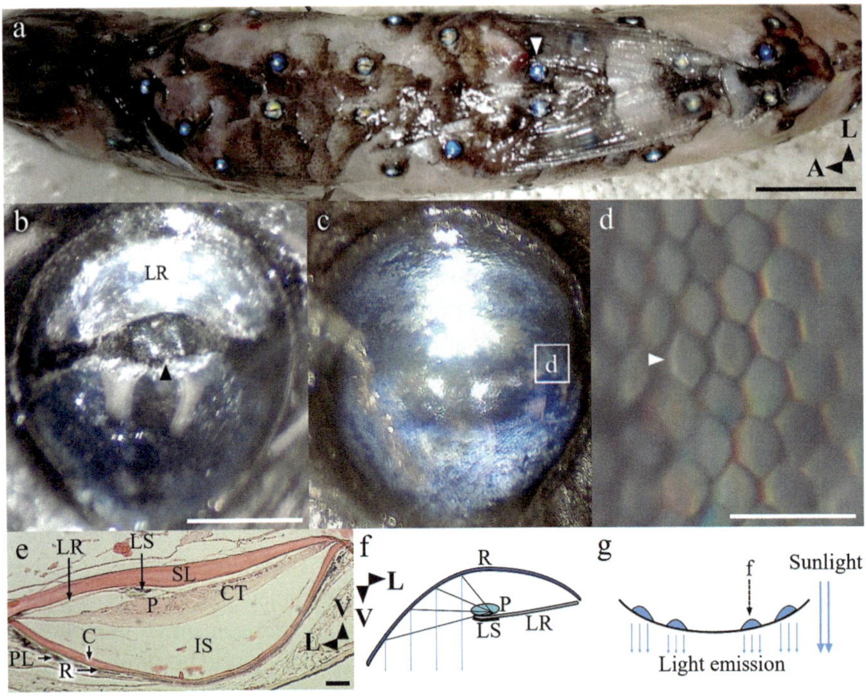

Fig. 5.17 Colored reflector in body photophores of lanternfishes. (**a**) Ventral view of *Diaphus* showing reflection spectra variation in photophores and position of photophore in **b**, **c** (arrowhead). (**b**) Ventral view of photophore. (**c**) Photophore without tissues between photocytes layer and scale lens (outlined white square indicates iridophores area in **d**). (**e**) Regular hexagonal arrangement of iridophores in photophore inner reflector. (**e**) Histological section of photophore. (**f**) Schematic view of light projection of body photophore. (**g**) Position of photophore on **f** (dashed black arrow) on ventral body of the fish and their light emission vertical angle relative to the downwelling sunlight. *L* lateral, *A* anterior, *V* ventral, *LR* lens reflector, *LS* lens septum, *SL* scale lens, *P* photocytes, *CT* connective tissue, *IS* internal space, *C* cup, *PL* pigment layer with melanocytes, *R* inner reflector. (**d**) Iridophore. Scale bars = (**a**) 1 cm, (**b**) 500 μm, (**d**) 50 μm; (**e**) 100 μm. Adapted from Paitio et al. (2020)

5.2.4 Fluorescence

The spectra of light produced by bioluminescent reaction can additionally be altered by the presence of fluorescent proteins. Perhaps the most famous example is the green fluorescent protein (GFP) in a luminous jellyfish, *Aequorea victoria*, (Hastings 1996; Shimomura 2006). Cnidarians commonly emit blue-green light; however, when GFP is present near the photoprotein it gets excited by the energy produced by the bioluminescent reaction and emits green photons. This mechanism is explained as follows: the light energy from aequorin (photoprotein, donor) is transferred to the GFP (fluorescent protein, acceptor) via fluorescence resonance energy transfer (FRET), defined by the closeness (less than 100 Å) and significant

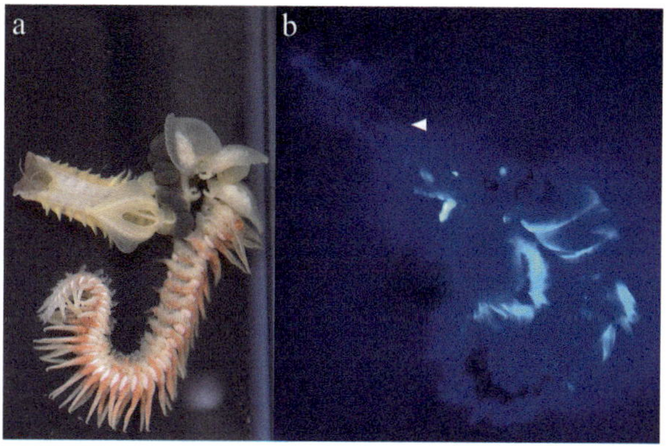

Fig. 5.18 Bioluminescence of the tubeworm *Chaetopterus*. (**a**) Under white light. (**b**) Body bioluminescence and mucus secretion (arrowhead), in the dark. Photographs: courtesy of Ikuhiko Kin

spectral overlap between the donor (emission) and acceptor (excitation) (Morin and Hastings 1971; Shimomura 2006). In contrast to GFP, fluorescent proteins on bioluminescent systems that do not meet the requisitions for FRET are considered as accessory emitters, as described in the following cases.

Millipedes and tubeworms are probably some of the most unknown animals that exhibit bioluminescence. The fluorescent cuticle of the millipede *Motyxia* is close to the bioluminescent spectra (Shimomura 1984; Kuse et al. 2001). The fluorescent spectra and bioluminescence spectra of the luminous slime of the tubeworm *Chaetopterus* (Fig. 5.18) are very similar; riboflavin is proposed as the light emitter in this case (Deheyn et al. 2013; Branchini et al. 2014; Rawat and Deheyn 2016). This observation suggests the involvement of fluorescent compounds in the luminous systems of the millipede and tubeworm; however, this theory has not been confirmed yet.

Luminous bacteria often emit blue-green light from the luciferin–luciferase reaction. The spectra may change because of fluorescent proteins that do not bind themselves to the luciferase molecules (Hastings 1996). Some strains of *Photobacterium* emits bluer light compared to other species of the same genera because of the presence of the fluorescent "lumazine protein," a protein that binds to with 6,7-dimethyl-8-(1-D-ribityl)lumazine, which shifts the emission wavelength and enhances the emission intensity (Koka and Lee 1979; Visser and Lee 1980; Vervoort et al. 1982). *Vibrio* Y-1 strain contains a fluorescent protein with a flavin mononucleotide ligand that is responsible for its yellow bioluminescent glow (Daubner et al. 1987; Macheroux et al. 1987). Although the alteration of the blue hue appears to be related to the sunlight spectrum in the sea where these bacteria live, the reason for the alteration of the yellow color in the bioluminescent spectra is unknown.

Juveniles of the *Erenna* siphonophores exhibit photophores on the tentacles with a photoprotein that emits blue-green light (Haddock et al. 2005). The adults possess an additional pigment that alters the bioluminescence color to red, for attracting prey. The fluorescent material involved in the alteration of the bioluminescence color appears similar to porphyrin-bound proteins.

The dragonfish *Malacosteus*, in addition to the lenticular red-light-absorbing pigment, possesses light-sensitive fluorophores described as phycobiliprotein-like (Campbell and Herring 1987). Similar fluorescent proteins extracted from the red-emitting photophores of the dragonfishes *Aristostomias* and *Pachystomias* suggest the involvement of related chemical compounds (Campbell and Herring 1987).

5.3 Perspectives

This chapter is, as far as the authors are aware, the first description that is focused solely on the diversity and role of pigments for bioluminescence. Although the phylogeny of organisms and chemistry of luminous reaction are quite diverse among bioluminescence organisms across the tree of life, it appears that the pigments and pigmented tissues exhibit a lower level of diversification. Considering the function of photophores and light organs to produce light, photogenic tissues would be the first to be developed, fulfilling the primary role for bioluminescence. Pigmented tissues must have developed later to modulate the light emitted by the photogenic tissues. One may then consider that pigmented tissues evolved secondarily and that they have a secondary function in the light organ, relative to the photogenic tissues. This might be explained by the low diversity of pigments in the light organs and photophores compared to the diversity of molecules involved in bioluminescent reactions. However, one should not discard the lack of clarity on this subject, as only a few reports are available on these pigments and several more studies need to be conducted.

The biological and ecological roles of pigments for bioluminescence are well studied. In contrast, few studies have focused on the chemical composition and microscopic arrangement of the pigments, and almost all available reports provide approximated suggestions. Such studies are needed to understand the physiological operation of the pigmented tissues, not only individually but also with an integrative perspective, because most light organs and photophores possess various pigmented tissues. Only this strategy will enable full comprehension of the role of each pigment in the entire functional light system that is a photophore.

Evolution of pigments and their functional roles remain unknown in bioluminescent animals. Modern technology such as molecular biology analyses should be utilized in the study of pigments on luminous species. The "When? How? Why?" for pigmentation tissues in light organs or photophores could then be answered.

Acknowledgments The authors are grateful to Dr. Keisuke Kawano of The Firefly Museum of Toyota Town, Shimonoseki, Japan for gently contributing to the photographs of the Genji firefly histological sections. We are thankful to Professor Hiromitsu Endo of Kochi University for contributing the photographs of bioluminescent fishes. We convey our kind regards to Mr. Takehito Miyatake of the Japan Professional Photographers Society for the photograph of the flashlight fish *Anomalops*. We thank Dr. Ikuhiko Kin for kindly contributing the *Chaetopterus* photographs. This chapter was written under financial support from JST CREST, Grant Number JPMJCRN16N1, Japan.

References

Abe K, Ono T, Yamada K et al (2000) Multifunctions of the upper lip and a ventral reflecting organ in a bioluminescent ostracod *Vargula hilgendorfii* (Müller, 1890). Hydrobiologia 419:73–82. https://doi.org/10.1023/A:1003998327116

Arnold JM, Young RE, King MV (1974) Ultrastructure of a cephalopod photophore. II. Iridophores as reflectors and transmitters. Biol Bull 147:522–534. https://doi.org/10.2307/1540737

Bagnara JT (1966) Cytology and cytophysiology of non-melanophore pigment cells. Int Rev Cytol 20:173–205. https://doi.org/10.1016/S0074-7696(08)60801-3

Bagnara JT, Hadley ME (1974) Chromatophores and color change. The comparative physiology of animal pigmentation. Prentice-Hall, Hoboken

Baldwin CC, Johnson GD, Paxton JR (1997) *Protoblepharon rosenblatti*, a new genus and species of flashlight fish (Beryciformes: Anomalopidae) from the tropical South Pacific, with comments on anomalopid phylogeny. Proc Biol Soc 110:373–383

Barraud J, Bassot J-M, Favard P (1959) Identification radiocristallographique et aspects cytologiques de la guanine dans le reflécteur des photophores chez *Maurolicus pennanti* (Téléostéen Maurolicidae) Walbaum. Compt Rend l'Académie Sci 249:2633–2635

Bassot J-M (1966a) On the comparative morphology of some luminous organs. In: Johnson FH, Haneda Y (eds) Bioluminescence in progress. Princeton University Press, Princeton

Bassot J-M (1966b) Une forme microtubulaire et paracristalline de reticulum endoplasmique dans les photocytes des annelides Polynoinae. J Cell Biol 31:135–158. https://doi.org/10.1083/jcb.31.1.135

Berry SS (1920) Light production in cephalopods, I. An introductory survey. Biol Bull 38:141–169. https://doi.org/10.2307/1536213

Bertelsen E (1958) A new type of light organ in the deep-sea fish *Opisthoproctus*. Nature 181:862–863

Best ACG, Bone Q (1976) On the integument and photophores of the alepocephalid fishes *Xenodermichthys* and *Photostylus*. J Mar Biol Assoc 56:227–236. https://doi.org/10.1017/S0025315400020567

Branchini BR, Behney CE, Southworth TL et al (2014) Chemical analysis of the luminous slime secreted by the marine worm *Chaetopterus* (Annelida, Polychaeta). Photochem Photobiol 90:247–251. https://doi.org/10.1111/php.12169

Buck JB (1948) The anatomy and physiology of the light organ in fireflies. Ann N Y Acad Sci 49:397–485. https://doi.org/10.1111/j.1749-6632.1948.tb30944.x

Butcher S, Dilly PN, Herring PJ (1982) The comparative morphology of the photophores of the squid *Pyroteuthis margaritifera* (Cephalopoda: Enoploteuthidae). J Zool 196:133–150. https://doi.org/10.1111/j.1469-7998.1982.tb03497.x

Campbell AK, Herring PJ (1987) A novel red fluorescent protein from the deep sea luminous fish *Malacosteus niger*. Comp Biochem Physiol 86:411–417. https://doi.org/10.1016/0305-0491(87)90314-2

Castle PHJ, Paxton JR (1984) A new genus and species of luminescent eel (Pisces: Congridae) from the Arafura Sea, Northern Australia. Copeia 1984:72–81. https://doi.org/10.2307/1445036

Cavallaro M, Mammola CL, Verdiglione R (2004) Structural and ultrastructural comparison of photophores of two species of deep-sea fishes: *Argyrpelecus hemigymnus* and *Maurolicus muelleri*. J Fish Biol 64:1552–1567. https://doi.org/10.1111/j.0022-1112.2004.00410.x

Claes JM, Mallefet J (2010) The lantern shark's light switch: turning shallow water crypsis into midwater camouflage. Biol Lett 6:685–687. https://doi.org/10.1098/rsbl.2010.0167

Clarke WD (1963) Function of bioluminescence in mesopelagic organisms. Nature 198:1244–1246. https://doi.org/10.1038/1981244a0

Cohen DM (1964) Bioluminescence in the Gulf of Mexico anacanthine fish *Steindachneria argentea*. Copeia 1964:406. https://doi.org/10.2307/1441034

Copeland D (1991) Fine structure of photophores in *Gonostoma elongatum*: detail of a dual gland complex. Biol Bull 181:144–157

Dahlgren U (1916) Production of light by animals. J Frankl Inst 181:525–556. https://doi.org/10.1016/S0016-0032(16)90461-7

Daubner SC, Astorga AM, Leisman GB, Baldwin TO (1987) Yellow light emission of *Vibrio fischeri* strain Y-1: purification and characterization of the energy-accepting yellow fluorescent protein. Proc Natl Acad Sci U S A 84:8912–8916. https://doi.org/10.1073/pnas.84.24.8912

Deheyn DD, Mallefet J, Jangoux M (2000) Cytological changes during bioluminescence production in dissociated photocytes from the ophiuroid *Amphipholis squamata* (Echinodermata). Cell Tissue Res 299:115–128. https://doi.org/10.1007/s004419900144

Deheyn DD, Enzor LA, Dubowitz A et al (2013) Optical and physicochemical characterization of the luminous mucous secreted by the marine worm *Chaetopterus* sp. Physiol Biochem Zool 86:702–715. https://doi.org/10.1086/673869

Delgado-Vargas F, Jiménez AR, Paredes-López O (2000) Natural pigments: carotenoids, anthocyanins, and betalains - characteristics, biosynthesis, processing, and stability. Crit Rev Food Sci Nutr 40:173–289. https://doi.org/10.1080/10408690091189257

Dennell R (1955) Observations on the luminescence of bathypelagic crustacea Decapoda of the Bermuda area. J Linn Soc 42:393–406. https://doi.org/10.1111/j.1096-3642.1955.tb02215.x

Denton EJ, Gilpin-Brown JB, Roberts BL (1969) On the organization and function of the photophores of *Argyropelecus*. J Physiol 204:38–39

Denton EJ, Gilpin-Brown JB, Wright PG (1970) On the "filters" in the photophores of mesopelagic fish and on a fish emitting red light and especially sensitive to red light. J Physiol 208:72–73

Denton EJ, Herring PJ, Widder EA et al (1985) The roles of filters in the photophores of oceanic animals and their relation to vision in the oceanic environment. Proc R Soc B Biol Sci 225:63–97. https://doi.org/10.1098/rspb.1985.0051

Dilly PN, Herring PJ (1974) The ocular light organ of *Bathothauma lyromma* (Mollusca: Cephalopoda). J Zool 172:81–100. https://doi.org/10.1111/j.1469-7998.1974.tb04095.x

Dilly PN, Herring PJ (1978) The light organ and ink sac of *Heteroteuthis dispar* (Mollusca: Cephalopoda). J Zool 186:47–59. https://doi.org/10.1111/j.1469-7998.1978.tb03356.x

Dilly PN, Herring PJ (1981) Ultrastructural features of the light organs of *Histioteuthis macrohista* (Mollusca: Cephalopoda). J Zool 195:255–266. https://doi.org/10.1111/j.1469-7998.1981.tb03463.x

Dunlap PV, Nakamura M (2011) Functional morphology of the luminescence system of *Siphamia versicolor* (Perciformes: Apogonidae), a bacterially luminous coral reef fish. J Morphol 272:897–909. https://doi.org/10.1002/jmor.10956

Dunlap PV, Urbanczyk H (2013) Luminous bacteria. In: Rosenberg E, DeLong E, Lory S et al (eds) The prokaryotes: prokaryotic physiology and biochemistry. Springer, Berlin

Fujii R (2000) The regulation of motile activity in fish chromatophores. Pigment Cell Res 13:300–319. https://doi.org/10.1034/j.1600-0749.2000.130502.x

Fulton BB (1941) A luminous fly larva with spider traits (Diptera, Mycetophilidae). Ann Entomol Soc Am 34:289–302. https://doi.org/10.1093/aesa/34.2.289

Goh K, Sheu H, Hua T-E et al (2013) Uric acid spherulites in the reflector layer of firefly light organ. PLoS One 8:e56406. https://doi.org/10.1371/journal.pone.0056406

Green LFB (1979) The fine structure of the light organ of the New Zealand glow-worm *Arachnocampa luminosa* (Diptera: Mycetophilidae). Tissue Cell 11:457–465. https://doi.org/10.1016/0040-8166(79)90056-9

Haddock SHD, Dunn CW, Pugh PR, Schnitzler CE (2005) Bioluminescent and red-fluorescent lures in a deep-sea siphonophore. Science 309:263. https://doi.org/10.1126/science.1110441

Haddock SHD, Moline MA, Case JF (2010) Bioluminescence in the sea. Annu Rev Mar Sci 2:443–493. https://doi.org/10.1146/annurev-marine-120308-081028

Haneda Y (1949) Luminous organs of fish which emit light indirectly. Pac Sci 4:214–227

Haneda Y (1951) The luminescence of some deep-sea fishes of the families Gadidae and Macrouridae. Pac Sci 5:372–378

Haneda Y (1952) Some luminous fishes from the genera *Yarrella* and *Polyipnus*. Pac Sci 4:13–16

Haneda Y (1963) Observations on the luminescence of the shallow-water squid, *Uroteuthis bartschi*. Sci Rep Yokosuka Mus 8:10–16

Haneda Y (1966) On a luminous organ of the Australian pine-cone fish, *Cleidopus gloria-maris* De Vis. In: Johnson FH, Haneda Y (eds) Bioluminescence in progress. Princeton University Press, Princeton

Haneda Y (1985) Pisces. In: Haneda Y (ed) Luminous organisms. Kouseisha-Kouseikaku, Tokyo

Haneda Y, Johnson FH (1962) The photogenic organs of *Parapriacanthus beryciformes* Franz and other fish with the indirect type of luminescent system. J Morphol 110:187–198. https://doi.org/10.1002/jmor.1051100206

Harvey EN (1952) Bioluminescence. Academic, New York

Harvey EN (1956) Evolution and bioluminescence. Q Rev Biol 31:270–287

Hastings JW (1996) Chemistries and colors of bioluminescent reactions: a review. Gene 173:5–11. https://doi.org/10.1016/0378-1119(95)00676-1

Hendry TA, de Wet JR, Dougan KE, Dunlap PV (2016) Genome evolution in the obligate but environmentally active luminous symbionts of flashlight fish. Genome Biol Evol 8:2203–2213. https://doi.org/10.1093/gbe/evw161

Herring PJ (1972) Depth distribution of the carotenoid pigments and lipids of some oceanic animals. J Mar Biol Assoc 52:179–189. https://doi.org/10.1017/S0025315400018634

Herring PJ (1977) Bioluminescence in an evermannellid fish. J Zool 181:297–307. https://doi.org/10.1111/j.1469-7998.1977.tb03244.x

Herring PJ (1981a) Studies on bioluminescent marine amphipods. J Mar Biol Assoc 61:161–176. https://doi.org/10.1017/S0025315400045999

Herring PJ (1981b) The comparative morphology of hepatic photophores in decapod crustacea. J Mar Biol Assoc 61:723–737. https://doi.org/10.1017/S0025315400048165

Herring PJ (1981c) Red fluorescence of fish and cephalopod photophores. In: DeLuca M, McElroy WD (eds) Bioluminescence and chemiluminescence. Basic chemistry and analytical applications. Academic, New York

Herring PJ (1983) The spectral characteristics of luminous marine organisms. Proc R Soc B Biol Sci 220:183–217. https://doi.org/10.1098/rspb.1983.0095

Herring PJ (1985) Bioluminescence in the crustacea. J Crustac Biol 5:557–573. https://doi.org/10.2307/1548235

Herring PJ (2000) Bioluminescent signals and the role of reflectors. J Opt A Pure Appl 2:R29–R38. https://doi.org/10.1088/1464-4258/2/6/202

Herring PJ, Cope C (2005) Red bioluminescence in fishes: on the suborbital photophores of *Malacosteus*, *Pachystomias* and *Aristostomias*. Mar Biol 148:383–394. https://doi.org/10.1007/s00227-005-0085-3

Herring PJ, Locket NA (1978) The luminescence and photophores of euphausiid crustaceans. J Zool 186:431–462. https://doi.org/10.1111/j.1469-7998.1978.tb03932.x

Herring PJ, Morin JG (1978) Bioluminescence in fishes. In: Herring PJ (ed) Bioluminescence in action. Academic, London, pp 273–329

Herring PJ, Munk O (1994) The escal light gland of the deep-sea anglerfish *Haplophryne mollis* (Pisces: Ceratioidei) with observations on luminescence control. J Mar Biol Assoc 74:747–763

Herring PJ, Clarke MR, von Boletzky S, Ryan KP (1981) The light organs of *Sepiola atlantica* and *Spirula spirula* (Mollusca: Cephalopoda): bacterial and intrinsic systems in the order Sepioidea. J Mar Biol Assoc 61:901–916. https://doi.org/10.1017/S0025315400023043

Herring PJ, Dilly PN, Cope C (1992) Different types of photophore in the oceanic squids *Octopoteuthis* and *Taningia* (Cephalopoda: Octopoteuthidae). J Zool 227:479–491. https://doi.org/10.1111/j.1469-7998.1992.tb04408.x

Herring PJ, Dilly PN, Cope C (2002) The photophores of the squid family Cranchiidae (Cephalopoda: Oegopsida). J Zool 258:73–90. https://doi.org/10.1017/S0952836902001279

Johnsen S (2014) Hide and seek in the open sea: pelagic camouflage and visual countermeasures. Annu Rev Mar Sci 6:369–392. https://doi.org/10.1146/annurev-marine-010213-135018

Johnson GD, Rosenblatt RH, Jolla L (1988) Mechanisms of light organ occlusion in flashlight fishes, family Anomalopidae (Teleostei: Beryciformes), and the evolution of the group. Zool J Linnean Soc 94:65–96. https://doi.org/10.1111/j.1096-3642.1988.tb00882.x

Karplus I (2014) The associations between fishes and luminescent bacteria. In: Karplus I (ed) Symbiosis in fishes: the biology of interspecific partnerships. Wiley, Hoboken

Kelsh RN (2004) Genetics and evolution of pigment patterns in fish. Pigment Cell Res 17:326–336

Kemp S (1910a) Decapoda natantia of the coasts of Ireland. Alexander Thom & Co, Dublin

Kemp S (1910b) Notes on the photophores of decapod crustacea. Proc Zool Soc 80:639–651. https://doi.org/10.1111/j.1096-3642.1910.tb01907.x

Koka P, Lee J (1979) Separation and structure of the prosthetic group of the blue fluorescence protein from the bioluminescent bacterium *Photobacterium phosphoreum*. Proc Natl Acad Sci U S A 76:3068–3072. https://doi.org/10.1073/pnas.76.7.3068

Kotlobay AA, Dubinnyi MA, Purtov KV et al (2019) Bioluminescence chemistry of fireworm *Odontosyllis*. Proc Natl Acad Sci U S A 116:18911–18916. https://doi.org/10.1073/pnas.1902095116

Kuse M, Kanakubo A, Suwan S et al (2001) 7,8-dihydropterin-6-carboxylic acid as light emitter of luminous millipede *Luminodesmus sequoiae*. Bioorg Med Chem Lett 11:1037–1040. https://doi.org/10.1016/s0960-894x(01)00122-6

Kuwabara S (1953) Ocurrence of luminous organs on the tongue of two scopelid fishes, *Neoscopelus macrolepidotus* and *N. microchir*. J Shimonoseki Coll Fish 3:283–287

Latz MI, Frank TM, Case JF (1988) Spectral composition of bioluminescence of epipelagic organisms from the Sargasso Sea. Mar Biol 98:441–446. https://doi.org/10.1007/BF00391120

Lawry JV (1973) Dioptric modifications of the scales overlying the photophores of the lantern fish, *Tarletonbeania crenularis* (Myctophidae). J Anat 114:55–63

Lubczak J, Cisek-Cicirko I, Myśliwiec B (2002) Preparation and applications of the products of reaction of uric acid with formaldehyde. React Funct Polym 53:113–124. https://doi.org/10.1016/S1381-5148(02)00167-0

Macheroux P, Schmidt KU, Steinerstauch P et al (1987) Purification of the yellow fluorescent protein from *Vibrio fischeri* and identity of the flavin chromophore. Biochem Biophys Res Commun 146:101–106. https://doi.org/10.1016/0006-291X(87)90696-6

McCosker JE (1977) Flashlight fishes. Sci Am 236:106–115

McFall-Ngai MJ, Dunlap PV (1983) Three new modes of luminescence in the leiognathid fish *Gazza minuta*: discrete projected luminescence, ventral body flash, and buccal luminescence. Mar Biol 73:227–237. https://doi.org/10.1007/BF00392247

McFall-Ngai MJ, Dunlap PV (1984) External and internal sexual dimorphism in leiognathid fishes: morphological evidence for sex-specific bioluminescent signaling. J Morphol 182:71–83. https://doi.org/10.1002/jmor.1051820105

McFall-Ngai M, Montgomery MK (1990) The anatomy and morphology of the adult bacterial light organ of *Euprymna scolopes* Berry (Cephalopoda:Sepiolidae). Biol Bull 179:332–339. https://doi.org/10.2307/1542325

McFall-Ngai M, Morin JG (1991) Camouflage by disruptive illumination in leiognathids, a family of shallow-water bioluminescent fishes. J Exp Biol 156:119–137

Meland K, Aas PØ (2013) A taxonomical review of the *Gnathophausia* (Crustacea, Lophogastrida), with new records from the northern mid-Atlantic ridge. Zootaxa 3664:199–225. https://doi.org/10.11646/zootaxa.3664.2.5

Morin JG (1974) Coelenterate bioluminescence. In: Muscatine L, Lenhoff HM (eds) Coelenterate biology: reviews and perspectives. Academic, New York

Morin JG (1983) Coastal bioluminiscence: patterns and functions. Bull Mar Sci 33:787–817

Morin JG, Hastings JW (1971) Energy transfer in a bioluminescent system. J Cell Physiol 77:313–318. https://doi.org/10.1002/jcp.1040770305

Morin JG, Reynolds GT (1974) The cellular origin of bioluminescence in the colonial hydroid *Obelia*. Biol Bull 147:397–410. https://doi.org/10.2307/1540457

Morin JG, Harrington A, Nealson K et al (1975) Light for all reasons: versatility in the behavioral repertoire of the flashlight fish. Science 190:74–76. https://doi.org/10.1126/science.190.4209.74
Munk O (1998) Light guides of the escal light organs in some deep-sea anglerfishes (Pisces; Ceratioidei). Acta Zool 79:175–186. https://doi.org/10.1111/j.1463-6395.1998.tb01156.x
Munk O (1999) The escal photophore of ceratioids (Pisces; Ceratioidei) - a review of structure and function. Acta Zool 80:265–284. https://doi.org/10.1046/j.1463-6395.1999.00023.x
Nicol JAC (1957) Observations on photophores and luminescence in the teleost. Porichthys 98:179–188. https://doi.org/10.1017/S002531540000574
Nicol JAC (1959) The regulation of light emission in animals. Biol Rev 30:1–40. https://doi.org/10.1111/j.1469-185X.1960.tb01321.x
Nicol JAC (1964) Special effectors: luminous organs, chromatophores, pigments, and poison glands. In: Wilburn K, Younge C (eds) Physiology of mollusca, 1st edn. Academic, New York
Nowel MS, Shelton PMJ, Herring PJ (1998) Cuticular photophores of two decapod crustaceans, *Oplophorus spinosus* and *Systellaspis debilis*. Biol Bull 195:290–307. https://doi.org/10.2307/1543141
Nuevo M, Milam SN, Sandford SA et al (2009) Formation of uracil from the ultraviolet photo-irradiation of pyrimidine in pure H_2O ices. Astrobiology 9:683–695. https://doi.org/10.1089/ast.2008.0324
Oba Y (2019) Living light list. https://www3.chubu.ac.jp/faculty/oba_yuichi/living_light_list/. Accessed 26 Oct 2019
Oba Y, Schultz DT (2014) Eco-evo bioluminescence on land and in the sea. Adv Biochem Eng Biotechnol 144:3–36. https://doi.org/10.1007/978-3-662-43385-0_1
Ohshima H (1911) Some observations on the luminous organs of fishes. J Coll Sci Imp Univ Tokyo 27:1–25
Okada Y (1935) Luminous apparatus in lampyrids, III. Bot Zool 3:1638–1648
Paitio J, Oba Y, Meyer-Rochow VB (2016) Bioluminescent fishes and their eyes. In: Thirumalai J (ed) Luminescence - on outlook on the phenomena and their applications. InTech, Rijeka
Paitio J, Yano D, Muneyama E et al (2020) Reflector of the body photophore in lanternfish is mechanistically tuned to project the biochemical emission in photocytes for counterillumination. Biochem Biophys Res Commun 521:821–826. https://doi.org/10.1016/j.bbrc.2019.10.197
Pickford G (1949) *Vampyroteuthis infernalis* Chun: an archaic dibranchiate cephalopod. II. External anatomy. Dana-Report 32:1–132
Poulsen JY, Sado T, Hahn C et al (2016) Preservation obscures pelagic deep-sea fish diversity: doubling the number of sole-bearing opisthoproctids and resurrection of the genus *Monacoa* (Opisthoproctidae, Argentiniformes). PLoS One 11:e01597. https://doi.org/10.1371/journal.pone.0159762
Rawat R, Deheyn DD (2016) Evidence that ferritin is associated with light production in the mucus of the marine worm *Chaetopterus*. Sci Rep 6:1–14. https://doi.org/10.1038/srep36854
Renwart M, Delroisse J, Claes JM, Mallefet J (2014) Ultrastructural organization of lantern shark (*Etmopterus spinax* Linnaeus, 1758) photophores. Zoomorphology 133:405–416. https://doi.org/10.1007/s00435-014-0230-y
Rigby LM, Merritt DJ (2011) Roles of biogenic amines in regulating bioluminescence in the Australian glowworm *Arachnocampa flava*. J Exp Biol 214:3286–3293. https://doi.org/10.1242/jeb.060509
Rosenblatt RH, Johnson GD (1991) *Parmops coruscans*, a new genus and species of flashlight fish (Beryciformes: Anomalopidae) from the South Pacific. Proc Biol Soc 104:328–334
Sano T, Kobayashi Y, Sakai I et al (2019) Ecological and histological notes on the luminous springtail, *Lobella* sp. (Collembola: Neanuridae), discovered in Tokyo, Japan. In: Suzuki H (ed) Bioluminescence - analytical applications and basic biology. IntechOpen, Rijeka
Schmidt J (1922) Live specimens of *Spirula*. Nature 110:788–790

Schweitzer-Stenner R (2014) Cytochrome c: a multifunctional protein combining conformational rigidity with flexibility. New J Sci 2014:484538. https://doi.org/10.1155/2014/484538
Shimomura O (1984) Porphyrin chromophore in *Luminodesmus* photoprotein. Comp Biochem Physiol:565–567. https://doi.org/10.1016/0305-0491(84)90367-5
Shimomura O (2006) Bioluminescence: chemical principles and methods. World Scientific Publishing Co., Singapore
Shinoda H, Shannon M, Nagai T (2018) Fluorescent proteins for investigating biological events in acidic environments. Int J Mol Sci 19:1548. https://doi.org/10.3390/ijms19061548
Sköld HN, Aspengren S, Cheney KL, Wallin M (2016) Fish chromatophores - from molecular motors to animal behavior. Int Rev Cell Mol Biol 321:171–219. https://doi.org/10.1016/bs.ircmb.2015.09.005
Somiya H (1981) On the bacterial-associated light organ in *Chlorophthalmus*. In: DeLuca M, McElroy WD (eds) Bioluminescence and chemiluminescence. Basic chemistry and analytical applications. Elsevier, New York
Sparks JS, Dunlap PV, Smith WL (2005) Evolution and diversification of a sexually dimorphic luminescent system in ponyfishes (Teleostei: Leiognathidae), including diagnoses for two new genera. Cladistics 21:305–327. https://doi.org/10.1111/j.1096-0031.2005.00067.x
Tebo BM, Linthicum DS, Nealson KH (1979) Luminous bacteria and light emitting fish: ultrastructure of the symbiosis. Biosystems 11:269–280
Thacker CE, Roje DM (2009) Molecular phylogenetics and evolution phylogeny of cardinalfishes (Teleostei: Gobiiformes: Apogonidae) and the evolution of visceral bioluminescence. Mol Phylogenet Evol 52:735–745. https://doi.org/10.1016/j.ympev.2009.05.017
Thompson EM, Rees J (1995) Origins of luciferins: ecology of bioluminescence in marine fishes. In: Hochachka P, Mommsen T (eds) Biochemistry and molecular biology of fishes. Elsevier Science, Amsterdam
Tsarkova AS, Kaskova ZM, Yampolsky IV (2016) A tale of two luciferins: fungal and earthworm new bioluminescent systems. Acc Chem Res 49:2372–2380. https://doi.org/10.1021/acs.accounts.6b00322
Vereshchaka AL, Olesen J, Lunina AA (2014) Global diversity and phylogeny of pelagic shrimps of the former genera *Sergestes* and *Sergia* (Crustacea, Dendrobranchiata, Sergestidae), with definition of eight new genera. PLoS One 9:e112057. https://doi.org/10.1371/journal.pone.0112057
Vervoort J, O'Kane DJ, Carreira LA, Lee J (1982) Identification of a lumazine protein from *Photobacterium leiognathi* by Coherent anti-strokes Raman spectroscopy. Photochem Photobiol 37:117–119. https://doi.org/10.1111/j.1751-1097.1983.tb04444.x
Visser AJWG, Lee J (1980) Lumazine protein from the bioluminescent bacterium *Photobacterium phosphoreum*. A fluorescence study of the protein-ligand equilibrium. Biochemistry 19:4366–4372. https://doi.org/10.1021/bi00559a033
Warrant EJ, Locket NA (2004) Vision in the deep sea. Biol Rev Camb Philos Soc 79:671–712. https://doi.org/10.1017/S1464793103006420
Watson M, Thurston EL, Nicol JAC (1978) Reflectors in the light organ of *Anomalops* (Anomalopidae, Teleostei). Proc R Soc Lond Biol Sci 202:339–351. https://doi.org/10.1098/rspb.1978.0071
Widder EA (2001) Marine bioluminescence. Biosci Explain 1:1–9
Widder EA (2010) Bioluminescence in the ocean: origins of biological, chemical, and ecological diversity. Science 328:704–708. https://doi.org/10.1126/science.1174269
Widder EA, Latz MI, Case JF (1983) Marine bioluminescence spectra measured with an optical multichannel detection system. Biol Bull 165:797–810. https://doi.org/10.2307/1541479
Widder EA, Latz MI, Herring PJ, Case JF (1984) Far red bioluminescence from two deep-sea fishes. Science 225:512–514. https://doi.org/10.1126/science.225.4661.512
WoRMS (2019) *Stylopandalus richardi* (Coutière, 1905). http://www.marinespecies.org/aphia.php?p=taxdetails&id=107665. Accessed 4 Nov 2019

Yaldwyn JC (1957) Deep-water crustacea of the genus *Sergestes* (Decapoda, Natantia) from Cook Strait, New Zealand. Victoria University of Wellington, Wellington

Young R (1975) *Leachia pacifica* (Cephalopoda, Teuthoidea): spawning habitat and function of the brachial photophores. Pac Sci 29:19–25

Young RE (1983) Oceanic bioluminescence: an overview of general functions. Bull Mar Sci 33:829–845

Young RE, Arnold JM (1982) The functional morphology of a ventral photophore from the mesopelagic squid, *Abralia trigonura*. Malacologia 23:135–163

Young RE, Mencher FM (1980) Bioluminescence in mesopelagic squid: diel color change during counterillumination. Science 208:1286–1288. https://doi.org/10.1126/science.208.4449.1286

Part II
Pigment Cell and Patterned Pigmentation

Chapter 6
Development of Melanin-Bearing Pigment Cells in Birds and Mammals

Heinz Arnheiter and Julien Debbache

Abstract Melanin-bearing pigment cells in birds and mammals, much as those in other vertebrates, have at least two developmental origins: the neural crest, giving rise to melanocytes of the integument (skin and its appendages) and of inner organs, and the optic neuroepithelium, giving rise to the retinal pigment epithelium and part of the iris. Both types of cells are derived from precursors that are guided towards their differentiation by complex signaling pathways and transcription factors, some common to both cell types and some unique. Numerous studies show that neural crest-derived melanocytes arise from precursors that emanate from the dorsal neural tube and migrate on a dorso-lateral pathway underneath the surface ectoderm. Others arise from poorly defined precursors that migrate on a ventro-medial pathway and also give rise to Schwann cells. Nevertheless, melanocyte precursors retain developmental plasticity for considerable time, potentially being capable of correcting developmental imbalances in an embryo's distinct cell populations. Some derivatives even exhibit stem cell features during adulthood, capable of replenishing melanocytes during hair and feather cycles. In fact, the study of the development of melanin-bearing pigment cells provides for fascinating insights into how specific cell types arise and maintain themselves or become abnormal or are lost in pathogenic processes.

Keywords Neural crest · Neuroepithelium · Cellular plasticity · Hair follicle · Feather follicle · Retinal pigment epithelium · MITF · KIT

H. Arnheiter (✉)
Scientist Emeritus, National Institute of Neurological Disorders and Stroke, National Institutes of Health, Bethesda, MD, USA

J. Debbache
Institute of Anatomy, University of Zürich, Zürich, Switzerland
e-mail: julien.debbache@anatomy.uzh.ch

H. Hashimoto et al. (eds.), *Pigments, Pigment Cells and Pigment Patterns*,
https://doi.org/10.1007/978-981-16-1490-3_6

6.1 Introduction

During development, cells exhibit a high degree of plasticity on their way to differentiation, but once differentiated, they tend to keep their differentiated state for prolonged periods of time. It usually requires exceptional conditions such as wound healing, experimental manipulations in vitro, or perhaps malignant transformation for a mature differentiated cell to lose its current state of differentiation and eventually assume a novel state. An interesting example illustrating these features is provided by the development and differentiation of melanin-bearing pigment cells. These cells, so visibly marked by their pigmentation when differentiated, have their origin in multipotent tissues and their states are initially unstable. One of these multipotent tissues is the neural crest, which forms at the dorsal lip of the closing neural tube and gives rise to all melanocytes of the skin, feather or hair bulbs, choroid of the eye, Harderian glands, stria vascularis of the inner ear (in mammals), meninges, heart septa, and mitral and tricuspid valves, in addition to a variety of other cell types (Fig. 6.1). Another multipotent tissue is the optic neuroepithelium, which buds out from the diencephalic neuroepithelium and gives rise to the retinal pigment epithelium (RPE) in addition to all neuronal and glial cell types in the retina (see below, Fig. 6.4). In some species including humans, a distinct melanin-type pigment, termed neuromelanin, is found in the brain, particularly in

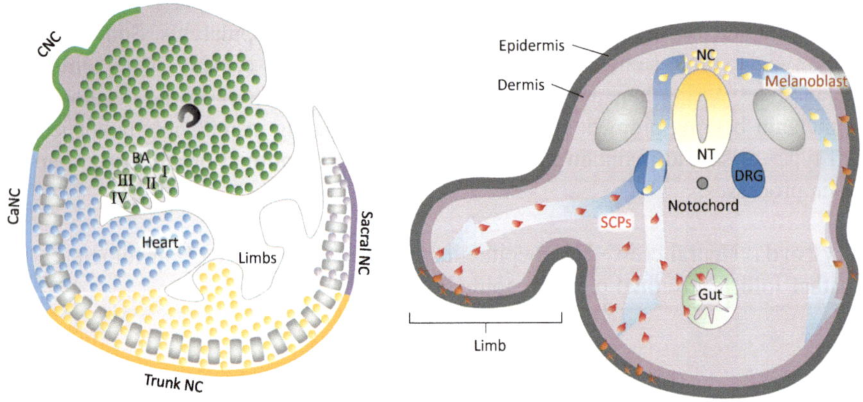

Fig. 6.1 Neural crest origin and migration of melanoblasts. The neural crest is a transient vertebrate tissue capable of giving rise to a multitude of cells including melanoblasts from which melanocytes are eventually generated. The crest can be divided into cranial (CNC), cardiac (CaNC), trunk, and sacral crest. Even though melanoblasts are generated at each axial level, other neural crest-derived cell types are specific to these regions. After delamination from the neural tube, crest cells in the trunk region migrate on two distinct pathways: a dorso-lateral pathway and a ventro-medial pathway. Classically, all melanoblasts/melanocytes in the trunk are derived from cells on the dorso-lateral pathway. Melanoblasts, however, can also be seen associated with nerves where they arise from the so-called Schwann cell precursors, which are neural crest-derived cells migrating on the ventro-medial pathway. It is possible that these two populations eventually populate distinct portions of the integument and inner organs. *BA* branchial arches

catecholaminergic neurons of the substantia nigra and the locus coeruleus. As much less is known about the function of the pigment in these cells, however, we here focus entirely on the development of melanocytes derived from the neural crest and the optic neuroepithelium and discuss their regulation by extrinsic signaling and intrinsic transcriptional pathways. We also discuss disorders associated with melanocyte deficiencies or aberrant growth to underscore the fact that the study of pigment cell development is not of mere academic interest.

6.2 The Development of Melanoblasts from the Neural Crest

The neural crest is a distinct tissue that evolutionarily is first encountered in vertebrates, though corresponding gene regulatory networks are already present in urochordates. Some of these urochordates, such as *Ecteinascidia turbinata*, even show intimations of neural crest-type cells that give rise to pigment cells (Nieto 2002). In birds and mammals, the neural crest contributes cells to the craniofacial skeleton, cranial ganglia, teeth, thyroid, enteric ganglia, smooth muscles, heart, dorsal root ganglia, sympathetic ganglia, and adrenal medulla, in addition to pigment cells in various locations of the body (Le Douarin and Kalcheim 1999; Dupin and Le Douarin 2014). Several of these derivatives originate in defined regions of the neural crest along the rostro-caudal axis, referred to as the cranial, vagal, trunk, and sacral crest, but neurons and pigment cells are produced at all axial levels, though in region-specific numbers and spatial distributions.

Based on experiments in frogs and other vertebrates, the neural crest is induced at the neural plate border mainly by a combination of the signaling factors WNT, FGF, and BMP, which activate the paired domain protein PAX3, the zinc finger proteins ZIC1/2, and others to converge onto the zinc finger transcription factor SNAI2 (SLUG) and the bHLH transcription factor TWIST1 among others (Pla and Monsoro-Burq 2018; Vandamme and Berx 2019). Delamination of neural crest cells from the neural folds depends on a variety of factors that include the Hippo pathway transcriptional co-activator YAP and the zinc finger proteins SNAI1 and ZEB2. Their action eventually leads to repression of epithelial and upregulation of mesenchymal markers, a process commonly referred to as epithelial-to-mesenchymal transition (EMT) (Vandewalle et al. 2009; Kumar et al. 2019; Vandamme and Berx 2019). Because unlike epithelial cells, mesenchymal cells are not fixed to each other via tight junctions, they are free to migrate, and they migrate extensively, aided by the action of specific metalloproteases that help them tunnel their way to their final destination.

Classical studies have shown that after delamination from the neural tube, neural crest cells first assemble in a so-called migration staging area (MSA) (Weston 1991). In chick embryos, entry of neural crest cells into the MSA occurs at around embryonic day (E) 2.5–3.0 (HH stages 15–19) and the first melanoblast markers appear at E3.5. In the mouse, neural crest cells enter the MSA at around E9.5 and the earliest melanoblast markers, such as the *microphthalmia*-associated transcription factor MITF and the receptor tyrosine kinase KIT (see below), appear at around E10.5, whereby the exact timing depends on the axial level.

After assembly in the MSA, neural crest cells migrate ventrally. Rostral to the first somite, they migrate underneath the surface ectoderm but also along the neural tube. In regions where somites are present, their migratory pathways are clearly separated into a dorso-lateral route between somite and surface ectoderm and a ventro-medial route between somite and neural tube. Interestingly, cells on the dorso-lateral pathway become melanocytes, and those on the ventro-medial pathway predominantly glial cells and neurons. These observations immediately beg the question of when and where precisely the cells are specified (Dupin et al. 2018). Are they specified before migration, and their specification selects them for a given path? Or are they specified only later by instructions from the microenvironment in which they find themselves by chance? This crucial question has been addressed in vitro and in vivo in multiple heterotopic and heterochronic transplantation experiments particularly using quail/chick chimeras in which quail cells can be identified selectively based on their unique nuclear morphology (Le Douarin 1974). It was found, for instance, that transplantation of quail cranial or vagal crest to the chick trunk region led to trunk crest derivatives, and transplantation of quail trunk crest to the chick vagal region to vagal derivatives, thus suggesting continuing cellular plasticity (Le Douarin 1974). On the other hand, transplantation of postmigratory melanoblasts (the cells normally destined to become melanocytes) from the dorso-lateral pathway to the premigratory MSA of younger embryos resulted in the preservation of their dorso-lateral migratory route, thus arguing against plasticity. Likewise, prospective neural progenitors diverted to the melanocytic pathway still adopted neural fates (Krispin et al. 2010).

What holds for chickens, though, may not necessarily hold for other species. One should not overlook the fact that developmental pathways and time constraints may differ between species and thus may lead to different demands on when precisely, and where exactly, a multipotent neural crest cell has to migrate and become fate-specified. In avian embryos, fate specification may occur at earlier developmental stages relative to those in other species. In mice, for instance, multipotentiality seems to be maintained even in migratory crest cells, as demonstrated by lineage tracing using R26R Confetti mice (Baggiolini et al. 2015). Still, even a simple species-centered view does not do full justice to the complexity of neural crest cell fate determination as the time points of specification also differ along the rostro-caudal axis within the same species. Nevertheless, conceptually it would seem easier to accept that specification occurred only when the cells are in an isolated microenvironment, as microenvironments that are separated from each other likely differ in availability and activity of specific extracellular cues. Distinct fates, however, might even be induced within the same microenvironment, for instance, when individual cells slightly differ in expression of specific receptors and so show differences in their temporal readiness to be specified. In avian embryos, for example, neurogenic progenitors seem to migrate out slightly earlier on their specific route than melanogenic progenitors on theirs.

Another question that has previously been addressed concerns the number and anatomical position of neural crest-derived melanocyte precursors. Answers to this question may help explain the size and distribution of pigment spots and congenital nevi of both mice and humans (Kinsler and Larue 2018). Nevertheless, the answers largely depend on the definition of what precisely constitutes a precursor and it is therefore not surprising that different studies came to different conclusions. Using chimeric mouse embryos experimentally created by fusion of blastocysts from pigmented and unpigmented mice and pigmentation of the resulting offspring as the readout, Mintz concluded that there are 34 precursors, 17 equally spaced on the right side of the embryo and 17 on the left (Mintz 1967). A small number, though not exactly 34, was also supported by studies in which a wildtype Tyrosinase (TYR) was expressed in *Tyr*-mutant (albino) mouse embryos using a retroviral vector (Huszar et al. 1991). Furthermore, staining in mouse embryos for the early melanoblast marker *Mitf* also indicated a small number of precursors, particularly in the trunk region (Opdecamp et al. 1997). Nevertheless, using a *Dopachrome tautomerase* (*Dct*) promoter-driven LacZ transgene, which marks melanocyte precursors but also glial cells, the numbers of cells on the dorso-lateral pathway were found to be much higher and the cells to extensively mix along the rostro-caudal axis (Wilkie et al. 2002). Without certainty about which cells are being counted as precursors, though, the question cannot be answered unequivocally.

After decades of research had established the above mentioned migratory pathways and cell type commitments, a more recent study by Adameyko et al. shook up the field by coming to a radically different conclusion. These authors used transgenic expression of a tamoxifen-regulatable Cre recombinase driven by the proteolipid protein promoter (Plp-CreER^{T2}) in mouse embryos to conditionally mark cells of the glial lineage. They found that rather than melanoblasts from the dorso-lateral pathway, it was the so-called Schwann cell precursors (SCPs) originating from the dorso-ventral pathway and found in close proximity to nerve endings that serve as the major source for coat melanocytes (Adameyko et al. 2009) (Fig. 6.1). This finding was supported by earlier studies showing that in the chick, peripheral nerves contain bipotent Schwann cell/melanocyte precursors and that nerve injury can cause lineage switching from glial cells to melanocytes (Dupin et al. 2000, 2003). Adameyko's et al. new interpretation has been challenged, though, mainly on the grounds that the Plp-CreER^{T2} transgene is not as strictly SCP/glial cell-specific as has been claimed because it tags melanocytes even when it is induced postnatally (Hari et al. 2012). Moreover, a *Desert hedgehog* (*Dhh*)-Cre transgene specifically tags SCPs and *not* melanocytes (Jaegle et al. 2003; Hari et al. 2012). Furthermore, with the notable exception of ADAR1 (Gacem et al. 2020), responsible for adenosine-to-inosine editing of RNA, most coat color mutants do *not also* affect SCPs (as would be manifested by disturbances in myelin production). The discrepancy to the Adameyko et al. study might be resolved if the Plp-CreER^{T2} and Dhh-Cre transgenes targeted different SCP subpopulations, but this is not very likely because a glial cell population not targeted by the Dhh driver has yet to be identified.

Alternatively, it is conceivable that Dhh-Cre targets principally the same SCP population but only after the cells have generated committed melanoblasts, while Plp-Cre would target them before. The different results might also be reconciled if melanoblasts on the dorso-lateral pathway eventually populated dermal tissues derived from the paraxial mesoderm while SCP-derived melanoblasts populated dermal tissues derived from the lateral plate mesoderm (Nitzan et al. 2013b). This distinction may be relevant when tracing the origins of melanocytes in different skin regions, such as in limb, trunk, and cranial skin (see also Fig. 6.1). In any event, it is still difficult to define an SCP in molecular terms as these cells are highly dynamic from the moment they originate in the neural crest and associate with nerves to the moment of their differentiation. Not only do they eventually generate Schwann cells but they can also give rise to enteric and parasympathetic neurons, neuroendocrine cells, endoneural fibroblasts, mesenchymal stem cells in teeth and bones, in addition to melanocytes (reviewed in Furlan and Adameyko (2018)). Hence, considerable efforts are required to further dissect the relationship between SCPs and melanocytes.

In sum then, the description of the initial stages of melanocyte development gives arguments for both premigratory and postmigratory commitment. Nevertheless, we have to keep in mind that fate maps as seen in vivo in normal embryos do not indicate fate restrictions. The determination of a definite fate restriction would require systematic challenges with conditions permissive for alternative fates. This point is of broad importance as a fate potential that is wider than what is encountered during normal development would provide an embryo with the opportunity to correct deficiencies in one cell population, be they caused genetically or epigenetically, by recruitment from another population and so still yield a healthy adult.

6.3 The Development of Melanocytes and Colonization of the Skin and Hair Follicles

Despite the fact that neural crest-derived melanoblasts migrate to, and differentiate in, various locations in the body, such as the eye, inner ear, brain, bone, and heart, most studies in birds and mammals focus on their colonization of the skin and feather or hair follicles. This is so because the relevant pathways have been elucidated mostly based on the analysis of mutant animals with visible coat color alterations, rather than, say, alterations in melanocytes in inner organs, which are not easily detectable.

At E11.5-E12 in mouse embryos, melanoblasts in the dermis of the trunk start to migrate across the basement membrane into the epidermis. The reduction in the number of dermal melanoblasts due to transmigration is compensated by cell division in the dermis. Once melanoblasts are in the epidermis, they speed up their division rate and eventually cluster in the developing hair follicles, with a *Tyrosinase*-negative subpopulation lodging in the upper part of the follicle and a differentiating population in the lower part. Here, they express the melanin biosynthetic enzymes and transfer their melanosomes to hair shaft keratinocytes and so are responsible for the pigmentation of the first hairs (Fig. 6.2). The cells in the upper part will finally reside, along with hair follicle stem cells, in a stem cell niche, the hair follicle bulge, which becomes anatomically distinct right after birth. A third population of melanoblasts remains in the interfollicular epidermis. In mice, this population is restricted to skin areas with few or no hairs such as the pinna of the ear, the tail, and the glabrous surface of the paws. In humans and some other mammals, however, most of the skin is free of hairs and interfollicular melanocytes are the norm. It is these cells that are responsible for skin pigmentation by transferring their melanosomes to keratinocytes to protect the skin against the damaging effects of UV irradiation.

Hair follicles are among few structures in the mammalian body that undergo lifelong cycles of loss and regeneration. They do this in specific stages called anagen (the growth phase), catagen (the degenerative phase), and telogen (the resting phase where degeneration is complete). Hair follicle loss during the hair cycle is partial, though, and one portion of the follicle, called the lower permanent portion (LPP), never degenerates. It is this portion from which regrowth is initiated by activation of hair stem cells. These stem cells give rise to all parts of the growing follicle, except melanocytes, which are regenerated from separate, non-pigmented stem cells derived from the melanoblasts that have originally populated the stem cell niche. Recently, it has been recognized that in telogen mouse hair follicles, melanocyte stem cells come in two distinct populations: a CD34-positive, likely multipotential population that resides in the hair follicle bulge region and the LPP and that is also capable of giving rise to glial cells; and a CD34-negative, seemingly unipotent population that resides in the secondary hair germ and may give rise only to melanocytes (Joshi et al. 2019). Melanocyte stem cells in the bulge have been characterized extensively after marking with a Dct-LacZ transgene (Nishimura et al. 2002). The cells proliferate very slowly, retain a 5-bromo-2′-deoxyuridine (BrdU) label for at least 2 months, lack (or nearly lack) expression of *Sox10*, *Mitf*, *Tyr*, *Tyrp1*, but retain endogenous expression of *Dct*, and can give rise to pigmented cells upon transplantation into albino skin. When a new anagen phase commences, melanocyte stem cells start dividing in synchrony with hair stem cells and differentiate. The expression of *Notch1* and *Tgf-ß,* combined with the absence of WNT/ß-catenin signaling, is responsible for stem cell quiescence, and expression of *Kit*, *Wnt*, and *Endothelins (EDNs)* is responsible for their activation, at least in the secondary hair germ (Moriyama et al. 2006; Rabbani et al. 2011). In fact, transient application of neutralizing KIT antibodies in neonates eliminates pigmentation during the first hair cycle but allows for restoration of pigmentation during the second hair cycle, suggesting that some melanocyte stem cells escape destruction by the antibody (Botchkareva et al. 2001). Expression of a dominant active form of ß-catenin in melanocyte stem cells initiates premature and excessive differentiation at the expense of self-renewal, leading to eventual hair graying during subsequent hair cycles (Rabbani et al. 2011). Conversely, conditional loss of WNT/ß-catenin signaling leads to absence of differentiated melanocytes in regenerating hairs

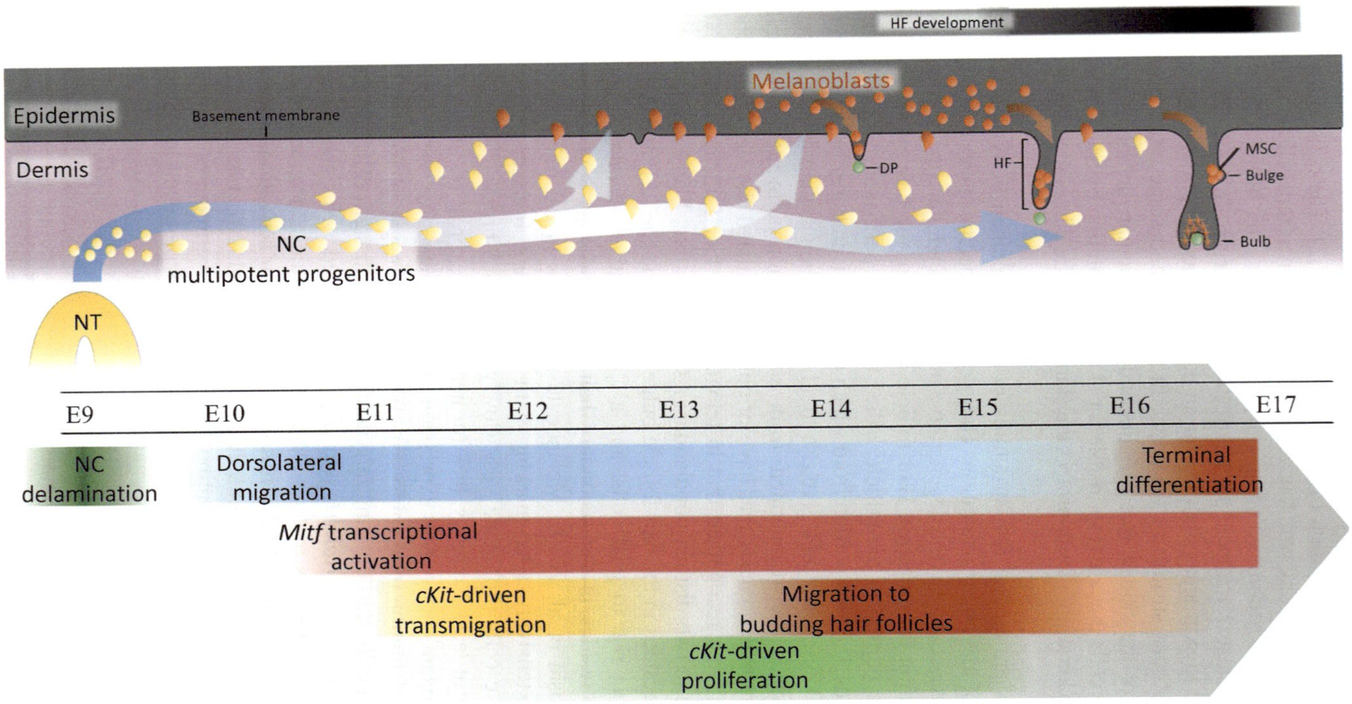

Fig. 6.2 Melanoblast migration into dermis and epidermis in mouse embryos. After delamination from the neural crest and migration (here indicated for the classical dorso-lateral pathway), the cells activate their signaling and transcriptional program marked by expression of *Mitf* and *Kit* (among other factors), begin to migrate into the epidermis at the indicated times and eventually into hair follicles where they give rise to differentiated melanocytes during the first hair cycle and to stem cells capable of replenishing melanocytes during the second and subsequent hair cycles. Schematic representation redrawn from (Vandamme and Berx 2019). *NT* neural tube, *NC* neural crest, *DP* dermal papilla, *HF* hair follicle, *MSC* melanocyte stem cell

(Rabbani et al. 2011; Zhou et al. 2016). Hence, WNT, which originates in the dermal papilla and hair stem cells (but not in melanocyte stem cells themselves), is crucial for the physiological regulation of pigmentation during normal hair cycling. The process is further aided by the expression of *Edns* as well as retinoic acid, which leads to increased expression of *Kit* in melanocyte stem cells (Lu et al. 2020). Nevertheless, melanocyte stem cell physiology is far more complex as the cells are also regulated by sympathetic innervation of the stem cell niche. It has been shown, for instance, that acute stress in mice leads to rapid proliferation of melanocyte stem cells followed by their rapid loss, thereby providing an explanation for stress-associated graying (Zhang et al. 2020).

Studies in mice have shown that while the first hair pigmentation is synchronous from head to tail, the second and subsequent hair cycles are not. As seen after hair clipping at postnatal day (P)21, when hairs are in telogen, the naked skin of both head and back appears pink, indicating lack of pigment-bearing follicular melanocytes. At P30, however, the shaved back skin appears black, indicating repigmentation of follicles during anagen, but the shaved head skin is still pink, indicating delayed initiation of anagen. Anagen and pigmentation in the head area can be advanced by hair plucking (epilation) at P21, which leads to mild inflammation, augmented levels of EDNs (in particular EDN3), and darker pigment patches. These changes are due to an EDN3-mediated activation of melanocyte stem cells (Li et al. 2017).

The regulation of melanocyte stem cells and their niche is critical for the physiology and pathophysiology of pigmentation. Physiological hair graying, for instance, appears to be associated with an age-dependent depletion of melanocyte stem cells associated with reduced self-renewal and the appearance of pigmented cells in the niche. In vitiligo, which is an immune-mediated patchy skin depigmentation, the depigmented interfollicular skin can sometimes be re-pigmented starting out from hair follicles that retain functional stem cells that have escaped immune destruction (reviewed in Nishimura (2011)). Nevertheless, while epithelial stem cells control the maintenance of melanocyte stem cells in the stem cell niche, the isolated absence of melanocyte stem cells does not impact the hair cycle.

As mentioned, in human skin and the parts of mouse skin with low density or complete absence of hair follicles, melanocytes reside in the interfollicular epidermis. The stem cells giving rise to these melanocyte populations are localized in part in the dermis and the epidermis, but their analysis lags behind that of melanocyte stem cells in hair follicles. In the mouse tail, stem cells have been characterized in a niche located at each inter-scale border region, but in human skin, they have not yet been characterized. Nevertheless, stem cells not associated with follicles may be of great importance in multiple human disorders such as human Waardenburg syndromes, whose common features are hearing deficiency and skin and eye pigment alterations; piebaldism; vitiligo, as mentioned above; and melanoma. In fact, heterogeneity in cutaneous melanoma, which is associated with a high mutational load and phenotypic switching, is potentially also linked to the still inherent plasticity of the neural crest cells from which melanoma cells are originally derived. Both in mice and humans, for instance, melanomas may acquire glial features. This may be of considerable interest as peripheral glial cells have increasingly been recognized in recent years not only to provide support for neurons and axons but also to promote survival and growth of stromal cells during tissue repair following injury (Parfejevs et al. 2018). Hence, it is conceivable that regardless of whether glial cells in melanomas are directly derived from melanoma cells themselves by phenotypic switching or are recruited into melanoma tissue from peripheral glial, glial cells may play important roles in supporting the stromal tumor microenvironment and hence the survival and growth of melanoma cells.

6.4 The Development of Melanocytes in Feather Buds and Follicles

As usual in biology, evolution selects for tight interplays between form and function. More so than mammalian hairs, feathers in different body regions and at different developmental stages perform vastly different functions, with some of them being downy to keep the bird warm, and others firm to provide lift for flight and resist the shear forces of the air or, in aquatic birds, the flow of water, or to provide water repellency. Furthermore, feathers may show intricate pigment patterns such as barring and lateral asymmetries.

The first intimations of feather distribution over different body regions are manifested in the formation of feather tracts, marking "feather-competence" fields, which in the chick, for instance, can be seen already at E6.5 of development. The molecular bases of tract formation, however, is only partially understood (Chen et al. 2015). Nevertheless, the formation of feather follicles shows some similarities to that of hair follicles, with early dermal condensations marking the future position of the respective appendage, and bud formation preceding the formation of the mature structure (Chen et al. 2015). Like hair follicles, feather follicles are divided into a proximal, lower part where the dermal papilla sits, and a more distal part, the collar bulge, that contains epithelial stem cells (Fig. 6.3). And like hair follicles, feather follicles undergo a regenerative cycling marked by growth, rest, and molting. During the resting phase, the width of the follicle shrinks along with the dermal papilla, and upon molting, when feathers are shed, the follicle is reduced in both length and width. In the growing phase, immature melanocytes that retain a BrdU label for 8 days (thus much shorter than mouse melanocyte stem cells in the bulge of hair follicles) are present in the lower bulge and may represent melanocyte progenitors. In the resting phase, such cells are found arranged in a ring in the papillar ectoderm just above the dermal papilla, and they are quiescent (Lin et al. 2013). Nevertheless, whether these quiescent cells represent the equivalent of melanocyte stem cells in the bulge or secondary hair germ of hair follicles requires further analysis (for a review, see Chen et al. (2015)).

Interestingly, as demonstrated by heterospecific transplantation in vivo and in vitro, purified melanocytes of quail form networks via filopodia and are capable of inducing expression of Agouti signaling protein (ASIP, responsible for switching black eumelanin to yellow pheomelanin production) in adjacent dermal cells. Hence,

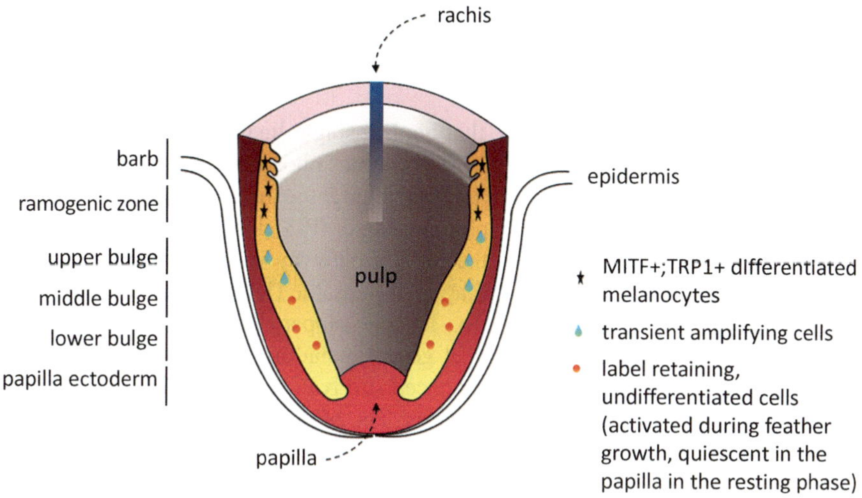

Fig. 6.3 Schematic depiction of melanocytes in growing feather follicle. The development of melanoblasts/melanocytes in birds is similar to that in mammals. Bird melanoblasts are neural crest-derived cells that migrate into feather follicles where they give rise to label retaining, undifferentiated cells, and transient amplifying cells that based on tracing techniques migrate distally and eventually differentiate into mature melanocytes in the ramogenic and barb zone of the feather. During the phase when the feather grows, the cells in the lower bulge retain a BrdU label for 8 days, shorter than the 70 days reported for mouse melanocyte stem cells in the hair follicle bulge. During the resting phase, the cells become quiescent and presumably retain label for a longer period, but this has not yet been analyzed in detail. Further analysis is also required to determine whether feather follicles harbor melanocyte stem cells equivalent to those in hair follicles. Redrawn after (Lin et al. 2013). For a review see (Chen et al. 2015)

by regulating ASIP, avian melanocytes can cell-autonomously generate striping patterns (Inaba et al. 2019).

6.5 The Interplay Between Signaling Pathways and Transcription Regulation During Development of Neural Crest-Derived Melanocytes

The combined analysis of in vivo genetic models and in vitro culture systems led to the discovery of many factors controlling melanocyte development. Many of these factors operate both in early and later stages of development, can either promote or inhibit melanocyte development, and are also instrumental in other organ systems. Hence, their mutations can lead to syndromic pigment alterations that affect not only skin and hair color and melanocytes in sensory organs but also, for instance, bone, heart, or brain development.

A list of factors implicated in melanocyte development is shown in Table 6.1, and we here highlight but a few of them. Among the earliest gene products marking future melanoblasts are the previously mentioned cell surface receptor KIT and the transcription factor MITF.

KIT is a receptor tyrosine kinase that binds a single ligand, KIT ligand (KITL). Upon ligand binding, KIT homodimerizes, becomes phosphorylated, and stimulates a wide array of downstream pathways, including the mitogen-activated protein kinase (MAPK) pathway, the phosphoinositide-3 (PI3) kinase pathway, and the JAK/STAT pathway. Interestingly, the MAPK pathway is critical for melanocyte development but the PI3 kinase pathway is dispensable (Kissel et al. 2000).

MITF is a basic helix–loop–helix leucine zipper factor that binds E-box sequences in the promoters of its target genes, including a wide variety of genes involved in metabolism, cell proliferation, cell survival, and lysosome biogenesis, in addition to genes involved in melanogenesis. The *Mitf* gene is comprised of at least 18 exons, with nine of them linked to their own promoters, leading to mRNAs with alternative 5′ ends and at least five protein isoforms differing at their amino termini. With multiple additional alternative splicing events, some of them affecting common exons, and numerous post-translational modifications, the *Mitf* gene may in fact give rise to a large number of distinct proteins. One type of these, M-MITF, represents the major isoform in melanoblasts and melanocytes (for a recent review, see Goding and Arnheiter (2019)).

Interestingly, in melanoblasts, KIT signaling is not required for the onset of MITF expression, nor is MITF required for the onset of KIT expression (Opdecamp et al. 1997; Hou et al. 2000). Nevertheless, the two genes interact in complex ways. As shown in melanoma cells, MAPK activation indirectly influences the nuclear/cytoplasmic distribution of M-MITF (and hence its transcriptional activity) (Ngeow et al. 2018) and MITF is needed for the maintenance of KIT expression in melanoblasts (Opdecamp et al. 1997). KIT and MITF are thus part of a feedback loop that further intersects with the above mentioned endothelin-3 (EDN3) that binds the G-coupled receptor EDNRB, activates Gαq and Gαi subunits, and stimulates melanoblast proliferation (Opdecamp et al. 1998). In fact, EDNRB-deficient embryos have few melanoblasts and adults derived from them display only a few small pigmented spots. EDNRB is, however, also expressed in the neural tube, and recombination experiments with EDNRB wildtype and mutant neural tubes have shown that EDNRB also stimulates melanoblast development in a cell non-autonomous way via other cells, likely by stimulating their KITL expression (Hou et al. 2004). KIT, MITF, and EDNRB are thus functionally interconnected.

As is well known, no gene acts in isolation, and small core regulatory networks are embedded in larger networks by which they are regulated and which they regulate. The M-promoter of *Mitf*, for instance, is positively regulated by a number of transcription factors, including SOX10, a member of the SRY-box containing high mobility group of proteins; PAX3, a paired domain transcription factor; LEF/TCF, another high mobility group protein; and CREB, a basic-leucine zipper protein. On the other hand, the forkhead box D3 protein (FOXD3) represses *M-Mitf* expression in neural crest cells and promotes the glial lineage at the expense of

Table 6.1 Factors implicated in the specification and development of neural crest-derived melanocytes[a]

Functional group	Symbol	Name	Function	Human neural crest-related disorders (OMIM No.)[b]
Extracellular matrix and cell adhesion	ADAMTS20	Adamts20 metalloprotease	Involved in ECM remodeling	
	ECM	Extracellular matrix (collagen, fibronectin, laminin)	Migration of melanoblasts	
	EFNB	Ephrin B	Early migration of melanoblasts	
	EPHB	Ephrin B receptor	Early migration of melanoblasts	
	INT	Integrins	Migration of melanoblasts	
Growth factors, growth factor receptors, and signaling-related molecules	BMP	Bone morphogenic protein	Promotes or inhibits melanoblast specification depending on conditions	
	CDH	Cadherins	Migration of melanoblasts regulated by various combinations of cadherin family members	
	EDN3	Endothelin-3 (a 20-residue peptide)	Melanoblast proliferation and survival	Waardenburg syndrome IV (277580), Hirschsprung disease and related disorders (142,623, 209,880)
	EDNRB	G-coupled endothelin receptor B	Melanoblast proliferation and survival	Waardenburg syndrome IV (277580), Hirschsprung disease and related disorders (142,623, 600,501)
	FGF	Fibroblast growth factors (various)	Neural crest induction	
	FZD	Frizzled family of G-coupled receptors for WNT	Activation of canonical WNT signaling, induction of neural crest, and promotion of melanoblast development	

(continued)

Table 6.1 (continued)

Functional group	Symbol	Name	Function	Human neural crest-related disorders (OMIM No.)[b]
	KITL	KIT ligand	Melanoblast survival, proliferation, migration	
	KIT	KIT receptor tyrosine kinase	Activates RAS/RAF/MAPK pathway, melanoblast survival, proliferation, migration	Piebaldism (172800)
	LRP	Family of receptors related to low-density lipoprotein receptors	Co-receptor for FZD, activation of canoncial WNT signaling, induction of neural crest, and promotion of melanoblast development	
	NOTCH	Family of receptors related to *Drosophila* Notch	Melanoblast survival and melanocyte stem cell maintenance	
	TGFß	Transforming growth factor	Stem cell quiescence	
	WNT	Wingless-related/ MMTV integration site	Induction of neural crest, promotion of melanoblast development	
	YAP1/2	YES-associated protein 1/2	Both cytoplasmic and nuclear localization (also acting as transcriptional co-factor). Neural crest cell delamination	
Transcription factors and co-factors	AP-2	AP-2 transcription factor	Melanoblast differentiation	
	CTNNB1	β-catenin	One of the effectors of WNT signaling	
	FOXD3	Forkhead box D3	Neural crest induction, can inhibit melanoblast specification	
	LEF1/TCF	Lymphoid enhancer factor/T cell-specific transcription factor	One of the effectors of WNT signaling	

(continued)

Table 6.1 (continued)

Functional group	Symbol	Name	Function	Human neural crest-related disorders (OMIM No.)[b]
	MITF	Microphthalmia-associated transcription factor, basic helix–loop–helix leucine zipper family	Melanoblast specification, survival, proliferation, migration	Waardenburg syndrome IIa (193510)
	PAX3	Paired domain protein-3	Melanoblast specification, survival, and proliferation; regulation of adult melanocyte stem cells	Waardenburg syndrome I, II, III (193,500, 193,510, 148,820)
	SNAI2/ SLUG	Snail homolog protein-2, zinc finger protein	Epithelial-to-mesenchymal transition, melanoblast specification downstream of MITF?	
	SOX2	SRY-box containing protein-2	Stem-cell maintenance, Glia specification, transcriptional repression of MITF	Microphthalmia, syndromic 3 (206900)
	SOX9	SRY-box containing protein-9	Induction and differentiation of the neural crest	
	SOX10	SRY-box containing protein-10	Melanoblast specification, survival, proliferation, and differentiation	Waardenburg syndrome IV (613266)
	ZEB1	Zinc finger E-box binding homeobox 1	Epithelial to mesenchymal transition	
	ZEB2	Zinc finger E-box binding homeobox 2	Epithelial to mesenchymal transition	
	ZIC1/2	Zinc finger protein of cerebellum 1/2	Neural crest induction	
RNA editing	ADAR1	Adenosine deaminase acting on RNA	Promoting survival of melanocytes *and* differentiation of Schwann cells. Mutations in ADAR1 in mice lead to upregulation of interferon-	

(continued)

Table 6.1 (continued)

Functional group	Symbol	Name	Function	Human neural crest-related disorders (OMIM No.)[b]
			stimulated genes. In vitro, elimination of MDA5, an RNA sensor and mediator of interferon production, rescues ADAR1-associated melanocyte and Schwann cell defects (Gacem et al. 2020).	

[a]Factors directly involved in melanin synthesis are excluded
[b]OMIM: Online Mendelian Inheritance in Man (https://omim.org)

pigment cells (Nitzan et al. 2013a). Interestingly, *Foxd3* is downregulated in melanoblasts on the dorso-lateral pathway earlier than in SCP-derived melanoblasts generated at nerve endings (Nitzan et al. 2013b). Based on in vitro experiments, each of these *Mitf*-regulating proteins is themselves regulated by extracellular signaling, for instance, CREB by cAMP activating signals, LEF1 by WNT/ß-catenin, PAX3 by PI3K and HIPPO signaling, and FOXD3 by BMP and WNT3A. None of these factors are melanocyte-specific, though, and it is only through their combination and temporal control that specificity might be generated (reviewed in Goding and Arnheiter (2019)).

As illustrated by the effects of altered activities of ß-catenin, the above mentioned factors need to be finely tuned both temporally and spatially to exert their normal function. In mice, for instance, expression of a stably active ß-catenin, provided it occurs within a specific temporal window, leads to the appearance of white belly spots, not by altering cell proliferation but by inhibition of melanoblast migration, in part by activation of MITF (Gallagher et al. 2013). In fact, it is still not entirely clear how even within the boundaries of its physiological levels, MITF exerts its many functions, which besides the control of migration includes the promotion of both proliferation and differentiation that are usually considered mutually exclusive states of a cell. It is conceivable that different target genes are accessed by selective cooperation with transcriptional co-factors (of which there are many, including p300/CBP and the SWI/SNF complex), by selective splicing or post-translational modifications of MITF, or by a combination of these mechanisms. Nevertheless, delineating the precise mechanisms of this differential control remains a challenge, mainly because co-factors and signaling pathways leading to post-translational modifications are not MITF-specific but affect a large number of other genes and pathways (for a deeper discussion of these issues, see (Goding and Arnheiter 2019).

6.6 The Development of RPE Cells

Like neural crest-derived melanocytes, RPE cells of vertebrates contain light-absorbing melanosomes. The functions of these cells, however, go far beyond their mere role to provide a shield preventing light from reaching deeper structures. RPE cells form a monolayer that on its apical side abuts the photoreceptors of the retina and on its basal side the choriocapillaris. They thus form a barrier, the blood–retina barrier, between the vasculature of the choriocapillaris and the blood vessel-free part of the retina. Aided by a multilayer basement membrane, Bruch's membrane, this barrier controls the bi-directional flow of water, ions, nutrients, and metabolites. RPE cells also provide trophic factors, are instrumental for the turnover of photoreceptor outer segments, catalyze the isomerization of all-trans retinal into 11-cis retinal in the so-called visual cycle, and have anti-oxidant properties (for reviews, see Strauss (2005), Ma et al. (2019)). In addition, the distal extension of the RPE forms the pigmented back layer of the iris. Without a functioning RPE, normal vision is impossible, and so it is not surprising that developmental and degenerative disorders of the retina, including age-related macular degeneration (AMD), may have their pathogenetic origin in alterations of the RPE (Ambati and Fowler 2012; Letelier et al. 2017).

Developmentally, the RPE is derived from the optic neuroepithelium that buds out from the neuroepithelium at the level of the diencephalon and also gives rise to the multiple neuronal and glial layers of the retina. Presumptive RPE and retina, however, soon show markedly different cellular proliferation rates that are linked, among other pathways, to the differential expression of the transcription factor MITF with its anti-proliferative activities (Nguyen and Arnheiter 2000; Bharti et al. 2006) (Fig. 6.4). This differential MITF expression is the result of exposition of the optic neuroepithelium to competing gradients of growth factors. On the distal side of the budding optic vesicle, the neuroepithelium comes in contact with the surface ectoderm, which secretes fibroblast growth factors (FGF1 and FGF2) and goes on to form the lens. The presumptive retina then expresses FGF8 in its central area. These FGFs induce the expression of the homeodomain protein VSX2, which in turn represses the expression of MITF and the related TFEC. In fact, *Vsx2* mutant mice show ectopic expression of MITF in the presumptive retina, which now develops as an RPE, but this does not happen in *Vsx2/Mitf* double mutants, which show a markedly improved retinal phenotype. This indicates that physiologically, MITF is the critical mediator of neuroepithelial hypoproliferation and development as an RPE. On the proximal side, the optic neuroepithelium is exposed to activin, a member of the TGFβ family of proteins; to BMPs, in particular BMP4 and BMP7, both critical for eye development; and to WNT/β-catenin signaling, which leads to upregulation of MITF and OTX2, another RPE transcription factor. Deletion of *Ctnnb1*, the gene encoding β-catenin, in the presumptive RPE leads to downregulation of *Mitf* and *Otx2* and concomitant upregulation of rctinal markers. On the other hand, expression of OTX2 or constitutively active β-catenin in the presumptive retina leads to upregulation of *Mitf* and expression of RPE markers. In

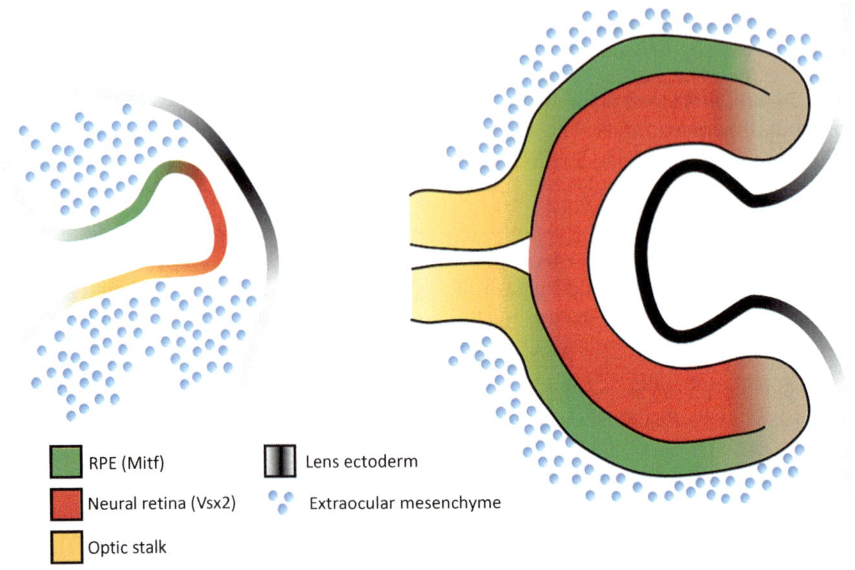

Fig. 6.4 Development of retinal pigment epithelial cells from the optic neuroepithelium. The eye buds out from the ventral neuroepithelium at the level of the diencephalon. Upon contact with the surface ectoderm and exposure to signaling molecules such as FGFs, the neuroepithelium differentiates into a distal portion that invaginates, activates a retinogenic transcriptional program, and eventually differentiates into the multilayered structure of the neuroretina. The proximal portion, in contrast, is exposed to BMPs and other factors. It remains a single layer of cells that activate a transcriptional program that in part is shared with neural crest-derived pigment cells. The cells become pigmented and form the retinal pigment epithelium (RPE) that proximally extends to form the back layer of the iris. RPE cells that abut the retina perform critical physiological functions for retinal photoreceptor cells, both for light reception and survival, and also act as a light shield. For a review, see (Bharti et al. 2006)

addition to the above mentioned factors, *Mitf* is also regulated by ZEB1, which inhibits *Mitf* expression (Liu et al. 2009), and *Klotho*, which promotes *Mitf* expression (Kokkinaki et al. 2013). Moreover, the LIM homeodomain protein LHX2 is required for BMP-mediated *Mitf* expression (Bharti et al. 2006; Yun et al. 2009; Ma et al. 2019).

An interesting role in the RPE is played by the paired domain protein PAX6. PAX6 has multiple roles in eye development across different phyla. In drosophila, for instance, ectopic expression of PAX6 in corresponding imaginal discs can induce functional eyes on legs, wings, or antennae (Halder et al. 1995). In frogs, ectopic PAX6 expression can lead to rudimentary additional eyes in the head region (Chow et al. 1999). Although PAX6 is normally responsible for the formation of the lens, iris, and retina, in mice it is also transiently expressed in the developing RPE. In fact, selective elimination of PAX6 in the RPE leads to RPE hypopigmentation, which is somewhat surprising as one might expect that elimination of a neuroepithelial factor, which acts to promote retina development, would further help RPE differentiation and not inhibit it (Raviv et al. 2014). Also, the introduction of a heterozygous *Pax6* mutation on an *Mitf* mutant background had an unexpected effect. In *Mitf* mutant RPE, the area showing RPE-to-retina transdifferentiation expresses PAX6 at high levels, but in *Mitf* mutants carrying a heterozygous *Pax6* null mutation, the transdifferentiation was exacerbated, not reduced as expected. Furthermore, PAX6 overexpression markedly improved the *Mitf* mutant RPE phenotype by reducing cell proliferation. Expression profiling and in vitro experiments then indicated that PAX6, together with MITF and its paralog TFEC, suppresses FGF15 and the Dickkopf WNT signaling pathway inhibitor-3 (DKK3), and that FGFs and DKK3 in combination inhibit canonical WNT/β-catenin signaling and stimulate the expression of retinogenic genes. Thus, PAX6 promotes either retinal or RPE development in a context-dependent manner (Bharti et al. 2012).

The above observations indicate that initially, the neuroepithelium is bipotential and can readily be converted from presumptive retina to presumptive RPE and vice versa. The central nexus in this developmental process seems to be MITF, not unlike what is seen in neural crest-derived melanocytes. The specific MITF isoforms found in the presumptive RPE, however, are different from those seen in melanoblasts and melanocytes. In the latter cells, M-MITF predominates, and other isoforms are negligible. Selective deletions of A- or D-Mitf, for instance, do not visibly affect coat pigmentation (Bharti et al. 2012; Phelep et al. 2017; Flesher et al. 2020). Conversely, in the developing RPE, several other isoforms, in particular A-, H-, and D-MITF, predominate (Bharti et al. 2008), but M-Mitf expression is negligible and its selective deletion does not affect RPE pigmentation (Flesher et al. 2020). Nevertheless, one consequence of the expression of multiple isoforms is the potential for compensation when one or several isoforms are missing. A selective elimination of D-MITF, for instance, just leads to a slight delay in onset of RPE pigmentation but is soon compensated for by the upregulation of other isoforms, in particular H-MITF, and the RPE soon appears normal (Bharti et al. 2012). Another consequence of the availability of different isoforms is the potential regulation of intracellular distribution. Unlike M-MITF (which is predominantly nuclear), the other isoforms contain in their aminotermini an SR-QL motif, which based on in vitro studies allows them to be regulated in their nuclear/cytoplasmic distribution in a nutrient and M-TOR-dependent fashion (for a review, see (Goding and Arnheiter 2019)).

While many factors and pathways are thus involved to various extents in RPE development, *Mitf* plays critical roles for RPE development and specification. In the absence of all major isoforms of *Mitf*, the presumptive RPE, though still discernible, remains free of pigment, hyperproliferates, and develops into a pseudostratified, retina-like epithelium that expresses retinal markers, particularly in its dorsal part. Eventually, this retina-like epithelium forms a multilayered second retina, which is, however, inverted compared to the normal retina, suggesting that cellular respecification is achieved without a change in apical-to-basal polarity. The second retina cannot form normal neuronal connections, however, and eventually degenerates (Nguyen and Arnheiter 2000). Also, the optic fissure does not properly close, and the eye remains small and malformed.

In humans, *MITF* mutations are associated with Waardenburg Syndrome type IIa, the more severe Tietz albinism-deafness syndrome (TADS), and a novel syndrome called COMMAD, which stands for **C**oloboma, **O**steopetrosis, **M**icrophthalmia, **M**acrocephaly, **A**lbinism, and **D**eafness and which is the result of compound heterozygosity for different mutant *MITF* alleles (George et al. 2016). Unlike COMMAD patients, Waardenburg IIa individuals usually retain one wildtype copy of *MITF*, which, however, may be insufficient to fully protect neural crest-derived melanocytes (Tassabehji et al. 1994) but sufficient to protect the RPE. The analysis of *Mitf* mutant mice has indeed shown that RPE cells are generally less sensitive to *Mitf* mutations compared to neural crest-derived melanoblasts/melanocytes. In fact, while melanoblasts/melanocytes cease development and presumably die in the total absence of functional copies of *Mitf*, mutant RPE cells hyperproliferate, "transdifferentiate" into retinal cells, and survive (for a recent review, see Ma et al. (2019)). It is conceivable that other bHLH-leucine zipper proteins from the same MIT subfamily, which share an overlapping set of target genes by binding to the same regulatory elements as MITF, are expressed at sufficient levels to partially compensate for the lack of MITF in RPE cells but not in melanocytes. Nevertheless, the mechanistic details for this differential sensitivity to *MITF/Mitf* mutations are still to be explored.

6.7 Concluding Remarks

The study of the development of neural crest-derived melanocytes in birds and mammals has given us many mechanistic and molecular insights that may eventually allow us to make the link to the many pigmentation patterns in the adult and help explain some of the pigmentation alterations seen in animals and humans. In fact, the easy visibility of the differentiated, pigmented cell, and in particular also the high dynamic range of the pigmentation output that allows us to see the slightest changes in cell number and/or performance, make this system ideal to study the phenotypic consequences of even mild genetic and epigenetic influences on individual cells.

Likewise, the study of the development of RPE cells has revealed insights into how neuroepithelial cells of the central nervous system can become separated into intricately connected neurons and glial cells, the neuroretina, and the seemingly simple monolayer of retinal pigmented cells. Although the developmental origin and functions of RPE cells are clearly different from those of melanocytes derived from the neural crest, RPE cells nevertheless share with the latter significant parts of developmentally important molecular pathways, such as the transcription factor MITF and the pigment biosynthetic enzymes. This begs the intriguing question of whether the different pigment cells might all be evolutionarily related (Arnheiter 1998). In fact, knowledge gained from the study of pigment cell development not only gives us insights into normal ontogenetic mechanisms and their aberrations in disease but also highlights interspecies differences that over evolutionary time may have been shaped to yield the magnificent abundance of pigmentary phenotypes we can observe today.

Acknowledgments We thank Drs. Ling Hou, Lionel Larue, and Lukas Sommer for critical suggestions and review of the manuscript. The authors' own work discussed in this paper was supported in part by NINDS, National Institutes of Health, United States, and the Kanton of Zürich, Switzerland.

References

Adameyko I, Lallemend F, Aquino JB, Pereira JA, Topilko P, Muller T, Fritz N, Beljajeva A, Mochii M, Liste I et al (2009) Schwann cell precursors from nerve innervation are a cellular origin of melanocytes in skin. Cell 139:366–379

Ambati J, Fowler BJ (2012) Mechanisms of age-related macular degeneration. Neuron 75:26–39

Arnheiter H (1998) Evolutionary biology. Eyes viewed from the skin. Nature 391:632–633

Baggiolini A, Varum S, Mateos JM, Bettosini D, John N, Bonalli M, Ziegler U, Dimou L, Clevers H, Furrer R et al (2015) Premigratory and migratory neural crest cells are multipotent in vivo. Cell Stem Cell 16:314–322

Bharti K, Nguyen MT, Skuntz S, Bertuzzi S, Arnheiter H (2006) The other pigment cell: specification and development of the pigmented epithelium of the vertebrate eye. Pigment Cell Res 19:380–394

Bharti K, Liu W, Csermely T, Bertuzzi S, Arnheiter H (2008) Alternative promoter use in eye development: the complex role and regulation of the transcription factor MITF. Development 135:1169–1178

Bharti K, Gasper M, Ou J, Brucato M, Clore-Gronenborn K, Pickel J, Arnheiter H (2012) A regulatory loop involving PAX6, MITF, and WNT signaling controls retinal pigment epithelium development. PLoS Genet 8:e1002757

Botchkareva NV, Khlgatian M, Longley BJ, Botchkarev VA, Gilchrest BA (2001) SCF/c-kit signaling is required for cyclic regeneration of the hair pigmentation unit. FASEB J 15:645–658

Chen CF, Foley J, Tang PC, Li A, Jiang TX, Wu P, Widelitz RB, Chuong CM (2015) Development, regeneration, and evolution of feathers. Annu Rev Anim Biosci 3:169–195

Chow RL, Altmann CR, Lang RA, Hemmati-Brivanlou A (1999) Pax6 induces ectopic eyes in a vertebrate. Development 126:4213–4222

Dupin E, Le Douarin NM (2014) The neural crest, a multifaceted structure of the vertebrates. Birth Defects Res C Embryo Today 102:187–209

Dupin E, Glavieux C, Vaigot P, Le Douarin NM (2000) Endothelin 3 induces the reversion of melanocytes to glia through a neural crest-derived glial-melanocytic progenitor. Proc Natl Acad Sci U S A 97:7882–7887

Dupin E, Real C, Glavieux-Pardanaud C, Vaigot P, Le Douarin NM (2003) Reversal of developmental restrictions in neural crest lineages: transition from Schwann cells to glial-melanocytic precursors in vitro. Proc Natl Acad Sci U S A 100:5229–5233

Dupin E, Calloni GW, Coelho-Aguiar JM, Le Douarin NM (2018) The issue of the multipotency of the neural crest cells. Dev Biol 444(Suppl 1):S47–S59

Flesher JL, Paterson-Coleman EK, Vasudeva P, Ruiz-Vega R, Marshall M, Pearlman E, Macgregor GR, Neumann J, Ganesan AK (2020) Delineating the role of MITF isoforms in pigmentation and tissue homeostasis. Pigment Cell Melanoma Res 33:279–292

Furlan A, Adameyko I (2018) Schwann cell precursor: a neural crest cell in disguise? Dev Biol 444 (Suppl 1):S25–S35

Gacem N, Kavo A, Zerad L, Richard L, Mathis S, Kapur RP, Parisot M, Amiel J, Dufour S, De La Grange P et al (2020) ADAR1 mediated regulation of neural crest derived melanocytes and Schwann cell development. Nat Commun 11:198

Gallagher SJ, Rambow F, Kumasaka M, Champeval D, Bellacosa A, Delmas V, Larue L (2013) Beta-catenin inhibits melanocyte migration but induces melanoma metastasis. Oncogene 32:2230–2238

George A, Zand DJ, Hufnagel RB, Sharma R, Sergeev YV, Legare JM, Rice GM, Scott Schwoerer JA, Rius M, Tetri L et al (2016) Biallelic Mutations in MITF Cause Coloboma, Osteopetrosis, Microphthalmia, Macrocephaly, Albinism, and Deafness. Am J Hum Genet 99:1388–1394

Goding CR, Arnheiter H (2019) MITF-the first 25 years. Genes Dev. https://doi.org/10.1101/gad.324657.119

Halder G, Callaerts P, Gehring WJ (1995) Induction of ectopic eyes by targeted expression of the eyeless gene in Drosophila. Science 267:1788–1792

Hari L, Miescher I, Shakhova O, Suter U, Chin L, Taketo M, Richardson WD, Kessaris N, Sommer L (2012) Temporal control of neural crest lineage generation by Wnt/beta-catenin signaling. Development 139:2107–2117

Hou L, Panthier JJ, Arnheiter H (2000) Signaling and transcriptional regulation in the neural crest-derived melanocyte lineage: interactions between KIT and MITF. Development 127:5379–5389

Hou L, Pavan WJ, Shin MK, Arnheiter H (2004) Cell-autonomous and cell non-autonomous signaling through endothelin receptor B during melanocyte development. Development 131:3239–3247

Huszar D, Sharpe A, Hashmi S, Bouchard B, Houghton A, Jaenisch R (1991) Generation of pigmented stripes in albino mice by retroviral marking of neural crest melanoblasts. Development 113:653–660

Inaba M, Jiang TX, Liang YC, Tsai S, Lai YC, Widelitz RB, Chuong CM (2019) Instructive role of melanocytes during pigment pattern formation of the avian skin. Proc Natl Acad Sci U S A 116:6884–6890

Jaegle M, Ghazvini M, Mandemakers W, Piirsoo M, Driegen S, Levavasseur F, Raghoenath S, Grosveld F, Meijer D (2003) The POU proteins Brn-2 and Oct-6 share important functions in Schwann cell development. Genes Dev 17:1380–1391

Joshi SS, Tandukar B, Pan L, Huang JM, Livak F, Smith BJ, Hodges T, Mahurkar AA, Hornyak TJ (2019) CD34 defines melanocyte stem cell subpopulations with distinct regenerative properties. PLoS Genet 15:e1008034

Kinsler VA, Larue L (2018) The patterns of birthmarks suggest a novel population of melanocyte precursors arising around the time of gastrulation. Pigment Cell Melanoma Res 31:95–109

Kissel H, Timokhina I, Hardy MP, Rothschild G, Tajima Y, Soares V, Angeles M, Whitlow SR, Manova K, Besmer P (2000) Point mutation in kit receptor tyrosine kinase reveals essential roles for kit signaling in spermatogenesis and oogenesis without affecting other kit responses. EMBO J 19:1312–1326

Kokkinaki M, Abu-Asab M, Gunawardena N, Ahern G, Javidnia M, Young J, Golestaneh N (2013) Klotho regulates retinal pigment epithelial functions and protects against oxidative stress. J Neurosci 33:16346–16359

Krispin S, Nitzan E, Kassem Y, Kalcheim C (2010) Evidence for a dynamic spatiotemporal fate map and early fate restrictions of premigratory avian neural crest. Development 137:585–595

Kumar D, Nitzan E, Kalcheim C (2019) YAP promotes neural crest emigration through interactions with BMP and Wnt activities. Cell Commun Signal 17:69

Le Douarin NM (1974) Cell recognition based on natural morphological nuclear markers. Med Biol 52:281–319

Le Douarin NM, Kalcheim C (1999) The neural crest. Cambridge, Cambridge University Press

Letelier J, Bovolenta P, Martinez-Morales JR (2017) The pigmented epithelium, a bright partner against photoreceptor degeneration. J Neurogenet 31:203–215

Li H, Fan L, Zhu S, Shin MK, Lu F, Qu J, Hou L (2017) Epilation induces hair and skin pigmentation through an EDN3/EDNRB-dependent regenerative response of melanocyte stem cells. Sci Rep 7:7272

Lin SJ, Foley J, Jiang TX, Yeh CY, Wu P, Foley A, Yen CM, Huang YC, Cheng HC, Chen CF et al (2013) Topology of feather melanocyte progenitor niche allows complex pigment patterns to emerge. Science 340:1442–1445

Liu Y, Ye F, Li Q, Tamiya S, Darling DS, Kaplan HJ, Dean DC (2009) Zeb1 represses Mitf and regulates pigment synthesis, cell proliferation, and epithelial morphology. Invest Ophthalmol Vis Sci 50:5080–5088

Lu Z, Xie Y, Huang H, Jiang K, Zhou B, Wang F, Chen T (2020) Hair follicle stem cells regulate retinoid metabolism to maintain the self-renewal niche for melanocyte stem cells. eLife 9: e52712

Ma X, Li H, Chen Y, Yang J, Chen H, Arnheiter H, Hou L (2019) The transcription factor MITF in RPE function and dysfunction. Prog Retin Eye Res 73:100766

Mintz B (1967) Gene control of mammalian pigmentary differentiation. I. Clonal origin of melanocytes. Proc Natl Acad Sci U S A 58:344–351

Moriyama M, Osawa M, Mak SS, Ohtsuka T, Yamamoto N, Han H, Delmas V, Kageyama R, Beermann F, Larue L et al (2006) Notch signaling via Hes1 transcription factor maintains survival of melanoblasts and melanocyte stem cells. J Cell Biol 173:333–339

Ngeow KC, Friedrichsen HJ, Li L, Zeng Z, Andrews S, Volpon L, Brunsdon H, Berridge G, Picaud S, Fischer R et al (2018) BRAF/MAPK and GSK3 signaling converges to control MITF nuclear export. Proc Natl Acad Sci U S A 115:E8668–E8677

Nguyen M, Arnheiter H (2000) Signaling and transcriptional regulation in early mammalian eye development: a link between FGF and MITF. Development 127:3581–3591

Nieto MA (2002) The snail superfamily of zinc-finger transcription factors. Nat Rev Mol Cell Biol 3:155–166

Nishimura EK (2011) Melanocyte stem cells: a melanocyte reservoir in hair follicles for hair and skin pigmentation. Pigment Cell Melanoma Res 24:401–410

Nishimura EK, Jordan SA, Oshima H, Yoshida H, Osawa M, Moriyama M, Jackson IJ, Barrandon Y, Miyachi Y, Nishikawa S (2002) Dominant role of the niche in melanocyte stem-cell fate determination. Nature 416:854–860

Nitzan E, Krispin S, Pfaltzgraff ER, Klar A, Labosky PA, Kalcheim C (2013a) A dynamic code of dorsal neural tube genes regulates the segregation between neurogenic and melanogenic neural crest cells. Development 140:2269–2279

Nitzan E, Pfaltzgraff ER, Labosky PA, Kalcheim C (2013b) Neural crest and Schwann cell progenitor-derived melanocytes are two spatially segregated populations similarly regulated by Foxd3. Proc Natl Acad Sci U S A 110:12709–12714

Opdecamp K, Nakayama A, Nguyen MT, Hodgkinson CA, Pavan WJ, Arnheiter H (1997) Melanocyte development in vivo and in neural crest cell cultures: crucial dependence on the Mitf basic-helix-loop-helix-zipper transcription factor. Development 124:2377–2386

Opdecamp K, Kos L, Arnheiter H, Pavan WJ (1998) Endothelin signalling in the development of neural crest-derived melanocytes. Biochem Cell Biol 76:1093–1099

Parfejevs V, Debbache J, Shakhova O, Schaefer SM, Glausch M, Wegner M, Suter U, Riekstina U, Werner S, Sommer L (2018) Injury-activated glial cells promote wound healing of the adult skin in mice. Nat Commun 9:236

Phelep A, Laouari D, Bharti K, Burtin M, Tammaccaro S, Garbay S, Nguyen C, Vasseur F, Blanc T, Berissi S et al (2017) MITF - A controls branching morphogenesis and nephron endowment. PLoS Genet 13:e1007093

Pla P, Monsoro-Burq AH (2018) The neural border: Induction, specification and maturation of the territory that generates neural crest cells. Dev Biol 444(Suppl 1):S36–S46

Rabbani P, Takeo M, Chou W, Myung P, Bosenberg M, Chin L, Taketo MM, Ito M (2011) Coordinated activation of Wnt in epithelial and melanocyte stem cells initiates pigmented hair regeneration. Cell 145:941–955

Raviv S, Bharti K, Rencus-Lazar S, Cohen-Tayar Y, Schyr R, Evantal N, Meshorer E, Zilberberg A, Idelson M, Reubinoff B et al (2014) PAX6 regulates melanogenesis in the retinal pigmented epithelium through feed-forward regulatory interactions with MITF. PLoS Genet 10:e1004360
Strauss O (2005) The retinal pigment epithelium in visual function. Physiol Rev 85:845–881
Tassabehji M, Newton VE, Read AP (1994) Waardenburg syndrome type 2 caused by mutations in the human microphthalmia (MITF) gene. Nat Genet 8:251–255
Vandamme N, Berx G (2019) From neural crest cells to melanocytes: cellular plasticity during development and beyond. Cell Mol Life Sci 76:1919–1934
Vandewalle C, Van Roy F, Berx G (2009) The role of the ZEB family of transcription factors in development and disease. Cell Mol Life Sci 66:773–787
Weston JA (1991) Sequential segregation and fate of developmentally restricted intermediate cell populations in the neural crest lineage. Curr Top Dev Biol 25:133–153
Wilkie AL, Jordan SA, Jackson IJ (2002) Neural crest progenitors of the melanocyte lineage: coat colour patterns revisited. Development 129:3349–3357
Yun S, Saijoh Y, Hirokawa KE, Kopinke D, Murtaugh LC, Monuki ES, Levine EM (2009) Lhx2 links the intrinsic and extrinsic factors that control optic cup formation. Development 136:3895–3906
Zhang B, Ma S, Rachmin I, He M, Baral P, Choi S, Goncalves WA, Shwartz Y, Fast EM, Su Y et al (2020) Hyperactivation of sympathetic nerves drives depletion of melanocyte stem cells. Nature 577:676–681
Zhou L, Yang K, Carpenter A, Lang RA, Andl T, Zhang Y (2016) CD133-positive dermal papilla-derived Wnt ligands regulate postnatal hair growth. Biochem J 473:3291–3305

Chapter 7
Pigment Cell Development in Teleosts

Hisashi Hashimoto, Makoto Goda, and Robert N. Kelsh

Abstract Fish generate their strikingly diverse body colouration with multiple pigment cell-types expressing different colours, in many species organized in complex and often beautiful patterns. Multiple pigment cell-types are derived from a transient embryonic tissue, the neural crest. Increasing genetic evidence suggests how the stem cell gives rise to the repertoire of pigment cell-types during development. Many genes encoding transcription factors and signaling molecules have been identified as key players regulating cell fate, proliferation and differentiation.

This chapter focuses on the genetic control of pigment cell development to generate melanophores, iridophores and xanthophores in zebrafish and also leucophores in medaka.

Keywords Fate specification · Intermediate progenitor · Transcription factor · Cell-type · Stem cell

7.1 Introduction

Skin colour patterns in vertebrates depend upon diverse pigments, either within organelles of pigment cells (chromatophores) or secreted from these cells into growing hair or feathers. In addition, some pigment cells rely on structural colour, whereby the physical properties of the organelles or their spacing affect the way they interact with light. Different combinations of chromatophores in different regions of the body generate integumental pigment pattern, providing an important component

H. Hashimoto
Graduate School of Science, Nagoya University, Nagoya, Japan

M. Goda
Hamamatsu University School of Medicine, Hamamatsu, Shizuoka, Japan

R. N. Kelsh (✉)
Department of Biology and Biochemistry, University of Bath, Bath, UK
e-mail: bssrnk@bath.ac.uk

H. Hashimoto et al. (eds.), *Pigments, Pigment Cells and Pigment Patterns*,
https://doi.org/10.1007/978-981-16-1490-3_7

of the adaptive mechanisms enabling survival (Fujii 1993b). Furthermore, in feathers, structural colour generated by the organization of keratin fibres and secreted pigment organelles play a major role in the formation of iridescent or intensely black or white colours. In fish, just as in amphibia and reptiles, skin pigment patterns depend upon a combination of (1) the multiple different chromatophore types, (2) their characteristic organelles containing pigments or generating structural colour, and (3) their distribution in the skin.

7.2 Pigment Cell-Types Identified in Fish

7.2.1 *Diversity of Pigment Cells in Poikilotherms (see Sect. 13.1.1)*

At least six kinds of chromatophores are present in poikilothermic vertebrates: namely, melanophores/melanocytes (black or brown), xanthophores (yellow), erythrophores (red), leucophores (whitish), cyanophores (blue) and iridophores (metallic or iridescent) (see Fig. 13.1, (Fujii 1993b; Fujii 1993a)). However, the situation is complex, because fish and amphibia have incredibly diverse pigment patterns, and some may well have evolved novel pigment cell-types, yet almost all species have never been assessed. Where other taxa are studied, new cell-types are identified. For example, recent examination of the genus *Danio* identified two populations of leucophore-like cells, each apparently derived by transdifferentiation of another cell-type, xanthophores and melanocytes (Lewis et al. 2019). Named xantholeucophores and melanoleucophores respectively, it remains to be seen how these relate to the genetically best-characterized leucophores, those of medaka, which appear to share an origin with xanthophores, but where transdifferentiation has never been proposed (Kimura et al. 2014; Nagao et al. 2014, 2018).

7.2.2 *Melanophore*

Melanophores in fish produce eumelanin, but apparently not pheomelanin. They are sometimes called melanocytes, since their developmental genetics shows strong homology to that of mammalian melanocytes. However, fish melanophores show two distinctive features compared to mammalian melanocytes: (1) they do not secrete melanin-containing melanosomes (see Fig. 13.2a) and (2) melanosomes in many species can be actively moved radially along the microtubules, causing them to either aggregate in the perikaryon or to disperse throughout the cytoplasm (aggregation-dispersion response) in response to environmental and/or nervous cues.

7.2.3 *Xanthophore and Erythrophore*

These two cell-types have been distinguished on the basis of their colour, and may be found in combinations within one species. However, in their biochemistry, xanthophores and erythrophores are highly similar, since both utilize pteridines and carotenoids as their pigmentary compounds (see Fig. 13.2b, c). Biochemical differences in the balance of their carotenoid composition appear to drive their specific colour. Since these colours range in a continuum through yellow to red, we consider these two chromatophores as variants of a single cell-type, which we will refer to as xanthophores in this chapter.

7.2.4 *Iridophore and Leucophore: Light Reflecting Chromatophores*

Iridophores and leucophores share the property of using a structural colouration mechanism to at least contribute to their appearance. Iridophores are abundant in iridescent parts of skin, especially silvery or whitish areas, but may also be blue. They are defined by transmission electron microscopy through having thin, membrane-bound guanine crystals, called reflecting platelets, in the cytoplasm (see Fig. 13.2e). Whether they look blue, silvery or whitish depends on the arrangement of the reflecting platelets, e.g. silver hue is dependent upon platelets being organized as uniformly-spaced stacks.

Leucophores look whitish under epi-illumination (Fujii 1993b). They are believed to contain crystalline uric acid in their cytoplasmic, membrane-bound leucosomes (see Fig. 13.2c), since this colouration becomes lost when they are treated with melamine (melamine is known to capture uric acid to form melamine–uric acid complex) (Hama 1975), but direct evidence for uric acid as a specific component of the leucosome has still to be provided. Their milky whitish colouration is thought to be dependent upon the less-organized arrangement of the leucosomes, which are more rounded and not organized as stacks, resulting in scattering of light.

Confusingly, iridophores can sometimes appear whitish, making them difficult to distinguish from leucophores. However, they can be distinguished by the following criteria: (1) Leucophores remain dendritic in their post-migratory positions, whereas iridophores are rounded in morphology (except when migrating). (2) Leucophores show motile activity of their leucosomes (aggregation-dispersion response), mostly in a direction contrasting with that of melanosomes and xanthosomes, e.g. under control of adrenergic neurons in medaka. (3) Iridophores contain 'reflecting platelets', stacks of long hexagonal guanine crystals, while leucophores contain leucosomes, which are rounded, when observed by transmission electron microscopy (see Fig. 13.2c, e). Iridophores are not dendritic nor motile in general, but an exception is the dendritic iridophores in a freshwater goby (*Odontobutis obscura*) with motile reflecting platelets (Iga and Matsuno 1986).

In taxonomic terms, iridophores are widespread, being found in teleosts, as well as in amphibians and reptiles. In contrast, to date leucophores have only been identified in a limited set of fish species, and are probably limited to the Osteichthyes: medaka and its related species (*Oryzias* species); guppy (*Lebistes reticulatus*); killifish, including Arabian killifish (*Aphanius dispar*) and mummichog (*Fundulus heteroclitus*); Darkbanded rockfish (*Sebastes inermis*); Japanese flounder (*Paralichthys olivaceus*) and zebrafish (*Danio rerio*) (see Sect. 7.2.6).

7.2.5 Cyanophore

Blue colouration in vertebrates, including fish such as the coral reef damselfish and surgeonfish, is mostly based on interference phenomena when incident light interacts with nanoscale structures (the so-called structural colour, see Chap. 13, Fig. 13.2d). As noted above, blue iridescence, such as in zebrafish dark stripes, but also widely in open water shoaling fish such as mackerel and tuna, is generated by light-reflection and interference on reflecting platelets within iridophores. However, novel blue chromatophores containing true blue pigments, known as cyanophores, have been discovered in two callionymid species, the mandarin fish, *Synchiropus splendidus*, and the psychedelic fish, *S. picturatus* (Goda and Fujii 1995, 1998). Cyanophores are dendritic and have blue organelles, cyanosomes, which are capable of an aggregation-dispersion response. The nature of the blue pigments is yet to be explored.

7.2.6 Dichromatic Chromatophores (Mosaic Pigment Cells)

Many pigment cells contain distinct organelles that together generate their colour. The erythrophores (xanthophores) of xiphophorine fish, such as the swordtail (*Xiphophorus helleri*) and platyfish (*X. maculates*), were found to have two kinds of pigmentary organelles, brownish-red pterinosomes and yellow carotenoid vesicles (Matsumoto 1965). Later it was reported that xanthophores in medaka also contained both pterinosomes and carotenoid granules (Hama 1967; Hama and Hasegawa 1967; Obika 1993). It is now generally accepted that most teleost xanthophores are dichromatic, having the two pigmentary organelles in each cell. Similarly, embryonic leucophores in medaka are dichromatic, containing both the white substance (presumably uric acid) and drosopterin (orange pteridine); in adults, the cells look white, presumably because pteridine pigments are decreased or no longer present.

In the mandarin fish, *Synchiropus splendidus*, where cyanophores were discovered, another extraordinary dichromatic chromatophore has been identified around the edge of blue regions in the dermis. These erythro-cyanophores contain both erythrosomes (carotenoid vesicles), as well as cyanosomes (Goda et al. 2013; Schartl et al. 2016) (see Fig. 13.2f). Another example of a dichromatic chromatophore is the erythro-iridophore, containing reddish carotenoid and reflecting platelets, which was discovered in the reddish-violet regions of the skin of bicolour dottyback *Pseudochromis diadema* (Goda et al. 2011).

Finally, as mentioned above (see Sect. 7.2.1), in addition to xanthophores, two novel, dichromatic chromatophore types have recently been reported from adult zebrafish and their relatives (Lewis et al. 2019). Xantholeucophores having orange and white hues are found in anal fins of seven Danio species and less prominently in dorsal fins of two species. Similar to xanthophores, these cells contained pteridines and carotenoids, and are thought to arise by transdifferentiation of xanthophores. Melanoleucophores, found at the distal edges of dorsal and caudal fins, and, within the anal fin of two species, have reflective but opaque white material, sometimes containing melanin but not pteridines or carotenoids. These cells arise directly by transdifferentiation of melanophores, accumulating white material and losing melanin over several days.

7.3 Embryonic/Larval Pigment Cells and Adult Pigment Stem Cells

Pigment cells are generated during two distinct stages of fish development, resulting in two distinct patterns. Production of embryonic/larval cells directly from neural crest cells (NCCs) generates the early larval pattern, whilst generation of new cells during metamorphosis from adult pigment stem cells (APSCs) generates the adult pattern (see Chap. 8). It should be noted that APSCs themselves are derived from the NCCs. Maintenance/regeneration of both patterns seems to also depend upon these APSCs. Pigment cell development is best-characterized in zebrafish, and we will focus our outline description on this species. It should be noted that aspects of the embryonic/early larval pattern are highly conserved across fish (and even amphibian) species, whereas adult patterns can differ dramatically even between sister species!

7.3.1 Development of Embryonic/Larval Pigment Cells

Differentiating melanophores first appear at approximately 24 h postfertilization (hpf) in the dorsolateral trunk and head (Kimmel et al. 1995). These cells begin to synthesize melanin even as they are migrating although they remain only partially pigmented until they settle in their final locations. Xanthophores containing pteridine pigments are first evident at approximately 42 hpf as pale yellow colouration on the dorsal aspect of the head. Around this same time iridophores are first observable around the choroid of the eye and later along the dorsal midline of the trunk and tail.

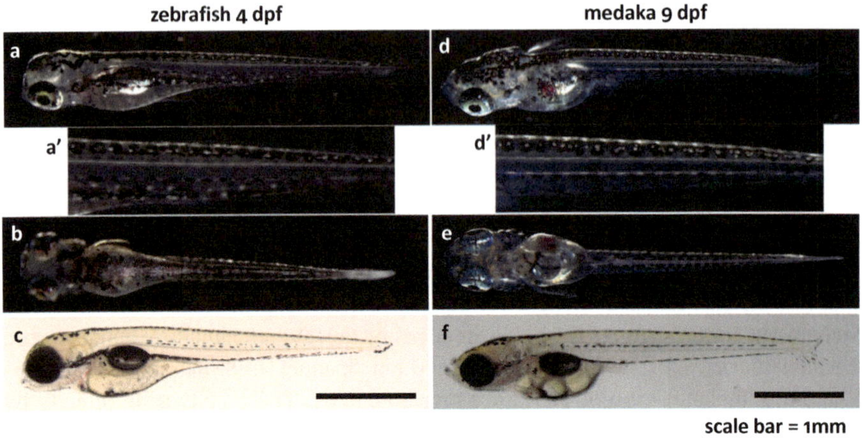

Fig. 7.1 Larval pigment pattern in zebrafish and medaka. (**a–c**) Zebrafish 4 dpf larva. (**d–f**) Medaka 9 dpf larva. Light-reflecting cells surrounded by black melanophores in the dorsal midline of zebrafish are iridophores (**a**, **a′**, dorsolateral view), while creamy cells in the corresponding region of medaka are leucophores (**d**, **d′**, dorsolateral view). Iridophores are also observed in the ventral midline of zebrafish (**b**, ventral view). Leucophores are localized in two lines of the ventral stripe of medaka (not yet very clear at this stage) as well as in a position ventral to the brain (**e**, ventral view). Xanthophore pigmentation has become more evident craniofacially and dorsally in zebrafish (**c**, lateral view) and in medaka (**f**, lateral view). The three horizontal stripes of melanophores are clearly seen in the trunk in both species in transmitted light (**c**, **f**)

Production of each cell-type continues and, by 6 days postfertilization (dpf), the embryonic pigment pattern is established: four horizontal melanophore stripes, three with associated iridophores, and xanthophores throughout the skin dorsal to the gut along the length of the body. This larval pattern remains unchanged until the start of adult pattern metamorphosis.

Medaka shows a similar early larval pattern (9 dpf) to zebrafish (Fig. 7.1), while the yolk melanophores are not organized as a 'stripe' but in a 'patch', thus instead are called the yolk patch in medaka.

Iridophores are found in the iris, on the yolk and along the dorsal and ventral stripes in zebrafish, and in medaka similarly in the former two regions but not along the dorsal stripe. Instead, leucophores, which exist in medaka but not in zebrafish, are closely associated with the melanophores along the dorsal stripe.

After metamorphosis, adult pigment cell development starts with iridophores becoming localized along the horizontal myoseptum. Xanthophores and melanophores appear over the iridophores spreading dorsally and ventrally to cover the body surface and to form horizontal stripes (see Sect. 8.2.3).

7.4 Origin of Skin Pigment Cells

7.4.1 Concepts of Stem Cells

It is generally assumed that all the pigment cell-types that can be seen in fish as well as in amphibians and reptiles originate from a NC-derived shared progenitor cell, a chromatoblast (Bagnara et al. 1979). However, there is currently no direct evidence that a single progenitor cell (chromatoblast) gives rise to all three or four (or more) different chromatophores, but not other neural crest derivatives such as neural cell-types, in fish or in other animals.

The 'chromatoblast' concept was rooted in observations of mosaic pigment cells having multiple types of pigmentary organelles, each of which is cell-type specific. Bagnara noted that, although distinct in their final appearance and biochemistry, these pigmentary organelles are initially similar, and all derived from the endoplasmic reticulum (ER, Fig. 7.2). Indeed, ultrastructural images show that, in medaka, vesicles at early formative stages in differentiating melanophores share striking similarities with those of pterinosomes, characteristic of xanthophores (Obika 1993). The implicit existence of a primordial organelle provides the key concept that the various pigment cell-types differentiate from a common stem cell, or chromatoblast. Bagnara and his colleagues assumed that the chromatoblast is formed in the neural crest and is partially restricted, able to differentiate exclusively into any pigment cell-type (Bagnara et al. 1979).

Mosaic pigment cells having dual pigmentary phenotypes have also been found in a wide range of vertebrates other than Osteichthyes. Well documented examples are: Iridophores in the iris of a dove contain reflecting platelets that seem to be partially melanized (Ferris and Bagnara 1970). The tapetum lucidum (choroid behind the retina) of a ray fish (*Dasyatis sabina*) contains dichromatic cells with melanosomes and reflecting platelets (Arnott et al. 1970). Erythrophores (xanthophores) having reflecting platelets and melanosomes, or more prevalently reflecting platelets, in addition to pterinosomes are present in the red central stripe

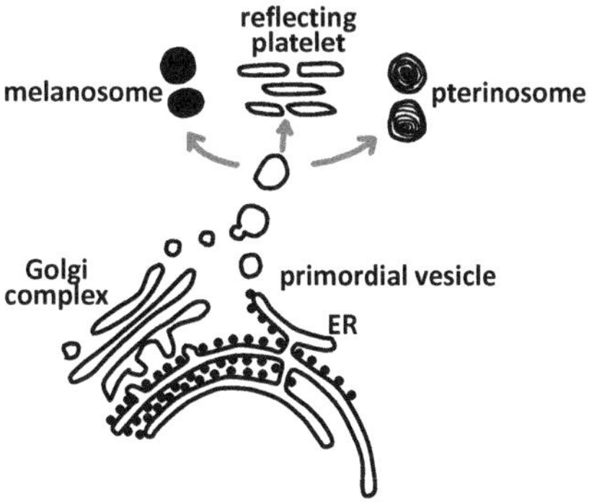

Fig. 7.2 Development of pigmentary organelles. Although the pigmentary organelles are specific to each of the multiple pigment cell-types, these are initially similar, and all derived from the endoplasmic reticulum (ER). Which pigmentary compounds are accumulated in these organelles is dependent on which synthetic enzymes and substrates are supplied to them during differentiation

area of the garter snake (*Thamnophis Proximus*) (Bagnara et al. 1978). Melanophores with an unusually large compound melanosome containing eumelanin surrounded by a fibrous mass of pteridine pigment can be found in the dermis of metamorphosing leaf frog (*Pachymedusa dacnicolor*) (Bagnara et al. 1973). The broad distribution of mosaic pigment cells in vertebrates suggests that the chromatoblast concept may be applicable to development and diversification of pigment cells throughout vertebrates. However, this concept demands rigorous testing, most readily in the zebrafish and medaka models.

7.4.2 Cell Migration of Embryonic/Larval Pigment Cells

In the embryo, pigment cell progenitors are thought to be specified early, prior to migration away from the neural tube, and then to migrate as fate-specified cells to the peripheral parts of the body where they differentiate. In mammals and birds, migration of melanoblasts is normally restricted to a pathway under the developing epidermis (dorsolateral pathway), whereas cells using the ventral pathway, between neural tube and somites, are mostly neural in fate. In zebrafish, the migration routes of specific chromatophores have been assessed: unlike in mammals, melanoblasts migrate on both the lateral and medial neural crest migration pathways, and these even begin to show some melanization (and thus should be described as melanophores) during migration. Xanthoblasts are restricted to the lateral pathway, and iridoblasts are restricted to the medial pathway.

7.5 Genetic Regulation of Pigment Cell Development from NC

Genetic control of pigment cell development has been best studied in zebrafish and medaka. The genetic resources for pigment cell biology in these two species (e.g. Johnson et al. 1995; Kelsh et al. 1996; Kelsh et al. 2004) are substantial. Furthermore, much work to date has focused on those mutants with phenotypes indicative of fate specification defects (Kelsh et al. 1996). New methods for reverse mutagenesis have now opened up the possibility of a direct comparison between the roles of homologous factors in pigment cell development in the two species. Melanophore/melanocyte fate specification mechanisms are highly conserved with mammals, but key players regulating iridophore, xanthophore and leucophore development have all been identified and characterized in the fish model systems. Epistasis experiments in model organisms have begun to assemble these genes into functional hierarchies. Some genes regulating specific aspects of pigment cell migration, survival and especially differentiation have also been identified, usually through phenotypic similarities to mutants for mammalian homologues. However, many known mutants remain unstudied, and RNA-seq studies have identified expression of numerous genes in one or more lineages, so that an integrated understanding remains an ambitious target. Here we begin by outlining a scheme for the genetic control of pigment cell development, before summarizing the key findings for particular genes where detailed investigations have been published.

7.5.1 *Intrinsic and Extracellular Factors*

Fate choice from multipotent progenitors requires the action both of intrinsic factors and extracellular signals. They act together to drive transcription of key fate-specific master switch transcription factors in premigratory or early migrating NCCs (Kelsh 2006). Thus, a set of 'master switch' transcription factor-encoding genes, acting separately or in combination, appear to be necessary and sufficient to drive development of specific fates (Kelsh 2006).

Extracellular factors signal to the cell through transmembrane receptors and determine which master switch transcription factor gene(s) become activated. Their downstream transcriptional effects act with endogenous transcription factors (e.g. Sox10) expressed in developing NC. Together these influences result in transcriptional activation of the master switch genes. Subsequently, the master switch transcription factors activate target genes that together result in survival, migration and differentiation of each pigment cell-type.

7.5.2 *The Genetics of Embryonic/Larval Pigment Cell Development*

A considerable literature documents detailed study of the embryonic roles of a set of genes with key roles in pigment cell development, most identified from cloning of mutations identified in the major zebrafish genetic screens. Here we summarize the key findings of that work, focusing on aspects related to pigment cells, and only more briefly commenting on effects on other NC derivatives. It should be noticed that some of these genes show pleiotropic effects involving non-NC-related phenotypes; we largely omit these considerations, in the interests of space. We organize these discussions based on the zebrafish studies, but making reference to medaka or other fish species where appropriate.

Chromatoblast (Cbl) Formation

The early NCCs express neural crest specifier genes such as *foxd3*, *sox9a*, *sox9b*, *sox10* and *tfap2a* (Simoes-Costa and Bronner 2015). NC-specific expression of *sox9a*, *sox9b* and *sox10* depends on *foxd3* and *tfap2a* function in zebrafish embryos (Ignatius et al. 2008). A subset of cells with *foxd3* and *tfap2a* expression downregulated emerge as early chromatoblasts, which have been proposed as progenitors for all the pigment cell-types.

The early Cbls express *tfec* in addition to *sox10* and *sox9b* (Petratou et al. 2018). *tfec* expression is initially activated in a Foxd3 and Tfap2a-dependent manner although it is not currently clear if this is through a direct or indirect mechanism. In early Cbls, establishment and maintenance of *tfec* expression are independent of Sox10. Subsequently, these cells become positive for *ltk* and *mitfa*, now being defined as late Cbls. The two genes are known to be functionally required for specification of the iridophore and melanophore lineages, respectively, but are expressed in multipotent progenitors (late Cbls) first.

Melanophore Lineage

Melanophore lineage specification begins in the late Cbls, and depends upon maintenance of *mitfa* expression. *sox10* is rapidly downregulated once the progenitor becomes specified to melanophore lineage, and appears not to be strictly necessary for melanophore differentiation (Greenhill et al. 2011). Mitfa activates differentiation through regulation of diverse genes, including melanization genes such as *dct* and *tyrp1* (indirectly) although the details of this have been little-studied in fish.

Iridophore Lineage

Iridophore lineage specification begins with maintenance and upregulation of *tfec* (Table 7.1). In contrast to the melanophore lineage, *sox10* expression needs to be maintained for *tfec* to remain expressed in the specified iridophore lineage (Petratou et al. 2018). Furthermore, a positive feedback loop between Tfec and Ltk is essential to maintain a stable expression of Tfec (and Sox10) and the resultant activation of key differentiation genes, such as *pnp4a*, in differentiating iridophores.

Xanthophore Lineage

A subset of chromatoblasts, which maintain expression of *pax3* and then *pax7* later, become specified to a xanthophore lineage (Minchin and Hughes 2008; Nord et al. 2016). Once specified, xanthophore differentiation markers such as *gch*, *xdh* and *aox* are upregulated by Pax7 rather than Pax3 activity.

Table 7.1 Genes encoding transcription factors important for fate specification

Gene	Protein type	Expression pattern	Mutant (species)	Loss-of-function phenotype in pigment cells	Deduced function in pigment cell lineages
tfap2a	Activator Protein 2 (AP2) family	eNCC-> Later in melanoblasts	*tfap2a/lock-jaw/mont blanc*	One-third reduction in M, with I and X less affected	Early NCC specification
			tfap2a;2c	Absence of NC and all NC derivatives	
			tfap2a;2e	More severe M defect than *tfap2a* single	
foxd3	Forkhead box family	eNCC-> Later, a subset of migrating NCCs on the medial pathway and Schwann cell precursors	*foxd3/sympathetic mutation 1* (frame shift in the coding sequence) *Mother superior* (lack of NC-specific regulatory element) *foxd3* enhancer trap line	Reduction of I, increase of M Or Reduction of an early subset of NC	I specification via mitfa suppression Or Early NCC specification
mitf	MiT/TFE family with bHLH	Late Cbl -> only M lineage	*mitfa/nacre*	Lack of M, reduction in X, increase in I	Specification and differentiation of M (master regulator of M)
tfec	MiT/TFE family with bHLH	eNCC -> Early Cbl -> late Cbl, -> MI -> only I lineage	*tfec* CRISPR	Lack of I, delayed development of M and X	Early Cbl specification, later specification and differentiation of I
sox10	Sry-related HMG box family E	eNCC -> early Cbl -> late Cbl -> -> only I lineage	*sox10/ colourless*	Lack of M, X, I	Early Cbl specification and activation of mitfa Maintenance of tfec in I lineage

(continued)

Table 7.1 (continued)

Gene	Protein type	Expression pattern	Mutant (species)	Loss-of-function phenotype in pigment cells	Deduced function in pigment cell lineages
			sox10a TALEN (medaka)	Severe reduction of M, X, I L present	Early Cbl specification Not required for L specification
			sox10b TILLING, TALEN (medaka)	No phenotype	
			sox10a;10b (medaka)	Lack of M, X, I L present	
sox5	Sry-related HMG box family D	Premigratory NCC (perhaps XL) and migrating Xb	*sox5/many leucophores-3* (medaka)	Lack of X, increase of L	Promotion of X lineage in XL
		Undetectable by in situ	*sox5* CRISPR	No phenotype Slight increase of X	Subtle inhibition of Sox10
pax3	Paired type homeobox family	Premigratory NCC and MX	*pax3a* MO	Lack of X, increase of M	Specification of MX and activation of mitfa
pax7		Premigratory NCC and perhaps migrating Xb	*pax7a* and *pax7b* single	Delayed development of X	Specification and differentiation of X
			pax7a;7b	Lack of X	
		Premigratory NCC and XL, and migrating Xb and Lb	*pax7a/ leucophore free-2* (medaka)	Lack of X, L	Specification and differentiation of X, L

M melanophore, *I* iridophore, *X* xanthophore, *L* leucophore, -> time passage

7.5.3 Roles of Intrinsic Factors

Tfap2a, Tfap2c and Tfap2e

Tfap2a appears to function together with Tfap2c in NC induction from the ectoderm (Greenhill et al. 2011), directly, but partially, regulating *sox10* (Van Otterloo et al. 2012). This early role for Tfap2a in NC specification is likely reflected in the phenotype of zebrafish *tfap2a/lockjaw/mont blanc* mutants, albeit weakly. Loss of Tfap2a results in severe reduction of *crestin* (a pan-neural crest maker) and *tfap2a*, but not *sox9b*, expression, suggesting that *tfap2a* is required for maintenance of some, but not all, characteristics of premigratory NC, consistent with the elevated death of premigratory NCCs and widespread defects in craniofacial cartilage, and neural derivatives (Hoffman et al. 2007; Li and Cornell 2007; O'Brien et al. 2004).

Tfap2a, later in development, functions redundantly with Tfap2e for embryonic melanophore differentiation (but not specification) (Table 7.1). Much of this role is focused on maintaining expression of *kit* in the melanocyte lineage, but they also act redundantly to maintain *mitfa* expression and Mitfa function in a Kit-independent manner too (Hoffman et al. 2007; Li and Cornell 2007; O'Brien et al. 2004). Consistent with melanoblast fate specification being unaffected, decrease of Tfap2a/2e activity does not result in reciprocal changes in numbers of other pigment cell-types nor elevated cell death beyond that attributable to Kit function (Van Otterloo et al. 2012).

Foxd3

Expression of *foxd3* in the premigratory NC begins shortly after *sox9b* (Lopes et al. 2008), and is apparently Sox9b-dependent (Yan et al. 2005). FoxD3 activity is required for initial expression of *sox10* in premigratory NCCs, which is delayed in mutants (Montero-Balaguer et al. 2006; Stewart et al. 2006). Recently it has been shown to be a crucial pioneer factor at this stage (Lukoseviciute et al. 2018). However, it has proven difficult to dissect its role in zebrafish chromatophore lineages, in part due to phenotypic differences in the initial morphant studies (Lister et al. 2006) and subsequently in three independent mutants (Curran et al. 2010, 2009; Hochgreb-Hagele and Bronner 2013; Ignatius et al. 2008; Lister et al. 2006; Montero-Balaguer et al. 2006; Stewart et al. 2006).

Differentiation of all three pigment cell-types is delayed in *foxd3* mutants and morphants, as is expression of earliest markers for these cell-types (Cooper and Raible 2009; Ignatius et al. 2008; Lister et al. 2006; Montero-Balaguer et al. 2006; Stewart et al. 2006). Correlated with these delays in fate specification, NCC migration is also delayed (Stewart et al. 2006). These observations suggest that this phenotype might best be attributed to problems with functions in premigratory NCCs, e.g. lineage priming (Lukoseviciute et al. 2018). Indeed, we note that the defective fate specification phenotypes have been suggested to reflect delayed NC induction (Hochgreb-Hagele and Bronner 2013; Montero-Balaguer et al. 2006; Stewart et al. 2006).

An alternative model for FoxD3 action in NC is that Foxd3 acts to prevent melanophore fate by restricting *mitfa* expression, similar to the proposal in chick that Foxd3 represses the formation of melanocytes (Thomas and Erickson 2009). There is some evidence that NCCs expressing FoxD3 protein or *mitfa* transcripts are mutually exclusive, and that FoxD3 expression represses Sox10-dependent expression of *mitfa,* implying that FoxD3 is required for iridophore fate but also inhibits melanophore fate (Curran et al. 2010, 2009; Lister et al. 2006; Stewart et al. 2006). It has been proposed that FoxD3 functioned in bipotent melanoiridoblasts to favour iridophore development through repression of *mitfa* expression and thus melanophore specification (Curran et al. 2009). Although in *foxd3* mutants, iridophore number is reduced, their number is elevated in *mitfa* mutants; intriguingly, *foxd3: mitfa* double mutants show partial rescue to a number closer to WT (Curran et al. 2009). However, it should be noted that these studies generally considered FoxD3 to function as a transcriptional repressor, so that reinterpretation of the data may now be warranted. A recent study focused on iridophore fate specification combined study of *tfec* mutant embryos with mathematical modeling of the core GRN, and concluded that a factor repressing *mitfa* expression/activity was a key component of this GRN (Petratou et al. 2018). However, they concluded that *foxd3* was unlikely to explain this repression of melanophore fate in developing iridophores, principally because expression levels were low and detectable only in a subset of iridophore lineage cells throughout 24–48 hpf (Petratou et al. 2018).

There is also evidence that FoxD3 interacts genetically with Kit (Petratou et al. 2018). Specifically, *kit* mutants (in which melanophore survival is reduced) can be partially rescued (50% increase) by homozygosity for a loss-of-function *foxd3* mutant. Conceivably, this effect could be explained by suggesting that simply producing more melanocytes results in higher numbers surviving, but examination of melanoblasts markers makes this seem unlikely (Cooper and Raible 2009). Instead, it may be that loss of FoxD3 function removes inhibition of *mitfa* transcription, and thereby promotes melanophore survival (Cooper and Raible 2009).

It seems likely that FoxD3 has a role in at least melanophore and iridophore development, but given its early role as a pioneer factor in lineage priming as well as a later role as a transcriptional repressor (Cooper and Raible 2009), it remains unclear which phenotypic aspects are caused by roles in early NCCs and which depends upon on-going roles in specified pigment cell lineages.

Tfec

Tfec was first suggested to be involved in iridophore development based upon the embryonic expression pattern: early in the premigratory NCCs, and later in a pattern reminiscent of iridophores (Lister et al. 2006). Furthermore, transcriptomic analysis of purified cells showed that *tfec* is one of several transcription factor genes with highly specific expression in the iridophore lineage (Higdon et al. 2013). Consistent with these observations, *tfec* mutant larvae show a complete loss of iridophores from the eyes, over the yolk, on the dorsal head and along the dorsal and ventral midline in the trunk and tail in addition to a non-inflated swim bladder, and anaemia (Mahony et al. 2016; Petratou et al. 2018). These suggest that Tfec functions as a master regulator of iridophores, equivalent to the role of Mitfa as a master regulator of melanophores.

Detailed characterization of the expression pattern showed that *tfec* is a valuable marker for the iridophore lineage at all stages: it is expressed rather early in premigratory NCCs and, through careful consideration of other markers, appears to be expressed throughout at least six phases of iridophore development from the NC, of which at least 4 were shared with melanophores (Petratou et al. 2018). Importantly, Tfec is expressed in the early NCCs prior to expression of *mitfa* and *ltk* and similarly to that of *sox10*; at these stages, *tfec* expression is independent of *sox10*, whereas initiation of *ltk* expression is dependent upon Tfec function (Petratou et al. 2018). Given the key role for Ltk in receiving an environmental iridophore fate specification signal, setting cells aside as specified iridoblasts, this Tfec-dependent expression of *ltk* is crucial in generating iridophore potential in these early cells. Subsequently, a positive feedback loop between Ltk and Tfec drives maintenance of iridophore specification and is likely crucial for iridophore commitment. At this stage, *sox10* continues to be co-expressed with *tfec* throughout iridophore differentiation and is required for maintenance of the Tfec-Ltk feedback loop and hence to maintain iridophore lineage specification.

Tfec function seems to have a subsidiary role in specification of xanthoblasts and iridoblasts (Petratou et al. 2021).

Mitfa

Microphthalmia-associated transcription factor or Melanocyte-inducing transcription factor a (Mitfa) has been shown to be necessary for specification, differentiation and survival of melanocytes/melanophores in zebrafish (Johnson et al. 2011; Lister et al. 1999), as reported in mice (Opdecamp et al. 1997). Indeed it is commonly considered a master regulator of melanophore development, in part because ectopic expression of Mitf drives embryonic, but also medaka ES cells, to adopt a melanophore phenotype (Bejar et al. 2003; Elworthy et al. 2003; Lister et al. 1999). This is likely to reflect the extensive and diverse array of melanophores genes that lie (often directly) downstream of this transcription factor although this has been much less extensively characterized in zebrafish compared to mammalian orthologues (Goding and Arnheiter 2019; Greenhill et al. 2011; Lister et al. 1999) (see Chap. 6).

mitfa is expressed initially in almost all NCCs and later specifically in the melanophore lineage (Curran et al. 2010, 2009; Lister et al. 1999). Maintenance of *mitfa* expression in a subset of NCCs is an important aspect of melanophore development and, indeed, likely represents a crucial part of the molecular basis for melanophore fate commitment. In contrast, cells in which *mitfa* expression is not maintained adopt alternative pigment cell fates, so that *mitfa* repression is a key, but overlooked, aspect of chromatophore fate specification in the NC (Petratou et al. 2018). Initial expression of *mitfa* depends upon both Wnt signaling and Sox10 (Dorsky et al. 2000; Elworthy et al. 2003). A role for miRNAs in repressing *mitfa* expression is evidenced by study of the *dicer1* mutant phenotype, in which there are subtle changes in melanophore number, morphology and pigmentation (Weiner et al. 2019).

Zebrafish *nacre/mitfa* homozygous mutant larvae have no trunk melanophores but more iridophores than WT, suggesting that the specification or differentiation of these two pigment cell-types may be coordinately regulated; it has been proposed that melanoblasts may switch fate to become iridoblasts in the absence of functional Mitf/Nacre protein (Curran et al. 2010, 2009; Lister et al. 1999).

Sox10, Sox9b

SoxE genes encode a key group of transcription factors, readily divided into subfamilies that share highly conserved DNA-binding HMG domains (Koopman et al. 2004; Wegner 1999). In vertebrates, the SoxE subfamily comprises Sox8, Sox9 and Sox10, and is characterized in part by a C-terminal transcriptional activation domain. Sox9 and Sox10 have been best studied, largely because they have been associated with specific congenital diseases, such as Waardenburg-Shah syndrome and Hirschsprung's disease, and as crucial in various aspects of NC development.

Sox10 expression is highly conserved across vertebrates (reviewed in Hong and Saint-Jeannet 2005): in zebrafish, for example, it begins very early in premigratory NCC, is maintained transiently in migrating cells (on both medial and dorsolateral migration pathways in the trunk), and then is lost as the cells differentiate with the exception of iridophores, and in all glial derivatives (Dutton et al. 2001b).

Sox9 expression begins in NC development even earlier than *sox10*, likely prior to delamination (Aoki et al. 2003; Cheng et al. 2000; Dutton et al. 2001b; Pusch et al. 1998; Southard-Smith et al. 1998). NC-specific expression of *sox9a*, *sox9b* and *sox10* has been shown to depend on both *foxd3* and *tfap2a* function in zebrafish embryos (Arduini et al. 2009). Reduction of function of *sox9a* and/or *sox9b* results in severe craniofacial cartilage defects, but pigment cell defects are restricted to a substantial reduction in iridophores in *sox9b* mutants (Yan et al. 2005).

Sox10 plays central roles in all non-ectomesenchymal fates, including development of all pigment cell-types (Dutton et al. 2001a, b; Kelsh and Eisen 2000). The phenotype is thus reminiscent of that in *Sox10* mutant mice, and in both species transplantation studies have shown that Sox10 acts cell autonomously, at least within the pigment cell and enteric neuron lineages (Kapur 1999; Kelsh and Eisen 2000).

In each of the affected fates, it has been proposed that the primary defect is in fate specification from multipotent progenitors, and that this role is conserved from lamprey to mammals (Dutton et al. 2001b; Kelsh 2006; Lee et al. 2016). This is best shown for melanocytes, where Sox10 directly regulates *mitfa* expression, and which is absent from the NC in *sox10* mutants (Elworthy et al. 2003) consequently, *sox10* mutants essentially lack all melanoblast marker expression (Kelsh and Eisen 2000) (Fig. 7.3). Furthermore, zebrafish *sox10* mutants appear to be fully rescued by expression of *mitfa* (Elworthy et al. 2003), with no evidence for an on-going function (Greenhill et al. 2011); this is in apparent contrast to the story in mice (Hou et al. 2006). Regulation of *mitfa* has been shown to be a direct transcriptional effect (Elworthy et al. 2003), and it has been argued that this may be the major role for Sox10 in melanocyte development, since *sox10* expression is rapidly downregulated in differentiating melanocytes and indeed this appears to be important for differentiation to occur, because Sox10 shows a weak, but detectable, repressive effect on melanin-synthesis genes (Greenhill et al. 2011). A role for miRNAs in repressing *sox10* expression is evidenced by study of the *dicer1* mutant phenotype and may at least contribute to this (Weiner et al. 2019); indeed, this may be a major part of the Mitfa-dependent repression of *sox10* transcription identified in an earlier study

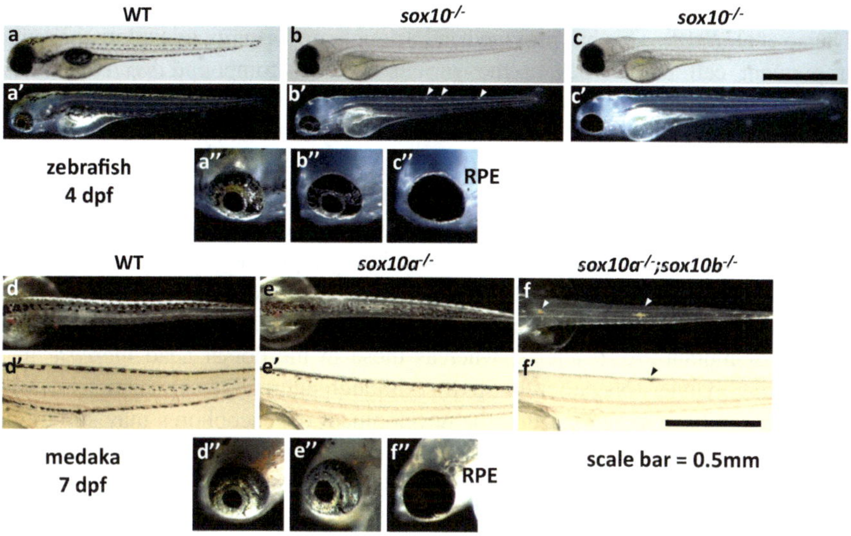

Fig. 7.3 Comparison of *sox10* mutant phenotypes between zebrafish and medaka. (**a–c**) zebrafish. (**a**) wild-type larva. (**b**, **c**) *sox10*$^{-/-}$ mutants. Zebrafish *sox10/colourless* homozygous mutants lack nearly all chromatophores except for melanophores in RPE (**a**, **a′**, **b**, **b′**, **c**, **c′**). In rare cases, the homozygotes retain iridophores in the dorsal midline (**a′**, **b′**, **c′**, arrowheads in **b′**) and in the eyes (**a″**, **b″**, **c″**). (**d–f**) medaka. (**d**) wild-type larva. (**e**) *sox10a*$^{-/-}$ mutant. (**f**) *sox10a*$^{-/-}$; *sox10b*$^{-/-}$ mutant. Medaka *sox10a* homozygotes partially lack chromatophores (**d**, **e**), particularly evident with melanophores being lost in the ventral two stripes of the trunk (**d′**, **e′**). The double mutant for *sox10a*$^{-/-}$; *sox10b*$^{-/-}$ mutant almost completely lacks chromatophores (**f**, **f′**), except for leucophores in the head (**f**) and in the trunk (**f**, **f′**). The eye iridophores are lost in the double whereas rather unaffected in the *sox10a* single (**d″**, **e″**, **f″**)

(Greenhill et al. 2011). The mechanisms driving early expression of *sox10* in the NC have only begun to be explored in zebrafish, but Wnt signaling-induced Lef1 binding, and, especially, Tfap2a best defined amongst multiple expected factors (Dutton et al. 2008; Trinh et al. 2017; Van Otterloo et al. 2012).

Melanocyte migration in zebrafish and mouse requires Kit signaling (Parichy et al. 1999; Wehrle-Haller and Weston 1995) and NCCs fail to express *kit* in *sox10* mutants of both species (Bondurand et al. 2000; Dutton et al. 2001b; Potterf et al. 2001). This is consistent with the role for Mitfa in driving *kit* expression in zebrafish melanoblasts (Lister et al. 1999).

Fate specification is also defective for the other two pigment cell-types in *sox10* mutants (Dutton et al. 2001b; Lopes et al. 2008; Petratou et al. 2018). In the case of iridophores, the key role is in expression of *tfec*, but here *sox10* expression is readily detectable in the iridophore lineage even long after differentiation, making an on-going role for Sox10 in maintaining the differentiated state more likely (Petratou et al. 2018).

Similarly, in medaka, specification of these three types requires Sox10 function, although here both duplicates, *sox10a* and *sox10b*, have been retained and act redundantly, but with Sox10a having stronger effects on pigment cell development (Nagao et al. 2018). This has provided an intriguing opportunity to explore the range of pigment cell phenotypes in the various medaka compound *sox10* mutants, whereas in zebrafish *sox10* mutants show fully recessive effects, and all alleles to date show very strong to full reduction in melanocytes (Carney et al. 2006; Dutton et al. 2001b; Kelsh and Eisen 2000) (Fig. 7.3).

Similar to mutants of the strongest zebrafish *colourless/sox10* alleles, medaka *sox10a; sox10b* double homozygous mutants lack all melanophores, iridophores and xanthophores (Nagao et al. 2018). However, intriguingly they retain a considerable number of leucophores (Nagao et al. 2018). The phenotype is surprisingly complex, with leucophores in the body and tail being lost in quantitative proportion to the number of mutant *sox10* alleles, whereas those in the dorsal head *increase* in a quantitative manner (Nagao et al. 2018). Although alternative explanations can be imagined, we have suggested that the development of leucophores may simply be achieved using lower levels of SoxE activity, perhaps maternally supplied SoxE and/or especially Sox9b; we note that a similar effect has been shown in zebrafish where additional loss of *sox9b* results in loss of the residual melanophores and sensory neurons of the DRGs in *colourless/sox10* mutants (Carney et al. 2006).

Sox5

The role of Sox5 in pigment cell development is best understood in medaka, where it is vital for xanthophore specification. *Many leucophores-3/sox5* mutant larvae and adults exhibit the absence and reduction, respectively, of xanthophores and instead the excessive formation of leucophores. Thus, Sox5 seems to promote xanthophore fate and suppress leucophore fate in medaka; in contrast, there is no melanophore and iridophore phenotype, suggesting that Sox5 is not required for fate specification of these cell-types.

In this context it is, at first sight, surprising that in zebrafish, Sox5 is dispensable for all pigment cell lineages, including xanthophores (Nagao et al. 2018). However, there is one aspect to Sox5 function in pigment cells that is conserved, not just in zebrafish, but even in mammals (Fig. 7.4). Thus, it should be noted that although loss of Sox5 does not result in any apparent abnormality in melanocyte development in mice, it partially rescues a severe reduction of melanoblasts and expression of *Mitf* and *Dct* in *Sox10* heterozygotes, suggesting that Sox5 slightly antagonizes the key role for Sox10 in melanocyte fate specification (Stolt et al. 2008). Through study of *sox5; sox10* double mutants in both zebrafish and medaka, we have shown that Sox5 plays a similar antagonistic role for Sox10-dependent specification of melanophores, iridophores and xanthophores in zebrafish, and of melanophores and iridophores in medaka (Fig. 7.4).

In contrast, Sox5 and Sox10 function collaboratively in specification of xanthophores in medaka. The unique function of Sox5, and its unusual co-activatory role with Sox10, in medaka xanthophores correlates with the novel presence of leucophores in medaka (Nagao et al. 2014, 2018). These observations

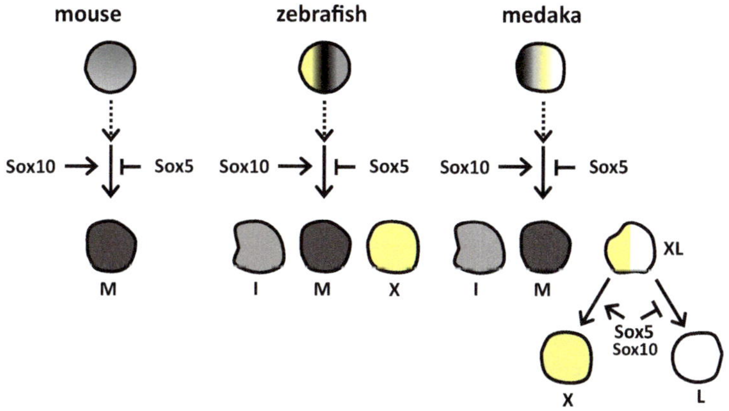

Fig. 7.4 Comparison of Sox10 and Sox5 roles. In mice, zebrafish and medaka, *sox10* is commonly required for pigment cell development and its activity is antagonistically modulated by *sox5* to generate melanocytes/melanophores (M) as well as iridophores (I) and xanthophores (X). In medaka, in the later phase when the common progenitor of xanthophore and leucophore (XL) is fate-specified to either of these cells, *sox10* and *sox5* cooperatively promote xanthophore fate (X) and suppress leucophore fate (L). The role for Sox5 in the first phase is relatively weak, whereas in the second phase Sox5 is the dominant player, which is unique to medaka

raise the possibility that leucophores originate from a common progenitor shared with xanthophores, and that Sox5 acts as a switch determining xanthophore versus leucophore fate choice within the shared progenitor; as such it would be unexpected for it to show a function conserved in zebrafish, which lack leucophores (Nagao et al. 2018). We note that Sox5 is expressed in that progenitor, but also maintained at high levels in the xanthophore lineage. The distinct modes of interaction between Sox5 and Sox10 can be reconciled by proposing a dual-phased action of Sox5 in pigment cell development (Fig. 7.4): firstly, a subtle Sox5 antagonism functions to modulate the Sox10-dependent fate specification. In the second phase, unique to medaka, Sox5 plays a dominant role in xanthophore fate choice, being absolutely required for xanthophore specification in the xanthophore progenitor, which has evolutionarily acquired xanthophore/leucophore bipotency. Consistent with this model, forced *sox5* expression in the shared xanthophore/leucophore progenitor (*pax7a*-expressing cells, see below) is sufficient to drive xanthophore fate and to repress leucophore fate in medaka.

Pax3/7

Pax3 and Pax7 are closely related to each other resulting from duplication of an ancestral gene (*pax3/7* in lamprey). The role of Pax3 in pigment cell development is barely known in fish, but in mammals *PAX3/Pax3* has a pivotal role in both melanocyte and enteric neuron development from the NC, as seen in Waardenburg syndrome type I, a human congenital pigmentation disorder, caused by a mutation in

PAX3 (Tassabehji et al. 1992). In mice, Pax3 promotes melanocyte fate in a synergistic fashion with Sox10 by binding to melanoblast-specific regulatory elements in *Mitf* gene and driving its expression in the melanoblasts, which, by contrast, can be inhibited by protein–protein interactions with FOXD3 (Kubic et al. 2008; Monsoro-Burq 2015).

In zebrafish, *pax3* is expressed in premigratory NCCs (Minchin and Hughes 2008; Seo et al. 1998). In contrast to mice, depletion of Pax3 by morpholino-mediated knocking down in zebrafish results, not in reduced melanocytes, but in decreased xanthophores and enteric neurons (Minchin and Hughes 2008). Furthermore, in these *pax3*-morphants, decreased xanthophores are accompanied by an increase in melanophores; this may indicate that Pax3 promotes xanthophore fate choice at the expense of melanophore fate within a shared multipotent progenitor. Early xanthophore lineage markers were essentially absent, suggesting a role in xanthophore specification. However, they also observed delayed expression of *mitfa* in migrating NCCs, suggesting that melanoblast specification was at least delayed in *pax3* morphants, partially reflecting the mammalian role (Minchin and Hughes 2008).

Pax7 functions in pigment cell lineages in fish, acting in a manner partially redundant to Pax3. There are two *pax7* orthologues in both zebrafish and medaka (Kimura et al. 2014; Minchin and Hughes 2008; Seo et al. 1998). Using a Pax3/7 antibody, but also *pax3* translation-blocking morpholinos to distinguish likely *pax3* from *pax7* expression, Minchin and Hughes showed that *pax3* expression was seen in the premigratory NC, but concluded that *pax7*, and not *pax3*, was expressed in migrating NCCs (Minchin and Hughes 2008). They did detect *pax7a* and *pax7b* mRNA in premigratory NCCs and in migrating cells on the lateral pathway (xanthoblasts), but not those on the medial pathway; the xanthoblasts were lost in *pax3* morphants (Minchin and Hughes 2008).

Medaka *leucophore free-2/pax7a* mutants exhibit a complete absence of xanthophores and leucophores throughout life, indicating that Pax7 is essential for xanthophore and leucophore lineages in medaka, perhaps with requirement for specification of bipotent intermediates of xanthophore and leucophore (see Sects. 7.5.3 and 7.6.1) (Kimura et al. 2014). In zebrafish, this role in xanthophores is conserved (Minchin and Hughes 2008), but *pax7a* and *pax7b* have a redundant role in xanthophore specification (Nord et al. 2016). Interestingly, melanocytes in the dorsal head were increased at 3 dpf and persisting to 6 dpf at least; this correlated with increased numbers of cells expressing melanocyte markers *mitfa* and *dct* at 24 hpf (Nord et al. 2016). This again hints at a shared origin from a progenitor, perhaps the melano-xanthoblast (Nord et al. 2016).

Hdac1

Hdac1 encodes a histone deacetylase. In *hdac1/colgate* mutants, melanophores are formed, but they tend to clump around the otic vesicle and ventral migration is impaired, reminiscent of the *kit* mutant phenotype (Ignatius et al. 2008). Using molecular markers, it was shown that there is a delay in *mitfa* and *dct* expression, but a more severe defect in migration, again correlating with a severe reduction in *kit* expression (Ignatius et al. 2008). The delayed melanophore development phenotype correlates with temporary persistence of *foxd3* expression in premigratory NCCs (Ignatius et al. 2008). There was no more than a slight delay in xanthophore differentiation and/or migration, although this was not reflected in early marker studies, but this recovers (Ignatius et al. 2008). When combined with decreased FoxD3 activity, melanophore numbers are increased, partially rescuing the *hdac1* melanophore defect; this phenotype correlated with partial rescue of the *mitfa* expression defect (Ignatius et al. 2008). The authors conclude that Hdac1 activity is required to downregulate *foxd3* expression in the premigratory NC, and consequently to promote *mitfa* expression; they provide evidence that FoxD3 might bind the *mitfa* promoter directly.

Gbx2

The *gastrulation brain homeobox 2 (gbx2)* gene encodes a transcription factor and was reported as upregulated in zebrafish iridophores (Higdon et al. 2013). It is expressed extensively in the premigratory NC and also in differentiated iridophores (Hozumi et al. 2018). Using morpholino-based knockdown, Gbx2 seems to be required for iridophore development, but not other pigment cell-types (Hozumi et al. 2018). Furthermore, *gbx2* morphants show absence of *ltk*$^{+}$ specified iridoblasts at 24 hpf, and of *pnp4a* expressing iridophores at 48 hpf, consistent with a key role in iridophore fate specification (Hozumi et al. 2018). These morpholino results are in strong contrast to a truncated mutant allele, which shows no phenotype, leading the authors to suggest that the key role of Gbx2 is repressive and mediated through the N-terminal part of the protein (Hozumi et al. 2018).

7.5.4 Extrinsic Signals

Wnt Signaling

An early finding in the study of zebrafish NC was the role for Wnt signaling in pigment cell fate choice. Early work on fate restriction in zebrafish indicated that, in the cranial NC, fate choice to a melanophore (in contrast to a neuron or glial fate) is determined according to the cell's mediolateral position within the NC (Kelsh and Raible 2002; Schilling and Kimmel 1994). Subsequent work identified the key factor as a Wnt signal, likely Wnt1/3a, arising from the dorsal midline of the ectoderm which, when active, biases cells to a melanophore fate (Dorsky et al. 1998). Later work showed that this effect results from direct transcriptional activation of *mitfa* via Lef1 binding (Dorsky et al. 2000).

Wnt signaling is also known to be required for NC induction from the ectoderm, and study of a transgenic line expressing an inducible inhibitor of Wnt signaling allowed confirmation that both NC induction, and later melanophore specification, were canonical Wnt signaling dependent (Lewis et al. 2004). In addition, they identified Wnt8, expressed in the prospective neural plate, as likely to be a key Wnt signal in NC induction (Lewis et al. 2004). Although members of the Frizzled family of Wnt receptors are strong candidates, to date the receptor mediating Wnt-dependent melanophore specification has remained elusive (Nikaido et al. 2013).

It is clear that Wnt signaling acts in conjunction with Sox10 to drive melanophore fate specification by activating *mitfa* expression. Wnt signaling alone is clearly insufficient for *mitfa* activation since *mitfa* expression is not detectable in *sox10* mutants. Once activated, however, *mitfa* can maintain its own expression dependently of positive feedback loop with Wnt signaling. However, there is an on-going role for Wnt signaling in differentiated melanophores, since both up- and downregulation of Wnt signaling in zebrafish embryos during melanophore differentiation results in altered melanophore morphology (Vibert et al. 2017). Thus, it is clear that Wnt signaling is involved in three temporally distinct aspects of melanophore development!

Kit Signaling

Kit is a type of receptor tyrosine kinase family, bound by Kit ligand, Kitlg/SCF/Steel factor. Kit signaling plays an important role in melanocyte/melanophore development as well as for a variety of physiological processes.

Requirement of Kit for embryonic melanophore development was revealed by studies with zebrafish *kita/sparse* (Parichy et al. 1999): the *kita/sparse* mutant larva exhibits accumulation of melanophores near the otocysts at their NC origin, reduction in formation of melanophores and apoptosis of the differentiated melanophores, suggesting a role for Kita in melanoblast migration, survival of melanophores. Expression of *kit* begins in premigratory NCCs, and is both Mitfa and Sox10-dependent (Greenhill et al. 2011; Parichy et al. 1999). Functions for Kit in melanoblast migration (before 2 dpf) and melanophore survival (between 2 and 4 dpf) are temporally separable, as shown by study of a temperature sensitive allele and of intriguing alleles which molecularly separate these functions (Rawls and Johnson 2003). Kita expression in the melanophore lineage is initially dependent on Tfap2a but later becomes dependent of Mitfa.

There are two zebrafish *kitlg* genes, but only *kitlga* is required for melanophore development (Hultman et al. 2007).

Medaka *few melanophore (fm)* mutants show a similar fully recessive melanophore phenotype to zebrafish *sparse*, suggesting that *fm* is likely to encode a key component of Kit signaling (Kelsh et al. 2004), as was confirmed by identification of *fm* as *kit-ligand a* (*kitlga*) (Otsuki et al. 2020). Interestingly, however, *fm* mutants exhibit a severe reduction in number of leucophores as well as melanophores, showing that Kit signaling is unexpectedly required for development of leucophores as well as melanophores. The recent suggestion that a subset of leucophores in *Danio* species might be melanophore-derived (melanoleucophores) suggests the hypothesis that loss of some leucophores in *fm* mutant medaka might reflect a failure of melanoleucophore formation (Lewis et al. 2019) although this cell-type has yet to be described in medaka and, if anything, the medaka leucophores would seem more similar to the xantholeucophores Lewis et al described (Kimura et al. 2014; Nagao et al. 2014, 2018). Alternatively, perhaps leucophores also express a Kit orthologue and are Kit signaling dependent for their survival.

Variation in Kitlg expression has been shown to underlie natural variation in stickleback pigmentation, associated with adaptation to freshwater environments. Analysis has shown that decreased adult skin pigmentation depends upon cis-regulatory changes in the *kitlg* gene that result in reduced *kitlg* expression (Miller et al. 2007).

fms/panther Signalling

Fms/Panther is a receptor tyrosine kinase closely related to Kit, that is expressed in many NCCs in a premigratory pattern and persists in the xanthophore lineage throughout its differentiation (Parichy et al. 2000). In the embryo, loss of Fms/Panther activity results in the apparent absence of xanthophores throughout the skin of early larva. However, using *xdh and gch* as early markers of xanthoblasts, reveals that the progenitors are formed, but fail to migrate on the lateral pathway. By 48 hpf, such cells are apparently absent. It seems that the role of Fms in xanthophores closely parallels that of Kit in melanophores.

Ltk Signaling

Leukocyte tyrosine kinase (Ltk) is a member of the insulin receptor superfamily, and within this is most closely related to anaplastic lymphoma kinase (ALK). The zebrafish gene identified as the *shady* locus was identified as an *ltk* homologue, but we note that as further Ltk and Alk sequences have been identified, it has been proposed that this gene may be better considered as *alk-1* (Mo et al. 2017). However, detailed assessment of this question has confirmed the original assignation as an *Ltk* orthologue (Dornburg et al. 2020). The *alk-2* gene appears not to have a role in iridophore (nor other pigment cell) development (Mo et al. 2017). At later embryonic stages, *ltk* shows an expression pattern reminiscent of iridophores, in the eyes, the lateral patches on the yolk and a series of spots along the dorsal and ventral midline, which suggests that Ltk may play a crucial role in differentiated iridophores, but it is also expressed in a subset of premigratory NCCs, deduced to be putative multipotent progenitors for pigment cell fates from study of the changed expression pattern in *sox10* mutants, and in migrating cells on the medial migration pathway, hinting at a role throughout iridophore development (Lopes et al. 2008). Iridophore fate specification is defective in both embryonic and adult *ltk/shady* mutants although no other NC phenotype was identified in this study (Lopes et al. 2008). This role was confirmed in subsequent work showing that constitutively activated Ltk signaling can drive NCCs to adopt an iridophore fate, as shown by overexpression of a constitutively activated chimaeric protein (Rodrigues et al. 2012) and by a spontaneous *ltk* gain-of-function allele (Fadeev et al. 2016). These studies also showed that there is a subsequent role for Ltk signaling, likely in survival and/or proliferation of differentiated iridophores (Fadeev et al. 2016; Rodrigues et al. 2012).

7.5.5 Pigmentation Genes (see Chap. 4 for detailed function)

dct, *tyr*, *tyrp1*: Melanophore Markers

dct is expressed in the melanophore lineage in zebrafish and medaka (Kelsh et al. 2000; Nagao et al. 2014). The expression is directly dependent on Sox10 and Mitfa. *tyr* and *tyrp1* expression marks the melanophore precursors, but different from *dct*, these are unlikely direct transcriptional targets of Sox10 (Greenhill et al. 2011). Mutant phenotypes have only been partially explored in fish. As expected the *tyrosinase/sandy* mutant shows a complete absence of melanin, but unpigmented melanophores survive and pattern normally. We are not aware of identified mutants for *tyrp1* or *dct*, but it would be interesting to see the relative importance of these enzymes on melanin synthesis in fish; based on the mouse phenotypes, we would expect their mutant phenotypes to be relatively subtle decreases in melanin.

gch, *xdh*: Xanthophore Markers

gch and *xdh* are often used as markers for xanthophore precursors in zebrafish (Parichy et al. 2000; Pelletier et al. 2001) while *gch* and *xdh* mark not only xanthophores but also leucophores in medaka (Nagao et al. 2014, 2018). Transcription factors that directly regulate these genes are not yet known. *gch* is supposed to be expressed also in melanophore lineage because Gch activity is required for the first reaction of pteridine pathway shared by melanophore to synthesize H4 biopterin, which catalyzes L-tyrosine production in the melanin synthesis. Indeed, *gch* is used as a marker for melanophore precursors in the Japanese flounder (Watanabe et al. 2008).

pnp4a: An Iridophore Marker

Purine nucleoside phosphorylase 4a is an enzyme converting guanosine monophosphate to guanine (Curran et al. 2010) and thereby generating a high concentration of guanine in iridophores to allow reflecting platelet formation. Consistently, medaka *pnp4a/guanineless* mutant lacks iridophore reflectivity (Kimura et al. 2017).

Although *pnp4a* is defined as an iridophore lineage marker, it is expressed rather widely in NCCs and long before iridophores differentiate (Curran et al. 2010). Also, it is expressed, albeit at lower levels, in melanophores (Petratou et al. 2018). It is notable that the expression pattern of *pnp4a* strikingly resembled that of *mitfa*, rather than that of *tfec*, in the early stages of NCCs. *pnp4a* expression is likely to be regulated independently by both Tfec and Mitfa. It appears to be transiently expressed in an Mitfa-dependent manner in chromatoblasts, but then on-going expression in iridoblasts requires Tfec.

7.5.6 Modeling of the Pigment Cell Gene Regulatory Network

As will be clear, there is considerable knowledge of a diversity of genes with key roles in chromatophore development in fish. However, as one starts to think about how these genes link together to form a gene regulatory network (GRN) driving fate specification and differentiation, it rapidly becomes clear how little we know!

Initial work from our lab has focused on identifying and validating 'core models' incorporating the key genes, defined as those with major phenotypic effects. We first developed an iterative systems biology approach to develop a core GRN for melanophore fate specification and differentiation (Greenhill et al. 2011). We noted that *sox10* expression is rapidly lost in differentiating melanophores, but also that *sox10* mutants have tiny amounts of residual melanin in a small number of cells, attributed to leaky expression of melanin synthesis genes (Greenhill et al. 2011). Building on a core model for sympathetic neuron development developed by Anderson's lab, we formulated a core model for the melanophore GRN, incorporating various novel features (Fig. 7.5). These features included (1) Sox10-dependent repression of Mitfa-dependent target gene expression; (2) feedback loop between Mitfa and Sox10; (3) Hdac-dependent repression of *sox10* in the melanophore lineage (Greenhill et al. 2011). Using dynamical mathematical modeling as a series of ordinary differential equations, we showed how this model was insufficient to explain the known gene expression patterns, and hypothesized further components to the core model that might solve these issues. Using this novel, iterative means to develop the core GRN, we first predicted, and then identified biological correlates of a number of novel features of the core GRN, including autoregulatory feedback by Mitfa on *mitfa* in a Sox10-independent manner, and Mitfa-dependent, Hdac1-mediated repression of *sox10* expression (Greenhill et al. 2011). We also predicted the low level, Mitfa-independent activation of several melanin synthesis genes and proposed that Sox9b was a strong candidate (Greenhill et al. 2011).

This initial melanophore GRN has subsequently been expanded by incorporating Wnt signaling (Vibert et al. 2017). We showed that Wnt signaling can be detected in the NC and throughout melanophore differentiation to around 72 hpf. Furthermore,

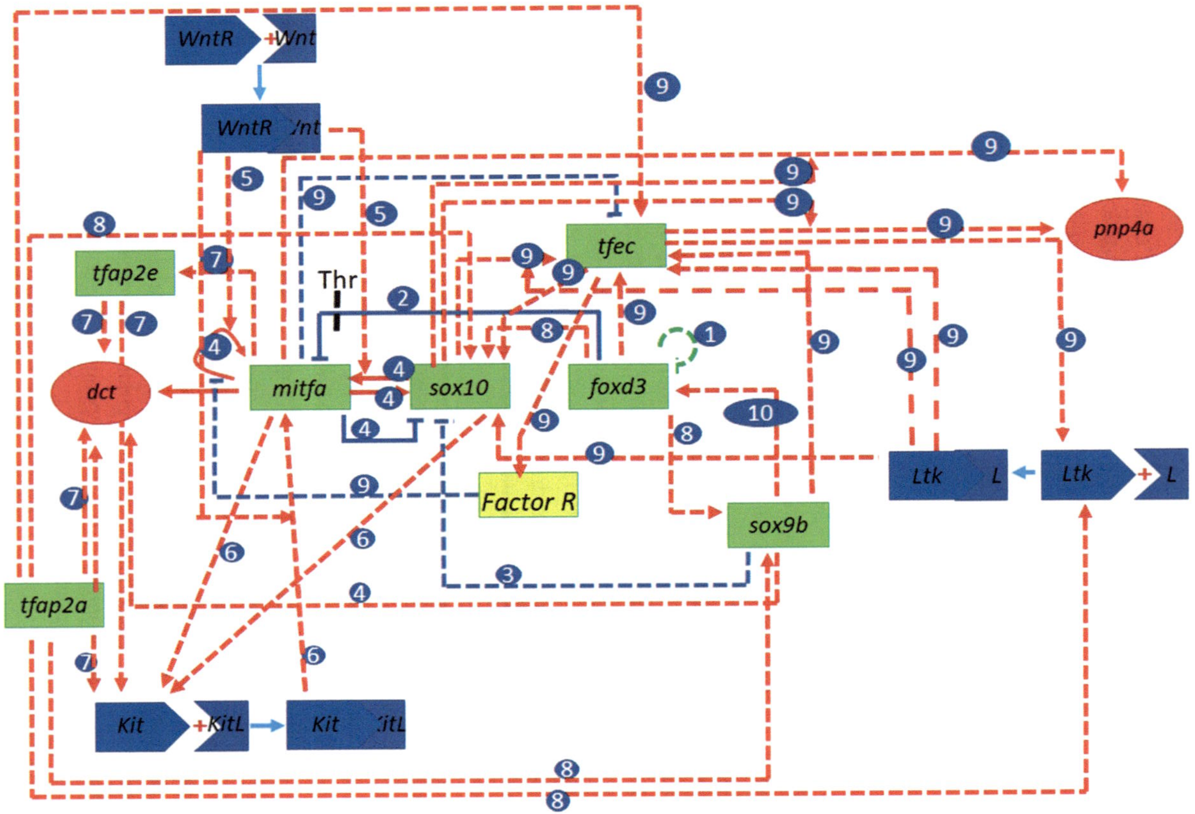

Fig. 7.5 Working model of melanophore and iridophore GRN based on published literature. Genes encoding transcription factors are indicated by green boxes, receptors and their ligands by compatible blue shapes, and differentiation genes (fate markers) in red ovals. Activation arrows in red and end in an arrowhead, repression arrows are in blue and are blunt-ended. Autoregulatory feedback loops are represented by curved arrows. Predicted genes shown in yellow. 'Thr' with

accompanying black line indicates threshold response governing activity of the associated input. Data supporting features are indicated by numbered circles, referencing the following papers: (1) (Pohl and Knochel 2001); (2) (Curran et al. 2009); (3) (Shakhova et al. 2015); (4) (Greenhill et al. 2011); (5) (Dorsky et al. 1998; Vibert et al. 2017; Vibert 2014); (6) (Lapedriza 2016); (7) (Knight et al. 2004; O'Brien et al. 2004; Van Otterloo et al. 2010); (8) (Arduini et al. 2009); (9) (Petratou et al. 2018, 2021); (10) (Yan et al. 2005)

using GSK3b inhibitor treatment to increase Wnt signaling and complementary reduction of Wnt signaling using expression of a dominant negative Tcf3 construct, our study identified two distinct phases of Wnt signaling, one driving initial fate specification (most likely early enhancement of Sox10-dependent expression of *mitfa*) and another involved in melanophore maintenance (likely via Sox10-independent autoregulatory positive feedback of Mitfa) (Vibert et al. 2017). We then used the same mathematical modeling approach to assess the impact of incorporating this biphasic Wnt signal, showing that it continues to match well the experimental observations on melanophore development (Vibert et al. 2017).

Due to their shared origin from a multipotent progenitor, a complete chromatophore GRN will need to integrate core networks for all pigment cell-types as well as other NC fates. Although this remains to be attempted for xanthophores, we recently extended our systems biology approach to develop a core GRN for iridophores (Petratou et al. 2018, 2021). We demonstrated that the on-going expression of *sox10* in the iridophore lineage was important for maintenance of *tfec* expression, and that Tfec and Ltk form a positive feedback loop maintaining Tfec expression in this lineage (Petratou et al. 2018). By integrating these data with evidence for independent regulation of *pnp4a* by Mitfa and Tfec, and previous data on Ltk, Mitfa and Sox10, we used the iterative modeling approach, but here with the innovation of using physiologically realistic parameter values based on the literature, to build a core GRN for iridophore development, in the context of a careful dissection of putative stages in the iridophore specification and differentiation process (Petratou et al. 2018). The modeling process proved valuable for distinguishing the likely relationship between Sox10, Tfec and Ltk signaling in establishing the iridoblasts, as well as identifying the predicted role for a currently unknown Factor R mediating repression of *mitfa* in the developing iridoblast as part of the fate restriction process (Petratou et al. 2018). In a subsequent series of experiments, we further expanded this model (Petratou et al. 2021).

7.6 Pigment Cells as an Evolutionary Novelty

Body colouration provides a key adaptive feature affecting many aspects of organismal biology. It may be highly effective in blending with the surrounding environments and hiding from predators. Species-specific colouration contributes to intra- and inter-species recognition, and may also function as a sexual attractant. Melanophores, xanthophores and iridophores are widely distributed amongst teleost lineages (Fujii 1993b). In contrast, to date leucophores and cyanophores have been identified in just a few species (see Sect. 7.2), and hence are likely an evolutionary novelty (Fig. 7.6).

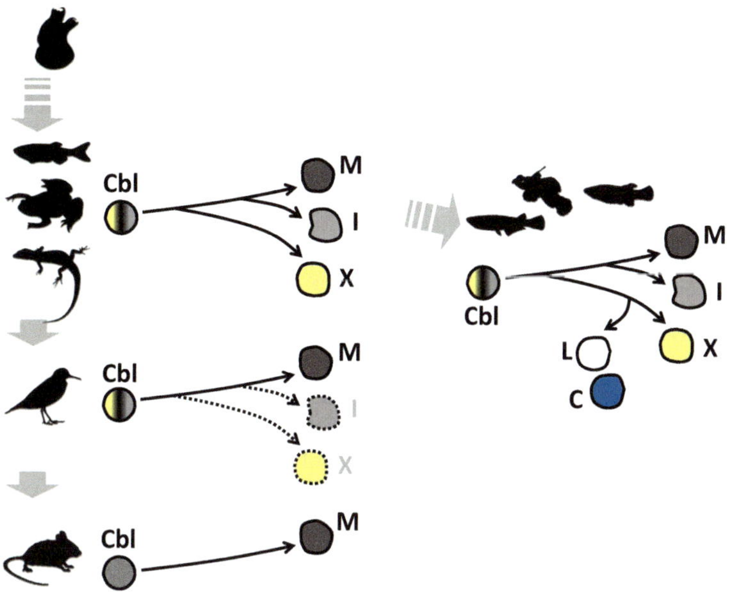

Fig. 7.6 Evolution of novel cell-types in the fish species. Within the neural crest, a chromatoblast (Cbl) is proposed to give rise to three different types of pigment cell, melanophores (M), iridophores (I) and xanthophores (X) in fish, amphibians and reptiles. During the evolution of birds and mammals, I and X disappeared, and thus these animals have a single type of melanocyte (M). It should be noted that rudimentary I and X are found in the iris of birds, implying that the composition of pigment cell-types in birds represents a transitional state of neural crest evolution. In some fish, novel pigment cell-types, including leucophores (L) and cyanophores (C), have evolved in specific lineages. Silhouettes were downloaded from www.phylopic.org, free of copyrights (CG Public Domain Dedication 1.0 license)

7.6.1 Medaka Leucophore

Although leucophores had been considered to be closely related to iridophores since the appearance of both cell-types is dominated by reflection of light due to intracellular purine crystal accumulation, we showed that leucophores share an ontogenetic origin with xanthophores, i.e. they share a progenitor (Kimura et al. 2014; Nagao et al. 2014). The genetic analyses with medaka mutants have suggested that leucophores develop via a common bipotent progenitor shared with xanthophores, which is consistent with a previous finding that leucophores contain a pteridine pigment (drosopterin) closely related to sepiapterin in xanthophores (Hama 1965, 1967; Oliphant and Hudon 1993). Fate choice between leucophore and xanthophore lineages is associated with *sox5* function as a molecular fate switch in the bipotent progenitor. This mode of Sox5 function to promote xanthophore fate is unique to medaka, but not seen in zebrafish, suggesting an evolutionary change in the role of Sox5. It is conceivable that medaka evolved the leucophore by expanding the potency of a xanthophore progenitor and using loss of Sox5 expression to drive accumulation of intracellular purine crystals in a xanthophore-like cell. Evolution of Sox5 function in medaka seems to be linked to the evolutionary acquisition of leucophores. Future work will explore Sox5 partner proteins and other transcriptional regulators of the leucophore lineage, and will be expected to clarify the evolution of this novel cell-type.

7.6.2 Zebrafish Leucophores

Leucophore-like pigment cells in zebrafish were documented by Irion et al. (Irion et al. 2016) as only being present in the dorsal and tail fins as well as, later by Lewis et al. (Lewis et al. 2019) in the anal fin. Similar cells are distributed in fins across nine species within *Danio*. These have been classified as two types, xantholeucophore in the anal and tail fins and melanoleucophore in the dorsal and tail fins. Lewis et al. reported behavioural data indicating that these cells contribute to zebrafish social interactions (Lewis et al. 2019).

Zebrafish xantholeucophore seems closely related to medaka leucophores (Lewis et al. 2019). During development, xantholeucophores first exhibit pronounced yellow-orange pigmentation, and later a white halo of pigmentation becoming gradually more distinct, consistent with some shared characteristics and inferred developmental origins of leucophores and xanthophores in medaka.

7.6.3 Arabian Killifish Leucophores

Arabian killifish (*Aphanius dispar*) larvae have numerous leucophores spreading over the body (Hamied et al. 2020). Hamied and his colleagues showed that these leucophores appear much earlier (during somitogenesis) in the killifish than those in medaka (after somitogenesis) and are extraordinarily fluorescent, thus named fluoroleucophores (Fig. 7.7). The autofluorescence of the leucophores depends on the pteridine synthetic pathway (see Chap. 4) since *gch* loss-of-function mutant larvae lack detectable autofluorescence, suggesting that the strong fluorescence occurs in pteridine compound(s). These *gch* mutant killifish, however, retain white colouration in the leucophores. These findings are reminiscent of the medaka *white leucophores* (*wl*) mutant, which exhibits white or dull but not orange embryonic/larval leucophores (Kelsh et al. 2004), implying that *wl/slc2a11b*, encoding a member of solute carrier family, is specific to pteridine substrates (Kimura et al. 2014). It would be interesting to know the adaptive role of these cells in the killifish; it seems likely that they may be an adaptation to life under the extreme environmental conditions, perhaps a response to high UV incidence and shallow water, transient pond environments.

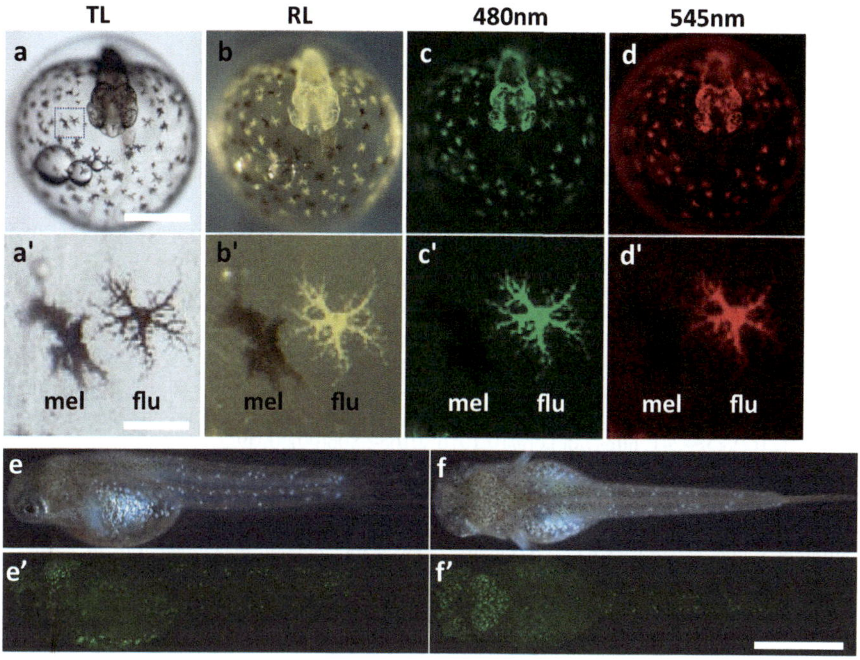

Fig. 7.7 Fluoroleucophores in Arabian killifish. The Arabian killifish embryo possesses strongly fluorescent cells, fluoroleucophores. A 3 dpf embryo was imaged under transmitted light (**a**, **a**′, TL), reflected light (**b**, **b**′, RL), fluorescent light with GFP filter (**c**, **c**′, 480 nm), or RFP filter (**d**, **d**′, 545 nm). (**a**–**d**) Dorsal views. (**a**′–**d**′) Melanophore (mel, left) and fluoroleucophore (flu, right). Scale bars, 500μm in (**a**–**d**) and 200 μm in (**a**′–**d**′). Photos (**a**–**d**, **a**′–**d**′) courtesy of Tetsuhiro Kudo. Larva has numerous flouroleucophores, which are equivalent to medaka leucophore, but highly autofluorescent and differentially distributed from those in medaka (see Fig. 7.1) (**e**, **e**′) dorsolateral views. (**f**, **f**′) dorsal views. (**e**, **f**) Reflected light. (**e**′, **f**′) 480 nm. Scale bar, 2 mm

Acknowledgements We are grateful to Tetsuhiro Kudo, University of Exeter for kindly providing the photos of Arabian killifish fluoroleucophores.

References

Aoki Y, Saint-Germain N, Gyda M, Magner-Fink E, Lee YH, Credidio C, Saint-Jeannet JP (2003) Sox10 regulates the development of neural crest-derived melanocytes in Xenopus. Dev Biol 259 (1):19–33. https://doi.org/10.1016/s0012-1606(03)00161-1

Arduini BL, Bosse KM, Henion PD (2009) Genetic ablation of neural crest cell diversification. Development 136(12):1987–1994. https://doi.org/10.1242/dev.033209

Arnott HJ, Best AC, Nicol JA (1970) Occurrence of melanosomes and of crystal sacs within the same cell in the tapetum lucidum of the stingaree. J Cell Biol 46(2):426–427. https://doi.org/10.1083/jcb.46.2.426

Bagnara JT, Taylor JD, Prota G (1973) Color changes, unusual melanosomes, and a new pigment from leaf frogs. Science 182(4116):1034–1035. https://doi.org/10.1126/science.182.4116.1034

Bagnara JT, Frost SK, Matsumoto J (1978) On the development of pigment patterns in amphibians. Am Zool 18(2):301–312

Bagnara JT, Matsumoto J, Ferris W, Frost SK, Turner WA Jr, Tchen TT, Taylor JD (1979) Common origin of pigment cells. Science 203(4379):410–415. https://doi.org/10.1126/science.760198

Bejar J, Hong Y, Schartl M (2003) Mitf expression is sufficient to direct differentiation of medaka blastula derived stem cells to melanocytes. Development 130(26):6545–6553. https://doi.org/10.1242/dev.00872

Bondurand N, Pingault V, Goerich DE, Lemort N, Sock E, Le Caignec C, Wegner M, Goossens M (2000) Interaction among SOX10, PAX3 and MITF, three genes altered in Waardenburg syndrome. Hum Mol Genet 9(13):1907–1917. https://doi.org/10.1093/hmg/9.13.1907

Carney TJ, Dutton KA, Greenhill E, Delfino-Machin M, Dufourcq P, Blader P, Kelsh RN (2006) A direct role for Sox10 in specification of neural crest-derived sensory neurons. Development 133 (23):4619–4630. https://doi.org/10.1242/dev.02668

Cheng Y, Cheung M, Abu-Elmagd MM, Orme A, Scotting PJ (2000) Chick sox10, a transcription factor expressed in both early neural crest cells and central nervous system. Brain Res Dev Brain Res 121(2):233–241. https://doi.org/10.1016/s0165-3806(00)00049-3

Cooper CD, Raible DW (2009) Mechanisms for reaching the differentiated state: Insights from neural crest-derived melanocytes. Semin Cell Dev Biol 20(1):105–110. https://doi.org/10.1016/j.semcdb.2008.09.008

Curran K, Raible DW, Lister JA (2009) Foxd3 controls melanophore specification in the zebrafish neural crest by regulation of Mitf. Dev Biol 332(2):408–417. https://doi.org/10.1016/j.ydbio.2009.06.010

Curran K, Lister JA, Kunkel GR, Prendergast A, Parichy DM, Raible DW (2010) Interplay between Foxd3 and Mitf regulates cell fate plasticity in the zebrafish neural crest. Dev Biol 344 (1):107–118. https://doi.org/10.1016/j.ydbio.2010.04.023

Dornburg A, Wang Z, Wang J, Mo ES, Lopez-Giraldez F, Townsend JP (2020) Comparative genomics within and across Bilaterians illuminates the evolutionary history of ALK and LTK proto-oncogene origination and diversification. Genome Biol Evol. https://doi.org/10.1093/gbe/evaa228

Dorsky RI, Moon RT, Raible DW (1998) Control of neural crest cell fate by the Wnt signalling pathway. Nature 396(6709):370–373. https://doi.org/10.1038/24620

Dorsky RI, Raible DW, Moon RT (2000) Direct regulation of nacre, a zebrafish MITF homolog required for pigment cell formation, by the Wnt pathway. Genes Dev 14(2):158–162

Dutton K, Dutton JR, Pauliny A, Kelsh RN (2001a) A morpholino phenocopy of the colourless mutant. Genesis 30(3):188–189. https://doi.org/10.1002/gene.1062

Dutton KA, Pauliny A, Lopes SS, Elworthy S, Carney TJ, Rauch J, Geisler R, Haffter P, Kelsh RN (2001b) Zebrafish colourless encodes sox10 and specifies non-ectomesenchymal neural crest fates. Development 128(21):4113–4125

Dutton JR, Antonellis A, Carney TJ, Rodrigues FS, Pavan WJ, Ward A, Kelsh RN (2008) An evolutionarily conserved intronic region controls the spatiotemporal expression of the transcription factor Sox10. BMC Dev Biol 8:105. https://doi.org/10.1186/1471-213X-8-105

Elworthy S, Lister JA, Carney TJ, Raible DW, Kelsh RN (2003) Transcriptional regulation of mitfa accounts for the sox10 requirement in zebrafish melanophore development. Development 130 (12):2809–2818. https://doi.org/10.1242/dev.00461

Fadeev A, Krauss J, Singh AP, Nusslein-Volhard C (2016) Zebrafish Leucocyte tyrosine kinase controls iridophore establishment, proliferation and survival. Pigment Cell Melanoma Res 29 (3):284–296. https://doi.org/10.1111/pcmr.12454

Ferris WR, Bagnara JT (1970) Iridophores in Iris of doves. J Invest Dermatol 54(1):86

Fujii R (1993a) Coloration and chromatophores. In: Evans DH (ed) The physiology of fishes. CRC Press, Boca Raton, pp 535–562

Fujii R (1993b) Cytophysiology of fish chromatophores. Int Rev Cytol 143(143):191–255

Goda M, Fujii R (1995) Blue chromatophores in two species of callionymid fish. Zool Sci 12 (6):811–813

Goda M, Fujii R (1998) The blue coloration of the common surgeonfish, Paracanthurus hepatus-II. Color revelation and color changes. Zool Sci 15(3):323–333. https://doi.org/10.2108/zsj.15.323

Goda M, Ohata M, Ikoma H, Fujiyoshi Y, Sugimoto M, Fujii R (2011) Integumental reddish-violet coloration owing to novel dichromatic chromatophores in the teleost fish, Pseudochromis diadema. Pigment Cell Melanoma Res 24(4):614–617

Goda M, Fujiyoshi Y, Sugimoto M, Fujii R (2013) Novel dichromatic chromatophores in the integument of the mandarin fish Synchiropus splendidus. Biol Bull 224(1):14–17. https://doi.org/10.1086/BBLv224n1p14

Goding CR, Arnheiter H (2019) MITF-the first 25 years. Genes Dev 33(15-16):983–1007. https://doi.org/10.1101/gad.324657.119

Greenhill ER, Rocco A, Vibert L, Nikaido M, Kelsh RN (2011) An iterative genetic and dynamical modelling approach identifies novel features of the gene regulatory network underlying melanocyte development. PLoS Genet 7(9):e1002265. https://doi.org/10.1371/journal.pgen.1002265

Hama T (1965) The relation between the pterins and chromatophores in the rnedaka Oryzias talipes. Proc Jpn Acad 41:305–309

Hama T (1967) Nouvelle demonstration de la coexistence de la drosopterine et de la purine dans le leucophore de Medaka (Oryzias latipes, teleosteen). C R Seances Soc Biol 161:1197–1200

Hama T (1975) Medaka (KILLFISH): biology and strains. Keigaku Publishing Company, Tokyo, pp 138–153

Hama T, Hasegawa H (1967) Studies on the chromatophores of Oryzias latipes (teleostean fish): behaviour of the pteridine, fat and carotenoid during xanthophore differentiation in the color varieties. Proc Jpn Acad 43:901–906

Hamied A, Alnedawy Q, Correia A, Hacker C, Ramsdale M, Hashimoto H, Kudoh T (2020) Identification and characterization of highly fluorescent pigment cells in embryos of the arabian killifish (Aphanius Dispar). iScience 23(11):101674. https://doi.org/10.1016/j.isci.2020.101674

Higdon CW, Mitra RD, Johnson SL (2013) Gene expression analysis of zebrafish melanocytes, iridophores, and retinal pigmented epithelium reveals indicators of biological function and developmental origin. PLoS One 8(7):e67801. https://doi.org/10.1371/journal.pone.0067801

Hochgreb-Hagele T, Bronner ME (2013) A novel FoxD3 gene trap line reveals neural crest precursor movement and a role for FoxD3 in their specification. Dev Biol 374(1):1–11. https://doi.org/10.1016/j.ydbio.2012.11.035

Hoffman TL, Javier AL, Campeau SA, Knight RD, Schilling TF (2007) Tfap2 transcription factors in zebrafish neural crest development and ectodermal evolution. J Exp Zool B Mol Dev Evol 308(5):679–691. https://doi.org/10.1002/jez.b.21189

Hong CS, Saint-Jeannet JP (2005) Sox proteins and neural crest development. Semin Cell Dev Biol 16(6):694–703. https://doi.org/10.1016/j.semcdb.2005.06.005

Hou L, Arnheiter H, Pavan WJ (2006) Interspecies difference in the regulation of melanocyte development by SOX10 and MITF. Proc Natl Acad Sci U S A 103(24):9081–9085. https://doi.org/10.1073/pnas.0603114103

Hozumi S, Shirai M, Wang J, Aoki S, Kikuchi Y (2018) The N-terminal domain of gastrulation brain homeobox 2 (Gbx2) is required for iridophore specification in zebrafish. Biochem Biophys Res Commun 502(1):104–109. https://doi.org/10.1016/j.bbrc.2018.05.128

Hultman KA, Bahary N, Zon LI, Johnson SL (2007) Gene Duplication of the zebrafish kit ligand and partitioning of melanocyte development functions to kit ligand A. PLoS Genet 3(1):e17. https://doi.org/10.1371/journal.pgen.0030017

Iga T, Matsuno A (1986) Motile iridophores of a freshwater goby, Odontobutis obscura. Cell Tissue Res 244(1):165–171

Ignatius MS, Moose HE, El Hodiri HM, Henion PD (2008) colgate/hdac1 repression of foxd3 expression is required to permit mitfa-dependent melanogenesis. Dev Biol 313(2):568–583. https://doi.org/10.1016/j.ydbio.2007.10.045

Irion U, Singh AP, Nusslein-Volhard C (2016) The developmental genetics of vertebrate color pattern formation: lessons from zebrafish. Curr Top Dev Biol 117:141–169. https://doi.org/10.1016/bs.ctdb.2015.12.012

Johnson SL, Africa D, Walker C, Weston JA (1995) Genetic control of adult pigment stripe development in zebrafish. Dev Biol 167(1):27–33. https://doi.org/10.1006/dbio.1995.1004

Johnson SL, Nguyen AN, Lister JA (2011) mitfa is required at multiple stages of melanocyte differentiation but not to establish the melanocyte stem cell. Dev Biol 350(2):405–413. https://doi.org/10.1016/j.ydbio.2010.12.004

Kapur RP (1999) Early death of neural crest cells is responsible for total enteric aganglionosis in Sox10(Dom)/Sox10(Dom) mouse embryos. Pediatr Dev Pathol 2(6):559–569. https://doi.org/10.1007/s100249900162

Kelsh RN (2006) Sorting out Sox10 functions in neural crest development. Bioessays 28(8):788–798. https://doi.org/10.1002/bies.20445

Kelsh RN, Eisen JS (2000) The zebrafish colourless gene regulates development of non-ectomesenchymal neural crest derivatives. Development 127(3):515–525

Kelsh RN, Raible DW (2002) Specification of zebrafish neural crest. Results Probl Cell Differ 40:216–236. https://doi.org/10.1007/978-3-540-46041-1_11

Kelsh RN, Brand M, Jiang YJ, Heisenberg CP, Lin S, Haffter P, Odenthal J, Mullins MC, van Eeden FJ, Furutani-Seiki M, Granato M, Hammerschmidt M, Kane DA, Warga RM, Beuchle D, Vogelsang L, Nusslein-Volhard C (1996) Zebrafish pigmentation mutations and the processes of neural crest development. Development 123:369–389

Kelsh RN, Schmid B, Eisen JS (2000) Genetic analysis of melanophore development in zebrafish embryos. Dev Biol 225(2):277–293. https://doi.org/10.1006/dbio.2000.9840

Kelsh RN, Inoue C, Momoi A, Kondoh H, Furutani-Seiki M, Ozato K, Wakamatsu Y (2004) The Tomita collection of medaka pigmentation mutants as a resource for understanding neural crest cell development. Mech Dev 121(7-8):841–859. https://doi.org/10.1016/j.mod.2004.01.004

Kimmel CB, Ballard WW, Kimmel SR, Ullmann B, Schilling TF (1995) Stages of embryonic development of the zebrafish. Dev Dyn 203(3):253–310. https://doi.org/10.1002/aja.1002030302

Kimura T, Nagao Y, Hashimoto H, Yamamoto-Shiraishi Y, Yamamoto S, Yabe T, Takada S, Kinoshita M, Kuroiwa A, Naruse K (2014) Leucophores are similar to xanthophores in their specification and differentiation processes in medaka. Proc Natl Acad Sci U S A 111(20):7343–7348. https://doi.org/10.1073/pnas.1311254111

Kimura T, Takehana Y, Naruse K (2017) pnp4a is the causal gene of the medaka iridophore mutant guanineless. G3 (Bethesda) 7(4):1357–1363. https://doi.org/10.1534/g3.117.040675

Knight RD, Javidan Y, Nelson S, Zhang T, Schilling T (2004) Skeletal and pigment cell defects in the lockjaw mutant reveal multiple roles for zebrafish tfap2a in neural crest development. Dev Dyn 229(1):87–98. https://doi.org/10.1002/dvdy.10494

Koopman P, Schepers G, Brenner S, Venkatesh B (2004) Origin and diversity of the SOX transcription factor gene family: genome-wide analysis in Fugu rubripes. Gene 328:177–186. https://doi.org/10.1016/j.gene.2003.12.008

Kubic JD, Young KP, Plummer RS, Ludvik AE, Lang D (2008) Pigmentation PAX-ways: the role of Pax3 in melanogenesis, melanocyte stem cell maintenance, and disease. Pigment Cell Melanoma Res 21(6):627–645. https://doi.org/10.1111/j.1755-148X.2008.00514.x

Lapedriza A (2016) Gene regulatory network of melanocyte development. University of Bath, Bath

Lee EM, Yuan T, Ballim RD, Nguyen K, Kelsh RN, Medeiros DM, McCauley DW (2016) Functional constraints on SoxE proteins in neural crest development: the importance of differential expression for evolution of protein activity. Dev Biol 418(1):166–178. https://doi.org/10.1016/j.ydbio.2016.07.022

Lewis JL, Bonner J, Modrell M, Ragland JW, Moon RT, Dorsky RI, Raible DW (2004) Reiterated Wnt signaling during zebrafish neural crest development. Development 131(6):1299–1308. https://doi.org/10.1242/dev.01007

Lewis VM, Saunders LM, Larson TA, Bain EJ, Sturiale SL, Gur D, Chowdhury S, Flynn JD, Allen MC, Deheyn DD, Lee JC, Simon JA, Lippincott-Schwartz J, Raible DW, Parichy DM (2019) Fate plasticity and reprogramming in genetically distinct populations of Danio leucophores. Proc Natl Acad Sci U S A 116(24):11806–11811

Li W, Cornell RA (2007) Redundant activities of Tfap2a and Tfap2c are required for neural crest induction and development of other non-neural ectoderm derivatives in zebrafish embryos. Dev Biol 304(1):338–354. https://doi.org/10.1016/j.ydbio.2006.12.042

Lister JA, Robertson CP, Lepage T, Johnson SL, Raible DW (1999) nacre encodes a zebrafish microphthalmia-related protein that regulates neural-crest-derived pigment cell fate. Development 126(17):3757–3767

Lister JA, Cooper C, Nguyen K, Modrell M, Grant K, Raible DW (2006) Zebrafish Foxd3 is required for development of a subset of neural crest derivatives. Dev Biol 290(1):92–104. https://doi.org/10.1016/j.ydbio.2005.11.014

Lopes SS, Yang X, Muller J, Carney TJ, McAdow AR, Rauch GJ, Jacoby AS, Hurst LD, Delfino-Machin M, Haffter P, Geisler R, Johnson SL, Ward A, Kelsh RN (2008) Leukocyte tyrosine kinase functions in pigment cell development. PLoS Genet 4(3):e1000026. https://doi.org/10.1371/journal.pgen.1000026

Lukoseviciute M, Gavriouchkina D, Williams RM, Hochgreb-Hagele T, Senanayake U, Chong-Morrison V, Thongjuea S, Repapi E, Mead A, Sauka-Spengler T (2018) From pioneer to repressor: bimodal foxd3 activity dynamically remodels neural crest regulatory landscape in vivo. Dev Cell 47(5):608–628. https://doi.org/10.1016/j.devcel.2018.11.009

Mahony CB, Fish RJ, Pasche C, Bertrand JY (2016) tfec controls the hematopoietic stem cell vascular niche during zebrafish embryogenesis. Blood 128(10):1336–1345. https://doi.org/10.1182/blood-2016-04-710137

Matsumoto J (1965) Studies on fine structure and cytochemical properties of erythrophores in swordtail, Xiphophorus helleri, with special reference to their pigment granules (Pterinosomes). J Cell Biol 27(3):493–504. https://doi.org/10.1083/jcb.27.3.493

Miller CT, Beleza S, Pollen AA, Schluter D, Kittles RA, Shriver MD, Kingsley DM (2007) cis-Regulatory changes in Kit ligand expression and parallel evolution of pigmentation in sticklebacks and humans. Cell 131(6):1179–1189. https://doi.org/10.1016/j.cell.2007.10.055

Minchin JE, Hughes SM (2008) Sequential actions of Pax3 and Pax7 drive xanthophore development in zebrafish neural crest. Dev Biol 317(2):508–522. https://doi.org/10.1016/j.ydbio.2008.02.058

Mo ES, Cheng Q, Reshetnyak AV, Schlessinger J, Nicoli S (2017) Alk and Ltk ligands are essential for iridophore development in zebrafish mediated by the receptor tyrosine kinase Ltk. Proc Natl Acad Sci U S A 114(45):12027–12032. https://doi.org/10.1073/pnas.1710254114

Monsoro-Burq AH (2015) PAX transcription factors in neural crest development. Semin Cell Dev Biol 44:87–96. https://doi.org/10.1016/j.semcdb.2015.09.015

Montero-Balaguer M, Lang MR, Sachdev SW, Knappmeyer C, Stewart RA, De La Guardia A, Hatzopoulos AK, Knapik EW (2006) The mother superior mutation ablates foxd3 activity in neural crest progenitor cells and depletes neural crest derivatives in zebrafish. Dev Dyn 235(12):3199–3212. https://doi.org/10.1002/dvdy.20959

Nagao Y, Suzuki T, Shimizu A, Kimura T, Seki R, Adachi T, Inoue C, Omae Y, Kamei Y, Hara I, Taniguchi Y, Naruse K, Wakamatsu Y, Kelsh RN, Hibi M, Hashimoto H (2014) Sox5 functions as a fate switch in medaka pigment cell development. PLoS Genet 10(4):e1004246. https://doi.org/10.1371/journal.pgen.1004246

Nagao Y, Takada H, Miyadai M, Adachi T, Seki R, Kamei Y, Hara I, Taniguchi Y, Naruse K, Hibi M, Kelsh RN, Hashimoto H (2018) Distinct interactions of Sox5 and Sox10 in fate specification of pigment cells in medaka and zebrafish. PLoS Genet 14(4):e1007260. https://doi.org/10.1371/journal.pgen.1007260

Nikaido M, Law EW, Kelsh RN (2013) A systematic survey of expression and function of zebrafish frizzled genes. PLoS One 8(1):e54833. https://doi.org/10.1371/journal.pone.0054833

Nord H, Dennhag N, Muck J, von Hofsten J (2016) Pax7 is required for establishment of the xanthophore lineage in zebrafish embryos. Mol Biol Cell 27(11):1853–1862. https://doi.org/10.1091/mbc.E15-12-0821

O'Brien EK, d'Alencon C, Bonde G, Li W, Schoenebeck J, Allende ML, Gelb BD, Yelon D, Eisen JS, Cornell RA (2004) Transcription factor Ap-2alpha is necessary for development of embryonic melanophores, autonomic neurons and pharyngeal skeleton in zebrafish. Dev Biol 265 (1):246–261. https://doi.org/10.1016/j.ydbio.2003.09.029

Obika M (1993) Formation of pterinosomes and carotenoid granules in xanthophores of the teleost oryzias-latipes as revealed by the rapid-freezing and freeze-substitution method. Cell Tissue Res 271(1):81–86

Oliphant LW, Hudon J (1993) Pteridines as reflecting pigments and components of reflecting organelles in vertebrates. Pigment Cell Res 6(4 Pt 1):205–208. https://doi.org/10.1111/j.1600-0749.1993.tb00603.x

Opdecamp K, Nakayama A, Nguyen MT, Hodgkinson CA, Pavan WJ, Arnheiter H (1997) Melanocyte development in vivo and in neural crest cell cultures: crucial dependence on the Mitf basic-helix-loop-helix-zipper transcription factor. Development 124(12):2377–2386

Otsuki Y, Okuda Y, Naruse K, Saya H (2020) Identification of kit-ligand a as the gene responsible for the medaka pigment cell mutant few melanophore. G3 (Bethesda) 10(1):311–319. https://doi.org/10.1534/g3.119.400561

Parichy DM, Rawls JF, Pratt SJ, Whitfield TT, Johnson SL (1999) Zebrafish sparse corresponds to an orthologue of c-kit and is required for the morphogenesis of a subpopulation of melanocytes, but is not essential for hematopoiesis or primordial germ cell development. Development 126 (15):3425–3436

Parichy DM, Ransom DG, Paw B, Zon LI, Johnson SL (2000) An orthologue of the kit-related gene fms is required for development of neural crest-derived xanthophores and a subpopulation of adult melanocytes in the zebrafish, Danio rerio. Development 127(14):3031–3044

Pelletier I, Bally-Cuif L, Ziegler I (2001) Cloning and developmental expression of zebrafish GTP cyclohydrolase I. Mech Dev 109(1):99–103. https://doi.org/10.1016/s0925-4773(01)00516-0

Petratou K, Subkhankulova T, Lister JA, Rocco A, Schwetlick H, Kelsh RN (2018) A systems biology approach uncovers the core gene regulatory network governing iridophore fate choice from the neural crest. PLoS Genet 14(10):e1007402. https://doi.org/10.1371/journal.pgen.1007402

Petratou KSS, Kelsh RN, Lister JA (2021) The MITF paralog tfec is required in neural crest development for fate specification of the iridophore lineage from a multipotent pigment cell progenitor. PLoS ONE 16:e0244794

Pohl BS, Knochel W (2001) Overexpression of the transcriptional repressor FoxD3 prevents neural crest formation in Xenopus embryos. Mech Dev 103(1-2):93–106. https://doi.org/10.1016/s0925-4773(01)00334-3

Potterf SB, Mollaaghababa R, Hou L, Southard-Smith EM, Hornyak TJ, Arnheiter H, Pavan WJ (2001) Analysis of SOX10 function in neural crest-derived melanocyte development: SOX10-dependent transcriptional control of dopachrome tautomerase. Dev Biol 237(2):245–257. https://doi.org/10.1006/dbio.2001.0372

Pusch C, Hustert E, Pfeifer D, Sudbeck P, Kist R, Roe B, Wang Z, Balling R, Blin N, Scherer G (1998) The SOX10/Sox10 gene from human and mouse: sequence, expression, and transactivation by the encoded HMG domain transcription factor. Hum Genet 103 (2):115–123. https://doi.org/10.1007/s004390050793

Rawls JF, Johnson SL (2003) Temporal and molecular separation of the kit receptor tyrosine kinase's roles in zebrafish melanocyte migration and survival. Dev Biol 262(1):152–161. https://doi.org/10.1016/s0012-1606(03)00386-5

Rodrigues FS, Yang X, Nikaido M, Liu Q, Kelsh RN (2012) A simple, highly visual in vivo screen for anaplastic lymphoma kinase inhibitors. ACS Chem Biol 7(12):1968–1974. https://doi.org/10.1021/cb300361a

Schartl M, Larue L, Goda M, Bosenberg MW, Hashimoto H, Kelsh RN (2016) What is a vertebrate pigment cell? Pigment Cell Melanoma Res 29(1):8–14. https://doi.org/10.1111/pcmr.12409

Schilling TF, Kimmel CB (1994) Segment and cell type lineage restrictions during pharyngeal arch development in the zebrafish embryo. Development 120(3):483–494

Seo HC, Saetre BO, Havik B, Ellingsen S, Fjose A (1998) The zebrafish Pax3 and Pax7 homologues are highly conserved, encode multiple isoforms and show dynamic segment-like expression in the developing brain. Mech Dev 70(1-2):49–63. https://doi.org/10.1016/s0925-4773(97)00175-5

Shakhova O, Cheng P, Mishra PJ, Zingg D, Schaefer SM, Debbache J, Hausel J, Matter C, Guo T, Davis S, Meltzer P, Mihic-Probst D, Moch H, Wegner M, Merlino G, Levesque MP, Dummer R, Santoro R, Cinelli P, Sommer L (2015) Antagonistic cross-regulation between Sox9 and Sox10 controls an anti-tumorigenic program in melanoma. PLoS Genet 11(1):e1004877. https://doi.org/10.1371/journal.pgen.1004877

Simoes-Costa M, Bronner ME (2015) Establishing neural crest identity: a gene regulatory recipe. Development 142(2):242–257. https://doi.org/10.1242/dev.105445

Southard-Smith EM, Kos L, Pavan WJ (1998) Sox10 mutation disrupts neural crest development in Dom Hirschsprung mouse model. Nat Genet 18(1):60–64. https://doi.org/10.1038/ng0198-60

Stewart RA, Arduini BL, Berghmans S, George RE, Kanki JP, Henion PD, Look AT (2006) Zebrafish foxd3 is selectively required for neural crest specification, migration and survival. Dev Biol 292(1):174–188. https://doi.org/10.1016/j.ydbio.2005.12.035

Stolt CC, Lommes P, Hillgartner S, Wegner M (2008) The transcription factor Sox5 modulates Sox10 function during melanocyte development. Nucleic Acids Res 36(17):5427–5440. https://doi.org/10.1093/nar/gkn527

Tassabehji M, Read AP, Newton VE, Harris R, Balling R, Gruss P, Strachan T (1992) Waardenburg's syndrome patients have mutations in the human homologue of the Pax-3 paired box gene. Nature 355(6361):635–636. https://doi.org/10.1038/355635a0

Thomas AJ, Erickson CA (2009) FOXD3 regulates the lineage switch between neural crest-derived glial cells and pigment cells by repressing MITF through a non-canonical mechanism. Development 136(11):1849–1858. https://doi.org/10.1242/dev.031989

Trinh LA, Chong-Morrison V, Gavriouchkina D, Hochgreb-Hagele T, Senanayake U, Fraser SE, Sauka-Spengler T (2017) Biotagging of specific cell populations in zebrafish reveals gene regulatory logic encoded in the nuclear transcriptome. Cell Rep 19(2):425–440. https://doi.org/10.1016/j.celrep.2017.03.045

Van Otterloo E, Li W, Bonde G, Day KM, Hsu MY, Cornell RA (2010) Differentiation of zebrafish melanophores depends on transcription factors AP2 alpha and AP2 epsilon. PLoS Genet 6(9):e1001122. https://doi.org/10.1371/journal.pgen.1001122

Van Otterloo E, Li W, Garnett A, Cattell M, Medeiros DM, Cornell RA (2012) Novel Tfap2-mediated control of soxE expression facilitated the evolutionary emergence of the neural crest. Development 139(4):720–730. https://doi.org/10.1242/dev.071308

Vibert L (2014) Experimental analysis of a gene regulatory network underlying stable melanocyte differentiation. University of Bath, Bath

Vibert L, Aquino G, Gehring I, Subkankulova T, Schilling TF, Rocco A, Kelsh RN (2017) An ongoing role for Wnt signaling in differentiating melanocytes in vivo. Pigment Cell Melanoma Res 30(2):219–232. https://doi.org/10.1111/pcmr.12568

Watanabe K, Washio Y, Fujinami Y, Aritaki M, Uji S, Suzuki T (2008) Adult-type pigment cells, which color the ocular sides of flounders at metamorphosis, localize as precursor cells at the proximal parts of the dorsal and anal fins in early larvae. Develop Growth Differ 50(9):731–741. https://doi.org/10.1111/j.1440-169X.2008.01071.x

Wegner M (1999) From head to toes: the multiple facets of Sox proteins. Nucleic Acids Res 27 (6):1409–1420. https://doi.org/10.1093/nar/27.6.1409

Wehrle-Haller B, Weston JA (1995) Soluble and cell-bound forms of steel factor activity play distinct roles in melanocyte precursor dispersal and survival on the lateral neural crest migration pathway. Development 121(3):731–742

Weiner AMJ, Scampoli NL, Steeman TJ, Dooley CM, Busch-Nentwich EM, Kelsh RN, Calcaterra NB (2019) Dicer1 is required for pigment cell and craniofacial development in zebrafish. Biochim Biophys Acta Gene Regul Mech 1862(4):472–485. https://doi.org/10.1016/j.bbagrm.2019.02.005

Yan YL, Willoughby J, Liu D, Crump JG, Wilson C, Miller CT, Singer A, Kimmel C, Westerfield M, Postlethwait JH (2005) A pair of Sox: distinct and overlapping functions of zebrafish sox9 co-orthologs in craniofacial and pectoral fin development. Development 132 (5):1069–1083. https://doi.org/10.1242/dev.01674

Chapter 8
Pigment Patterning in Teleosts

Jennifer Owen, Christian Yates, and Robert N. Kelsh

Abstract Pigment patterns in adult fish show astonishing diversity, reflecting their likely roles in camouflage, kin- and mate-selection, and display. Adult pigment pattern formation has only been explored extensively in the zebrafish, *Danio rerio*, and here we summarize the experimental studies that help to define a myriad of cell–cell interactions that result in this pattern. We begin by summarizing zebrafish stripe morphology and the general process whereby the adult pattern forms. We then provide a detailed summary of the genetic, cell biology, and ablation studies that inform our knowledge of the biological bases for stripe formation.

Keywords Melanophore · Xanthophore · Iridophore · *Danio* · Zebrafish · Cell–cell signaling

8.1 Introduction

Pigment patterns in adult fish show astonishing diversity, reflecting their likely roles in camouflage, kin- and mate-selection, and display. These patterns are highly labile, differing prominently even in closely-related species, suggesting that they are subject to rapid evolution. Adult pigment pattern formation has only been explored extensively in the zebrafish, *Danio rerio*, and here we summarize the experimental studies that help to define a myriad of cell–cell interactions that result in this pattern. We begin by summarizing zebrafish stripe morphology and the general process whereby the adult pattern forms. We then provide a detailed summary of the genetic, cell biology, and ablation studies that inform our knowledge of the biological bases for stripe formation. This chapter should, therefore, be considered a complement to those summarizing theoretical studies of fish pigment pattern formation (see Chap. 9) and the evolution of these patterns (see Chap. 10).

J. Owen · R. N. Kelsh (✉)
Department of Biology and Biochemistry, University of Bath, Bath, UK
e-mail: bssrnk@bath.ac.uk

C. Yates
Department of Mathematical Sciences, University of Bath, Bath, UK

H. Hashimoto et al. (eds.), *Pigments, Pigment Cells and Pigment Patterns*,
https://doi.org/10.1007/978-981-16-1490-3_8

8.2 Pigment Pattern Formation of Adult Zebrafish

8.2.1 *The Composition of the Adult Striped Pattern*

The adult zebrafish pigment pattern is comprised of 4–5 dark blue stripes, interspersed with 4 golden interstripes across the body tail and fins. The pattern comprises the superimposition of three layers of pigment cell-types (Hirata et al. 2003). Starting from the deepest layer of the hypodermis just above the muscle and moving to the dermis, the dark stripes consist of consecutive layers of L-iridophores, melanophores, and loose S-iridophores. Similarly, light interstripes are made up of layers of S-iridophores and xanthophores. It is the iridescent blue coloration of the loose S-iridophores lying over black melanophores that gives the blue-black color to the dark stripes, whilst the yellow xanthophores lying over the iridescent silver dense S-iridophores give the yellow appearance to the pale interstripes (Fig. 8.1).

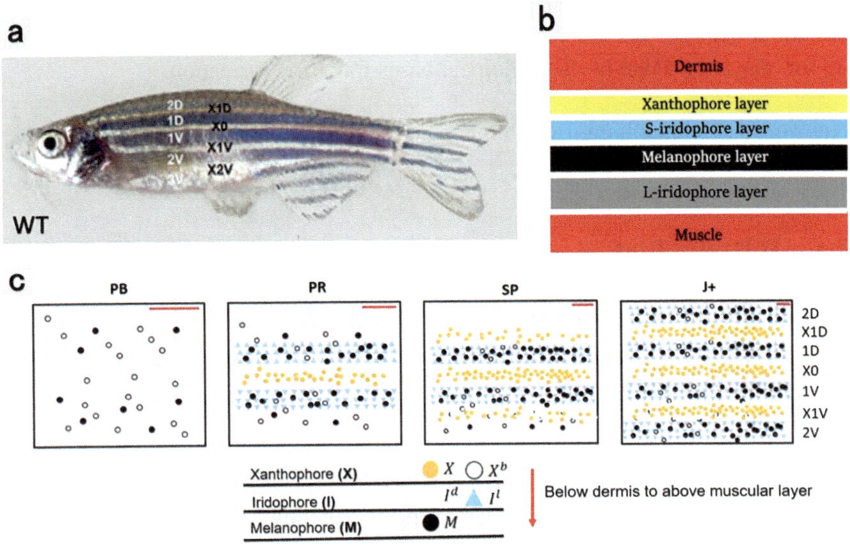

Fig. 8.1 Zebrafish stripe composition and development. (**a**) An adult wild type (WT) fish. Stripes and interstripes are labeled according to their order of temporal appearance and position on the body. X0 is the first interstripe to appear. 1D and 1 V (D-Dorsal, V-Ventral) are the first two stripes to appear. X1D and X1V are the next two interstripes to appear and so on. (**b**) Summary of pigment cell distribution in adult zebrafish. The cells in the xanthophore, S-iridophore, melanophore, and L-iridophore layers consist of xanthophores and xanthoblasts, S-iridophores, melanophores, and L-iridophores, respectively. (**c**) Schematic of WT patterns on the body of zebrafish. Stages PB, PR, SP, J+ correspond to developmental stages described in Table 8.1. Patterns form sequentially outward from the central X0 interstripe. Colour dots indicate melanocytes (M, black), xanthophores (X, yellow), xanthoblasts (X^b, open circle), the unpigmented precursor cell to xanthophores and S-iridophores in either dense or loose form (I^d (white), I^l (blue triangle), respectively). Images reproduced from Owen et al. (2020) and licensed under CC-BY 3.0 (http://creativecommons.org/licenses/by/3.0); published by The Company of Biologists Ltd

8.2.2 Adult Zebrafish Pattern Development

Developmental Staging

During adult metamorphosis the zebrafish grows significantly (Parichy et al. 2009; Singh and Nusslein-Volhard 2015). Developmental staging of zebrafish is typically measured using standard length (SL) or standardized standard length (SSL). Standard length is defined as the distance from the tip of the snout to the caudal peduncle. Standard length is used as a proxy for developmental stage in zebrafish as it can explain significantly more developmental variance than days postfertilization (dpf) and hence is preferred over dpf as an indicator of developmental progress (Parichy et al. 2009). The standardized standard length is the predicted standard length based on developmental progress, i.e. normalized for individual variation or strain differences. A list of developmental stages and their approximate times are given in Table 8.1.

8.2.3 Adult Zebrafish Pigment Stem Cells and their Niche

Whereas embryonic/larval pigment cells are directly derived from neural crest cells, most adult pigment cells are not generated from migrating neural crest cells but rather from a reserved population of postembryonic neural crest-derived progenitors (Fig. 8.2). In zebrafish, melanophores and iridophores are newly generated in the adult: embryonic/larval cells are replaced with postembryonic stem cells during metamorphosis. In contrast, adult xanthophores are derived from two distinct sources of progenitors: one is postembryonic multipotent APSCs and the other is embryonic/larval xanthophores which undergo a remarkable process of dedifferentiation, proliferation, and then xanthophore differentiation.

The location of APSCs in the fish is only partially resolved. There is strong evidence that at least some are located in the peripheral nervous system, most clearly shown for the dorsal root ganglia (DRGs), but likely also to include the spinal nerves (Dooley et al. 2013; Camargo-Sosa et al. 2019; Budi et al. 2011). In the DRGs, an

Table 8.1 Developmental stages of development as determined by Parichy et al. (2009)

Stage	Acronymn	Size (SSL)	~Age (dpf)
Pelvic bud	PB	7.2	21
Pelvic ray	PR	8.6	30
Squamation onset posterior	SP	9.6	39
Squamation through anterior	SA	10.4	44
Juvenile	J	11	51
Juvenile +	J+	13.5	70
Juvenile ++	J++	16	90

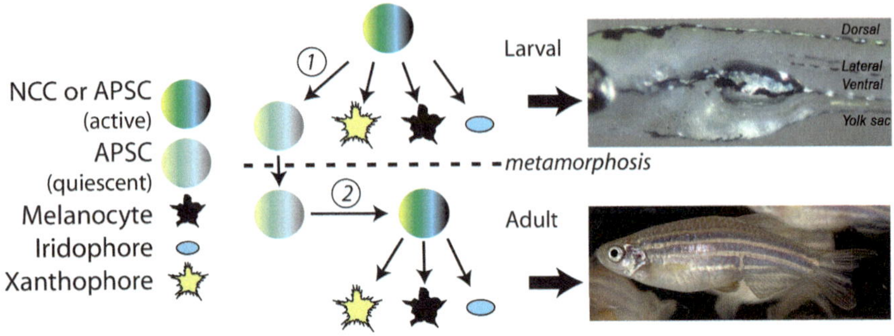

Fig. 8.2 Zebrafish pigment patterns in the early larva and adult are derived independently, but share a neural crest origin. Production of each pigment cell-type from neural crest cells (NCCs) in the embryo begins from c 27 hpf and, by 6 dpf, the embryonic pigment pattern is established (1): four horizontal melanocyte/melanophore stripes, three with associated iridophores, and xanthophores throughout the skin dorsal to the gut along the length of the body (Upper panel). This larval pattern remains unchanged until the start of adult pattern metamophosis at which point metamorphic melanophores, iridophores, and some xanthophores differentiate in the skin from Adult Pigment Stem Cells (APSCs), set aside from the embryonic neural crest and which remain quiescent until activated at metamorphosis (2). Note that many xanthophores are likely to form by dedifferentiation, proliferation, and then redifferentiation of early larval xanthophores (not shown). These cells together form the final adult pattern (Lower panel)

mitfa:GFP reporter transgene labels a small subset of ganglial cells. APSCs were initially shown indirectly to be set aside during a defined, embryonic time-window, in a manner that was ErbB-dependent, with adult pigment pattern being severely disrupted in adult *erbb3b* mutants. Consistent with this, in the absence of ErbB signaling formation of DRGs is defective and *mitfa:GFP* labeled cells are absent. Melanophores are delivered to the adult skin via the peripheral nerves, and do so in a manner that initially places them in widely-scattered locations throughout the flank.

Adult iridophores are also derived from APSCs in the DRGs, and are largely lost in *erbb3b/picasso* mutants. Iridoblasts appear to migrate through the horizontal myoseptum where they continue to migrate and spread after arrival in the skin. Iridoblasts proliferate dramatically in the skin, in contrast to melanoblasts, which are less proliferative after fate-specification. Melanoblasts proliferate while migrating along the peripheral nerve.

Lineage tracing has shown that these APSCs in the DRGs remain broadly multipotent and give rise to some xanthophores, neurons, and glia, in addition to the metamorphic melanophores and iridophores (see below).

Consistent with the different cellular origins of the larval and adult pigment cells, there are differences (but also strong similarities) in the genetic control of each set. The idea that embryonic/larval and adult pigment cell populations are genetically separable was first indicated by study of zebrafish *sparse/kita* mutants (Johnson et al. 1995; Parichy et al. 1999): melanophores differentiate in *sparse* mutants but fail to migrate and subsequently die within the first week. Ultimately the *sparse* larvae lack melanophores until a new wave of melanophores, which are *kita*-independent, arises during metamorphosis and contributes to stripe formation.

A second wave of metamorphic melanophores was suggested by the analyses of zebrafish *puma/tuba8l3* and *picasso/erbb3b* mutants (Budi et al. 2008; Parichy et al. 1999; Parichy and Turner 2003; Parichy et al. 2003): these mutants normally develop embryonic/larval melanophores but barely develop additional melanophores later than metamorphosis.

Adult Pattern Development

The adult pattern is transformed from an early larval pattern formed by 3 mm SSL (~4dpf) (Kelsh et al. 2009). Transformation of the early larval pattern to the adult pigment pattern starts at the onset of metamorphosis. Thyroid hormone acts as a trigger for metamorphosis and also regulates xanthophore proliferation and differentiation (McMenamin et al. 2014). Our understanding of the molecular basis for pattern formation is very limited, although identification of a diverse set of mutants affecting the process provides clues to drive future understanding. Our understanding of the cell–cell interactions that drive patterning is much better, and we will focus on this in the following summary.

The start of adult stripe development is marked by the differentiation of metamorphic S-iridophores around 7.2 mm SL (3wpf) along the horizontal myoseptum, which has been considered to serve as a morphological prepattern determining the orientation of the adult stripes (Patterson and Parichy 2013; Singh and Nusslein-Volhard 2015). From the horizontal myoseptum, the iridophores migrate and proliferate dramatically (doubling time of 3–4 days (Walderich et al. 2016), performing patterned shape transitions between dense form (interstripe regions) and loose form (stripe regions) regulated by tight junction proteins (Mahalwar et al. 2014; Singh et al. 2014). Tight junction protein 1a, for example, regulates the transition from dense to loose and in fish where this is disrupted, dense iridophores extend to form a coherent sheet over the entire stripe region (Fadeev et al. 2015). Iridophores move without directionality. However, the availability of space along the dorsal–ventral axis coupled with competition from other iridophores along the anteroposterior axis appears to orient the population of migrating iridophores bidirectionally along the dorso–ventral axis (Singh et al. 2014). By local proliferation, iridophores spread to typically 2–4 segments (87% of 163 labeled iridophore clones by Singh et al. spanned at least two interstripes; Singh et al. 2014). Not all iridophores in the interstripe of one segment are clonally related, indicating a mutual mixing of iridophores from neighboring segments. However, most iridophore clones span at least two interstripes, including the first one, suggesting that most of the later arising iridophores are clonally derived from iridophores of the initial interstripe X0 (Singh et al. 2014). A combination of extensive migration, shape transition, and proliferation produce the striping observed.

Most metamorphic xanthophores arise from larval xanthophores (Haddad et al. 2013). Larval xanthophores suddenly de-differentiate in response to the release of thyroid hormone, before dividing rapidly (Mahalwar et al. 2014). Iridophores express colony stimulating factor-1 (Csf1), which promotes local differentiation of some xanthophores in the interstripes (Patterson and Parichy 2013). Many larval xanthophores do not differentiate and instead remain invisible as unpigmented xanthoblasts in the prospective stripes (Haddad et al. 2013). Xanthophores migrate slowly, at a rate approximately three times slower than melanophores (~33um per week) in vitro (Yamanaka and Kondo 2014), and are generally observed to be slower than melanophores in vivo. Long-term time lapse imaging of labeled xanthophore clusters shows that differentiated xanthophores maintain their relative positions during stripe formation. Though differentiated xanthophores proliferate continuously (a doubling time of about 1 week; Mahalwar et al. 2014) and interact with newly differentiated cell-types during stripe morphogenesis, there is no global reorganization of xanthophores during stripe formation (Mahalwar et al. 2014).

Larval melanophores play only a minor role in *Danio rerio* adult pattern development, though they are more important in other species of *Danio*. Residual larval melanophores occupying the ventral and dorsal position persist, shifting slightly dorsally and ventrally during pattern metamorphosis but otherwise remain unchanged (Parichy and Turner 2003). A minority of the lateral early larval melanophores contribute to adult primary stripe 1D (Fig. 8.1), most die in the interstripe following its development of dense iridophores and xanthophores (Parichy and Turner 2003; Takahashi and Kondo 2008). The adult melanophore population is instead dominated by the appearance of metamorphic melanophores. During stripe development, progenitors of metamorphic melanophores migrate to the skin and begin to differentiate widely over the flank, though most melanophores differentiate in prospective stripe regions. Melanophores migrate slowly (a rate of approximately 100μm per week in vivo) (Takahashi and Kondo 2008). Furthermore, melanoblasts and melanophores rarely proliferate (Singh et al. 2014). As the stripes form, melanophores present in prospective interstripes move into nearby dark stripes or die in the interstripes generating a stripe/interstripe distinction (Takahashi and Kondo 2008).

During pattern development, new stripes and interstripes form sequentially, and seemingly as the result of the stripes and interstripes that appear before them. Starting from the onset of metamorphosis, at the *Pelvic Bud* stage (PB ~ 7.2 mm SSL) iridophores are visible in dense form along the horizontal myoseptum, marking the first X0 interstripe. Larval de-differentiated xanthophores re-differentiate in the X0 interstripe giving the X0 interstripe its yellow color. From the X0 interstripe, iridophores proliferate and migrate to the edge of the X0 interstripe where they then transition from silver/dense to a blue/loose form. Meanwhile, metamorphic melanophores begin to appear either side of the initial interstripe, forming stripes 1D and 1 V. By the *Pelvic fin Ray* stage (PR ~ 8.6 mm SSL), the X0 interstripe is complete and the adult primary melanophore stripes are partially formed: melanophores and loose iridophores occupy the 1D and 1 V regions. From the stripes 1D and 1 V, iridophores continue to migrate and proliferate to the edges of 1D and 1 V where they next transition to dense form forming the beginning of interstripes X1D and X1V. By stage *Squamation onset Posterior* (SP ~ 9.5 mm SSL), stripes 1D and 1 V are complete and interstripes X1D and X1V are partially formed: occupied by xanthophores and dense iridophores. This process of generating stripes and interstripes continues until all stripes and interstripes are formed at *Juvenile* + stage (J + ~13 mm SSL).

8.3 Genetics of Pigment Pattern Formation in Zebrafish

Understanding pigment pattern formation has come from various studies, but key amongst these have been careful analysis of the pigment pattern defects in a series of adult pigment pattern mutants, defined as those where the distribution (as opposed to the coloration) of pigment cells in the adult is disrupted. Here we will discuss these, dividing them into classes by phenotype.

8.3.1 The Initial Iridophore Interstripe Provides a Morphological Landmark for Pattern Development

The *choker* (*cho*) mutant larvae, which lack a horizontal myoseptum, fail to show the initial focused production of iridophores marking the propsective X0 at the onset of metamorphosis (Svetic et al. 2007; van Eeden et al. 1996). Instead, iridophores, which usually appear along the horizontal myoseptum at stage PB in WT fish, differentiate later in dispersed clusters above the myotomes throughout the body at stage PR. Following iridophore differentiation, xanthophores emerge in patches associated with dense iridophore regions across the domain, interspersed with melanophores (Frohnhofer et al. 2013). As pattern development ensues, melanophore and xanthophore regions develop and separate, until a pattern emerges across the body by stage J+. This pattern consists of stripes and interstripes of the same spacing as normal WT fish, but which branches and twists in a labyrinthine manner (van Eeden et al. 1996).

The adult pattern indicates that whilst cues at the horizontal myoseptum are important for orienting the first interstripe, stripe and interstripe width is dependent on the interactions between melanophores, xanthophores, and iridophores (Fig. 8.3).

8.3.2 All Three Pigment Cell-Types Are Required to Generate the Adult Pigment Pattern

One key class of pigment pattern mutants is those missing a single chromatophore type: *ltk/shd* which lack iridophores, *csfr1a/pfe* which lack xanthophores, and *mitfa/nac* which lacks melanophores. In all cases where a pigment cell-type is missing,

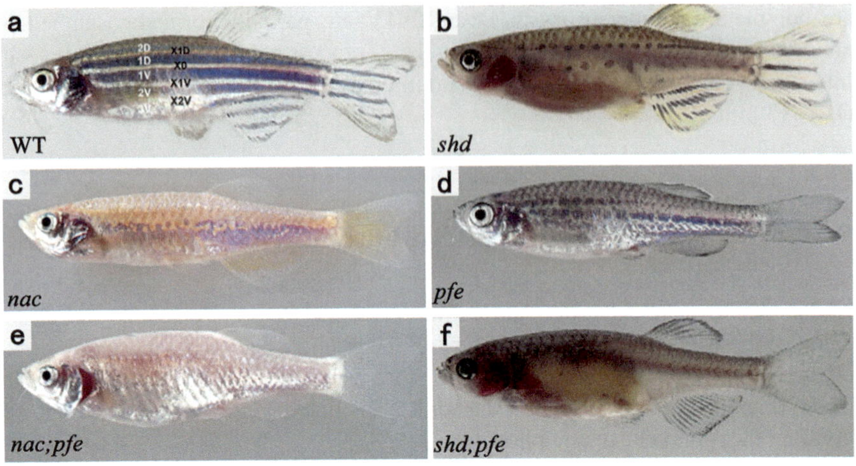

Fig. 8.3 Mutations causing loss of one or more cell-types. (**a**) WT fish. (**b**) *shd* mutant fish lack iridophores and display spots of melanophores surrounded by xanthophores. (**c**) *nac* mutant fish lack melanophores and exhibit an extension of the X0 interstripe into blebs of xanthophores and dense iridophores ventrally, surrounded by loose blue iridophores. (**d**) *pfe* mutant fish lack xanthophores and exhibit melanophore spots as well as ectopic melanophores in interstripe regions. (**e**) *nac;pfe* mutants lack melanophores and xanthophores and display a coherent sheets of dense iridophores (**f**) *shd;pfe* mutants lack xanthophores and iridophores and exhibit melanophores scattered uniformly over the fish. Images reproduced from Fronhofer et al. (Frohnhofer et al. 2013) and licensed under CC-BY 3.0 (http://creativecommons.org/licenses/by/3.0); published by The Company of Biologists Ltd

normal adult pigment stripe formation does not occur, indicating that all three pigment cell-types are required to generate the pattern. Study of the single mutants in this class has been revealing regarding the roles of individual cell-types.

No Iridophores (*ltk/shady*)

The strongest *leukocyte receptor tyrosine kinase* (*ltk*, also known as *shady/shd*) mutant alleles are early larval lethal, but escaper adults lack iridophores and display two stripes of melanophore spots surrounded by xanthophores (Lopes et al. 2008). In *shd* mutants, fewer metamorphic melanophores emerge during pattern metamorphosis and they remain more homogeneously dispersed over the entire flank. Later, following the widespread differentiation of xanthophores, melanophores aggregate into spots roughly in the positions of 1D and 1 V. Since metamorphic melanophore number is reduced, this suggests that iridophores may play a role in promoting melanophore differentiation. Furthermore, aggregation of melanophores into spots in the absence of iridophores suggests that iridophores are not necessary for enforcing strict "stripe–interstripe" boundaries. Inhibiting Ltk signaling during postembryonic stages results in the loss of adult iridophores, whereas ectopic iridophores are induced by Aklal2a overexpression and develop in *moonstone*, an *ltk* gain-of-function mutant (Fadeev et al. 2018; Mo et al. 2017; Fadeev et al. 2016).

No Xanthophores (*csf1a/c-fms/fms/panther/pfeffer/salz*)

In homozygous *colony stimulating factor 1 receptor 1a* (*csf1a*, also known as *pfeffer/pfe* and *salz/sal*) mutants, the development of xanthophores is strongly suppressed in larvae and abolished in adults (Haffter et al. 1996; Kelsh et al. 1996; Odenthal et al. 1996). The adult homozygous *pfe* mutant displays spots of melanophores associated with dense iridophores most aligned with the normal stripe positions with loose iridophores and scattered ectopic melanophores in the interstripe regions. They also show an increase in S-iridophores and a 50% reduction in melanophore number. Iridophores appear in the X0 region as in WT, around stage PB although sometimes earlier, followed by the appearance of melanophores in stripe regions 1D and 1 V in reduced numbers, suggesting that xanthophores are also important for melanophore differentiation. Iridophores proliferate and dense iridophores spread without transitioning into loose form, into 1D and 1 V splitting the stripes into melanophore spots (Frohnhofer et al. 2013). Removal of melanophores from the interstripes is mediated by xanthophores, as melanophores persist in the light stripe in *pfe* mutants. Since metamorphic melanophore number is reduced, albeit at a lower number than in the *shd* mutant, this suggests that xanthophores also play a role in promoting melanophore differentiation and/or survival. Furthermore, the appearance of melanophores in the interstripes suggests that xanthophores are important for ensuring removal of melanophores from interstripe regions. As iridophores are able to transition between loose and dense forms in the absence of xanthophores, this suggests that melanophores may provide signaling cues for iridophore transitions independent of xanthophores. Meanwhile, fish heterozygous for *pfe* have a reduced number of xanthophores when compared with WT and show irregular and interrupted stripes (Odenthal et al. 1996). Otherwise, stripes and interstripes are more clearly distinguished than those observed on fish homozygous for *pfe,* further suggesting that xanthophores likely play a role in maintaining strict stripe/interstripe boundaries.

No Melanophores (*mitfa/nacre*)

The *melanocyte inducing transcription factor a (mitfa,* also known as *nacre/nac)* mutant lacks both larval and adult melanophores (Lister et al. 1999). Adult *mitfa* mutants display an expanded X0 interstripe of dense iridophores and xanthophores with spots of xanthophores and dense iridophores more ventrally in a sea of loose iridophores (Frohnhofer et al. 2013). Pattern development differs from metamorphosis. Instead of changing from dense to loose at the boundaries of the X0 interstripe, dense iridophores continue to spread further. Xanthophores differentiate associated with the dense iridophore regions. More ventrally, dense iridophores change to the loose blue form. Since in the absence of melanophores, xanthophores still localize in dense iridophore "interstripe" regions, this suggests that xanthophores may also be attracted to dense iridophores. As iridophores are able to transition between loose and dense form in the absence of melanophores, this suggests that xanthophores likely provide signaling cues for iridophore transitions independent of melanophores.

Double Mutants of *shd, nac*, and *pfe*

Through combination of two of these missing cell mutants, mutants lacking two of the three chromatophore types have been generated (Frohnhofer et al. 2013). In all of these double mutant cases, the remaining chromatophores homogeneously fill the domain, arguing against a patterning mechanism based on morphogens or physical prepatterning, and indicating that it is indeed the interactions between the three chromatophores that generate the pattern.

Interestingly *nac;pfe*, which lacks melanophores and xanthophores, only display iridophores in dense form, whilst mutants lacking melanophores or xanthophores (*nac* and *pfe*, respectively) display iridophores in both dense and loose form. From this we can infer that both melanophores and xanthophores regulate the transitions of iridophores between loose and dense states (Table 8.2).

8.3.3 Regulation of Cellular Processes

Thyroid Hormone

Thyroid hormone—best known for coordinating disparate cellular processes during amphibian metamorphosis as well as its functions in mammalian metabolism and neural development— has profound effects on all three major classes of adult pigment cells (McMenamin et al. 2014). Zebrafish that lacks thyroid hormone owing to transgenic ablation of the thyroid gland, or loss-of-function alleles of thyroid-stimulating hormone receptor, develops about twice as many adult melanophores as WT fish and lacks visible xanthophores. No investigation has been made into the effects on iridophores. In these hypothyroid fish, melanophores proliferate inappropriately, whereas xanthophores are present but fail to acquire pigmentation. Adult hypothyroid fish observe broken stripes which become spots moving ventrally and dorsally away from the initial 1D and 1 V stripes (McMenamin et al. 2014). Conversely, a hyperthyroid mutant develops too many pigmented xanthophores and too few melanophores. Adult hypothyroid mutants display an expanded X0 interstripe with no stripes. Outside the initial interstripe are interspersed melanophores and loose iridophores, in a pattern similar to that observed in *D. albolineatus*. Recent findings suggest that the role of thyroid hormone is a key change between *D. albolineatus* and *D. rerio* (McMenamin et al. 2014). Despite having opposite

Table 8.2 Mutations causing loss of pigment cell-types. M, X, I abbreviate melanophore, xanthophores, and iridophores, respectively

Mutant	Pigment cells present	Homozygous pigment cell phenotype	Adult pattern	Deduced cell-cell interactions involved in pattern formation
ltk/shd	X and M only	Lack of S- and L-iridophores on the body of the fish. A reduction to 10-20% of metamorphic M in the adult fish	Spots of M aggregates, not necessarily in usual stripe region, surrounded by X.	M are more greatly repulsed by X than by M.
csf1a/ pfe	M and I only	None or very few visible X. Furthermore, the area covered by M is strongly reduced	S-iridophores spread into stripes 1D and 1V and split the melanophore field into patches, which decrease in an antero-posterior gradient. M and I patches mix in the ventral half of the flank from stage SP onwards	S-iridophores and M mutually repulse each other in the short range.
mitfa/ nac	I and X only.	Complete lack of M. Increased number of I. Number of X is variable	An extended interstripe of X and dense S-iridophores with irregular borders in the region of X0, accompanied by spots of X and I ventrally separated by regions composed of a thin net of blue S-iridophores, as well as L-iridophores	X promotes the localisation of I forming part of the aggregation process.
nac; shd	X only.	Lack of M and I	X homogeneously fill most of the flanks in a dense layer	X can differentiate independently of the other cell-types
shd; pfe	M only.	Lack of I and X	M homogeneously cover the flank. Density lowers with age	M can differentiate independently of the other cell-types, however, survival is likely dependent on X and M
pfe; nac	I only.	Lack of X and M	A dense layer of S-iridophores covers the trunk region, replaced towards anterior region by L-iridophores	M and X mediate I shape transitions from dense to loose

Fig. 8.4 Mutations causing a reduction in the number of iridophores in the adult. All fish (except for **b**) share a similarity with the phenotype of mutant *shd* which lacks both L- and S-iridophores. (**a**) *rse* fish lack all L-iridophores and some S-iridophores and exhibit two of five broken stripes at adulthood. (**b**) Edn3a mutant fish only lack L-iridophores and exhibit normal pattern development. (**c**) Edn3b mutant fish exhibit a dramatic decrease in the proliferation of S-iridophores and display two broken stripes. (**d**) *tra* mutant fish exhibit an unusually high death of differentiated iridophores and display two broken stripes at adulthood. Image (**a**) is reproduced from Krauss et al. (Krauss et al. 2014), images (**b**)–(**c**) are reproduced from Spiewak et al. (Spiewak et al. 2018) and image (**d**) is reproduced from Krauss et al. Figures 8.2 and 8.3f (Krauss et al. 2013). Images (**a**)–(**d**) are licensed under CC-BY 3.0 (http://creativecommons.org/licenses/by/3.0); published by The Company of Biologists Ltd

effects on melanophore and xanthophore population sizes, thyroid hormone promotes the maturation of both cell-types, as revealed by single-cell transcriptomics coupled with mutational and other analyses (Spiewak et al. 2018). For melanophores, thyroid hormone promotes a terminally differentiated, binucleated state in which proliferation ceases. For xanthophores, it drives acquisition of carotenoid-based pigmentation, making previously cryptic cells visible. Without thyroid hormone, melanophores remain "young" and continue to divide, whereas xanthophores stay hidden. Whilst, these results indicate the importance of the cellular process for pattern formation, more information is required about thyroid hormone effect on iridophores to discern pattern effects. For example, recent modeling studies indicate that the *D. albolineatus* phenotype, which has been identified as having evolutionary decoupling of the effects of thyroid hormone, is likely caused by changes to requirements for S-iridophore shape change during pattern development (Owen et al. 2020; Volkening and Sandstede 2018) (Fig. 8.4).

8.3.4 Reduced Iridophores in the Adult

Another class of mutants has a partial reduction of iridophores. This consists of mutants affecting endothelin signaling, (*endrb1/rse, edn3a, edn3b*), as well as *mpv17/tra.* In the absence of endothelin signaling, iridophores differentiate near the horizontal myoseptum but fail to populate other regions of the hypodermis

(Spiewak et al. 2018; Parichy et al. 2000a; Krauss et al. 2014; Frohnhofer et al. 2013). Despite having differences in the underlying development—*tra* mutants develop iridophores which are subsequently lost, whilst *rse* mutants exhibit defective proliferation of iridophores—mutants of this class exhibit a similar adult phenotype to *ltk/shd:* reduction in the number of metamorphic melanophores and a reduced number of stripes, broken into spots. This suggests that iridophores play a role in pattern maintenance.

Alkal2a; Alkal2b

Ltk was an orphan receptor, but two candidate ligands for Ltk were identified recently. The orthologs of FAM150A and FAM150B (ALK and LTK Ligand 1 (ALKAL1) and ALKAL2 or augmentors, Aug) encode small secreted proteins (Fadeev et al. 2018; Mo et al. 2017). In zebrafish there are three augmentor orthologs, designated *alkal2a*, *alkal2b*, and *alkal1*. Loss-of-function phenotypes in *alkal2a*; *alkal2b* double homozygous embryos are identical to those in *ltk/shady* mutants: the mutant embryos completely lack iridophores, suggesting that *alkal2a* and *alkal2b* are the primary Ltk ligands driving iridophore development in embryogenesis (Mo et al. 2017). Prominent expression of all three *aug* genes in iridophores at 72 hpf equivalent to the expression of *ltk* itself (Fadeev et al. 2018), and of *alkal2a* and *alkal2b* in NCCs, e.g. near the eye, at 24 hpf (Mo et al. 2017), suggests that these ligands activate the receptor Ltk in a paracrine and/or autocrine fashion, although further characterization of the extent of expression is required. Embryonic overexpression of each, but especially Alkal2a, stimulates numerous ectopic iridophores in an Ltk-dependent manner, indicating that Alkal2a is perhaps more potent in stimulating Ltk activity in iridophores (Fadeev et al. 2018).

alkal2a; *alkal2b* double homozygous adult fish have severe iridophore reduction in the body like *ltk/shady* mutants do while a partial iridophore defect in the eye. In contrast, *alkal2b*; *alkal1* double mutants exhibit a severe reduction of eye iridophores with no significant body stripe defects, suggesting a larger contribution of *alkal1* in the adult eye pigmentation.

Ednrb1/Ednrb1a/Rse

The *endothelin b1a receptor/rose* (*ednrb1a*, also known as *rose/rse*) mutants are defective for L-iridophores and some S-iridophores (Parichy et al. 2000a). Furthermore, there is a reduction to 30% of metamorphic melanophores. In strong mutant alleles of *rse* only the first stripes, 1D and 1 V, develop, broken into spots. In weak alleles, stripes 2D and 2 V may also develop, with stripes broken into spots more ventrally.

In adult zebrafish, *rse* is expressed by melanophore and iridophore lineages but becomes restricted to iridophores later (Parichy et al. 2000a; Krauss et al. 2014). Moreover, cell transplantations have shown that *rse* acts cell-autonomously in iridophores, but not in melanophores, suggesting that iridophores supply an unknown factor required for melanophore development (Krauss et al. 2014).

Edn3

Ednrb1a ligand endothelin 3b (Edn3b) mutants exhibit a dramatic decrease in the proliferation of S-iridophores in the skin (but not L-iridophores) whereas Edn3b overexpression results in iridophore population expansion and, indirectly, more melanophores (Spiewak et al. 2018). Edn3b has been shown to be specifically required for iridophore but not for melanophore development during early larval stages (Krauss et al. 2014). Edn3b mutants have S-iridophore and melanophore deficiencies but normal L-iridophores. The adult displays a similar phenotype to *shd;* two stripes of melanophores broken into spots, surrounded by xanthophores, except unlike *shd,* the stripes are more coherent and there are fewer breaks, especially towards the tailfins. Edn3a mutants have a normal pigment pattern, including stripes and iridophore interstripes but lack L-iridophores.

Mpv17/Tra/tra

The mitochondrial protein mitochondrial inner membrane protein (Mpv17, also known as *transparent/tra*) mutant phenotype is caused by a deletion in the Mpv17 (Krauss et al. 2013) leading to the death of differentiated iridophores and a reduction in iridophore number throughout development. The adult *tra* mutant displays a phenotype similar to the *shd* mutant; two stripes of melanophores broken into spots in a sea of xanthophores.

The reduction in iridophore pigmentation during pattern metamorphosis occurs gradually. The first X0 interstripe forms normally, and unlike *rse* or *shd*, *tra* mutants initially develop two full stripes of melanophores (seen at 2 months). However, soon the silvery pigmentation is gradually lost and these stripes subsequently dissolve into spots by 6 months (Krauss et al. 2013).

Cell transplantations have shown that *tra* acts cell-autonomously in iridophores, and the subsequent reduction in melanophores in the body occurs as a consequence of this loss. This is consistent with the idea that iridophores supply an unknown factor required for melanophore maintenance.

8.3.5 *Genes Affecting the Pigment Cell Environment*

In the pigment pattern field, the emphasis has been on how pigment cell interactions result in the emergence of the pigment pattern. However, another class of mutants has been shown likely to act non-cell-autonomously to the pigment cells. These mutations affect cell development indirectly by affecting aspects of the tissue environment during pattern formation. In particular, analysis of *bnc2* (Lang et al. 2009) and *karneol* mutants (Krauss et al. 2014) points towards a crucial role of the tissue environment surrounding the pigment cells during stripe pattern formation. However, the apparent paradox of a cell–cell interaction system showing environmental influences is readily reconciled, since the two examples to date both control provision of environmental ligands acting through receptors expressed in the pigment cells.

Bnc2

The gene *basonuclin2 (bnc2)* encodes a highly conserved, nuclear-localized zinc finger protein of unknown function that is required for normal stripe formation, but does not act in pigment cells themselves (Patterson and Parichy 2013). Strong mutant alleles of *basonuclin2* (e.g., $bnc2^{bonaparte}$) lack iridophores and almost all melanophores. They further display a few residual iridophores near the horizontal myoseptum, but a complete loss of stripes and interstripes on the body; in weak alleles (e.g., $bnc2^{poppy}$), the X0 interstripe is reduced and more tortuous, and 1D and 1 V are broken up into asymmetric spots. During development, *bnc2* mutants exhibit extreme reductions in metamorphic melanophores and, unlike in WT fish acquire "stripe" melanophores from persisting early larval lateral melanophores. This suggests that without local signals from differentiating metamorphic chromatophores, many more early larval melanophores can persist in the skin to adulthood. The *bnc2* gene regulates both *kitlga* and *csf1* expression, which are necessary for melanophore and xanthophore survival, respectively, but it is difficult to be sure whether other targets are also important contributions to the phenotype (Patterson and Parichy 2013).

Ece2/Kar

Endothelin-converting enzyme, (Ece2 also known as *karneol/kar*) mutants, has a phenotype similar to mutants with weak alleles of *rse* (Krauss et al. 2014). Adults display two stripes of spots of melanophores, with a reduction in iridophore and melanophore numbers. *Kar* activates endothelin pro-ligands by proteolytic cleavage in the tissue environment, readily explaining the lack of cell-autonomous requirement for *kar* activity in any of the chromatophores. *Kar* promotes iridophore development in a non-cell autonomous manner (Krauss et al. 2014) via a paracrine mode of endothelin signaling during pigment patterning facilitating the production of active Edns. Iridophores express the receptor *rse*, which in *kar* mutant embryos, cannot be fully activated due to the lack of processed ligand. Since the *kar* mutation affects iridophore production and melanophore numbers are also depleted, these results provide further evidence that iridophores supply an unknown factor required for melanophore development and maintenance.

Somatolactin

Somatolactin (SL) is a pituitary hormone of the growth hormone/prolactin family that to date has been found only in fish. SL function in pigment cell development was first suggested through study of the medaka *color interfere* (*ci*) mutant, which was shown to encode *SLa* gene (Fukamachi et al. 2004). The *ci* adult fish show pale gray skin color as a result of increase in leucophores in number and size and concomitant decrease in xanthophores while the *ci* larvae have a normal number of those pigment cells despite their abnormally-pale body coloration. These phenotypes indicate that SL predominantly acts after metamorphosis to promote xanthophore differentiation and to repress leucophore proliferation in medaka (Fukamachi et al. 2004; Yu et al. 2006; Fukamachi et al. 2009). Consistently, overexpression of SL results in more yellowish body coloration due to increased xanthophores and decreased leucophores.

Receptors for SL (SLRs) have not yet been identified, except that salmon SLR has been identified as being highly similar to growth hormone receptors but biochemically specific to SL rather than to growth hormone. The contribution of this SLR to skin pigmentation has yet to be investigated. Likewise, it is not clear whether SL is likely to have a pigmentation role in zebrafish.

8.3.6 Genes Required for Early Metamorphic Melanophores

Early studies of metamorphic melanophore development distinguished two subsets of these cells, according to the time of their appearance: early-appearing, which differentiates during the first week of metamorphosis, and late-appearing, which differentiates in the latter stages of metamorphosis. Single- and double-mutant analyses indicate that early-appearing metamorphic melanophores are dependent on *kit/sparse (spa)* and late-appearing metamorphic melanophores are dependent on *rse* and *pfe*.

As expected, both early and late melanophores are absent when fish are doubly mutant for *spa* and either *rse* or *pfe* (Spiewak et al. 2018; Parichy et al. 2000b; Parichy et al. 1999), whilst *pfe;rse* doubles show no significant further reduction in melanophore number beyond that seen in each of the single mutants.

However, as new evidence brings attention to the importance of iridophores and xanthophores in the generation of metamorphic melanophores, it becomes increasingly uncertain that late-appearing melanophores are directly dependent on *edrnb1* and *csf1ra*. The essential functions of both *ednrb1* and *csf1ra* are nonautonomous to melanophores, and implied mechanisms through which they act are only vaguely understood (Parichy et al. 2000a; Frohnhofer et al. 2013).

We propose that the early and late populations are more likely distinguished by the indirect effect of a reduction of iridophores (*rse*) or the complete lack of xanthophores (*pfe*). Indeed, recent mathematical modeling studies show that these reduced melanophore phenotypes in the cases of *rse* and *pfe* can be explained by simply reducing the number of iridophores and a complete lack of xanthophores, respectively (Owen et al. 2020; Volkening and Sandstede 2018). Hence, we speculate that whilst early, *kit*-requiring metamorphic melanophores can be distinguished from late-appearing ones, the latter are those cells whose differentiation is stimulated by xanthophores and iridophores.

kit/spa* and *kitla/slk

The adult loss-of-function *receptor tyrosine kinase* (*kit*, also known as *sparse/spa*) and *kit ligand a (kitlga*, also known as *sparse-like/slk*) mutants are almost completely devoid of melanophores at the beginning of metamorphosis. Nevertheless, 1D and 1 V stripes made up of a reduced number of melanophores appear later either side of the X0 interstripe. Adult stripes appear faded and more tortuous than those of WT fish. Furthermore, stripes contain only 15% of the normal number of melanophores on the body (Parichy et al. 2000b). Differentiation, migration, and survival of early-appearing adult melanophore precursors are promoted by paracrine factor Slk signaling through the receptor tyrosine kinase Kita. During metamorphosis, *slk* is expressed in the hypodermis, and its ectopic expression induces melanophore differentiation in extrahypodermal locations (Dooley et al. 2013). These results indicate that timely and numerous melanophore differentiation is required for pattern formation, but ultimately a very minimal number of metamorphic melanophore differentiation is required for some semblance of stripe development.

8.3.7 Genes Required for Metamorphic Chromatophore Development

Mutants with reductions in melanophores in the adult pattern but not during embryonic pattern development are important in identifying factors required uniquely to establish, maintain, or recruit the latent precursors to metamorphic melanophores. Zebrafish *tuba8l3/puma* and *erbb3b/picasso* mutants both show reduction in the number of metamorphic melanophores, with only partial compensation through retention of early larval melanophores, and display disrupted striping in adulthood. Both mutants display reductions in other chromatophores too, and it is currently unclear to what extent the metamorphic melanophore reduction results from cell-autonomous function of these genes.

tuba8l3/puma

The tubulin, alpha 8 like 3 (*tuba8l3*, also known as *puma*) mutant exhibits substantial reductions in the numbers of metamorphic melanophores and xanthophores and an expanded X0 interstripe region with disrupted stripes. Whilst the early larval pattern is unchanged compared to WT, the *puma* mutant exhibits 73% as many xanthophores and only 31% as many melanophores compared to WT (Parichy et al. 2003). Tubulins are key components of the microtubule cytoskeleton, but the exact role of this variant is still unclear. The *puma* gene has been shown to play a role in promoting the development of late-appearing *pfe*- and *rse*-dependent metamorphic melanophores as well as the numbers and organization of early-appearing *kit*-dependent metamorphic melanophores (Parichy et al. 2003). Analysis suggests that *puma* acts cell-autonomously to promote the expansion of pigment cell lineages mostly during metamorphosis but possibly also at even earlier embryonic stages, to neural crest-derived cells and possibly other lineages as well (Parichy et al. 2003). Pattern development in *puma* differs from the start of metamorphosis. Early lateral larval melanophores, which in WTs would usually die in response to local xanthophore signals (Takahashi and Kondo 2008), survive more frequently than WT early larval melanophores. Metamorphic melanophore births are negligible during early metamorphosis with a gradual increase towards the end of pattern metamorphosis 3 weeks later (Parichy and Turner 2003). Metamorphic melanophores in *puma* differentiate close to the early larval melanophores and rarely migrate. More research is required to fully understand the effects of this mutation and how it aids our understanding with regard to pattern development.

erbb3b/picasso

The *erb-b2 receptor tyrosine kinase 3b* (Erbb3b, also known as *picasso*) mutant displays two undulating stripes with breaks in the mid-trunk. *Picasso* encodes Erbb3, and the phenotype identifies a set of Erbb-dependent progenitors of adult melanophores that are established in association with the peripheral nervous system. Partial loss of DRG cells through ablation of ErbB signaling by either pharmacological inhibitors or *picasso* knockdown results in defective formation of adult melanophores while formation of the embryonic/larval melanophores appears normal, suggesting that ErbB signaling is required at a very early embryonic stage for much later development of metamorphic melanophores. Pattern development in *picasso* is similar to *puma*: The *picasso* mutant develops very few metamorphic melanophores, particularly in the mid-trunk and instead many embryonic/early larval melanophores persist into the adult during pattern metamorphosis.

8.3.8 Genes Relating to Cell Shape

Mutations which affect the cell shape need not impact pattern development as such but may change the appearance of the stripes. However, the one example that has been characterized to date changes both the appearance of the stripes and the distribution of melanophores in the skin.

bace2/wanderlust

The adult *beta-secretase 2* (*bace2*, also known as *wanderlust*) mutant exhibits poorly defined stripes across the body tail and fins. Pattern development has not been described. These mutants have melanophores that are hyperdendritic and found ectopically in interstripes and other locations (van Bebber et al. 2013). *Bace2* is expressed by melanophores and encodes a sheddase that cleaves insulin receptors from the plasma membrane, thereby curtailing insulin/PI3K/mTOR signaling. The salient insulin gene, *insulin b*, is expressed in the head at EL stages and subsequently in muscle and other tissues, suggesting the likelihood of endocrine roles early and possibly endocrine and paracrine roles later.

8.4 Cell Communication During Pattern Development

As seen in Sects. 8.3.1–8.3.4, all three pigment cell-types are required to generate the striped pattern. Increasing evidence suggests that this is due to the fact that the pigment cells rely on each other for signals relating to differentiation, death, and cell movement. As a result, many cell, ablation and genetic studies have been conducted into cell communication during pattern development. Cell communication between zebrafish pigment cells is divided into short range interactions, defined as interactions between neighboring cells that are likely regulated by filopodia or dendrites, and long range interactions, which span around half a stripe width distance and are regulated by filopodia or airinemes.

8.4.1 Cell-Studies into Cell Communication

Studies of cell–cell interactions in vivo and in vitro indicate that whilst there appears to be little to no homotypic cell–cell interaction (although this has not been assessed in iridophores), heterotypic interaction of xanthophores and iridophores with melanophores is likely to influence the movement and differentiation of metamorphic melanophores.

Melanophore–Melanophore Interactions

Despite the widespread aggregation of melanophores during pattern development, all cell studies indicate that melanophores do not interact in the short range. In vitro studies of extracted fin melanophores suggest that there are no obvious cell–cell interaction responses between melanophores in the short range (Yamanaka and Kondo 2014). In vivo ablation studies show that melanophores do not aggregate together (Takahashi and Kondo 2008) but instead may repulse each other in the short range.

Xanthophore–Xanthophore Interactions

There is also no evidence that xanthophores interact in the short range. Both in vitro and detailed electron microscopy analysis in vivo suggest that there are no obvious interaction responses between xanthophores in the short range (Takahashi and Kondo 2008; Yamanaka and Kondo 2014).

Xanthophore–Melanophore Interactions

In contrast, many cell–cell studies suggest that melanophores and xanthophores interact in the short range facilitated by filopodia. Firstly, in vivo studies into melanophore–xanthophore interactions during pattern metamorphosis indicated that when xanthophores appear in the interstripe regions at the onset of metamorphosis, melanophores initially distributed randomly across the flank, either migrated away to the black stripe regions or died (Nakamasu et al. 2009; Takahashi and Kondo 2008). In contrast, when xanthophores nearby isolated melanophores in interstripes were ablated, melanophores moved randomly, were less likely to leave the interstripe region and did not die, suggesting that melanophores are repelled by xanthophores in the short range and that xanthophores can promote the death of melanophores in the short range.

This is consistent with findings of in vitro analysis of melanophores and xanthophores extracted from the fins. They observed a "run-and-chase" behavior in the short range between melanophores and xanthophores whereby xanthophore contact-dependent depolarization of melanophores caused xanthophores to chase melanophores (surprisingly, in a specifically anticlockwise direction) as melanophores attempt to move away from the offending xanthophore. Since the melanophores are approximately four times faster than xanthophores in vitro (and faster in general in vivo (Singh et al. 2014), this gave rise to the idea that "slow xanthophores chase fast melanophores" (Yamanaka and Kondo 2014).

Melanophores have also been shown to influence the shape of xanthophores. For example, it has been observed that xanthophores reorganize their protrusions, and become stellate when they encounter melanophores, even in the absence of iridophores (Mahalwar et al. 2016; Mahalwar et al. 2014). In some cases, xanthophores will locally rearrange to accommodate their increasing numbers (Mahalwar et al. 2014).

Xanthoblast–Melanophore Interaction

During pattern formation, unpigmented xanthophore precursors differentiate in the prospective interstripe region, but also occur at lower densities in developing and completed stripes where they remain unpigmented or lightly pigmented as incompletely differentiated xanthoblasts (Haddad et al. 2013). Experiments by Eom et al. have shown that these xanthoblasts exhibit long (up to 5–6 cell diameters), thin projections with distinctive, membranous vesicles at their tips, that extend and retract quickly (individual cells could extend several projections within 18 hours) (Eom et al. 2015). These fast projections, named “airinemes,” are almost unique to the xanthophore lineage, being rare for melanophores and not observed for iridophores. More remarkably, the airinemes associated with xanthophores appear to be regulated by macrophages (Eom and Parichy 2017). Macrophages recognize airinemes by phospholipid phosphatidyl-serine (PS), an “eat me” signal that is commonly found on the outer leaflet of the plasma membrane on apoptotic cells.

The current working model is that xanthoblast blebs present PS to macrophages, which drag melanophore-specific targeted airenemes out from the xanthoblast. Contact between the airinemes and the target cell activates Notch-delta signaling, which ultimately promotes the consolidation of melanophores into stripes. Without airineme production in the xanthophore lineage, melanophores persist ectopically in the interstripe instead of migrating into the stripe regions (Eom et al. 2015).

Iridophore–Melanophore Interactions

An increasing number of studies indicate that iridophores, in particular iridophores in dense form, also affect melanophores independently of xanthophores. During pattern formation, the center of melanophore cells on the melanophore layer is rarely seen directly above a dense-form iridophore, however, melanophores frequently settle adjacent to dense-form iridophores. Melanophores will migrate to or differentiate in dense iridophore-free sites, suggesting short range repulsion of melanophores by dense-form iridophores and that dense-form iridophores may inhibit the differentiation of melanophores in the short range via some unknown signal (Patterson and Parichy 2013).

Iridophore–Xanthophore Interactions

Arrival of iridophores during metamorphosis leads to a compaction of the overlying xanthophores and in general xanthophore filopodia become shorter upon arrival of iridophores (Mahalwar et al. 2014), suggesting that there may be short range attraction of xanthophores by iridophores.

8.4.2 Ablation Studies to Assess Cell–Cell Communication

Ablation experiments, whereby a subset of cells are ablated on a developing or fully developed (and hence, strictly speaking, then addressing pattern maintenance rather than pattern formation) zebrafish and the subsequent differentiation of new cells is documented, can give insight into how cells, local to or in the long range, can promote or inhibit the differentiation of new cells.

Ablation of a rectangular subsection of the adult zebrafish stripes and interstripes, including all visible melanophores, xanthophores, and the majority of iridophores, leads to the subsequent generation of a labyrinthine pattern whereby stripes and interstripes of the same width stemming from the remaining stripes and interstripes curve in varying orientations. This regeneration indicates that stripe and interstripe widths are determined primarily by cell–cell interactions, but that orientation in the normal pattern formation process is driven by another, prepatterning process.

More specific ablations restricted to individual cell-types can help us to understand specific cell–cell interactions during pattern development.

Melanophore–Melanophore Interactions

Ablation experiments, whereby in a fully developed stripe pattern, a subsection of interstripe is fully ablated as well as partial ablation of neighboring stripes, suggest that melanophores may inhibit the differentiation of new melanophores in the long range (Nakamasu et al. 2009). When melanophores occupied neighboring stripe regions there was less differentiation of melanophores in the ablated interstripe region than when the melanophores occupying the neighboring stripe region at a half stripe width away were also ablated.

Melanophore–Xanthophore Interactions

Xanthophores promote the differentiation and survival of melanophores in the long range (Nakamasu et al. 2009). When melanophores occupied stripe regions, fewer melanophores differentiated in the neighboring interstripe region than when melanophores occupying the stripe region were ablated.

Iridophore–Xanthophore Interactions

Ablation experiments indicate that iridophores promote the development of xanthophores. Ablation of interstripe iridophores prior to xanthophore development results in fewer xanthophores in regions from which iridophores were lost, although both iridophores and xanthophores can be recovered gradually during later development, indicating compensation mechanisms for early deficiencies (Patterson and Parichy 2013).

8.4.3 Genetic Studies into Cell Communication

Mutations which affect cell-signaling but not development of chromatophores are of particular interest, as they can provide insights into the molecular mechanisms of cell–cell interactions underlying stripe formation. Mutants of this form generate robust patterns that deviate from the usual stripes. These mutations indicate the importance of cell communication during pattern development for stripe formation.

Several mutants have been described in which dark stripes are broken into spots. These consist of mutants *Cx41.8/leo, Cx39.4/luchs, Tjp1a/sbr, and Igsf11/seu.* Other mutations have been described where the widths of the stripes are notably different to WT fish: *Kir7.1/obe*, *asterix/ase*, and *Tspan3c/dali* mutants display fewer and broader stripes. In contrast, the *srm/ide* mutant shows notably thinner stripes.

cx41.8/leo

The adult Connexin 41.8 (Cx41.8, also known as *leopard/leo)* mutant is characterized by a spotted pattern of melanophores and loose iridophores in a bed of dense iridophores and xanthophores on the trunk (Irion et al. 2014).

Pattern development in *leo* mutants begins to deviate from WT patterning from stage PR. Melanophores either side of the initial interstripe (which usually gather in the stripe regions in WTs) remain dispersed in *leo* mutants. By stage J, dense iridophores and associated xanthophores are dispersed in an irregular manner so that the X0 interstripe branches into the usual striped region breaking the melanophores stripes into patches. By stage J++ (~16 mm SSL) spots can be observed instead of stripes on the body of the fish.

Cell autonomy studies demonstrate that expression of *leo* is required in melanophores and xanthophores but not iridophores (Irion et al. 2014; Maderspacher and Nusslein-Volhard 2003). It has been suggested that *leo* forms heteromeric gap junctions, promoting cell–cell interactions between melanophores and xanthophores. Double mutants for *leo* with missing cell mutants (*shd, nac*, and *pfe*) reveal disturbed patterning in all cases, suggesting that the defective interactions of *leo* melanophores and xanthophores may subsequently influence the patterning of iridophores (Irion et al. 2014). For example, *leo;nac* double mutant which lacks melanophores displays dense iridophores covered by xanthophores and the absence of loose iridophore areas compared to *nac*. *Leo;pfe* double mutants lack xanthophores and, instead of the spotted pattern observed in *pfe,* show melanophores randomly distributed across the body of the fish. *Leo;shd* double mutants lack iridophores and also have an almost complete loss of melanophores on the body of the fish. It has been suggested that the *leo* mutant phenotype is largely caused by iridophores failing to undergo shape transitions from dense to loose form at the correct time and place (Irion et al. 2014).

In vitro studies have demonstrated defective cell–cell interactions between *leo* mutant melanophore and WT xanthophore cells. Unlike in WT cells, there is no melanophore repulsive response or xanthophore chasing movement observed, suggesting that *leo* is important for mediating these interactions. Interestingly, in vivo studies have shown that expression of *leo* in mutant *leo* melanophores (only) can partially recover striping, albeit thinner than normal stripes, (Watanabe and Kondo 2011), indicating that *leo* is particularly important for heterotypic interactions involved with melanophores.

Ccx39.4/luc

The heterozygous Connexin 39.4 (Cx39.4, also known as *luchs/luc)* mutant zebrafish exhibit wavy stripes with breaks, whilst in homozygotes, stripes are completely dissolved into very small spots and individual melanophores. Pattern development in *luc* mutants begins to deviate from WT patterning from stage PR. Iridophores at the edges of the interstripes fail to change to the loose form and suppress melanophores. At stage J, dense iridophores form patches around aggregated melanophores. By stage J+, two undulating stripes either side of X0 can be observed with patches of dense iridophores more ventrally.

luc is required in melanophores and xanthophores but not iridophores (Irion et al. 2014). It has been suggested that *luc* also forms heteromeric gap junctions, promoting interactions of melanophores and xanthophores. Mutant crosses of *luc* with missing cell mutants (*tra, nac*, and *pfe*) reveal disturbed patterning in all cases. For example, in *luc;nac* double mutants, which lack melanophores, both xanthophores and iridophores are more homogenously distributed than in *nac* alone, although the difference is subtle. In *luc;pfe* double mutant, which lacks xanthophores, melanophores are distributed more evenly and only rarely form small spots. In *luc;tra* double mutants, which lack iridophores, the spotted pattern of melanophore of *tra* is reduced, and only a few small melanophore spots are present, although this is a weaker phenotype when compared with *leo;tra*.

igsf11/utr15e1/seu

The adult Immunoglobulin superfamily member 11 (Igsf11, also known as *seurat/seu)* mutant displays an irregular spotted pattern on the body of the fish and a spotted pattern on the tailfins (Eom et al. 2012). *Igsfl1* is required autonomously by the melanophore lineage and mutants exhibit defects in both melanophore survival and migration during pattern metamorphosis. Whilst melanophores in WT fish move at a rate of approximately 6μm/h, *Igsfl1* mutant melanophores move at a rate of 2μm/h. Whilst the survival rate of melanophores in WT fish is >95%, the survival rate of *Igsfl1* melanophores is around 70%. It has been hypothesized that *seu* could also play a role in mediating adhesion between differentiated melanophores, however, this has been disproven by time lapse imaging which showed that melanophores do not adhere to each other in vivo (Takahashi and Kondo 2008).

tjp1a/sbr

The adult Tight Junction Protein 1a (ZO-1a/Tjp1a, also known as *schachbrett/sbr*) mutant, which exhibits delayed changes in iridophore shape and organization caused by truncations in *sbr*, displays a checkerboard pattern of rectangular melanophore spots associated with loose iridophores, organized into stripes within a bed of loose iridophores and xanthophores.

Pattern development in Tjp1a mutants deviates at stage PR. As new melanophores differentiate either side of the initial interstripe, iridophores, which usually transition from dense to loose at the interstripe boundary, remain dense for longer and spread over the melanophore stripe regions 1D and 1 V. As a result the melanophore stripe breaks into a rectangular spotted pattern. Later in development, some dense iridophores transition to the loose form in regions densely populated with melanophores.

Tjp1a is expressed in dense iridophores but down-regulated in the loose form. It is postulated that Tjp1a regulates the transition of dense iridophores to loose type. Though there is also a reduction in the number of melanophores, it has been shown that the phenotype is not caused by this reduction, but rather that the reduction is caused by the delayed transitions of dense iridophores to loose type (Fadeev et al. 2015).

Double mutant crosses with missing cell mutants demonstrate that functional Tjp1a is important for iridophore shape changes. For example, *sbr;shd* double mutants lack iridophores, but the pattern is indistinguishable from *shd*. However, *sbr;nac* and *sbr;pfe,* which lack melanophores and xanthophores, respectively, display a greater spread of the dense iridophores across the domain. As a result, *sbr;nac* fish show a reduced blue iridophore section on the flank compared to *nac*, and *sbr;pfe* displays smaller melanophore spots than in *pfe*.

kir 7.1/jag/obe

The adult *inwardly rectifying potassium channel 7.1* (*kir7.1*, also known as *jaguar/jag* and formerly known as *obelix/obe*) (Iwashita et al. 2006) mutant has fewer (2–3 instead of 5) and broader stripes across the body tails and fins than WT *D. rerio*, and displays ambiguous stripe–interstripe boundaries (Iwashita et al. 2006).

Pattern development in *kir7.1* mutants deviates from WT development from stage PB onwards. Melanophores which appear either side of the initial interstripe fail to aggregate into stripes and instead persist in more ventral regions as well as within the initial interstripe itself. Xanthophores, which initially appear in the interstripe, later also emerge between the scattered melanophores (Iwashita et al. 2006).

kir7.1 is expressed in melanophores and xanthophores. Transplantation and in vitro studies suggest that *kir7.1* is only required in melanophores and not xanthophores (Iwashita et al. 2006). Currently, it is not clear how the function of Kir7.1 can be linked to the *jag* phenotypes. However, since the role of the Kir protein family is to maintain membrane potential and sustain ionic composition (Hibino et al. 2010; Kofuji and Newman 2004) it is likely that Kir7.1 functions in cell–cell signaling. In particular, it is thought to be responsible for managing cell–cell interactions which enforce the clear boundaries between melanophore and xanthophores during pattern development (Iwashita et al. 2006).

Related to this, it has been postulated that the *jag* gene functions to mediate melanophore cell–cell interactions. In vitro studies have shown that the directed movement of melanophores away from xanthophores is inhibited in *jag* (Yamanaka and Kondo 2014). The plasma membrane of melanophores derived from *jag* mutants showed no electrical polarization and failed to retreat from xanthophores, indicating that Kir7.1 may set the resting membrane potential of melanophores. The loss of this repulsive interaction in *kir7.1* mutants appears to contribute to a failure of cell sorting in vivo (Yamanaka and Kondo 2014).

tspan3c/dali

Heterozygous tetraspanin 3c (*tspan3c*, also known as *dali)* mutants have a reduced number of interrupted block-like stripes of melanophores. The *dali* homozygous mutants show thinner stripes than normal, with melanophores intermingled abnormally with xanthophores. The development of the *dali* mutant has not been described in the literature.

tspan3c is expressed in melanophores and xanthophores. It is not clear exactly how the function of *tspan3c* can be linked to the *dali* phenotype. The role of the tetraspanin family is to act as scaffolds to anchor proteins at cell and organelle membranes within tetraspanin-enriched microdomains and hence to contribute to cell adhesion, migration, proliferation, and cell–cell fusion. In vitro studies have shown that the *dali* mutants have reduced melanophore motility and disrupted melanophore–xanthophore interactions, but *dali* is not required for melanophore depolarization (Inoue et al. 2014). More information is needed to understand the role of *dali* in pattern development.

srm/idefix

Homozygous *spermidine synthase* (*srm*, also known as *idefix/ide*) mutants have fewer stripes, and these are less regular, usually narrower, and often interrupted; the X0 interstripe is also expanded (Frohnhöfer et al. 2016). The *ide* phenotype develops during adult metamorphosis. The mutant phenotype becomes clearly visible at stage SP, in which the first interstripe is already noticeably wider than in WT. The full phenotype is developed by stage J++. The phenotype of the *ide* mutant is caused by a severe reduction in spermidine levels. *Ide* is likely to influence the behavior of the pigment cells by providing enough spermidine for the regulation of cell–cell communication (Frohnhofer et al. 2016).

Potential Molecular Mechanisms Mediating Pigment Cell Interactions

Analyses of the *leo* and *luc* mutants suggest that Cx41.8 and Cx39.4 interact to form heteromeric gap junctions and that these channels are responsible for some of the interactions that take place among melanophores, xanthophores, and iridophores during pattern metamorphosis. Whilst *leo* and *luc* single mutants retain an organized but spotted pattern on the body of the fish, the body of the *leo;luc* double mutant is covered with dispersed melanophores associated with loose iridophores (Irion et al. 2014). The rest of the flank around the melanophores is covered with dense iridophores and xanthophores. This suggests that in the absence of one connexin-partner, homomeric channels form, but retain only some functionality, whereas the absence of both connexin proteins in the double mutants results in a complete block to activity.

In *jag;leo* double mutant, there is a residual 1D melanophore stripe and X0 interstripe. Some melanophores are interspersed with xanthophores and iridophores elsewhere (Maderspacher and Nusslein-Volhard 2003). For *jag*, previous analyses indicate that this gene is required in the melanophores to control their homotypic clustering, although it may also be involved in other cell–cell interactions (Maderspacher and Nusslein-Volhard 2003). The *jag;leo* double mutant suggests that at least one of the two cell–cell signaling pathways functions redundantly with a third signaling pathway to form coherent patterns.

Tjp1a in iridophores may also be involved in cell communication with xanthophores, and/or melanophores. The *sbr* mutation enhances the phenotypes of both *luc* and *leo* mutants. The *leo;sbr* double mutant displays a decrease in the size of the spots compared to both *sbr* and *leo* (Fadeev et al. 2015). The *luc;sbr* mutant exhibits an almost complete loss of melanophore clustering; the upper part of the body is covered with a layer of dense iridophores (Fadeev et al. 2015). The very ventral region of the fish is covered in loose iridophores and some melanophores.

Analysis of these double mutants suggest that Tjp1a and connexins do not act in a linear pathway to regulate pattern formation, but most likely work through different mechanisms. Absence or truncation of Tj1pa results in a delay between dense iridophores (1) receiving the signal from melanophores to switch to the loose form and (2) actually changing morphology. The combination of this delay with the disrupted communication between melanophores, xanthophores, and iridophores in *leo* and *luchs* leads to an enhanced phenotype in both *leo;sbr* and *luc;sbr* double mutants.

In double mutants where *ide* is combined with *leo, luc* or *jag*, a superimposition of both phenotypes is visible. That is, the characteristic spots (*leo, luc*) and widened stripes (*jag*) are combined with an extended X0 interstripe and slight reduction in melanophore stripe/spot width. Likely targets for spermidine are the gap junction components *leo*, *luc*, and *jag*. The phenotypic analysis of *ide* double mutants with *leo, luc,* or *jag* suggests that *ide* does not exclusively affect one of these channels, and that it may affect all or even a different yet unknown channel. Interestingly, it has been shown that if Cx41.8 is expressed only in melanophores, but not in xanthophores, a phenotype very similar to *ide* is produced (Watanabe and Kondo 2011), suggesting that *ide* may play a function in heterotypic polarized gap junctions requiring Cx41.8 (Table 8.3).

8.5 Pigment Pattern Formation in the Zebrafish Fins

The adult zebrafish caudal and anal fins show a striking pattern of pigmented stripes and interstripes that run parallel to the body stripes and interstripes. The caudal and anal fins also comprise melanophores, xanthophores, however, unlike in the body of the fish, pattern formation in the fins do not require iridophores (Hirata et al. 2005). Consistent with this, of the missing cell mutants, only *shd* (lacks iridophores) forms normal caudal fin development, indicating that here, in contrast to the body, iridophores are not required for fin pattern formation. *Pfe* and *nac* mutants show a random distribution of melanophores and xanthophores, respectively, across the fins suggesting that melanophore–xanthophore communication is important for caudal fin development. Consistent with this, the majority of in vitro studies identifying key short range melanophore–xanthophore interactions were performed using cells extracted from the caudal fins (Yamanaka and Kondo 2014).

The pigment pattern of the caudal fin forms simultaneously to fin outgrowth. During early stages of development (<6.4 mm SSL), a scattered arrangement of first melanophores and later xanthophores appears on the caudal fin. By ~7 mm SSL, melanophores become more concentrated in the center of the base of the caudal fin, initializing the first stripe in the middle of the caudal fin. The following interstripes and stripes form sequentially either side of the already established stripes and interstripes. Initially stripes and interstripes are offset from body stripes and interstripes, but eventually they become continuous (Parichy et al. 2011).

Table 8.3 Mutations affecting body stripe pigment pattern formation

Gene name	Protein type	Expression pattern in pigment cells	*Mutant allele*	Predicted allele type	Type defect	Adult stripe pattern phenotype (Body)	Pattern development deviations from WT	Deduced function of gene in pigment cell lineages	Deduced features important for pattern formation
asterix	Unknown	Unknown	*ase*^{dtt269}	Unknown	Unknown	(Het) exhibit fewer and broader stripes (Hom) exhibit a stronger phenotype, with a further reduction in the number of stripes (Haffter et al. 1996)	Not documented	Unknown	Not clear
basonuclin-2	Zinc finger protein (Bnc2)	N/A (Hypodermal cells in contact with M,X and I)	*poppy*^{j10e1}	Partial loss of function	Fewer metamorphic M,X and I. Differentiated pigment cells die at high frequency	(Hom) Two partial stripes of M (arranged in very small spots)	Not documented	Promotes M and X development by regulating expression of the Kitlga and Csf1 (Patterson and Parichy 2013)	Not clear
			bonaparte^{utr16e1}	Loss of function	Fewer metamorphic M,X and I. Differentiated pigment cells die at high frequency.	(Hom) Lack stripes on the body	Lack early appearing metamorphic I, but develop dispersed metamorphic M later. (Lang et al. 2009)		
choker	meox1	N/A	*cho*^{tm26}	Loss of function	Loss of the horizontal myoseptum (Svetic et al. 2007)	Stripes and interstripes are normal width but occupy a labyrinthine pattern.	M appear before I and X, evenly dispersed over the flank. I and X emerge in patches interspersed with M, but then	N/A	I occupying the horizontal myoseptum play a role in aligning the stripes.

(continued)

Table 8.3 (continued)

							aggregate into stripe-like arrangements, and separate from M regions (Frohnhofer et al. 2013)		
dali	Transmembrane scaffolding protein Tetraspanin 3c (Inoue et al. 2014)	M and X	*dali*^{wpr21e1}	Missense substitution that causes ER-retention and incomplete glycosylation of Tspan3c protein	Reduction in M movement M-X interaction impaired	(Het) broken stripes with clear boundaries between M and X (Hom) fewer, broader stripes that contain M and X	Not documented	Cell-cell interactions between M and X	M-X cell-cell interactions are important for pattern formation
duchamp	Unknown	Unknown	*duchamp*$^{ut.r19e1}$	Unknown	Unknown	(Het) fingerprint-like spots across the body. 45% of M remain. (Hom) exhibit more dispersed M than het (Quigley et al. 2005)	Not documented	Unknown	Unclear
Edn3	Ligand endothelin signalling	I	*edn3a*^{b1282}	Loss of function	Lack peritoneal iridophores	Normal stripes and interstripes. Rosy cast to the ventrum. (Spiewak et al. 2018)	Not documented	edn3 promotes iridophore development in Danio (Spiewak et al. 2018). Required for larval iridophore development (Krauss et al. 2014)	Unclear
			edn3b^{b1283}	Loss of function	Deficiency in hypodermal iridophores	Two stripes are broken into spots.	Not documented		

idefix	Enzyme spermidine synthase (Frohnhofer et al. 2016)	Not required in pigment cells	*ide*^{t26743}	Loss of function	Likely cell signalling interrupted	(Hom) The light stripe area is wider than in WT and only two dark stripes develop narrower than WT with frequent interruptions	Boundaries between the first interstripe and stripes are less regular from stage SP	Spermidine likely required for gap junctions and hence cell-cell signalling	Unclear
jaguar	K+ channel kcnj13; (previously Kir7.1) mutation. (Iwashita et al. 2006)	M and X lineage	*jag*^{b230}, *obe*^{tc271d}, *and obe*^{td15}	Loss of function	M-X interaction impaired. M constantly depolarized and is not repelled by X (Yamanaka and Kondo 2014)	(Het) fewer and broader stripes on the fins and some incomplete stripes on the body (Hom) broader interstripes and thinner stripes. Boundaries between M and X are blurred	M fail to aggregate into stripes and instead persist into more ventral regions as well as within the initial interstripe itself. X later emerge between the scattered melanophores	Cell-cell interactions between M - X	M-X cell-cell interactions are important for maintaining stripe/ interstripe boundaries
karneol	Enzyme Endothelin-converting enzyme 2	No chromatophores (Krauss et al. 2014)	*kar*^{ltNO046}	Premature stop and the loss of the C-terminal peptidase domain containing the catalytic centre of the enzyme.	Reduction of I and M (Krauss et al. 2014)	Reductions in iridophores and melanophores (Krauss et al. 2014)	At stages PB and PR, kar mutants show only slight reductions in I, they still form a continuous sheet just ventral to the horizontal myoseptum. From PR to SP reduction of I becomes more apparent. They do not extend properly into dorsal and ventral	kar promotes iridophore development in a non-cell autonomous manner	Unclear

(continued)

Table 8.3 (continued)

							interstripe regions, occasionally, remnants or patches develop. From stage SP onwards the number of M is reduced compared to WT		
kit	Receptor Kit signalling	eNCC and M	*spa*^{b5}	Loss of function	M precursors divide on average only ~1.75 times between leaving the NC and establishing the early larval pattern (Raible and Eisen 1994)	One-third reduction in M, with I and X less affected Stripes appear faded and more tortuous than those of WT fish Stripes contain only ~15% of the normal number of M	For the first week of pattern metamorphosis M are almost completely absent (Raible and Eisen 1994). M only begin to differentiate during middle and late stages of metamorphosis.	Promote survival, migration and differentiation of melanoblasts at all stages of development (Hultman et al. 2007; Johnson et al. 1995; Raible and Eisen 1994)	Timely M differentiation important for pattern formation.
kit lga	Ligand kit signalling	eNCC and M	*sparse like*^{tc244b}	Loss of function	Reduction of I, increase of M. Late larval and metamorphic M are absent (Dooley et al. 2013)	(Het) wave-like boundaries between stripes and interstripes, sometimes break into spots (Watanabe et al. 2006) (Hom) spotted pattern (Watanabe et al. 2006)	Almost completely devoid of M at the beginning of metamorphosis.	Early NCC specification and suppression of mitfa	Metamorphic M are important for generating the stripes.

leopard	Gap Junction Protein Connexin 41.8 (Watanabe et al. 2006)	M and X	*leo*^{tq270}	Missense mutation		(Het) Elongated spots instead of stripes/ some spots (Hom) Spot pattern	Not documented	Gap-junction formation between M and X (Irion et al. 2014; Watanabe et al. 2006)	M-X interactions are important for maintaining M-X boundaries
			leo^{t1}	Nonsense mutation in the first extracellular loop domain, which reduces the predicted molecular size of Cx41.8 encoded. (Recessive)	M scarcely show repulsive response to X (Dooley et al. 2013; Yamanaka and Kondo 2014)	(Het) No change (Hom) Spots	Not documented		
			leo^{tw28}	Missense mutation		(Het) Undulating stripes (Hom) Spot pattern	M stay more dispersed instead of arranging in a stripe like pattern either side of the interstripe. Dense I covered with X disperse in an irregular manner in the mutants to form a wider first X0 interstripe, branching into the stripe region and suppressing M dorsally and ventrally to form irregular patches		
			leo^{tk3}	Loss of function		(Het) Undulating and disrupted stripes. (Hom) Individual M that hardly cluster together are still present, mostly associated with loose I			

(continued)

Table 8.3 (continued)

luchs	Gap Junction Protein Connexin, 39.4 (Irion et al. 2014)	M and X, but not I	*luc^{tXA9}* and *luc^{tXG1}*	Loss of function	I appear almost entirely in dense form	(Het) undulating and disrupted stripes on their flanks (Hom) Individual M that hardly cluster together are still present, mostly associated with loose I	M ventral to the first interstripe are more scattered at stage PR. The second interstripe fails to form in the ventral region. Later at stage J++ irregular stripes and spots become visible. (Irion et al. 2014)	Cx39.4 form heteromeric gap junction channels in the plasma membranes of M and X which may facilitate communication with I	Suggest M and X interact with I to signal shape transitions
oberon	Unknown	Unknown	*oberon^{j198e1}*	Not clear	Unknown	Pale stripes and interstripes (Engeszer et al. 2008)	Not documented	Unknown	Not clear
rose	Receptor Endothelin receptor B (Ednrb1)	X precursors and I	*rse^{tAN17X}*	Weak (partial loss)	Deficit of I	2 of 5 stripes develop. Fewer I and M than WT (Krauss et al. 2014)	Appearance of X in rse mutants is delayed in X0. M differentiate near to X0 interstripe in reduced numbers, X fill the space between melanophore aggregates	rse may be responsible for the expansion of I	Suggest that dense S-iridophores of the interstripes may exert a long-range effect on the aggregation and support of melanophores in the neighbouring stripes (Frohnhofer et al. 2013). Lack of iridophores, and to a lesser extent of xanthophores, is correlated with a strong reduction of melanophore numbers
			rse^{tLF802}	Strong loss of function	Severe deficit of I and a severe reduction in M number compared to WT there are around ~10–20% of normal M in adult (Frohnhofer et al. 2013; Parichy et al. 2000a)	2 of 5 stripes develop. Fewer I and M than WT. Stripes are broken into spots (Krauss et al. 2014)			
			rose^{b140}	Premature stop codon	Fewer I and M	2 of 5 stripes develop. 1D stripe forms normally. 1V stripe is broken into spots. (Parichy et al. 2000a)			

			$rose^{j3e1}$	Missense substitution was detected in the phenotypically weaker allele	Fewer I and M	Four stripes develop. 1D and 1V form normally. 2D and 2V are broken into spots (Parichy et al. 2000a)			
seurat	Cell surface protein that mediates adhesive interactions Immunoglobulin superfamily member 11 (Igsf11) (Eom et al. 2012)	M	$seurat^{utr15e1}$	Missense substitution (strong)	M migrate at 1/3 of normal speed, survival of M is around 70% (instead of 95%)	(Hom) Irregular spots of M	Not documented	Promote survival and migration of M lineage (Eom et al. 2012)	M survival is important for pattern formation
			$seurat^{twp15e2}$ *(weak)*	Mutations at the boundary between the predicted signal sequence and the beginning of the first immunoglobulin domain	Reduction in number of M	Two stripes of M with one stripe broken into spots ventrally.	Not documented		
schachbrett	Tight Junction Protein 1a (Fadeev et al. 2015)	Dense I	$sbr^{tnh009b}$, sbr^{twl1}, sbr^{twl4}, sbr^{twc2}	Partial loss of function	Delayed transition of I from loose to dense	Interupted undulating dark stripes of normal arrangement and width. M stripes are broken into rectangular spots	From the initial interstripe, I spread in a coherent sheet over M. Later some of the dense I change to loose type	Promotes transition of dense I to loose form	Cell shape transition of I from dense to loose are important for pattern formation

(continued)

Table 8.3 (continued)

picasso	Receptor Epidermal growth factor receptor (EGFR)-like tyrosine kinase ErbB signalling (Lyons et al. 2005)	eNCC and M	*picasso*$^{wp.r2e2}$	Loss of function	Persistence of early larval melanophores. Deficient in metamorphic M particularly in the mid-trunk (Budi et al. 2008)	(Het) WT pattern development ensues (Budi et al. 2008) (Hom) Two (of 5) incomplete stripes form (Budi et al. 2008)	Early larval M persist and are incorporated into an incomplete stripe. Additional metamorphic melanophores differentiate where persisting embryonic/early larval M contributed to the adult stripe	erbb3b promotes the development of latent precursors (Budi et al. 2008)	Metamorphic M are important for pattern development
puma	tuba8l3	eNCC, M	*puma*^{j115el}	Recessive loss of function	Persistence of early larval melanophores. Reduction in M birth particularly during early metamorphosis with a gradual increase during terminal pigment pattern metamorphosis. (Parichy and Turner 2003)	(Hom) Fewer stripes with irregular borders and many breaks. 73% of X and 31% of M (Parichy et al. 2003)	Persistence of early larval M which move to form irregular striped patterns. During late metamorphosis, generation of metamorphic M into disorganised stripes	Early M precursor proliferation	Timely M differentiation important for pattern formation

Cell communication is important and may be partly conserved in its role for establishing stripe patterns on the fins. All cell communication mutants display defects in caudal fins suggesting that the melanophore–xanthophore communication mechanisms are shared in the body and fins. For example, the spotted *leo* mutant also displays an elongated spotted pattern in the caudal and anal fins. *Luc* mutants display slightly more tortuous stripes and interstripes in both caudal and anal fins. "Stripe width" mutants *dali* and *jag* also exhibit a reduced number of stripes in the caudal fin but are subtly different from each other. *Dali* exhibits the remnants of the middle stripe, in the form of a "stub" of melanophores at the base of the caudal fin surrounded by xanthophores. *Jag* mutants in contrast show two stripes (instead of the usual five stripes) at the very posterior and anterior region of the caudal fin. The rest of the space is filled with xanthophores (Inoue et al. 2014). *ide* mutants combine these two phenotypes (Frohnhofer et al. 2016).

Meanwhile plenty of mutants which have adult pigment pattern defects do not affect striping in the caudal fins, suggesting that some parts of development are fundamentally different. This list includes, but is not limited to, *cho, puma, picasso, bnc-2,* and *spa*.

8.5.1 *Dhrsx/Pyewacket*

The *dehydrogenase/reductase (SDR family) X-linked* (Mellgren and Johnson 2006) (*dhrsx*, also known as *pyewacket/pye*) mutant is unusual in the sense that whilst mutants show normal pattern development on the body, development of the adult caudal and anal fin pigment pattern is aberrant. The adult *pye* mutant has "stripes" in the anal and caudal fin that extend from the base, but are aligned with the fin rays and only extend a third of the full fin length. Caudal fin pattern development deviates from WT from very early stages. At 3 weeks, *pye* mutants display fewer xanthophores in early stages on the caudal fins. This deficiency persists and by 5 weeks, when in WT stripe pattern on the caudal fins is almost complete, *pye* fins continue to have fewer xanthophores and melanophores are still dispersed throughout the fin. The fins are fully developed by 7 weeks. The mutant phenotype is postulated to be related to xanthophore development in the fins, and seems to indicate that, at some level at least, fin pigment pattern formation is distinct from that in the body.

8.6 Dorsal Ventral Countershading

Alongside the striping pattern, a more general aspect of the pigment pattern common to almost all vertebrates, and widespread across the animal kingdom, is displayed in zebrafish. Thus, the dorsal region is noticeably darker and the ventrum paler, a phenomenon known as Dorsal Ventral Countershading (DVC). This evolutionary patterning mechanism is thought to be a basic camouflage mechanism from predators or prey positioned directly above or below the animal. From above, fish exhibiting DVC camouflage with the dark sea and from below, the bright sky.

Countershading in fish is achieved by a patterned distribution of the pigment cells, with the light-absorbing and light-reflecting chromatophores mostly distributed in the dorsal and ventral areas, respectively (Fukuzawa and Ide 1988; Zuasti et al. 1992; Joseph et al. 2007). Experimental data in fish and amphibian species suggest that DVC is regulated by a combination of a melanization simulating hormone (MSH) with a melanization inhibition factor (MIF) (Zuasti 2002). MSH is a hormone produced in both the ventral and dorsal regions that promotes the differentiation of melanophores and possibly inhibits iridophore proliferation. The identity of MIF has long been unknown, but MIF is thought to be locally produced by cells in the ventral skin and acts as an antagonist to MSH, inhibiting melanoblast differentiation and stimulating iridophore proliferation in this region (Fukuzawa and Ide 1988; Zuasti 2002). Recent studies have identified Agouti-signaling protein 1 (asip1) as the fish MIF (Guillot et al. 2012; Ceinos et al. 2015). A priority for future research will be to understand how the balance of Asip1 and MSH results in the differential abundance of pigment cells in different body regions.

8.6.1 Scale Development

Contributing to the darker dorsal surface is the association of pigment cells with dorsal scales. Development of scales begins posteriorly over the lateral surfaces of the tail at ~8.1 mm SL and proceeds from this location towards the anterior (Sire et al. 1997). By juvenile stages around 11 mm SL, scales are fully formed. Pigment cells generated from adult pigment stem cells, and which likely travel along the nerves lining the scales (Rasmussen et al. 2018), form two rows along each scale edge (Parichy et al. 2011). By adulthood multiple rows of pigment cells line the distal edges of the scales. The distribution of the pigment cells correlates with the spatial position dorso-ventrally on the fish. Thus, scales located dorsally contain all of melanophores, xanthophores, and iridophores. As a result the dorsum appears dark. In contrast, the ventral scales do not contain any pigment cells (Ceinos et al. 2015). Since melanophores are also virtually absent in the belly skin, the ventral region appears white as a result of the high number of iridophores in the abdominal wall.

8.6.2 Genes Relating to Dorso-Ventral Countershading (DVC)

Agouti Signaling

The melanocortin system consists of four endogenous antagonists of MSH signaling, which together form the agouti signaling protein family; *agouti signaling protein 1* (*asip1*), *agouti signaling protein 2* (*asip2*), *agouti related protein 1* (*agrp1*), and *agouti related protein 2* (*agrp2*).

Of these, the most well understood Agouti signaling molecule in zebrafish is *asip1*. *asip1* is a secreted protein that serves as an important component of dorsal ventral countershading by acting as an endogenous antagonist for MC1R (Cal et al. 2017a; Matsuda et al. 2018; Ollmann and Barsh 1999). Expression of *asip1* represses melanization and *mitfa* expression ventrally, reducing the number of melanophores cells, although analyses of *asip1* mutants indicate it may also have roles in inhibiting the production of other cells too.

The zebrafish mutant *asip1*, which likely results in a partial loss of *asip1* function, displays a wider 2 V stripe and extra 3 V stripe ventrally. The belly, branchiostegal, jaw, ventral skin, and operculum regions are hyperpigmented with melanophores and xanthophores and hence appear darker than in WT fish (Ceinos et al. 2015). The internal abdominal wall, which usually contains a high number of iridophores, has a reduced number and appears more yellow (due to increased xanthophores) than WT. Pigment cells on the scales are also affected. At 210 dpf, ventral scales in *asip1* mutants show ectopic melanophores, xanthophores, and iridophores; in contrast, dorsal scales are unaffected.

In contrast, transgenic fish ubiquitously overexpressing *asip1* have an almost normal striped pattern, although stripe 2D is thinner. However, the DVC is reduced, with fish showing a hypopigmented dorsum, and retaining a light belly. In 210 dpf *asip1* transgenics dorsal scales are hypopigmented compared to WT and *asip1* KO. This is consistent with *asip1* overexpression phenotypes in other teleost fish; e.g. in turbot and sole, injection of *asip1* caused localized lightening of pigment (Guillot et al. 2012). Analyses of these mutants indicate that aside from inhibiting melanophore differentiation, *asip1* may also play a role in inhibiting the production of other pigment cells ventrally.

Melanocortin Receptors

The teleost fish melanocortin system consists of five melanocortin receptor subtypes (MC1R, MC2R, MC3R, MC4R, MC5R) (Cal et al. 2017c). Zebrafish has six, because Mc5r is duplicated, as MC5Ra and MC5Rb (Cortés et al. 2014).

The melanocortin 1 receptor (MC1R) is the central melanocortin receptor involved in vertebrate pigmentation. The zebrafish loss-of-function mutant of MC1R displays a similar phenotype to the *asip1* mutant. It is characterized by a darker ventrum than WT. The branchiostegal, jaw, ventral skin, and operculum regions are hyperpigmented with increased numbers of melanophores and xanthophores. The abdominal wall appears overall more yellowish than WT (Cal et al. 2019). Whilst most of the striped pigment pattern is not changed compared to WT, MC1R KO fish develop a thicker 2 V stripe and an additional 3 V stripe in the ventrum compared to WT. The scales are also affected in MC1R KO. The dorsal scales exhibit a strong reduction in the number of melanophores. The ventral scales which usually lack pigment cells, exhibit both melanophores and xanthophores.

The MC1R KO still retains pigmentation ventrally, revealing that *asip1* interaction is more complex than simply blocking of the constitutive activity of MC1R. Rather, this data suggests that melanophore lineage specification and differentiation in the dorsal skin are not dependent solely on MC1R activity, and thus, some other receptors, presumably other melanocortin receptors, interact with *asip1*.

MC2R, MC3R, MC4R, MC5Ra, and MC5Rb bind many MSH ligands with varying affinity (Cerda-Reverter et al. 2005).

MSH and MCH

Physiological color change is controlled by both sympathetic (noradrenalin) and endocrine (α-melanocyte stimulating hormone (α-MSH), and melanin concentrating hormone (MCH) systems; Fujii 1993) (see Chap. 13). Noradrenalin and MCH induce pigment-aggregation on light backgrounds (Kawauchi et al. 1983; Logan et al. 2006; Sugimoto 2002), whereas α-MSH induces pigment dispersion on dark backgrounds (Logan et al. 2006; Mizusawa et al. 2013).

Suggested Molecular Mechanisms

Asip1 is an endogenous antagonist and inverse agonist of melanocortin 1 receptor (MC1R). Many of the effects of dorsal-ventral countershading are likely mediated through MC1R, via ligands *asip1* and *agrp2* (Cal et al. 2017b; Guillot et al. 2012). Different theories exist regarding the result of *asip1* antagonism of the MC1R pathway in fish. Thus, it may (1) promote melanin synthesis and/or (2) switch cell fate from iridophores to melanophores, with further work needed to assess these definitively. Future work is also required to assess the roles of the other MCRs in this process.

8.7 Conclusion

While there has been enormous progress in our understanding of pigment pattern formation in the zebrafish, and recent modeling studies have indicated that the biological mechanisms identified are nearly sufficient to explain pigment pattern formation up to juvenile stages at least (Owen et al. 2020; Volkening and Sandstede 2018), there do remain some key aspects deserving of further experimental investigation, in particular the likely role of L-iridophores in pattern maintenance. Furthermore, further investigation of the mutants defined here, to directly quantitate the interactions between cell-types and underlying molecular mechanisms that they define, promises a remarkable understanding of this fascinating process. As noted in the complementary chapter by Parichy (see Chap. 10), understanding of the evolution of these pigment patterns is the next major challenge.

References

Budi EH, Patterson LB, Parichy DM (2008) Embryonic requirements for ErbB signaling in neural crest development and adult pigment pattern formation. Development 135(15):2603–2614

Budi EH, Patterson LB, Parichy DM (2011) Post-embryonic nerve-associated precursors to adult pigment cells: genetic requirements and dynamics of morphogenesis and differentiation. PLoS Genet 7(5):e1002044

Cal L, Megias M, Cerdá-Reverter JM, Postlethwait JH, Braasch I, Rotllant J (2017a) BAC Recombineering of the Agouti loci from spotted gar and Zebrafish reveals the evolutionary ancestry of dorsal-ventral pigment asymmetry in fish. J Exp Zool Part B 328(7):697–708

Cal L, Suarez-Bregua P, Braasch I, Irion U, Kelsh R, Cerda-Reverter JM, Rotllant J (2019) Loss-of-function mutations in the melanocortin 1 receptor cause disruption of dorso-ventral countershading in teleost fish. Pigment Cell Melanoma Res 32(6):817–828. https://doi.org/10.1111/pcmr.12806

Cal L, Suarez-Bregua P, Cerda-Reverter JM, Braasch I, Rotllant J (2017b) Fish pigmentation and the melanocortin system. Comp Biochem Physiol A Mol Integr Physiol 211:26–33. https://doi.org/10.1016/j.cbpa.2017.06.001

Cal L, Suarez-Bregua P, Cerdá-Reverter JM, Braasch I, Rotllant J (2017c) Fish pigmentation and the melanocortin system. Comp Biochem Physio Part A: Mol Integr Physiol 211:26–33. https://doi.org/10.1016/j.cbpa.2017.06.001

Camargo-Sosa K, Colanesi S, Muller J, Schulte-Merker S, Stemple D, Patton EE, Kelsh RN (2019) Endothelin receptor aa regulates proliferation and differentiation of Erb-dependent pigment progenitors in zebrafish. PLoS Genet 15(2):e1007941. https://doi.org/10.1371/journal.pgen.1007941

Ceinos RM, Guillot R, Kelsh RN, Cerda-Reverter JM, Rotllant J (2015) Pigment patterns in adult fish result from superimposition of two largely independent pigmentation mechanisms. Pigment Cell Melanoma Res 28(2):196–209. https://doi.org/10.1111/pcmr.12335

Cerda-Reverter JM, Haitina T, Schioth HB, Peter RE (2005) Gene structure of the goldfish agouti-signaling protein: a putative role in the dorsal-ventral pigment pattern of fish. Endocrinology 146(3):1597–1610

Cortés R, Navarro S, Agulleiro MJ, Guillot R, García-Herranz V, Sánchez E, Cerdá-Reverter JM (2014) Evolution of the melanocortin system. Gen Comp Endocrinol 209:3–10. https://doi.org/10.1016/j.ygcen.2014.04.005

Dooley CM, Mongera A, Walderich B, Nusslein-Volhard C (2013) On the embryonic origin of adult melanophores: the role of ErbB and kit signalling in establishing melanophore stem cells in zebrafish. Development 140(5):1003–1013. https://doi.org/10.1242/dev.087007

Engeszer RE, Wang G, Ryan MJ, Parichy DM (2008) Sex-specific perceptual spaces for a vertebrate basal social aggregative behavior. Proc Natl Acad Sci U S A 105:929–933

Eom DS, Bain EJ, Patterson LB, Grout ME, Parichy DM (2015) Long-distance communication by specialized cellular projections during pigment pattern development and evolution. Elife 4: e12401. https://doi.org/10.7554/eLife.12401

Eom DS, Inoue S, Patterson LB, Gordon TN, Slingwine R, Kondo S, Watanabe M, Parichy DM (2012) Melanophore migration and survival during zebrafish adult pigment stripe development require the immunoglobulin superfamily adhesion molecule Igsf11. PLoS Genet 8(8): e1002899. https://doi.org/10.1371/journal.pgen.1002899

Eom DS, Parichy DM (2017) A macrophage relay for long-distance signaling during postembryonic tissue remodeling. Science 355(6331):1317–1320

Fadeev A, Krauss J, Frohnhofer HG, Irion U, Nusslein-Volhard C (2015) Tight junction protein 1a regulates pigment cell organisation during zebrafish colour patterning. Elife 4:e06545. https://doi.org/10.7554/eLife.06545

Fadeev A, Krauss J, Singh AP, Nusslein-Volhard C (2016) Zebrafish leucocyte tyrosine kinase controls iridophore establishment, proliferation and survival. Pigment Cell Melanoma Res 29 (3):284–296. https://doi.org/10.1111/pcmr.12454

Fadeev A, Mendoza-Garcia P, Irion U, Guan J, Pfeifer K, Wiessner S, Serluca F, Singh AP, Nusslein-Volhard C, Palmer RH (2018) ALKALs are in vivo ligands for ALK family receptor tyrosine kinases in the neural crest and derived cells. Proc Natl Acad Sci U S A 115(4):E630–E638. https://doi.org/10.1073/pnas.1719137115

Frohnhofer HG, Geiger-Rudolph S, Pattky M, Meixner M, Huhn C, Maischein HM, Geisler R, Gehring I, Maderspacher F, Nusslein-Volhard C, Irion U (2016) Spermidine, but not spermine, is essential for pigment pattern formation in zebrafish. Biology Open 5(6):736–744. https://doi.org/10.1242/bio.018721

Frohnhofer HG, Krauss J, Maischein HM, Nusslein-Volhard C (2013) Iridophores and their interactions with other chromatophores are required for stripe formation in zebrafish. Development 140(14):2997–3007. https://doi.org/10.1242/dev.096719

Fujii R (1993) Cytophysiology of fish chromatophores. Int Rev Cytol 143:191–255

Fukamachi S, Kinoshita M, Aizawa K, Oda S, Meyer A, Mitani H (2009) Dual control by a single gene of secondary sexual characters and mating preferences in medaka. BMC Biol 7:64. https://doi.org/10.1186/1741-7007-7-64

Fukamachi S, Sugimoto M, Mitani H, Shima A (2004) Somatolactin selectively regulates proliferation and morphogenesis of neural-crest derived pigment cells in medaka. Proc Natl Acad Sci U S A 101(29):10661–10666

Fukuzawa T, Ide H (1988) A ventrally localized inhibitor of melanization in Xenopus laevis skin. Dev Biol 129(1):25–36

Guillot R, Ceinos RM, Cal R, Rotllant J, Cerdá-Reverter JM (2012) Transient ectopic overexpression of Agouti-Signalling protein 1 (Asip1) induces pigment anomalies in flatfish. PLoS One 7(12):1–10

Haddad RI, Clark JR, Wein RO, Grillone GA (2013) Thyroid hormone-dependent adult pigment cell lineage and pattern in zebrafish. Biomarkers 72(19):5004–5013. https://doi.org/10.1158/0008-5472.CAN-11-3277

Haffter P, Odenthal J, Mullins MC, Lin S, Farrell MJ, Vogelsang E, Haas F, Brand M, van Eeden FJ, Furutani-Seiki M, Granato M, Hammerschmidt M, Heisenberg CP, Jiang YJ, Kane DA, Kelsh RN, Hopkins N, Nusslein-Volhard C (1996) Mutations affecting pigmentation and shape of the adult zebrafish. Dev Genes Evol 206(4):260–276. https://doi.org/10.1007/s004270050051

Hibino H, Inanobe A, Furutani K, Murakami S, Findlay I, Kurachi Y (2010) Inwardly rectifying potassium channels: their structure, function, and physiological roles. Physiol Rev 90 (1):291–366. https://doi.org/10.1152/physrev.00021.2009

Hirata M, Nakamura K, Kanemaru T, Shibata Y, Kondo S (2003) Pigment cell organization in the hypodermis of zebrafish. Dev Dyn 227(4):497–503

Hirata M, Nakamura K, Kondo S (2005) Pigment cell distributions in different tissues of the zebrafish, with special reference to the striped pigment pattern. Dev Dyn 234(2):293–300

Hultman KA, Bahary N, Zon LI, Johnson SL (2007) Gene duplication of the zebrafish kit ligand and partitioning of kit ligand a. PLoS Genet 3:e17

Inoue S, Kondo S, Parichy DM, Watanabe M (2014) Tetraspanin 3c requirement for pigment cell interactions and boundary formation in zebrafish adult pigment stripes. Pigment Cell Melanoma Res 27(2):190–200. https://doi.org/10.1111/pcmr.12192

Irion U, Frohnhofer HG, Krauss J, Colak Champollion T, Maischein HM, Geiger-Rudolph S, Weiler C, Nusslein-Volhard C (2014) Gap junctions composed of connexins 41.8 and 39.4 are essential for colour pattern formation in zebrafish. Elife 3:e05125. https://doi.org/10.7554/eLife.05125

Iwashita M, Watanabe M, Ishii M, Chen T, Johnson SL, Kurachi Y, Okada N, Kondo S (2006) Pigment pattern in jaguar/obelix zebrafish is caused by a Kir7.1 mutation: implications for the regulation of melanosome movement. PLoS Genet 2(11):e197

Johnson SL, Africa D, Walker C, Weston JA (1995) Genetic control of adult pigment stripe development in zebrafish. Dev Biol 167(1):27–33

Joseph T, Bagnara A, Matsumoto J (2007) Comparative anatomy and physiology of pigment cells in nonmammalian tissues. In: The pigmentary system: physiology and pathophysiology, 2nd edn. Wiley, Hoboken, NJ, pp 11–59

Kawauchi H, Kawazoe I, Tsubokawa M, Kishida M, Baker BI (1983) Characterization of melanin-concentrating hormone in chum salmon pituitaries. Nature 305(5932):321–323. https://doi.org/10.1038/305321a0

Kelsh RN, Brand M, Jiang YJ, Heisenberg CP, Lin S, Haffter P, Odenthal J, Mullins MC, van Eeden FJ, Furutani-Seiki M, Granato M, Hammerschmidt M, Kane DA, Warga RM, Beuchle D, Vogelsang L, Nusslein-Volhard C (1996) Zebrafish pigmentation mutations and the processes of neural crest development. Development 123:369–389

Kelsh RN, Harris ML, Colanesi S, Erickson CA (2009) Stripes and belly-spots: a review of pigment cell morphogenesis in vertebrates. Semin Cell Dev Biol 20(1):90–104. https://doi.org/10.1016/j.semcdb.2008.10.001S1084-9521(08)00096-7

Kofuji P, Newman EA (2004) Potassium buffering in the central nervous system. Neuroscience 129 (4):1045–1056. https://doi.org/10.1016/j.neuroscience.2004.06.008

Krauss J, Astrinidis P, Frohnhofer HG, Walderich B, Nusslein-Volhard C (2013) Transparent, a gene affecting stripe formation in zebrafish, encodes the mitochondrial protein Mpv17 that is required for iridophore survival. Biology Open 2(7):703–710. https://doi.org/10.1242/bio.20135132

Krauss J, Frohnhofer HG, Walderich B, Maischein HM, Weiler C, Irion U, Nusslein-Volhard C (2014) Endothelin signalling in iridophore development and stripe pattern formation of zebrafish. Biology Open 3(6):503–509. https://doi.org/10.1242/bio.20148441

Lang MR, Patterson LB, Gordon TN, Johnson SL, Parichy DM (2009) Basonuclin-2 requirements for zebrafish adult pigment pattern development and female fertility. PLoS Genet 5(11): e1000744. https://doi.org/10.1371/journal.pgen.1000744

Lister JA, Robertson CP, Lepage T, Johnson SL, Raible DW (1999) Nacre encodes a zebrafish microphthalmia-related protein that regulates neural-crest-derived pigment cell fate. Development 126(17):3757–3767

Logan DW, Burn SF, Jackson IJ (2006) Regulation of pigmentation in zebrafish melanophores. Pigment Cell Res 19(3):206–213

Lopes SS, Yang X, Muller J, Carney TJ, McAdow AR, Rauch GJ, Jacoby AS, Hurst LD, Delfino-Machin M, Haffter P, Geisler R, Johnson SL, Ward A, Kelsh RN (2008) Leukocyte tyrosine kinase functions in pigment cell development. PLoS Genet 4(3):e1000026. https://doi.org/10.1371/journal.pgen.1000026

Lyons DA, Pogoda HM, Voas MG, Woods IG, Diamond B, Nix R, Arana N, Jacobs J, Talbot WS (2005) erbb3 and erbb2 are essential for schwann cell migration and myelination in zebrafish. Curr Biol 15:513–524

Maderspacher F, Nusslein-Volhard C (2003) Formation of the adult pigment pattern in zebrafish requires leopard and obelix dependent cell interactions. Development 130(15):3447–3457
Mahalwar P, Singh AP, Fadeev A, Nüsslein-Volhard C, Irion U (2016) Heterotypic interactions regulate cell shape and density during color pattern formation in zebrafish. Biol Open 5 (11):1680–1690
Mahalwar P, Walderich B, Singh AP, Nusslein-Volhard C (2014) Local reorganization of xanthophores fine-tunes and colors the striped pattern of zebrafish. Science 345 (6202):1362–1364. https://doi.org/10.1126/science.1254837
Matsuda N, Kasagi S, Nakamaru T, Masuda R, Takahashi A, Tagawa M (2018) Left-right pigmentation pattern of Japanese flounder corresponds to expression levels of melanocortin receptors (MC1R and MC5R), but not to agouti signaling protein 1 (ASIP1) expression. Gen Comp Endocrinol 262:90–98. https://doi.org/10.1016/j.ygcen.2018.03.019
McMenamin SK, Bain EJ, McCann AE, Patterson LB, Eom DS, Waller ZP, Hamill JC, Kuhlman JA, Eisen JS, Parichy DM (2014) Thyroid hormone-dependent adult pigment cell lineage and pattern in zebrafish. Science 345(6202):1358–1361. https://doi.org/10.1126/science.1256251
Mellgren EM, Johnson SL (2006) Pyewacket, a new zebrafish fin pigment pattern mutant. Pigment Cell Res 19(3):232–238
Mizusawa K, Kobayashi Y, Yamanome T, Saito Y, Takahashi A (2013) Interrelation between melanocyte-stimulating hormone and melanin-concentrating hormone in physiological body color change: roles emerging from barfin flounder Verasper moseri. Gen Comp Endocrinol 181:229–234. https://doi.org/10.1016/j.ygcen.2012.09.026
Mo ES, Cheng Q, Reshetnyak AV, Schlessinger J, Nicoli S (2017) Alk and Ltk ligands are essential for iridophore development in zebrafish mediated by the receptor tyrosine kinase Ltk. Proc Natl Acad Sci U S A 114(45):12027–12032. https://doi.org/10.1073/pnas.1710254114
Nakamasu A, Takahashi G, Kanbe A, Kondo S (2009) Interactions between zebrafish pigment cells responsible for the generation of Turing patterns. Proc Natl Acad Sci U S A 106 (21):8429–8434. https://doi.org/10.1073/pnas.0808622106
Odenthal J, Rossnagel K, Haffter P, Kelsh R, Vogelsang E, Brand M, van Eeden F, Furutani-Seiki M, Granato M, Hammerschmidt M, Heisenberg C, Jiang Y, Kane D, Mullins M, Nusslein-Volhard C (1996) Mutations affecting xanthophore pigmentation in the zebrafish, Danio rerio. Development 123:391–398
Ollmann MM, Barsh GS (1999) Down-regulation of melanocortin receptor signaling mediated by the amino terminus of Agouti protein in Xenopus melanophores. J Biol Chem 274 (22):15837–15846
Owen JP, Kelsh RN, Yates CA (2020) A quantitative modelling approach to zebrafish pigment pattern formation. elife 9:e52998. https://doi.org/10.7554/eLife.52998
Parichy DM, Elizondo MR, Mills MG, Gordon TN, Engeszer E (2011) Normal table of post-embryonic zebrafish development: staging by externally visible anatomy of the living fish. Dev Dyn 238(12):2975–3015
Parichy DM, Elizondo MR, Mills MG, Gordon TN, Engeszer RE (2009) Normal table of postembryonic zebrafish development: staging by externally visible anatomy of the living fish. Dev Dyn 238(12):2975–3015. https://doi.org/10.1002/dvdy.22113
Parichy DM, Mellgren EM, Rawls JF, Lopes SS, Kelsh RN, Johnson SL (2000a) Mutational analysis of endothelin receptor b1 (rose) during neural crest and pigment pattern development in the zebrafish Danio rerio. Dev Biol 227(2):294–306. https://doi.org/10.1006/dbio.2000.9899
Parichy DM, Ransom DG, Paw B, Zon LI, Johnson SL (2000b) An orthologue of the kit-related gene fms is required for development of neural crest-derived xanthophores and a subpopulation of adult melanocytes in the zebrafish, Danio rerio. Development 127(14):3031–3044
Parichy DM, Rawls JF, Pratt SJ, Whitfield TT, Johnson SL (1999) Zebrafish sparse corresponds to an orthologue of c-kit and is required for the morphogenesis of a subpopulation of melanocytes, but is not essential for hematopoiesis or primordial germ cell development. Development 126 (15):3425–3436

Parichy DM, Turner JM (2003) Zebrafish puma mutant decouples pigment pattern and somatic metamorphosis. Dev Biol 256(2):242–257

Parichy DM, Turner JM, Parker NB (2003) Essential role for puma in development of postembryonic neural crest-derived cell lineages in zebrafish. Dev Biol 256(2):221–241

Patterson LB, Parichy DM (2013) Interactions with iridophores and the tissue environment required for patterning melanophores and xanthophores during zebrafish adult pigment stripe formation. PLoS Genet 9(5):e1003561. https://doi.org/10.1371/journal.pgen.1003561

Quigley IK, Manuel JL, Roberts RA, Nuckels RJ, Herrington ER, MacDonald EL, Parichy DM (2005) Evolutionary diversification of pigment pattern in Danio fishes: differential fms dependence and stripe loss in D. albolineatus. Development 132:89–104

Raible DW, Eisen JS (1994) Restriction of neural crest cell fate in the trunk of the embryonic zebrafish. Development 120:495–503

Rasmussen JP, Vo NT, Sagasti A (2018) Fish scales dictate the pattern of adult skin innervation and vascularization. Dev Cell 46(3):344–359. https://doi.org/10.1016/j.devcel.2018.06.019

Singh AP, Nusslein-Volhard C (2015) Zebrafish stripes as a model for vertebrate colour pattern formation. Curr Biol 25(2):R81–R92. https://doi.org/10.1016/j.cub.2014.11.013

Singh AP, Schach U, Nusslein-Volhard C (2014) Proliferation, dispersal and patterned aggregation of iridophores in the skin prefigure striped colouration of zebrafish. Nat Cell Biol 16 (6):607–614. https://doi.org/10.1038/ncb2955

Sire JY, Allizard F, Babiar O, Bourguignon J, Quilhac A (1997) Scale development in zebrafish (Danio rerio). Int J Dev Biol 48(2–3):233–247

Spiewak JE, Bain EJ, Liu J, Kou K, Sturiale SL, Patterson LB, Diba P, Eisen JS, Braasch I, Ganz J, Parichy DM (2018) Evolution of endothelin signaling and diversification of adult pigment pattern in Danio fishes. PLoS Genet 14(9):e1007538. https://doi.org/10.1371/journal.pgen.1007538

Sugimoto M (2002) Morphological color changes in fish: regulation of pigment cell density and morphology. Microsc Res Tech 58(6):496–503. https://doi.org/10.1002/jemt.10168

Svetic V, Hollway GE, Elworthy S, Chipperfield TR, Davison C, Adams RJ, Eisen JS, Ingham PW, Currie PD, Kelsh RN (2007) Sdf1a patterns zebrafish melanophores and links the somite and melanophore pattern defects in choker mutants. Development 134(5):1011–1022

Takahashi G, Kondo S (2008) Melanophores in the stripes of adult zebrafish do not have the nature to gather, but disperse when they have the space to move. Pigment Cell Melanoma Res 21 (6):677–686. https://doi.org/10.1111/j.1755-148X.2008.00504.x

van Bebber F, Hruscha A, Willem M, Schmid B, Haass C (2013) Loss of Bace2 in zebrafish affects melanocyte migration and is distinct from Bace1 knock out phenotypes. J Neurochem 127 (4):471–481. https://doi.org/10.1111/jnc.12198

van Eeden FJ, Granato M, Schach U, Brand M, Furutani-Seiki M, Haffter P, Hammerschmidt M, Heisenberg CP, Jiang YJ, Kane DA, Kelsh RN, Mullins MC, Odenthal J, Warga RM, Allende ML, Weinberg ES, Nusslein-Volhard C (1996) Mutations affecting somite formation and patterning in the zebrafish, Danio rerio. Development 123:153–164

Volkening A, Sandstede B (2018) Iridophores as a source of robustness in zebrafish stripes and variability in Danio patterns. Nat Commun 9(1):3231. https://doi.org/10.1038/s41467-018-05629-z

Walderich B, Singh AP, Mahalwar P, Nusslein-Volhard C (2016) Homotypic cell competition regulates proliferation and tiling of zebrafish pigment cells during colour pattern formation. Nat Commun 7:11462. https://doi.org/10.1038/ncomms11462

Watanabe M, Kondo S (2011) Changing clothes easily: connexin41.8 regulates skin pattern variation. Pigment Cell Melanoma Res 25:375–383. https://doi.org/10.1111/j.1755-148X.2011.00989.x

Yamanaka H, Kondo S (2014) In vitro analysis suggests that difference in cell movement during direct interaction can generate various pigment patterns in vivo. Proc Natl Acad Sci U S A 111 (5):1867–1872. https://doi.org/10.1073/pnas.1315416111

Yu JF, Fukamachi S, Mitani H, Hori H, Kanamori A (2006) Reduced expression of vps11 causes less pigmentation in medaka, Oryzias latipes. Pigment Cell Res 19(6):628–634

Zuasti A (2002) Melanization stimulating factor (MSF) and melanization inhibiting factor (MIF) in the integument of fish. Microsc Res Tech 58(6):488–495. https://doi.org/10.1002/jemt.10167

Zuasti A, Johnson WC, Samaraweera P, Bagnara JT (1992) Intrinsic pigment-cell stimulating activity in the catfish integument. Pigment Cell Res 5(5 Pt 1):253–262

Chapter 9
Theoretical Studies of Pigment Pattern Formation

Seita Miyazawa, Masakatsu Watanabe, and Shigeru Kondo

Abstract A wide variety of patterns can be observed in multicellular organisms. How these various spatial regularities are generated from the seemingly homogeneous field of an egg cell has been a great mystery, and various mathematical models have been proposed to provide a conceptual explanation of the underlying mechanism. Unlike other types of morphogenesis that involve three-dimensional processes, the patterns produced by pigment cells are simple in that they occur on the two-dimensional surface of the body. Therefore, they are relatively easy to reproduce using mathematical models. In addition, the ease with which the developmental process can be traced on a cell-by-cell basis without using special visualization techniques is also advantageous for studies linking experimental data to the models. Thus, pigment patterns have contributed significantly to the development of mathematical models of pattern formation and to the widespread recognition and acceptance of these models among experimental biologists. In this chapter, we discuss some of the important models that have been proposed and developed to understand the mechanisms of pigment pattern formation, including their relationship to patterns found in real organisms.

Keywords Pigment pattern · Mathematical model · Reaction–diffusion · Agent-based model · Cell–cell interaction · Gap junction

9.1 Turing's Reaction–Diffusion System

Let us start by considering a hypothetical diffusible factor. If various developmental processes such as cell differentiation and organogenesis are dependent upon the concentration of this factor in each part of the body, the morphogenesis of an organism can be schematically understood as a formation process of the

S. Miyazawa (✉) · M. Watanabe · S. Kondo
Graduate School of Frontier Biosciences, Osaka University, Osaka, Japan
e-mail: seita@fbs.osaka-u.ac.jp; watanabe-m@fbs.osaka-u.ac.jp; skondo@fbs.osaka-u.ac.jp

H. Hashimoto et al. (eds.), *Pigments, Pigment Cells and Pigment Patterns*,
https://doi.org/10.1007/978-981-16-1490-3_9

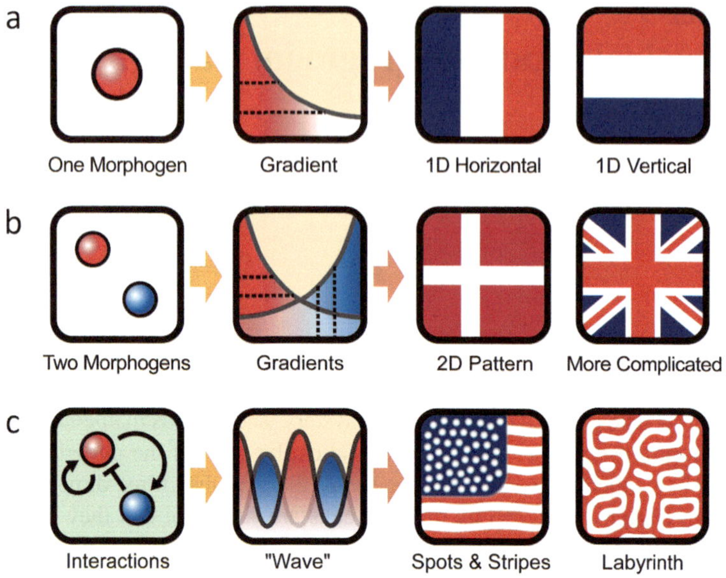

Fig. 9.1 A scheme showing the difference between the morphogen gradient model and the Turing's reaction–diffusion model. (**a**) Morphogen molecules produced at one end of an embryo form a gradient by diffusion. Cells "know" their position because of the concentration of molecules. The gradient is totally dependent on the pre-pattern of the morphogen source (boundary conditions). (**b**) The addition of a second morphogen yields a somewhat more complex pattern; however, with no interactions between the morphogens, the system is not self-regulating and still heavily dependent on the pre-patterns. (**c**) The addition of interactions between morphogens results in a self-regulating system, which can form a variety of patterns independent of the pre-patterns. (Figure modified from Kondo and Miura 2010)

concentration pattern of this factor (often called a morphogen) (Wolpert 1969; Meinhardt 1982) (Fig. 9.1a, b).

In 1952, the British mathematician Alan Turing considered a system in which multiple factors diffuse while interacting with each other (commonly called a reaction–diffusion system) and found that under certain conditions, a stable spatial pattern can be formed autonomously (Turing 1952) (Fig. 9.1c). The most revolutionary feature of Turing's reaction–diffusion system is the introduction of the "reaction" that produces morphogens. If only "diffusion" is at work, a source and sink of morphogens is needed for the formation of a stable concentration gradient of morphogens, which specifies positional information. In such a case (i.e., a morphogen gradient model), the positional information produced by the system depends on the pre-pattern that determines the location of the source and sink (Wolpert 1969; Meinhardt 1982). Introducing reactions into the system causes it to exhibit a variety of interesting behaviors from a nearly uniform initial state without relying on any pre-patterns.

Turing theoretically derived the following six different patterns that emerge from slight fluctuations in the initial state, depending on various conditions (Turing 1952;

Kondo and Miura 2010): (a) a static pattern that is uniform across the entire region, (b) a pattern in which the entire region oscillates uniformly, (c) a static pattern in which adjacent cells are inverted from one another, (d) a static pattern in which concentration peaks appear in the form of equally spaced waves, (e) an oscillating pattern in which equally spaced waves move, and (f) an oscillating pattern in which adjacent cells are inverted from one another. The equally spaced static pattern (d, the so-called Turing pattern) is of particular interest in relation to biological patterning. Herein, we will attempt to elucidate the mechanism by which the Turing pattern arises. Various models have been proposed in terms of the number of factors and the properties of their interactions (Gierer and Meinhardt 1972; Meinhardt 1982; Gray and Scott 1985). However, here we focus on the activator–inhibitor type model, as represented by the Gierer–Meinhardt model (Gierer and Meinhardt 1972) (Fig. 9.2a).

9.2 Activator–Inhibitor Model

Let us assume two diffusible factors: an activator and an inhibitor (Fig. 9.2a). Suppose that, in a concentration-dependent manner for each, the activator activates the production of both the activator itself and the inhibitor, and the inhibitor inhibits the activator. Both activator and inhibitor can spread diffusively to the periphery; however, the inhibitor can spread faster than the activator. Let us consider the time evolution from the initial state with an almost uniform concentration distribution in a one-dimensional system (Fig. 9.2b). (1) First, if some fluctuation causes a slightly higher activator concentration in one region, (2) the activator increases in a self-amplifying manner in that region (local activation). At the same time, the inhibitor is also increased by the action of the activator. (3) The activator and inhibitor diffuse to the periphery according to the concentration gradient. However, as the rate of diffusion of the inhibitor is greater, the inhibitor spreads over a broader area than the activator and inhibits the activator in the periphery (long-range inhibition). (4) If similar processes occur here and there in space, a stable and equally spaced concentration pattern (Turing pattern), which looks like a steady wave as a whole, is produced.

The wave properties (wavelengths and amplitudes) of the final concentration pattern are determined by the system parameters (e.g., the rate coefficients of each reaction and the balance with the diffusion rate) and are independent of the initial state. Gierer and Meinhardt demonstrated that, for a Turing pattern to be produced, the system needs only to contain a network that combines "local activation (positive feedback) and long-range inhibition (negative feedback)" (Gierer and Meinhardt 1972; Meinhardt and Gierer 2000). A similar understanding is possible in linear models that approximate activator inhibitor type dynamics (Fig. 9.2c) and substrate–consumer type models, such as the Gray–Scott model (Gray and Scott 1985) (Fig. 9.2d). In each model, by changing the parameters of the system, we can

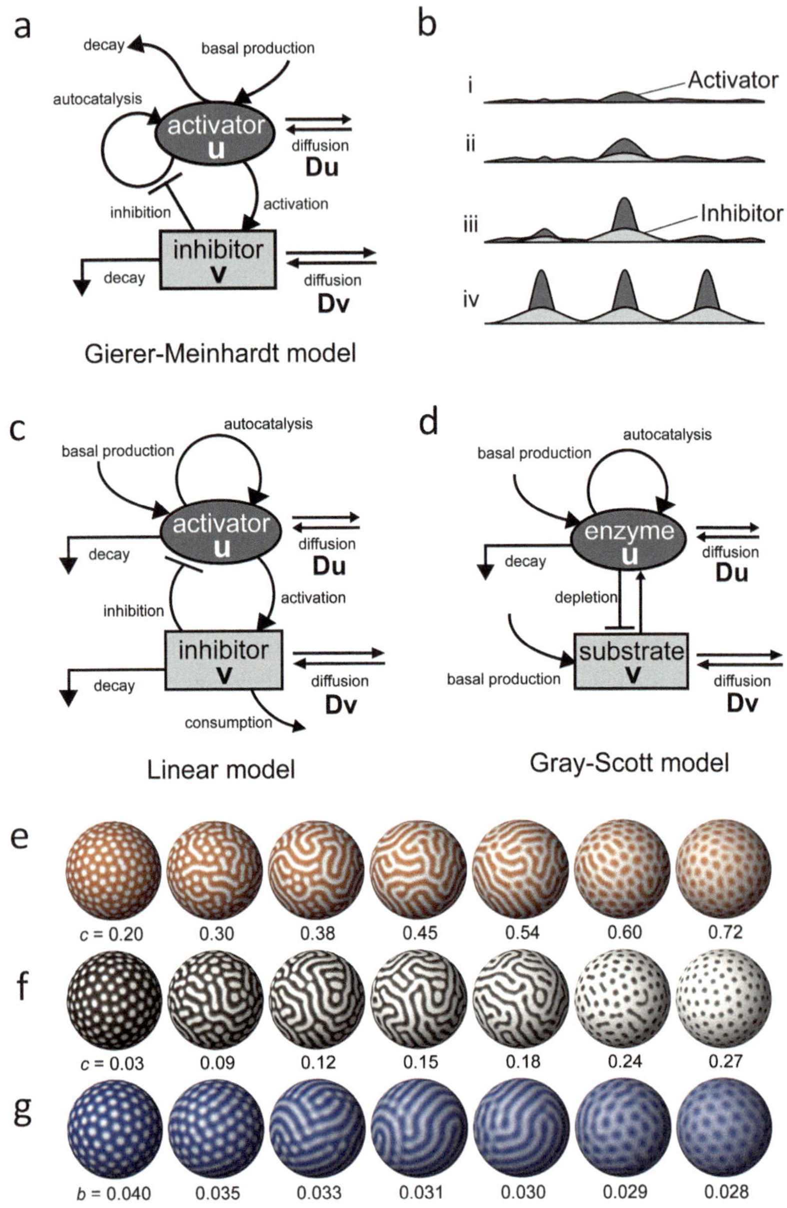

Fig. 9.2 Reaction–diffusion models. (**a**) A diagram of the Gierer–Meinhardt model. The diffusible activator activates the production of itself and the inhibitor. The inhibitor inhibits the activator and diffuses faster than the activator. (**b**) An intuitive overview of the pattern formation process in the activator–inhibitor model. (i) First, some fluctuation (or inherent heterogeneity) causes a slightly higher concentration of the activator in a region. (ii) In that region, the activator increases by self-activation (local activation) and also produces the inhibitor. (iii) Both factors diffuse. Since the inhibitor diffuses faster than the activator, it spreads over a wider area and inhibits the activator in the periphery (long-range inhibition). (iv) Similar processes occur elsewhere, resulting in a stable and equally spaced concentration pattern for both factors. (**c**) The linear model. (**d**) The Gray–Scott model. (**e**–**g**) Patterns generated by numerical simulations based on the Gierer–Meinhardt model

observe the pattern's continuous transition, producing a variety of motifs, such as spots, labyrinths, and reticulated patterns (Fig. 9.2e–g).

9.3 Application to Pattern Formation in Real Organisms

Attempts to understand the patterning processes of real organisms through models based on Turing's reaction–diffusion system have been made since the 1970s, mainly by mathematical biologists such as Meinhardt and Murray (Meinhardt 1982; Murray 2002). Meinhardt, together with Gierer, explained positional patterns in hydra head regeneration with the activator–inhibitor model described above (Gierer and Meinhardt 1972). Murray showed that body surface patterns found in various mammals, such as spotted patterns in leopards, could be well reproduced using two-dimensional reaction–diffusion models (Murray 1981). Although these results were not widely accepted by experimental biologists at the time, in 1995, Kondo and Asai succeeded in predicting the dynamic changes in the stripe pattern of the emperor angelfish *Pomacanthus imperator* by simulations based on the reaction–diffusion system and confirmed them by experimental observations (Kondo and Asai 1995). This led to the reemergence of the reaction–diffusion system as an effective model for pattern formation studies among experimental biologists.

Here, we introduce an example of a non-trivial pattern formation process revealed by the prediction based on the reaction–diffusion model: pattern blending by hybridization (Miyazawa et al. 2010). Let us now consider the crossing of two animal species that have different patterns (Fig. 9.3a–c). In general, when crossing occurs between two animal species with different characteristics, the phenotype exhibited by the hybridized individuals is expected to be one of the following: (1) a phenotype similar to either of the parents, (2) a phenotype that combines the traits of both parents, (3) an intermediate phenotype that is a mixture or blend of both parents' traits, or (4) an extreme phenotype that goes beyond the phenotype of the parents.

It is easy to imagine that case (1) represents situations where one or a few responsible genes define a trait and follow Mendelian inheritance. Case (2) represents co-dominance. It has been shown that the patterns observed on the wings of certain insects are determined by a combination of expression patterns of several transcription factors (Prud'homme et al. 2007; Werner et al. 2010) (see Chap. 12). In this case, individuals resulting from crossing between parents with different sets of cis-regulatory regions (and thus different wing patterns) have multiple sets of cis-regulatory regions, and therefore exhibit a phenotype that is similar to the superposition of the parents' wing patterns (Mavárez et al. 2006) (pattern addition, Fig. 9.3a, b). Cases (3) and (4) represent quantitative traits. When traits are defined

Fig. 9.2 (continued) (**e**), linear model (**f**), and Gray–Scott model (**g**). Each color represents the concentration of the core factor in each model (the activator or autocatalytic enzyme). Lighter colors indicate higher concentrations. (Figure modified from Miyazawa et al. 2010)

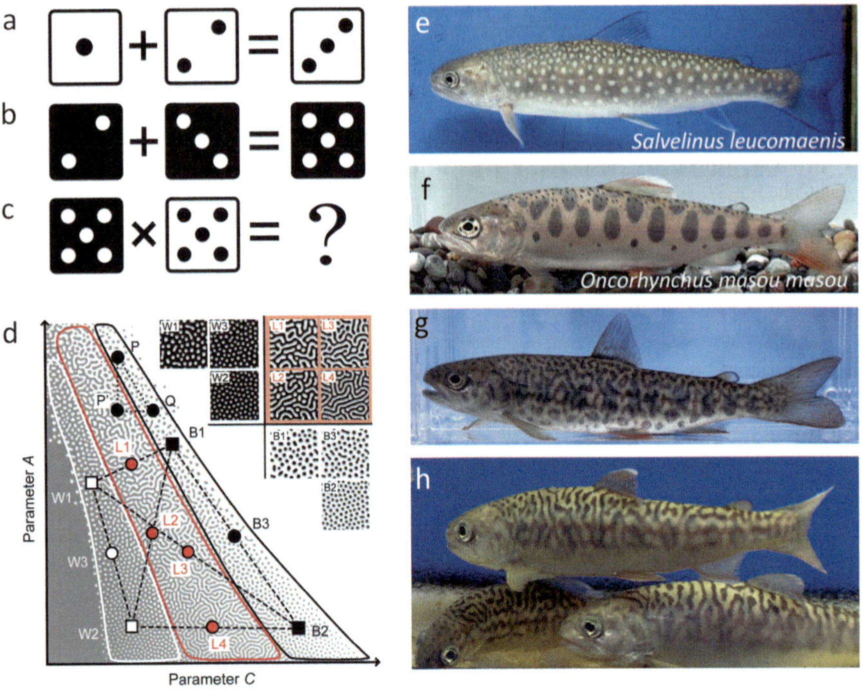

Fig. 9.3 Pattern blending by hybridization. (**a**–**c**) Schematic representation of pattern "addition" (**a** and **b**) and pattern "crossing" (**c**). (**d**) Pattern dynamics in the linear model reaction–diffusion system. Each point (x, y) represents a parameter set ($C = x$, $A = y$) for the model equations. Broken line segments joining the colored squares (W1–2 and B1–2) denote "in silico hybridization" between distinct parameter sets. The midpoints of these segments (colored circles; W3, B3, and L1–4) correspond to the "hybrids." The color patterns calculated using the parameter sets of the indicated points are shown in the insets. (**e**–**h**) Body patterns of salmonid fishes. (**e**) White-spotted charr (*Salvelinus leucomaenis*), (**f**) masu salmon (*Oncorhynchus masou masou*), and (**g**) their natural and (**h**) artificial intergeneric hybrids. (Figure modified from Miyazawa et al. 2010)

by the additive effects of a large number of genes, the progeny's phenotype often exhibits intermediate features, such as a mixture of traits from both parents (3). However, in certain rare cases, a set of alleles with extreme effects can be assembled (4), which produce an extreme phenotype that exceeds that of the parents, the so-called transgressive segregation (Rieseberg et al. 1999).

Let us further consider two simple patterns (Fig. 9.3c): "white spots on black" and "black spots on white." What kind of pattern can be expected after hybridization between animal species with similar yet different color patterns?

Assuming that patterns are quantitative traits, it is predicted that the color patterns appearing in hybrids would be intermediate (3) or transgressive phenotypes (4), according to the above classification. When a quantitative trait is represented as a one-dimensional variable, an "intermediate phenotype" can easily be imagined, for example, body length, weight, or the number of organs. Hybrids between large and small species should be of medium size, and hybrids between parents with different numbers of teeth or fin rays should have an average number of teeth or fin rays (Hubbs 1955; Grant and Grant 1992). However, in the case of color patterns that are essentially two-dimensional traits, it is difficult to imagine an "intermediate pattern" in which two different patterns are mixed together.

Therefore, Miyazawa et al. attempted to "mix" the patterns with an in silico hybridization experiment based on the reaction–diffusion model (Miyazawa et al. 2010). Continuously changing one parameter of the model caused various patterns to be produced (Fig. 9.2e–g). Regardless of the model selected, a labyrinthine pattern appeared between "white spots on black" and "black spots on white" (Fig. 9.2e–g). This was also the case when multiple parameters were changed simultaneously (Fig. 9.3d).

If the results of this in silico hybridization are correct, we can predict that the crossing of an animal with "white spots on black" and an animal with "black spots on white" will always produce hybrid offspring with strange camouflage labyrinthine patterns. This non-trivial prediction was confirmed by observing the pattern of hybrids in real animals. The white-spotted charr (*Salvelinus leucomaenis*) (Fig. 9.3e) and masu salmon (*Oncorhynchus masou masou*) (Fig. 9.3f) are both species found in Japanese rivers. The former has light spots on a dark background, while the latter has dark spots on a light background. These two species have different habitats, but natural hybrids have occasionally been observed. Examining the patterns of natural and artificial hybrids (Fig. 9.3g, h) revealed that they all had mysterious labyrinth patterns exactly as predicted by in silico hybridization. This has also been confirmed in other salmonid hybrids, such as intergeneric hybrids between the genera *Salvelinus* (light spots on a dark background) and *Oncorhynchus* and *Salmo* (dark spots on a light background), in which labyrinthine patterns were always present in all parental combinations examined.

Although both parents have very simple spotted patterns, the hybrids exhibit a labyrinthine pattern with a high degree of pattern complexity. It is important to note that the patterns of the hybrids are "intermediate" but also "transgressive" compared to the parent species. In terms of the four phenotypes of hybridization mentioned above, a phenotype can simultaneously be both a "intermediate phenotype" (3) and a "transgressive phenotype" (4) when color patterns are mixed. This can be regarded as a type of "transgressive segregation" that differs from conventional definition in that all hybrid individuals, rather than a rare occurrence of individuals, exhibit a transgressive phenotype. Miyazawa et al. pointed out that the emergence of transgressive traits through this kind of "pattern blending" may contribute to hybrid speciation and suggested that mathematical models of developmental pattern formation may also provide insights into evolutionary processes (Miyazawa et al. 2010; Miyazawa 2020).

9.4 The Kernel-Based Turing Model

Meinhardt and Gierer stated that the condition of "long activation and local inhibition" (LALI) is sufficient for stable pattern formation (Gierer and Meinhardt 1972; Meinhardt and Gierer 2000). This suggestion is very important because it implies that the effects of diffusion in the classical Turing model may be replaced by other effects (e.g., cell migration, physical stress, or neural signals). Indeed, a number of mathematical models have been proposed that depend on effects other than diffusion (Swindale 1980; Belintsev et al. 1987; Budrene and Berg 1995; Wearing et al. 2000). However, in all these models, LALI is the anticipated set of conditions sufficient to form a periodic pattern, and the patterning ability is similar. Therefore, these models are also called LALI models (Oster 1988).

Meanwhile, experimental researchers have pointed out the problems that arise when the LALI model is used as a working hypothesis. For example, even when different LALI models are based on different cellular and molecular functions, the patterns produced are nearly identical (Oster 1988). As such, model simulations rarely help to identify detailed molecular mechanisms (Economou and Green 2014). Even if the pattern-forming phenomenon is successfully reproduced by the simulation of a reaction–diffusion system, this in itself does not guarantee the involvement of diffusion. In most experimental settings, the key molecular event governing the phenomenon is unknown at the start of the study, which presents a serious problem.

It has also become clear that the existing LALI models are, in some cases, not suitable to represent real biological phenomena. Cell migration and apoptosis induced by direct physical interactions of cell projections are key factors in the formation of skin pigmentation patterns in zebrafish (Inaba et al. 2012; Yamanaka and Kondo 2014; Hamada et al. 2014). This is not the only case in which the key signals for pattern formation are transferred not by diffusion, but by fine cell projections such as filopodia (De Joussineau et al. 2003; Sagar et al. 2015; Vasilopoulos and Painter 2016), which may be essentially different from signaling by diffusion. During diffusion, the concentration of the substance is highest at the position of the source cell and rapidly decreases depending on the distance from the source. Therefore, it is difficult to use diffusion to model the condition in which the functional level of the signal has a sharp peak at a location distant from the source (Fig. 9.4a). As each LALI model is restricted by its assumed signaling mechanism, it is difficult to adapt a model to an arbitrary stimulation distance profile of a real system.

To address these issues, Kondo proposed an extension of the Turing model using kernel functions (Kondo 2017). In this alternative model, the dynamics of the system are represented in the form of kernels rather than partial differential equations (Fig. 9.4b). A kernel is a quantification of the distance-dependent effect of adjacent regions, and the kernel's shape determines how the system will behave. The Fourier transform of the kernel's shape reveals the stable (amplified) wavelengths, so that the patterns can be identified without actually calculating the model in a two-dimensional plane. The LALI condition can be considered a kernel shape that

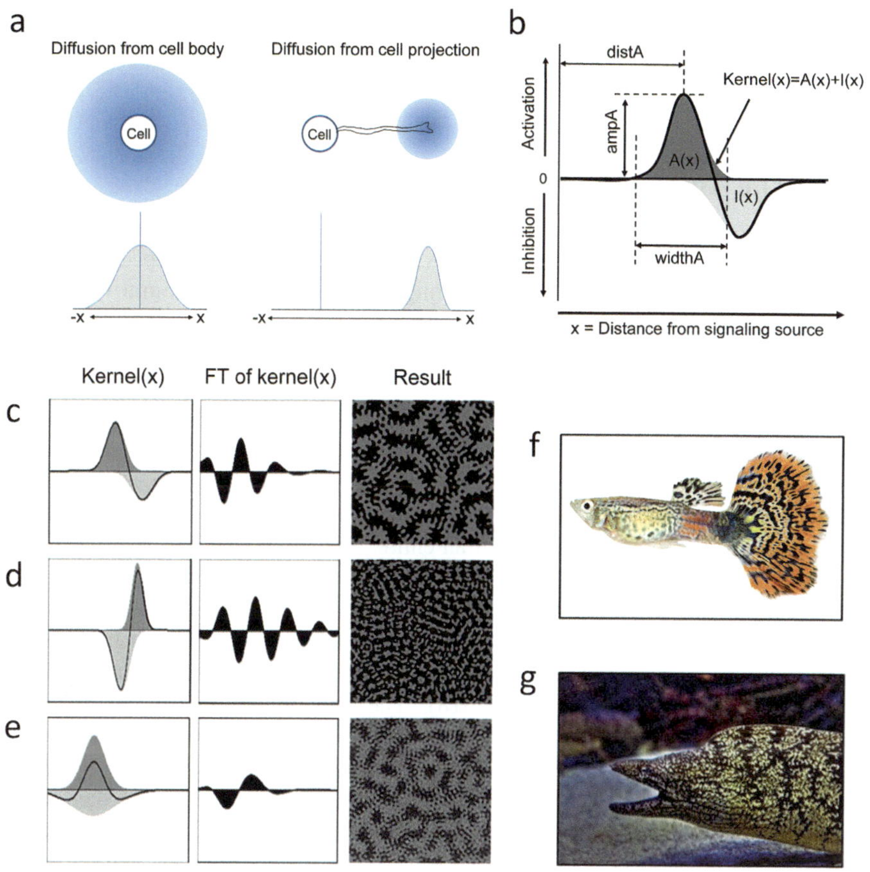

Fig. 9.4 Kernel-based Turing model. (**a**) Interaction strength profiles depend on the method of signal transfer. (**b**) Definition of the kernel shape. (**c–e**) Three different types of kernels, their Fourier transforms, and the resulting patterns. The model can generate complex spatial patterns (e.g., nested patterns) that conventional reaction–diffusion models cannot. (**f**) An artificial line of guppy. (**g**) A moray eel. (Figure modified from Kondo 2017)

makes stable waves. Thus, one simple method to generalize the conventional Turing or LALI models would be to directly input an arbitrary kernel shape not based on the assumption of any concrete molecular or cellular events. Kondo called this model the kernel-based Turing model (KT model).

The KT model can generate standard variations of stable two-dimensional patterns (spots, stripes, and networks) and more complex patterns that are difficult to generate with conventional mathematical models (Fig. 9.4c–g). As the KT model is not based on any specific behavior of molecules or cells, it is more abstract than the pre-existing mathematical models. However, it is useful in practice because the shape of the interaction kernel can be easily measured by a simple experiment, even when the detailed mechanisms are not well understood. These properties of the KT model complement the shortcomings of conventional models and may contribute to the understanding of biological pattern formation.

9.5 The Agent-Based Model

In the KT model introduced in the previous section, arbitrarily shaped kernels are used to investigate what patterns might emerge from arbitrarily defined (or experimentally detected) interaction dynamics. The agent-based model (Railsback and Grimm 2019) and similar models represent different approaches to explore what patterns might emerge from arbitrarily configured (or experimentally determined) interactions. In agent-based models, situations in which multiple agents (individuals, cells, etc.) act and interact with each other are simulated. Only the properties of the individual agents and the interactions between agents and/or between agents and fields are set, with an attempt to reproduce, understand, and predict how the complex behavior of an entire population emerges from these rules in a bottom-up manner.

Since the late 2000s, several studies have used agent-based and similar models aimed at reproducing the formation of pigment patterns in zebrafish (Caicedo-Carvajal and Shinbrot 2008; Bullara and De Decker 2015; Volkening and Sandstede 2015; Volkening and Sandstede 2018). The agent-based model developed by Volkening and Sandstede defined three types of cellular agents (melanophores, xanthophores, and iridophores) and their subclasses, and set complex interaction rules depending on the type of cellular agents and the distance between them based on empirical data and assumptions (Volkening and Sandstede 2018) (Fig. 9.5a). The model reproduces not only the developmental process of the pigment patterns seen in the zebrafish, but also the changes that occur when disturbances are applied, such as cell removal by laser ablation (Fig. 9.5b, c). A variety of pattern mutants are known in zebrafish, and the genes responsible for these mutations have been successively identified (see Chap. 8). Volkening and Sandstede have shown that the patterns found in these mutant lines can be reproduced by modifying or constraining the model conditions in various ways, such as eliminating one or more cell types (Fig. 9.5d). They also successfully simulated different pigment patterns (a central light stripe, light spots) found in closely related species of zebrafish, such as *Danio albolineatus* and *Danio margaritatus* (Fig. 9.5d).

The fact that the model is able to reproduce the patterns found in real animals may suggest that rules other than those set on the basis of empirical data, that is, arbitrary rules added to the model based on assumptions, are also valid and effective for pattern formation of real organisms. Agent-based models are easy to use and friendly mathematical models for experimental biologists, because they describe specific rules for each discrete agent such as cells. This makes it straightforward to understand how these rules correspond to experiments on actual organisms. If, as Volkening and Sandstede attempted, the model goes beyond reproducing known

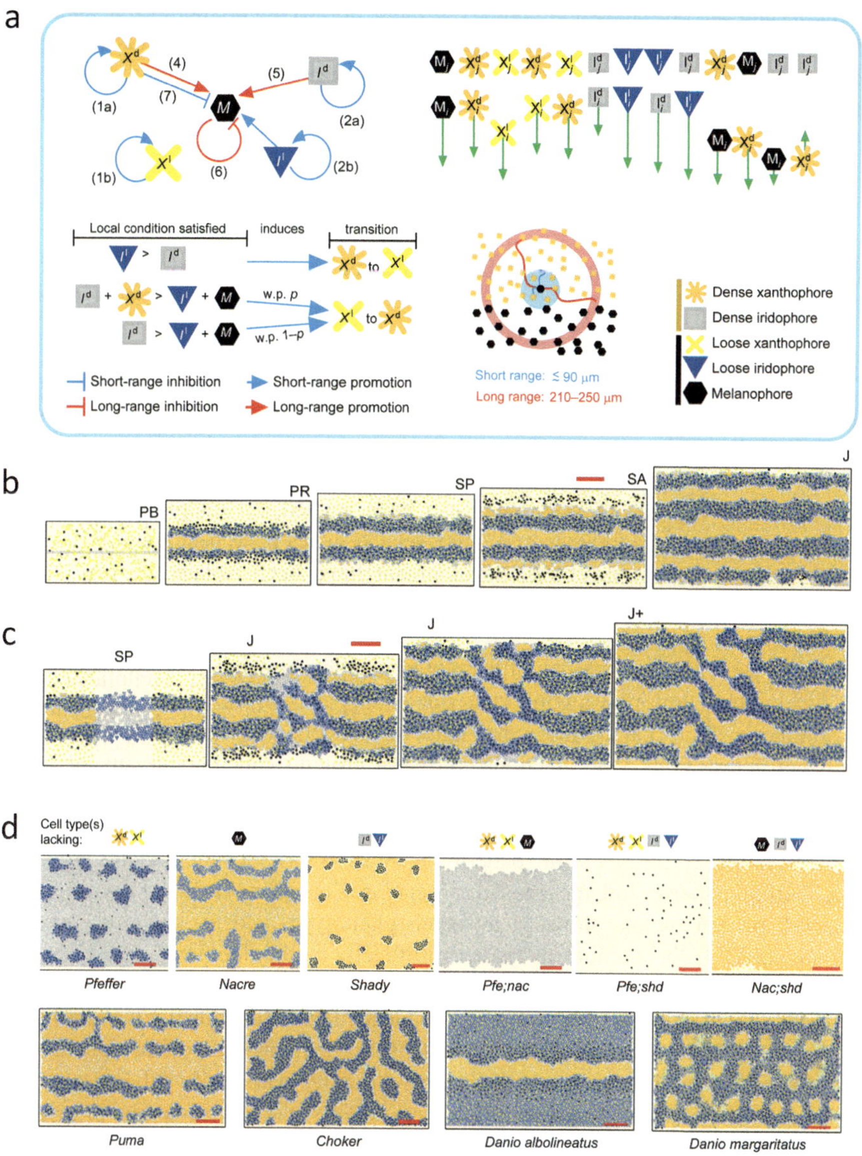

Fig. 9.5 An agent-based model describing the pigment pattern formation of zebrafish. (**a**) Model overview. The model defines five types of cell agents and sets rules of interaction according to the cell types and the distance between them. (**b**) Model simulation of the development of wild-type zebrafish. (**c**) Model simulation of the cell ablation experiment. (**d**) Patterns generated by the model under various conditions. The model reproduces the patterns of the mutant lineages and closely related species of zebrafish. (Figure modified from Volkening and Sandstede 2018)

patterns and can present experimentally testable, non-trivial predictions, it could be a very useful approach to address ongoing debates among experimental biologists, such as whether iridophores, or a subclass of them, play a central role in the formation of zebrafish stripe patterns (Mahalwar et al. 2014; Singh et al. 2015; Watanabe and Kondo 2015a, b).

9.6 Connecting Mathematical Models and Molecular Biology

With the development of molecular biology in the last 30 years, many events during growth and development have been clarified at the molecular level. Regarding pigment cells, as described in Chaps. 1–8, 10 many genes involved in development, differentiation, migration, cell death, and lesions have been isolated and analyzed. Molecular biological analyses have also been carried out on pigment cells and pattern formation in zebrafish, and many related genes have been identified. Some genes involved in pigment cell development have been isolated as mammalian homologs, whereas many molecules and genes involved in pattern formation are rather unique.

Here, we will introduce one of them, gap junction proteins. As described above, cell–cell interactions play an important role in the pattern formation of pigment cells. Gap junctions are channel molecules that connect cells and are directly involved in cell–cell interactions. Gap junctions consist of proteins called connexins, of which two types, Connexin 41.8 and Connexin 39.4, are expressed in zebrafish pigment cells (Watanabe et al. 2006; Irion et al. 2014). Zebrafish usually show a striped pattern; however, the "*leopard*" pattern mutant shows a spot pattern that is typical of the Turing pattern, and the causative gene is Connexin 41.8. Additionally, the "*luchs*" pattern mutant shows a labyrinth phenotype, and the causative gene is Connexin 39.4. An attempt was made to artificially modify the function of Connexin 41.8 as a channel using molecular techniques. Gene modifications produced three types of gap junction mutants: a mutant that lost channel function but could function as an intercellular adhesion factor, a mutant that exhibited abnormal opening and closing of the channel, and a mutant that had difficulty in controlling the opening and closing of the channel. Expressing these artificially mutated connexins in pigment cells caused pattern changes (Fig. 9.6). The fish showed typical Turing patterns on their bodies, as predicted by simulations (Watanabe and Kondo 2012). As can be seen from the figure, the cell–cell interaction via the gap junction is closely related to the formation of the Turing pattern.

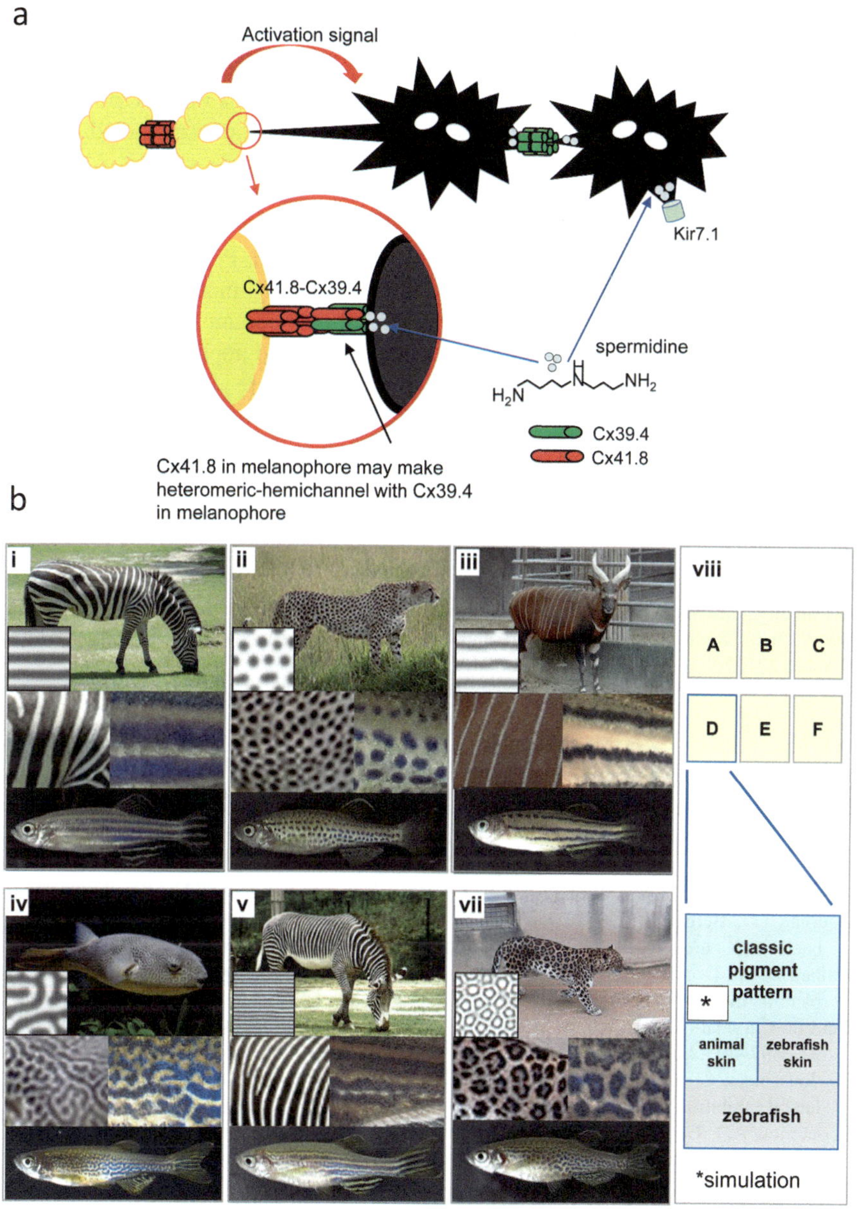

Fig. 9.6 (**a**) Gap junction network model. Stripe patterns may be formed if Connexin 39.4 is expressed in melanophores and Connexin 41.8 is expressed in xanthophores. Spermidine in melanophores inhibits outward flow from melanophores through gap junctions (Modified from Usui et al. 2019). (**b**) Artificially produced Turing pattern on fish skin (Modified from Watanabe and Kondo 2012)

9.7 Conclusions

In this chapter, we have introduced a few theoretical models of pattern formation mechanisms that are considered to be of particular importance, including their relationship to pigment patterns found in real organisms.

Models that reproduce the exquisite and diverse patterns found in organisms are beautiful in their own right. However, as the often-quoted and famous saying goes, "All models are wrong, but some are useful" (George E. P. Box). A model that reproduces every minute detail is not necessarily a "true" model. Nor is building such a model in and of itself our main goal. What is important is how the model is used to obtain new knowledge and testable predictions and implications. To achieve this, it is essential to keep the model as simple as possible, to always be aware of the limitations of the model, and for theoretical biologists to work closely with experimental biologists. In this sense, the most desirable and ideal situation is that a researcher is both an experimental biologist and a theoretical biologist. In the end, we hope that this chapter will help those who are interested in the experimental and theoretical study of pigment patterns to take their first step towards this goal.

Acknowledgements This work was supported by JSPS KAKENHI Grant Numbers JP15H04415 and JP19H03283 (to SM) and 15KT0079 and 17H03683 (to MW), by the MEXT KAKENHI Grant Number 22127003 (to SK), and by the CREST program (to SK).

References

Belintsev BN, Beloussov LV, Zaraisky AG (1987) Model of pattern formation in epithelial morphogenesis. J Theor Biol 129:369–394. https://doi.org/10.1016/S0022-5193(87)80019-X

Budrene EO, Berg HC (1995) Dynamics of formation of symmetrical patterns by chemotactic bacteria. Nature 376:49–53. https://doi.org/10.1038/376049a0

Bullara D, De Decker Y (2015) Pigment cell movement is not required for generation of Turing patterns in zebrafish skin. Nat Commun 6:1–7. https://doi.org/10.1038/ncomms7971

Caicedo-Carvajal CE, Shinbrot T (2008) In silico zebrafish pattern formation. Dev Biol 315:397–403. https://doi.org/10.1016/j.ydbio.2007.12.036

De Joussineau C, Soulé J, Martin M et al (2003) Delta-promoted filopodia mediate long-range lateral inhibition in Drosophila. Nature 426:555–559. https://doi.org/10.1038/nature02157

Economou AD, Green JBA (2014) Modelling from the experimental developmental biologists viewpoint. Semin Cell Dev Biol 35:58–65. https://doi.org/10.1016/j.semcdb.2014.07.006

Gierer A, Meinhardt H (1972) A theory of biological pattern formation. Biol Cybern 12:30–39

Grant PR, Grant BR (1992) Hybridization of bird species. Science 256:193

Gray P, Scott SK (1985) Sustained oscillations and other exotic patterns of behavior in isothermal reactions. J Phys Chem 89:22–32

Hamada H, Watanabe M, Lau HE et al (2014) Involvement of Delta/Notch signaling in zebrafish adult pigment stripe patterning. Development 141:1418. https://doi.org/10.1242/dev.108894

Hubbs CL (1955) Hybridization between fish species in nature. Syst Zool 4:1–20

Inaba M, Yamanaka H, Kondo S (2012) Pigment pattern formation by contact-dependent depolarization. Science 335:677. https://doi.org/10.1126/science.1212821

Irion U, Frohnhöfer HG, Krauss J et al (2014) Gap junctions composed of connexins 41.8 and 39.4 are essential for colour pattern formation in zebrafish. eLife 3:e05125. https://doi.org/10.7554/eLife.05125

Kondo S (2017) An updated kernel-based Turing model for studying the mechanisms of biological pattern formation. J Theor Biol 414:120–127. https://doi.org/10.1016/j.jtbi.2016.11.003

Kondo S, Asai R (1995) A reaction-diffusion wave on the skin of the marine angelfish Pomacanthus. Nature 376:765–768

Kondo S, Miura T (2010) Reaction-diffusion model as a framework for understanding biological pattern formation. Science 329:1616–1620. https://doi.org/10.1126/science.1179047

Mahalwar P, Walderich B, Singh AP, Volhard CN (2014) Local reorganization of xanthophores fine-tunes and colors the striped pattern of zebrafish. Science 345:1362–1364. https://doi.org/10.1126/science.1254837

Mavárez J, Salazar CA, Bermingham E et al (2006) Speciation by hybridization in *Heliconius* butterflies. Nature 441:868–871

Meinhardt H (1982) Models of biological pattern formation. Academic Press, London

Meinhardt H, Gierer A (2000) Pattern formation by local self-activation and lateral inhibition. BioEssays 22:753–760

Miyazawa S (2020) Pattern blending enriches the diversity of animal colorations. Sci Adv 6(49): eabb9107. https://doi.org/10.1126/sciadv.abb9107

Miyazawa S, Okamoto M, Kondo S (2010) Blending of animal colour patterns by hybridization. Nat Commun 1:66. https://doi.org/10.1038/ncomms1071

Murray JD (1981) A pre-pattern formation mechanism for animal coat markings. J Theor Biol 88:161–199. https://doi.org/10.1016/0022-5193(81)90334-9

Murray JD (2002) Mathematical biology, 3rd edn. Springer-Verlag, New York

Oster GF (1988) Lateral inhibition models of developmental processes. Math Biosci 90:265–286. https://doi.org/10.1016/0895-7177(89)90248-3

Prud'homme B, Gompel N, Carroll SB (2007) Emerging principles of regulatory evolution. Proc Natl Acad Sci U S A 104(Suppl):8605–8612. https://doi.org/10.1073/pnas.0700488104

Railsback SF, Grimm V (2019) Agent-based and individual-based modeling: a practical introduction, 2nd edn. Princeton University Press, Princeton, NJ

Rieseberg LH, Archer MA, Wayne RK (1999) Transgressive segregation, adaptation and speciation. Heredity 83(Pt 4):363–372

Sagar PF, Wiegreffe C, Scaal M (2015) Communication between distant epithelial cells by filopodia-like protrusions during embryonic development. Development 142:665–671. https://doi.org/10.1242/dev.115964

Singh AP, Frohnhöfer HG, Irion U, Nüsslein-Volhard C (2015) Response to comment on "Local reorganization of xanthophores fine-tunes and colors the striped pattern of zebrafish". Science 348:297. https://doi.org/10.1126/science.aaa2804

Swindale NV (1980) A model for the formation of ocular dominance stripes. Proc R Soc Lond Biol Sci 208:243–264. https://doi.org/10.1098/rspb.1980.0051

Turing AM (1952) The chemical basis of morphogenesis. Philos Trans R Soc Lond B 237:37–72. https://doi.org/10.1007/BF02459572

Usui Y, Aramaki T, Kondo S, Watanabe M (2019) The minimal gap-junction network among melanophores and xanthophores required for stripe pattern formation in zebrafish. Development 146, dev181065. https://doi.org/10.1242/dev.181065

Vasilopoulos G, Painter KJ (2016) Pattern formation in discrete cell tissues under long range filopodia-based direct cell to cell contact. Math Biosci 273:1–15. https://doi.org/10.1016/j.mbs.2015.12.008

Volkening A, Sandstede B (2015) Modelling stripe formation in zebrafish: an agent-based approach. J R Soc Interface 12(112):20150812. https://doi.org/10.1098/rsif.2015.0812

Volkening A, Sandstede B (2018) Iridophores as a source of robustness in zebrafish stripes and variability in Danio patterns. Nat Commun 9:1–14. https://doi.org/10.1038/s41467-018-05629-z

Watanabe M, Kondo S (2012) Changing clothes easily: connexin41.8 regulates skin pattern variation. Pigment Cell Melanoma Res 25:326–330. https://doi.org/10.1111/j.1755-148X.2012.00984.x

Watanabe M, Kondo S (2015a) Is pigment patterning in fish skin determined by the Turing mechanism? Trends Genet 31:88–96. https://doi.org/10.1016/j.tig.2014.11.005

Watanabe M, Kondo S (2015b) Comment on "Local reorganization of xanthophores fine-tunes and colors the striped pattern of zebrafish". Science 348:297. https://doi.org/10.1126/science.aaa2804

Watanabe M, Iwashita M, Ishii M et al (2006) Spot pattern of leopard Danio is caused by mutation in the zebrafish connexin41. 8 gene. EMBO Rep 7:893–897

Wearing HJ, Owen MR, Sherratt JA (2000) Mathematical modelling of juxtacrine patterning. Bull Math Biol 62:293–320. https://doi.org/10.1006/bulm.1999.0152

Werner T, Koshikawa S, Williams TM, Carroll SB (2010) Generation of a novel wing colour pattern by the Wingless morphogen. Nature 464:1143–1148

Wolpert L (1969) Positional information and the spatial pattern of cellular differentiation. J Theor Biol 25:1–47

Yamanaka H, Kondo S (2014) In vitro analysis suggests that difference in cell movement during direct interaction can generate various pigment patterns in vivo. Proc Natl Acad Sci U S A 111:1867–1872. https://doi.org/10.1073/pnas.1315416111

Chapter 10
Evolution of Pigment Pattern Formation in Teleosts

David M. Parichy and Yipeng Liang

Abstract Teleosts comprise roughly half of all vertebrates and exhibit a diverse array of pigment cells and pigment patterns that function in behaviors ranging from camouflage to mate choice. This chapter examines recent advances in understanding the developmental genetics of teleost pigment cells and pattern formation, and how comparative studies across species and populations are providing insights into fundamental aspects of vertebrate development and evolution.

Keywords Chromatophore · Neural crest · Convergence · Diversification · Heterochrony · Melanophore · Iridophore · Xanthophore · Leucophore · Cichlid fishes · Zebrafish · Cavefish · Stickleback · Clownfish

10.1 Introduction

No vertebrates have pigment patterns more spectacularly diverse than those of teleost fishes (Fig. 10.1). They come in a wide variety of forms and can evolve over short periods of time. Long known for their beauty, these patterns are increasingly recognized for the insights they provide into mechanisms of differentiation and morphogenesis. Substantial recent progress also has been made towards understanding the developmental evolution of pigment patterns, and these efforts allow for interesting comparisons to other vertebrates while pointing the way towards exciting areas of future exploration.

This chapter addresses several issues fundamental to understanding pigment pattern evolution, namely the functional significance of these patterns, and generalities of how they develop, including cell lineage origins and patterning mechanisms. With this background, the chapter then considers some of the outstanding questions

D. M. Parichy (✉) · Y. Liang
Department of Biology and Department of Cell Biology, University of Virginia, Charlottesville, VA, USA
e-mail: dparichy@virginia.edu

H. Hashimoto et al. (eds.), *Pigments, Pigment Cells and Pigment Patterns*,
https://doi.org/10.1007/978-981-16-1490-3_10

Fig. 10.1 Diverse pigment patterns of teleost fishes. Shown are a variety of species from marine and freshwater environments, exhibiting many different pattern features that depend on the numbers and arrangements of melanophores, xanthophores, iridophores, and other chromatophore classes (Schartl et al. 2016)

about pigment cell and pigment pattern evolution and how recent studies, especially those of emerging model species, are beginning to provide answers.

10.2 Pigment Pattern Function

The specific functions of teleost pigment patterns are likely to be almost as diverse as the patterns themselves. Broadly, the contribution of a pattern to individual fitness can be mediated through effects on survival or reproduction (Endler 1978; Rosenthal 2007; Price et al. 2008; Marshall et al. 2018). Some patterns help to avoid—or accomplish—predation by providing camouflage. Some warn, confuse, or misdirect predators. In other instances, patterns are visual signals for conspecifics, aiding in species recognition and social aggregation, aggressive displays, or choice of mate

(Hemingson et al. 2019). Often patterns change ontogenetically to match alterations in environment or social factors. Such changes can be seasonal, and either sex-specific or sex-independent. Reflecting alterations in pigment quantity and pigment cell number, and occurring over relatively long periods of time (hours to months), these modifications are often referred to as "morphological color change." Visual appearance also can be regulated at even finer temporal scales (milliseconds to hour), as some pigment cells—chromatophores—of fishes can disperse or contract their pigment granules in response to physiological and neuroendocrine stimuli, a phenomenon referred to as "physiological color change" (Fujii 1993). These changes can affect overall brightness and hue, enabling crypsis in the context of particular light intensities and wavelengths, as well as substrates or backgrounds.

Selection on pigmentation is well known, but in many instances response to selection—and phenotypes observed—likely depends not only on color and pattern but also how these features interact with individual behavior and specifics of the environment in which they are displayed. A classic example is the complex set of tradeoffs found for guppies, in which conspicuousness to potential mates is balanced against conspicuousness to predators and other factors (Endler 1983; Houde 1997). This implies that functional consequences, and therefore selective bases, can be difficult to ascertain for any specific pattern. Moreover, it seems likely there are often different, but equally good, pigmentation "solutions" to any given set of functional roles. Any particular outcome may therefore depend on both chance and the starting developmental and genetic milieu, and this makes inferences about functional significance of specific pattern features difficult at best. Of course some patterns also may be entirely non-adaptive, arising as byproducts of other developmental processes and gene activities.

These considerations suggest caution when interpreting pigmentation phenotypes. But they also highlight the richness of the system for work at multiple levels of biological hierarchy. With the tremendous diversity of teleost pigmentation, and the rapidity of its evolution, mechanistic studies of how genes and cell behaviors evolve to produce divergent, or convergent, phenotypes can provide important insights into the origins of form, and how such forms have been selected (or not) in the natural world.

10.3 Chromatophores: Development and Pattern Formation

There is a long history of research on the pigmentation of aquatic vertebrates [e.g., reviewed by: (Parichy 1996; Parichy et al. 2006)]. Indeed, studies of aquatic larvae—salamanders rather than fishes—provided early indications that vertebrate skin pigment cells are derived from embryonic neural crest cells (Dushane 1934), which arise at the border between neural and non-neural ectoderm and then migrate throughout the embryo, also contributing to bone and cartilage of the craniofacial skeleton, glia and neurons of the peripheral nervous system, and a variety of other tissues and organs (Hörstadius 1950; Dupin et al. 2018; Le Douarin and Dupin 2018). Developmental studies of amphibian pigmentation continue to provide insights, including some potentially relevant to understanding pigmentation in fishes (Fukuzawa 2010; Nakayama et al. 2017; Woodcock et al. 2017). Nevertheless, most recent progress on mechanisms of pigmentation in ectothermic vertebrates comes from teleosts, especially the minnow zebrafish (*Danio rerio*) and the killifish medaka (*Oryzias latipes*).

10.3.1 Chromatophores and Pigmentation of Early Larvae

Newly hatched zebrafish (Fig. 10.2a) and medaka have similar pigmentation (Kimmel et al. 1995; Kelsh et al. 1996; Odenthal et al. 1996b; Lamoreux et al. 2005). In both species, migratory neural crest cells begin to differentiate as chromatophores of several classes (Fujii 1993; Schartl et al. 2016). Melanophores contain melanin in specialized organelles (melanosomes) that are typically retained by the fully differentiated cell, in contrast to mammalian and avian melanocytes, the homologous cells to melanophores, that often transfer melanin granules to keratinocytes for deposition in feathers or hair (Tadokoro et al. 2019). By 3 days post-fertilization, embryonic/early larval melanophores are found along the dorsal and ventral myotomes, above and below the yolk sac, on the head, and sparsely along the horizontal myosepta in the middle of the flank.

Iridophores are a second class of chromatophore. These cells are iridescent owing to stacks of membrane-bound reflecting platelets formed from guanine crystals that confer different hues depending on their arrangement and the angle of observation (Gur et al. 2016, 2020; Kimura et al. 2017). At early larval stages of zebrafish, iridophores are found principally in association with melanophores.

Xanthophores, a third chromatophore class, are yellow or yellow-orange and can be difficult to discern individually. But as a population, xanthophores confer a yellowish cast to the flank where they are abundant. The color of xanthophores at embryonic/early larval stages of zebrafish depends on pteridine pigments that are localized in pterinosomes (Odenthal et al. 1996a; Ziegler et al. 2000; Ziegler 2003; Lister 2019). Additional pteridines are colorless in the visible spectrum but might be apparent to fish, which can be seen in the ultraviolet range (Losey et al. 1999). Xanthophore color also can arise from dietary carotenoids concentrated in lipid droplets, or "carotenoid vesicles." In zebrafish, the yellow-orange color of adult xanthophores results principally from carotenoids rather than pteridines (Granneman et al. 2017; Saunders et al. 2019).

Finally, leucophores comprise a fourth class of chromatophore and have a white or yellow-white appearance. These cells occur in early larvae of medaka though not early larvae of zebrafish [(Lamoreux et al. 2005; Kimura et al. 2014); but see below].

How and when different chromatophore classes develop from neural crest progenitors has been studied extensively, and features of the essential gene regulatory

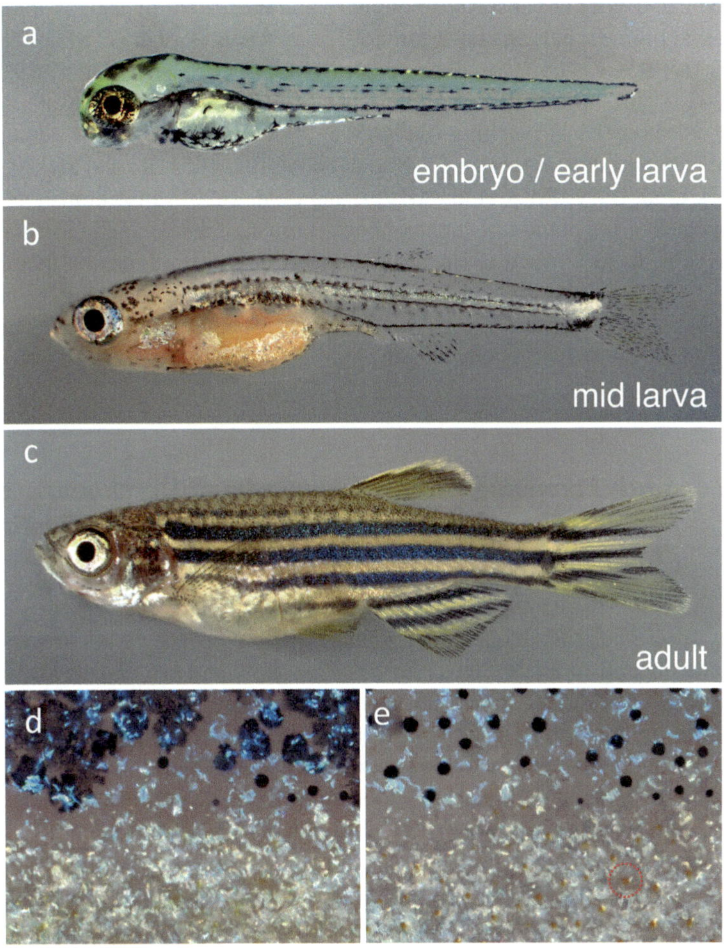

Fig. 10.2 Development of pigment pattern in zebrafish. (**a**) Embryo/early larva. (**b**) Middle stage larva during larva-to-adult transformation when new chromatophores arise from latent neural crest derived progenitors. (**c**) Adult illustrating alternating light interstripes and dark stripes. (**d**, **e**) Closeup of boundary between stripe and interstripe. Shown is the same view before (**d**) and after (**e**) treatment with epinephrine to retract pigment granules towards cell centers. Individual melanophores can be discerned by spots of melanin, iridophores by their iridescence (bluish in stripe regions, yellowish in interstripes), and xanthophores by concentrations of orange carotenoids (dashed red circle) (Images courtesy Emily Bain)

networks are beginning to be uncovered, including signaling pathways and transcription factors that contribute to fate specification and differentiation (Parichy et al. 2000b; Dutton et al. 2001; Elworthy et al. 2003; Minchin and Hughes 2008; Curran et al. 2009, 2010; Greenhill et al. 2011; Nagao et al. 2014, 2018; Nord et al. 2016; Petratou et al. 2018). What is considerably less well understood are factors that promote the migration and localization of different chromatophore classes at embryonic and early larval stages, though a small handful of mutants with defective early larval pigment patterns have provided some clues (Parichy et al. 1999; Svetic et al. 2007; Fadeev et al. 2016; Zhang et al. 2018; Camargo Sosa et al. 2019; Eskova et al. 2020). Although patterns reminiscent of zebrafish or medaka are common among other species at hatching (Quigley et al. 2004; Li et al. 2015), many other types of patterns occur as well [e.g., (Budd 1940; Orton 1953; Ré and Meneses 2008; Hendrick et al. 2019; Roux et al. 2019; Liang et al. 2020)]. In zebrafish and likely other species, a population of melanophores protects hematopoietic stem and progenitor cells from ultraviolet irradiation during early development (Kapp et al. 2018). It will be interesting to learn if neural and other stem cells are similarly protected, what other adaptive significance these patterns might have, and how different pigment cell arrangements develop.

10.3.2 Adult Pigment Pattern Formation and Developmental Features that May Contribute to Evolutionary Lability of Pattern

In many species, pigmentation changes markedly between embryonic or early larval stages and juvenile or adult stages (Parichy et al. 2009; Baldwin 2013; McMenamin and Parichy 2013; Holzer et al. 2017; Hendrick et al. 2019; Roux et al. 2019; Liang et al. 2020). In principle, such remodeling could depend on changes in the arrangements or numbers of chromatophores that developed from the embryonic neural crest, if these cells reinitiate migration to localize at new sites, proliferate differentially across chromatophore classes or body regions, or undergo programmed cell death to sculpt new patterns out of old. Alternatively, development of post-embryonic patterns could require contributions from newly differentiating chromatophores, not present at embryonic or early larval stages.

In zebrafish, both old and new chromatophores contribute to the adult pattern, which forms between 1 and 6 weeks post-fertilization (Parichy and Turner 2003c; Patterson and Parichy 2013, 2019; Patterson et al. 2014). As early larvae begin to feed and grow, embryonic xanthophores continue to proliferate across the flank (Walderich et al. 2016) but also gradually lose their yellow color (McMenamin et al. 2014). New adult iridophores begin to differentiate near the horizontal myoseptum where they, too, proliferate to form an epithelial-like mat that will contribute to the first light “interstripe” of the adult (Fig. 10.2b) (Patterson and Parichy 2013; Patterson et al. 2014; Singh et al. 2014). Newly differentiating melanophores also become visible during these stages, some in the nascent interstripe and others in prospective stripe regions.

As the larva-to-adult transformation progresses, some new melanophores, as well as melanophores that had differentiated in the embryo, migrate from the interstripe to prospective stripe regions [reviewed in (Patterson and Parichy 2019)]. Other melanophores initially in the interstripe die or are covered by proliferative iridophores

(Parichy et al. 2000b; Parichy and Turner 2003c; Takahashi and Kondo 2008). Simultaneously, some new iridophores develop within the darker stripes, where they remain sparse (Singh et al. 2014; Spiewak et al. 2018; Gur et al. 2020), and "depigmented" embryonic xanthophores begin to accumulate yellow-orange carotenoids while also increasing their densities within the interstripes (Mahalwar et al. 2014; McMenamin et al. 2014; Saunders et al. 2019). Some "new" xanthophores differentiate as well. These events of interstripe and stripe formation are then reiterated as zebrafish grow to an adult size, ultimately yielding 4–5 dark stripes with intervening light interstripes (Fig. 10.2c–e).

The source of "new" chromatophores during the larva-to-adult transformation of zebrafish is a population of latent progenitor cells that are themselves neural crest derived. During early neural crest migration, on the first day of embryonic development, some neural crest cells localize within dorsal root ganglia and likely other sites within the peripheral nervous system, where they proliferate but remain undifferentiated (Budi et al. 2008, 2011; Tryon et al. 2011; Dooley et al. 2013; O'Reilly-Pol and Johnson 2013). These progenitors express genes typical of embryonic neural crest (Saunders et al. 2019) and at least some are multipotent, in being able to contribute progeny to more than a single chromatophore fate (Singh et al. 2016). As post-embryonic pigmentation develops, some of these "deep" progenitors are recruited to the skin where they differentiate as adult chromatophores. Chromatophore progenitors also occur within the skin at later stages, where they may contribute to pattern homeostasis as well as regeneration of pattern after injury (Parichy and Turner 2003b; Yamaguchi et al. 2007; O'Reilly-Pol and Johnson 2008; Nakamasu et al. 2009; Iyengar et al. 2015). Newly differentiating pigment cells arise during larva-to-adult transformations of other teleosts and amphibians as well, presumably from similar pools of latent progenitors (Parichy 1998; Watanabe et al. 2008; Yamada et al. 2010; Hendrick et al. 2019; Roux et al. 2019; Liang et al. 2020).

It is tempting to speculate that distinct pools of embryonic and post-embryonic progenitors have allowed patterns to diversify across life cycle stages in ways that might not otherwise have been evolutionarily attainable, owing to pleiotropy and developmental or selective tradeoffs across stages (Parichy 1998; Quigley et al. 2004). Consistent with this idea, mutational analyses have revealed some genes required by chromatophores of one pool but not the other (Parichy et al. 2000a; Iwashita et al. 2006; Watanabe et al. 2006; Budi et al. 2008; Eom et al. 2012; Irion et al. 2014). Nevertheless, transcriptomic signatures within chromatophore classes are largely indistinguishable between cells of embryonic and post-embryonic origins, at least at the resolutions thus far attained (Saunders et al. 2019). The degree to which genes and gene regulatory programs are decoupled between chromatophores derived directly from the neural crest and chromatophores derived from peripheral progenitors remains an open question, but one whose answer should be broadly informative for understanding mechanisms of pattern diversification.

Beyond the cell lineage origins of chromatophores, considerable attention has been given to the manner in which these cells achieve a pattern of interstripes and stripes. In general, such mechanisms can be conceptualized as depending on interactions between chromatophores and non-chromatophore cell types in their tissue environment, or interactions between chromatophores and other chromatophores. That cells other than chromatophores are essential for morphogenesis and differentiation is not surprising and is demonstrated by aberrant patterns of zebrafish mutants deficient for specific types of skin cells or signaling factors of the tissue environment (Lang et al. 2009; Dooley et al. 2013; Patterson and Parichy 2013; McMenamin et al. 2014; Mo et al. 2017; Fadeev et al. 2018; Spiewak et al. 2018), as well as mutants that have malformations in anatomical features (myosepta and nerves) that normally guide deep progenitors to the skin (Frohnhofer et al. 2013; Parichy and Spiewak 2015). Effects of tissue environment documented so far appear to be permissive; i.e., allowing pattern to form, but not specifying particular features of the pattern itself. It remains to be seen whether more "instructive" cellular or molecular cues contribute to pattern initiation or reiteration.

By contrast, interactions within and between chromatophore classes play roles that are both essential and instructive to pattern outcome. The requirement for interactions of some sort between chromatophore classes is evident from mutants that lack any one class, as the remaining cells fail to arrange themselves into normally positioned interstripes and stripes (Fig. 10.3a, b) (Lister et al. 1999; Parichy et al. 2000b; Lopes et al. 2008; Frohnhofer et al. 2013; Patterson and Parichy 2013). Similar outcomes are observed when chromatophore types are ablated using conditional genetic approaches or laser-targeting (Parichy and Turner 2003a; Yamaguchi et al. 2007; Nakamasu et al. 2009; Gur et al. 2020).

Recent analyses have started to identify mechanistic bases for these interactions and have implicated signaling pathways, cell adhesive interactions, and electrophysiological gap junctional communication within and between chromatophore types [e.g., (Eom et al. 2012, 2015; Inaba et al. 2012; Watanabe et al. 2012; Hamada et al. 2014; Inoue et al. 2014; Irion et al. 2014; Fadeev et al. 2015; Mahalwar et al. 2016; Eom and Parichy 2017; Usui et al. 2019); reviewed in: (Watanabe et al. 2012; Watanabe and Kondo 2015; Irion et al. 2016; Patterson and Parichy 2019)]. Some aspects of these interactions have dynamics resembling those predicted for Turing mechanisms of pattern formation (Nakamasu et al. 2009). Interactions have also been modeled using agent-based and other approaches (Caicedo-Carvajal and Shinbrot 2008; Bullara and De Decker 2015; Volkening and Sandstede 2015, 2018; McGuirl et al. 2020; Owen et al. 2020; Volkening et al. 2020).

Collectively, these studies indicate the presence of a dynamic "engine" of interactions, that drives pattern formation during normal development, is able to generate regulative patterns under aberrant conditions in vivo (Parichy and Turner 2003b; Yamaguchi et al. 2007; Nakamasu et al. 2009; Patterson and Parichy 2013; Sawada et al. 2018) and can even yield predictable cellular behaviors in vitro (Inaba et al. 2012; Inoue et al. 2014; Yamanaka and Kondo 2014). The degree to which specific cellular and molecular interactions are altered in the patterning engines of other species remains an open question. The increasingly thorough understanding of pattern-forming mechanisms in zebrafish should provide testable hypotheses for interrogating patterning mechanisms in other species and how they may have diverged from zebrafish.

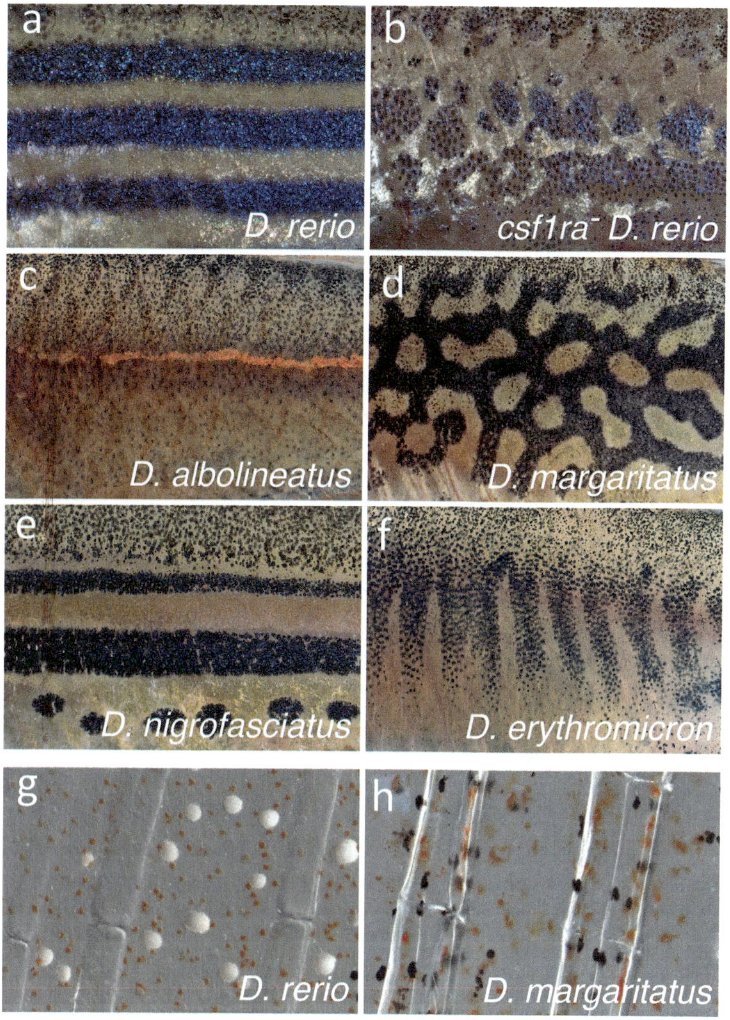

Fig. 10.3 Pattern and chromatophore variation in *Danio*. (**a**) Detail of interstripes and stripes in wild-type zebrafish, *D. rerio*. (**b**) A disrupted pattern develops in zebrafish that lack xanthophores owing to a mutation in the xanthogenic gene *colony stimulating factor 1 receptor a* (*csf1ra*). (**c**) Relatively uniformly distributed chromatophores and residual interstripe in *D. albolineatus*. (**d**) Reticulated pattern with light stripes in *D. margaritatus*. (**e**) Attenuated interstripes and stripes of *D. nigrofasciatus*. (**f**) Vertical bars of *D. erythromicron*. (**g**, **h**) Differential complements of white melanoleucophores in dorsal fins of zebrafish and *D. margaritatus* (Images courtesy Emily Bain, Samantha Sturiale)

10.4 Evolution of Teleost Pigment Cells and Patterns

Pigment pattern formation has been examined at the cellular and genetic levels in only a handful of the >33,000 species of fishes described (fishbase.org): generalities about pattern evolution are likely premature. Nevertheless, recent studies have provided insights into some basic questions relevant to the evolution of teleost pigmentation and highlight areas deserving of additional study. Some representative studies are presented below according to broader insights provided, though some studies might have been placed equally well under multiple subheadings.

10.4.1 Changes Within Pigment Cell Lineages Underlying Adult Pattern Evolution

Because neural crest cells and their progeny migrate so extensively and encounter so many different tissue environments, a long-standing question has been whether evolutionary transformations in neural crest derived traits result from changes that are autonomous or non-autonomous to the lineage itself (Twitty and Bodenstein 1939; Rawles 1948; Schneider and Helms 2003; Quigley et al. 2004; Inaba et al. 2019). Examples of autonomous changes might include the loss of gain of receptors expressed at the surface of these cells, thereby modifying their competence to respond to patterning cues in the tissue environment. Autonomous changes might also correspond to changes in the expression of genes encoding transcription factors that specify particular chromatophore fates.

Cichlid fishes provide a nice example in which a presumptively chromatophore-autonomous change has contributed to naturally occurring pattern variation. Cichlids of the East African Great Lakes are a classic example of adaptive radiation, exhibiting an extraordinary range of behaviors and morphologies, including pigment patterns (Seehausen et al. 2008; Maan and Sefc 2013; Brawand et al. 2014). Many of these species are sexually dimorphic, with males having bright coloration and visually distinctive patterns (like vertical bars), and females having duller, less distinctive patterns. Yet, several species are polymorphic, with some females exhibiting a typical dull pattern (brown barred, *BB*) and other females having conspicuous coloration, comprising dark patches of melanophores on bright backgrounds of orange xanthophores (orange-blotch, *OB*) (Fig. 10.4a). These phenotypes result from alternative alleles of the transcription factor gene *Pax7*, located on the female sex chromosome (Streelman et al. 2003; Roberts et al. 2009, 2017). In zebrafish, Pax7 genes are expressed by cells of the neural crest–xanthophore lineage and are required for xanthophore development (Minchin and Hughes 2008; Nord et al. 2016), consistent with a cell-autonomous role for *Pax7* in the cichlid orange-blotch phenotype. Indeed, cichlid *Pax7* was found to be transcribed at a higher level from the orange-blotch allele than the brown barred allele, strongly suggesting that a *cis*-regulatory variant drives more *Pax7* expression. Remarkably, this same allele

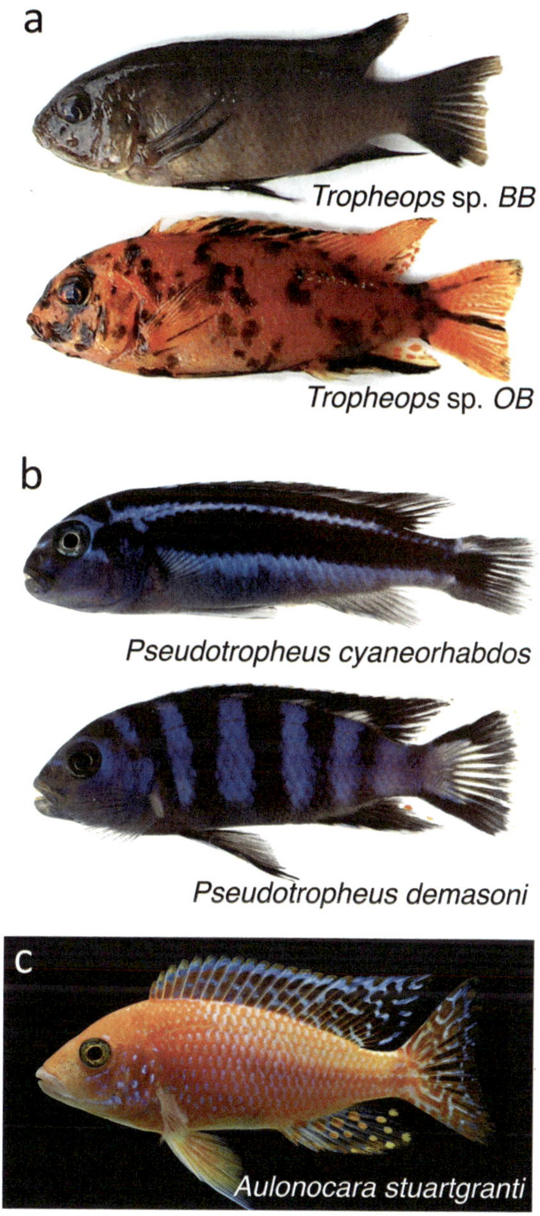

Fig. 10.4 Pigment patterns of selected East African cichlids. (**a**) Examples of brown barred (BB) and orange-blotch (OB) variants. (Courtesy R. Roberts). (**b**) Examples of species with and without horizontal stripes (Courtesy C. Kratochwil). (**c**) Ornamentation in the form of anal fin egg spots

was associated with orange-blotch phenotypes in multiple cichlid species, implying the variant arose once and was either retained as a polymorphism in some phylogenetic lineages or was introduced to lineages by hybridization and introgression while the adaptive radiation was underway (Roberts et al. 2009).

An additional intriguing—but still hypothetical—instance of changes that may have occurred within chromatophore lineages comes from *Danio*. The network of pattern-forming interactions among chromatophores that has been described in zebrafish is a potentially rich source of variation for the diversification of pattern phenotypes. Indeed, mathematical simulations starting with the zebrafish chromatophore interactions network can generate patterns outcomes resembling those of *D. albolineatus* and *D. margaritatus* (Fig. 10.3d, e) when specific features of interactions between xanthophores and iridophores are altered (Volkening and Sandstede 2018). One might imagine that other *Danio* patterns depend on additional tuning of interactions, perhaps in conjunction with alternative tissue-environmental cues for pattern initiation (Fig. 10.3f). Because *Danio* species other than zebrafish are also amenable to genetic and transgenic manipulation (Mills et al. 2007; McMenamin et al. 2014; Patterson et al. 2014; Eom et al. 2015; Spiewak et al. 2018), it should be possible to test empirically whether changes within pigment cell lineages have generated these divergent patterns within the genus.

10.4.2 Role of Tissue Environment in Pattern Evolution

In addition to chromatophore-autonomous mechanisms, changes in the tissue environments experienced by pigment cells or their progenitors might underlie some natural pattern variants. Examples of chromatophore non-autonomous changes might include alterations in the abundance of trophic factors needed for chromatophore survival or population expansion, or specific signals for guiding migration, localization, or in situ differentiation.

An example of such variation comes from stickleback, *Gasterosteus aculeatus*, a species that has evolved morphologically, physiologically, and behaviorally distinct morphs after repeated colonization of freshwater benthic or limnetic environments by a marine ancestor. One such phenotypic transition has been reduced melanization of gills and skin in freshwater stickleback as compared to marine stickleback, and genetic mapping identified *Kit ligand* (*Kitlg*) as a major-effect locus underlying this variation (Miller et al. 2007). Kitlg is expressed in the environment of chromatophores, and in zebrafish, guppy, and mammals, the Kitlg receptor, Kit, is expressed by melanized cells (melanophores or melanocytes); signaling through this pathway promotes proliferation, survival, migration, and differentiation (Besmer et al. 1993; Parichy et al. 1999; Rawls and Johnson 2003; Dooley et al. 2013; Kottler et al. 2013). In stickleback, the freshwater *Kitlg* allele is transcribed at lower levels than the marine allele, even in a common hybrid background, indicating a *cis*-regulatory change. The reduced melanization of freshwater stickleback is thus explicable by reduced transcription of Kitlg in the tissue environment, causing a less marked expansion of the melanophore population as compared to the marine fish.

A second example of non-autonomous change is found in *D. nigrofasciatus*, which has a derived pattern of fewer stripes (and interstripes) than zebrafish (Fig. 10.3e). Close inspection reveals that even in the stripes that are present, fewer melanophores are found, owing to a reduced number of cells contributed by post-embryonic progenitors. This deficit is partially offset by more neural crest derived melanophores persisting from the embryo/early larva pattern into the adult. Thus, the cellular composition of stripes in *D. nigrofasciatus* differs markedly from zebrafish (Quigley et al. 2004). Interspecific cell transplantation revealed these differences in melanophore complements to depend on factors extrinsic to the melanophores themselves, raising the possibility of tissue-derived trophic or other factors. A clue to one such mechanism came from zebrafish mutants of *endothelin receptor b1a* (*ednrb1a*), which encodes a seven pass transmembrane receptor expressed by pigment cells (Parichy et al. 2000a). Endothelin signaling is well-studied for its roles in promoting the development of melanocytes in birds and mammals and also melanophores, xanthophores and iridophores in salamanders (Kelsh et al. 2009; Saldana-Caboverde and Kos 2010; Woodcock et al. 2017). Zebrafish *ednrb1a* mutants have a pigment pattern resembling *D. nigrofasciatus*, suggesting this pathway might be involved. Yet the chromatophore non-autonomous nature of the difference suggested Ednrb1a ligands as candidates, rather than Ednrb1a itself.

A contributing role for ligand-encoding *endothelin 3b* (*edn3b*) was revealed by mutational analyses in zebrafish that resulted in a phenotype similar to *D. nigrofasciatus*. Moreover, *edn3b* of *D. nigrofasciatus* is transcribed at a lower level in skin than the zebrafish allele, owing to a *cis*-regulatory difference (Spiewak et al. 2018). Genetic and transgenic analyses further supported the inference that *D. nigrofasciatus* is fixed for a hypomorphic *edn3b* allele.

These observations might suggest a direct effect on melanophore populations, analogous to Kitlg in stickleback, yet this seems not to be the case. Rather, endothelin signaling in *Danio* appears to act primarily on iridophores, not melanophores. Iridophore population expansion is markedly reduced in *D. nigrofasciatus* and *edn3b* mutant zebrafish compared to wild-type zebrafish, as revealed by time-lapse imaging of iridophore proliferation, as well as vital labeling to assess clonal expansion. These cellular phenotypes translate into a curtailment of interstripe reiteration, which in turn leads to fewer melanophore stripes owing to the loss of essential interactions between iridophores and melanophores. Conversely, when higher levels of Edn3b are expressed in *D. nigrofasciatus*, more iridophores are generated, more interstripes are produced, and melanophores are reallocated to additional stripes. Although factors besides Edn3b certainly to contribute to the pattern difference—and especially the fewer melanophores in *D. nigrofasciatus*—this example illustrates how signaling from a tissue environmental factor can lead to an early "halt" to the pattern-forming engine of chromatophore interactions, with cascading effects on the pattern that is produced.

10.4.3 Heterochrony and Heterotopy as Factors in Diversification

Changes in the relative timing of events—heterochronies—have provided a classic framework for understanding morphological diversification (Gould 1977; Alberch et al. 1979; McKinney and McNamara 1991). With the advent of methods to identify molecular and cellular mechanisms, even in non-traditional model organisms, heterochronic explanations of character state changes have fallen out of favor. Nevertheless, evolutionary changes in timing are real, and elucidating these changes, especially using modern methods, seem likely to provide new insights into how modifications in rate and timing of developmental processes yield novel, yet still functionally integrated phenotypes at the whole-organism level. In the preceding example, for instance, a quantitative change in *edn3b* transcript abundance led to retention of an ancestrally immature phenotype (fewer stripes) into the adult—a case of paedomorphosis. Of course heterochronies also can be combined with heterotopies, in which the locations of events change relative to an ancestral state.

Consequences of changes in both time and place are evident in another *Danio*, *D. albolineatus* (Fig. 10.3c). In this species, a nearly uniform pattern is found in which chromatophore classes are broadly intermingled with one another; only in the posterior, middle region of the flank does a vestigial interstripe, free of melanophores, occur (Quigley et al. 2005). Interestingly, at least some latent signals for melanophore stripe formation remain, despite the lack of overt stripes: in *D. albolineatus* mutagenized and then screened to identify an inactivating mutation in the Kit gene, *kita*, most melanophores were ablated, yet melanophores that differentiated and persisted did so in an apparently atavistic stripe pattern (Mills et al. 2007). This and other observations indicated that a uniform pattern develops at least in part because—unlike zebrafish—initially dispersed melanophores fail to organize into stripes, obscuring the ancestral stripe pattern.

A likely explanation for this phenotype lies in alterations to the time and place of cues that influence melanophore localization. In zebrafish, melanophores are repelled at short range by fully differentiated xanthophores in the interstripe, and these interactions provide a directional cue (i.e., away from the interstripe) that helps confine melanophores to stripes (Fig. 10.2d, e). In *D. albolineatus*, fully differentiated xanthophores develop not only in the residual interstripe, but scattered all across the flank (and even internally), and they differentiate at a much earlier stage than in zebrafish (Quigley et al. 2005). Indeed, xanthophore differentiation in *D. albolineatus* is so exuberant that it occurs to some extent even without an endocrine factor, thyroid hormone, that is essential for the differentiation of xanthophores in zebrafish (McMenamin et al. 2014). This difference in xanthophore production results from a *cis*-regulatory change affecting the expression of xanthogenic factor Colony stimulating factor-1a (Csf1a), which acts through the Csf1 receptor to promote xanthophore differentiation, survival, and migration (Parichy et al. 2000b; Parichy and Turner 2003b). In zebrafish, *csf1a* is expressed by iridophores and this promotes the differentiation of xanthophores in the interstripe, where iridophores are abundant (Patterson and Parichy 2013). In *D. albolineatus*, however, *csf1a* is expressed more broadly, at higher levels, and at earlier stages (Patterson et al. 2014). Together these findings suggested that a Csf1a-dependent change in the time and place of xanthophore differentiation might be sufficient to drive the development of a uniform, rather than striped, pattern. Indeed, when Csf1a was expressed broadly, at higher levels and at earlier stages in zebrafish, to match *D. albolineatus*, extra xanthophores developed and melanophores failed to coalesce into stripes, phenocopying *D. albolineatus*. This outcome suggests that when xanthophores appear early and over a wider area, melanophores are deprived of directional cues they require to form stripes. More generally, these results illustrate how changes in the relative times and places of chromatophore differentiation can generate very different patterns, even when the pattern-forming engine of chromatophore interactions is itself unaltered.

Mesoamerican cichlids are another group in which heterochronies are suspected to play a role in pattern diversification. Extensive description of pattern ontogenies across multiple species has allowed the association of specific pattern types with differences in rates of pattern formation: patterns of lateral stripes develop slowly; patterns of lateral stripes interrupted by vertical bars develop at an intermediate rate; and patterns of vertical bars without lateral stripes develop rapidly (Rican et al. 2005, 2016).

One factor that can influence rate of somatic development overall—and that can differentially impact specific traits—is thyroid hormone. For example, zebrafish lacking thyroid hormone exhibit marked retardations in growth and development. But they also develop an excess of melanophores because these cells fail to enter a proliferatively arrested, terminally differentiated state, and they lack differentiated xanthophores, owing to a failure of these cells to accumulate the carotenoids required for their yellow-orange color (McMenamin et al. 2014; Saunders et al. 2019). Conversely, hyperthyroid zebrafish develop far too many xanthophores and mostly lack melanophores.

Pharmacological manipulation to generate hypothyroid or hyperthyroid states in convict cichlid (*Amatitlania nigrofasciata*) delayed or accelerated the appearance of specific pigment pattern elements, respectively, and caused defects in pigment cell abundance, reminiscent of thyroid hormone phenotypes in zebrafish (Prazdnikov and Shkil 2019). Interestingly, these disrupted patterns resembled the fast or slow developing patterns of other species, lending some support to the notion that heterochronic changes may underlie this range of phenotypes. It will be interesting to learn if thyroid hormone signaling itself, or other factors, contributes to differences in the rate of pattern formation in this group.

Finally, another example of heterotopic change can be found in sand-dwelling cichlids from Lake Malawi of East African (Hendrick et al. 2019). *Dimidiochromis compressiceps* develops a horizontal stripe of melanophores in the middle of the flank, whereas *Copadichromis azureus* develops a series of vertical bars of melanophores. Anatomical and histological comparisons of these during pattern development suggest that melanophore precursors that will contribute to stripes emerge from deeper regions at the horizontal myoseptum and subsequently differentiate in situ.

By contrast, melanoblasts that will form bars emerge from the same location and are thought to migrate more extensively in dorsal and ventral directions, before they differentiate. In another Lake Victoria cichlid species, *Haplochromis latifasciatus*, with fixed bar number and thicker bars compared to *C. azureus*, the formation of vertical bars is driven by the formation of new melanophores via differentiation of melanoblasts in situ rather than rearrangement of existing mature melanophores (Liang et al. 2020). Thus, heterotopy in final site of melanophore differentiation, owing to inferred differences in melanoblast migration, potentially contributes to the different patterns of these cichlid species.

10.4.4 Convergence and Parallelism in Phenotype Evolutions

In many groups of organisms, similar phenotypes have arisen independently across phylogenetic lineages (Hall 2003; Wake et al. 2011). Classic examples of such homoplasy include wings of birds and bats, complex eyes of vertebrates and cephalopod mollusks, and the evolution of multicellularity itself. An important and largely unresolved question about such homoplasies is how often they arise through the same (parallel) or different (convergent) mechanisms at the levels of molecular interactions and cellular behaviors. If mechanisms are often the same that would suggest biases on the sorts of variants that arise in populations and become available for selection. If mechanisms are often different, then sources of variation are apparently unconstrained, allowing many developmental genetic "solutions" to any given selective problem.

Examples of homoplastic pigmentary phenotypes abound in fishes, with horizontal stripes, vertical bars, uniform patterns, and even specific ornaments, like caudal eyespots, arising repeatedly and independently within and among groups. For instance, ultra-black skin that can help to reduce reflected bioluminescence has evolved multiple times independently in several orders of deep-sea fishes (Davis et al. 2020). Histological analyses revealed a continuous layer of melanosomes immediately beneath the epidermal basement membrane, optimally sized and arranged to minimize light reflectance. Another particularly compelling example can be found in the East African cichlids (Fig. 10.4b). Although horizontal stripes and vertical bars of *D. compressiceps*, *C. azureas* and *H. latifasciatus* were mentioned above, horizontal stripes have in fact evolved repeatedly from ancestors lacking stripes (Seehausen et al. 1999). The genetic basis for this transition has been examined in the closely related species *Haplochromis sauvagei* (striped) and *Pundamilia nyererei* (non-striped), for which genetic mapping and additional analyses identified *agouti-related peptide 2* (*agrp2*) as the locus responsible (Kratochwil et al. 2018). Agouti-signaling peptides antagonize Melanocortin receptors, leading to reductions in eumelanin in melanocytes of mammals and birds (Fitch et al. 2003; Manceau et al. 2011; Inaba et al. 2019) and melanin biosynthesis, melanosome dispersion within melanophores, and melanophore number in fishes (Cal et al. 2017b). In cichlids, a *cis*-regulatory difference leads to reduced expression of *agrp2* in the striped species, evidently reflecting the loss of a repressive signal for stripe formation. This inference was confirmed by targeted knock-out of *agrp2* in a normally non-striped species, which led to the development of stripes. Remarkably, stripes in another species (*H. chilotes*) also mapped to *agrp2*, and comparisons across a dozen striped species (including *D. compressiceps*) and a dozen non-striped species—representing three independent adaptive radiations in three different lakes—revealed concordant associations between low *agrp2* transcript abundance and presence of horizontal stripes. Interestingly, however, *agrp2* haplotypes differed across striped species (Urban et al. 2020), implying the repeated, independent evolution of hypomorphic *agrp2* alleles (a situation contrasting with the orange-blotch phenotype, above, in which a common allele was associated with repeated instance of the phenotype). Thus—as seems often the case when homoplasy is examined at mechanistic level—similarity or difference lies in the eye of the beholder. Here, a common phenotype has arisen independently by changes in the same gene (parallelism, *sensu* Hall 2003; Wake et al. 2011), but alterations at this locus are likely to differ from one another (convergence).

A different sort of homoplasy, at the cellular level, is illustrated by white cells of teleosts. Besides the most abundant and commonly studied chromatophores—melanophores, xanthophores, and iridophores—zebrafish have long been known to exhibit white cells at the distal edges of their dorsal and caudal fins (Johnson et al. 1995) (Fig. 10.3g). In other teleosts, chromatophores with white reflective, but non-iridescent, material are termed leucophores; those present in the embryo/early larva pattern of medaka share a close lineage relationship with xanthophores (Kimura et al. 2014; Nagao et al. 2014, 2018) and their visual appearance is believed to depend on crystalline deposits of uric acid (Hama 1975).

In the first detailed characterization of leucophores in adult zebrafish (Lewis et al. 2019), cells harboring white deposits were found not only in the dorsal and caudal fin (Johnson et al. 1995), but also in the anal fin where they additionally contained orange carotenoids. These latter cells developed from yellow-orange progenitors, resembling xanthophores, and so were termed xantholeucophores. By contrast, clonal analysis, fate mapping, and direct observation of dorsal and caudal fin white cells revealed an origin directly from melanophores, and so these were named melanoleucophores. In this natural example of transdifferentiation, melanophores lose their melanin and simultaneously acquire deposits of crystalline guanine, with associated changes in the expression of genes required for melanin synthesis and guanine production. Although crystalline guanine is also found in iridophore reflecting platelets, the crystal-containing organelles of melanoleucophores are irregularly shaped and disordered in their arrangement, leading to a matte white, rather than iridescent phenotype. The presence or absence of melanoleucophores impacts the propensity of fish to associate with one another in laboratory shoaling assays, and, remarkably, the complements of both melanoleucophores and xantholeucophores vary among *Danio* species (Fig. 10.3h), raising the possibility of different phenotypes in response to different selective regimes.

A different sort of white cell is found in the distinctive bars of clownfish, *Amphiprion* (Salis et al. 2019) (Fig. 10.5). Rather than developing from

Fig. 10.5 Pattern variation among clownfish *Amphiprion*. White patches consist of non-iridescent iridophores and patches have been lost evolutionarily in a posterior-to-anterior sequence

melanophores or xanthophores, these cells are a distinct class of iridophore, as indicated by transcriptomic similarities to zebrafish iridophores, as well as a requirement for common signaling pathway (not required by zebrafish melanoleucophores) and shared ultrastructural features. Unlike the parallel stacks of reflecting platelets in iridophores of other species, however, clownfish iridophores have reflecting platelets arranged radially, again leading to a matte, rather than iridescent appearance. Still more distant phylogenetically, squamate reptiles also exhibit white iridophores, but these cells exhibit irregular and disordered reflecting platelet—more like melanoleucophores—despite arising apparently independently of melanophores (Saenko et al. 2013). Collectively, these observations reveal many ways to make a white cell, with convergence in overt phenotype from cells of different lineages, as well as convergence and parallelism in ultrastructure (organized vs. disorganized platelets) and even chemistry [guanine, uric acid, and likely colorless pteridines as well (Oliphant and Hudon 1993)].

10.4.5 Breaking Phenotypes by Regressive Evolution

A special case of morphological change is that occurring during regressive evolution, in which structures are either lost or traits are retained by their states revert to an ancestral condition (Cronk 2009). For teleost pigment patterns, regressive evolution can include the loss of specific pattern elements, examples of which include *D. nigrofasciatus* (lost stripes and interstripes relative to the inferred ancestral state; Fig. 10.3e). An additional example comes from clownfish, *Amphiprion* (Salis et al. 2018) (Fig. 10.5). In this genus, a condition of three vertical white bars is inferred to be ancestral yet bars have been lost repeatedly, with evolutionarily

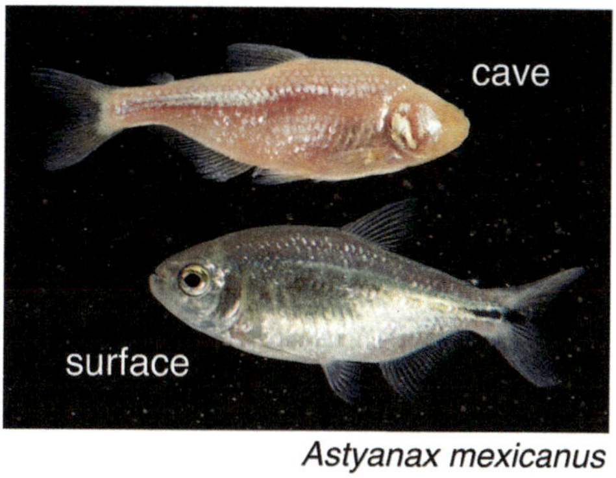

Fig. 10.6 Cave and surface forms of *Astyanax mexicanus* (Courtesy, Nicholas Rohner)

derived states including anywhere from 0 to 2 bars. Bars develop ontogenetically in a roughly anterior to posterior sequence, but are lost in a posterior-to-anterior sequence; the last bar to develop is the first to be lost. Moreover, in some species bars form ontogenetically only to be lost during later stages. These various patterns seem not to be correlated with body size or growth rate.

Regressive evolution also applies to the loss of chromatophore types (e.g., melanoleucophores from some *Danio*; Fig. 10.3h) and the loss of ancestral pigments. An especially well characterized example is *Astyanax mexicanus*, in which ~30 cave-dwelling populations, arising from a small number of founder events, have evolved distinctive morphologies that include reduction or loss of melanin, as well as reduced or absent eyes, enhancement of other sensory systems, and profound changes in physiology (Jeffery 2009; Riddle et al. 2018) (Fig. 10.6). Fully albino phenotypes are attributable to inactivation of *ocular and cutaneous albinism II* (*Oca2*) as revealed by genetic mapping of two cave populations relative to surface fish, as well as functional assays (Protas et al. 2006), and subsequent validation by targeted mutagenesis (Klaassen et al. 2018). Other comparative studies have also suggested an association between *oca2* and the amelanistic morph of medaka (Fukamachi et al. 2004) and the golden color of the Malawi golden cichlid (Kratochwil et al. 2019). *oca2* corresponds to the gene most often implicated in albinism in humans and thus represents one of the several examples in which natural or induced variants affecting teleost pigmentation mirror genetic and phenotypic variation in human populations (Lamason et al. 2005; Lang et al. 2009; Jacobs et al. 2013; Crawford et al. 2017). Two different cavefish populations, arising from distinct introduction of surface fish, share a common *oca2* allele, implying segregation of this variant in the ancestral populations (Gross and Wilkens 2013), similar to observations for the Pax7 OB allele of cichlids (above). Parallelism in the genetic basis for an albino phenotype suggests a selective advantage to melanin loss and perhaps this specific allele; what this advantage might be remains mysterious. *oca2* mutation is not the only route to melanin reduction, however, as some cavefish with a brown, rather than albino, phenotype harbor variant and independently arisen, alleles of Melanocortin-1 receptor (Mc1r) (Gross et al. 2009, 2018; Espinasa et al. 2018), well known for contributions to pigmentary phenotypes of birds and mammals (Mundy et al. 2003; Sturm et al. 2003; Hoekstra et al. 2006). Other examples of regressive evolution will be interesting to document, and particularly the roles of selection in actively "deconstructing" phenotypes vs. genetic drift leading to the degeneration of phenotypes that are no longer advantageous.

10.4.6 Genome Duplication and the Evolution of Novelty

A particularly exciting discovery regarding tempo and mode of genome evolution has been that two whole-genome duplications occurred at the base of the vertebrate lineage, and an additional whole-genome duplication (teleost-specific genome duplication) occurred at the base of the teleost lineage (Dehal and Boore 2005; Braasch et al. 2015). These duplications seem to have been followed by massive losses of newly redundant loci: non-teleost vertebrates do not, today, retain four times as many genes as non-vertebrate chordates, and teleosts do not retain twice as many genes as other vertebrates. Actual numbers of protein coding genes are similar between human and the non-vertebrate chordate amphioxus, and only ~1.3-fold greater in zebrafish than human (Howe et al. 2013; Marletaz et al. 2018). Despite this eventual rediploidization of genomes, the periods immediately following duplications must have provided substantial opportunities for diversification, as copy-specific mutational events allowed some newly duplicated genes to become subfunctionalized relative to original loci, thereby alleviating selective constraints owing to pleiotropy, or allowing the acquisition of functions that were entirely new (Ohno 1970; Force et al. 1999; Prince and Pickett 2002).

Many examples of divergent expression and function between paralogous gene copies have now been described (McClintock et al. 2001; Kleinjan et al. 2008; Lambert et al. 2015; Braasch et al. 2016; Saunders et al. 2019), but an outstanding question remains whether the transient evolutionary opportunities afforded by these bursts of genetic abundance were translated into realized diversifications in morphology. Interestingly, genes with roles in pigmentation have been retained at higher rates than genes lacking known functions without a pigmentary association (Braasch et al. 2009a; Lorin et al. 2018), certainly consistent with duplications providing raw material for adaptively significant subfunctionalization or neofunctionalization in pigment cells and pattern formation. Comparative genomic analyses have further documented the diversification of pigment-associated gene families (Braasch 2006; Braasch et al. 2007, 2008, 2009c; Selz et al. 2007; Braasch and Schartl 2014) and positive (diversifying) selection on some family members (Braasch et al. 2006; Salzburger et al. 2007). Developmental analyses have also indicated divergence in expression and function between paralogous loci having roles in chromatophore and pattern development (Lister et al. 1999; Hultman et al. 2007; Braasch et al. 2009b; Spiewak et al. 2018).

One remarkable pigment-associated gene family is that of agouti, which contains four members in teleosts owing to two rounds of whole-genome duplication, the teleost-specific genome duplication, and the subsequent loss of some copies. One member of this gene family, *agouti-signaling protein 1* (*asip1*), has a phylogenetically conserved dorsoventral expression gradient that locally inhibits melanization over the abdomen, thereby contributing to pigment countershading (Cerda-Reverter et al. 2005; Kurokawa et al. 2006; Guillot et al. 2012; Ceinos et al. 2015; Cal et al. 2017a, 2019). By contrast, a teleost-specific paralog of *asip1*, *agouti-related protein 2* (*agrp2*) of cichlids, acts a global stripe repressor, as noted above. The relatively broad domain of *agrp2* expression seems to obscure, rather than provide, instructive cues for pattern formation (Kratochwil et al. 2018; Liang et al. 2020). The corresponding gene of zebrafish (originally designated *agrp2* and now designated *asip2b*) is expressed in the pineal gland and functions in physiological color change and background adaptation by promoting melanosome contraction within melanophores that allows for blanching of the skin (Zhang et al. 2010). Though the evolutionary history of agouti family genes continues to be debated (Braasch and Postlethwait 2011; Shainer et al. 2019), observations to date are consistent with subfunctionalization or neofunctionalization of gene duplicates facilitating repeated evolution or diversification of pattern.

If extra genes have indeed allowed for pigmentary diversification, one might expect associations between paralogue functions and novel pattern features. One potential example comes from egg spots (or egg "dummies") on the anal fins of male East African cichlids, an evolutionary innovation present in ~1500 of these species that contributes to mouth brooding behaviors of these fishes (Salzburger et al. 2007) (Fig. 10.4c). Egg spots consist of concentrations of xanthophores and iridophores, sometimes surrounded by melanophores. Paralogous genes, *fhl2a* and *fhl2b*, encoding four and a half LIM domain protein 2 transcription factors, are expressed in iridophores of egg spots and recruitment of *fhl2b* expression to these cells is associated with a transposable element insertion into its upstream regulatory region, as assayed by comparing regulatory regions from cichlids with and without egg spots in zebrafish reporter assays (Santos et al. 2014). *fhl2a* and *fhl2b* are expressed by iridophores of zebrafish and clownfish (Salis et al. 2019; Saunders et al. 2019) and it will be interesting to learn how evolutionary changes in gene regulation within cichlid lineages might be causally associated with egg spot origins.

10.5 Conclusions

As chromatophore and pigment pattern development becomes increasingly understood, and technical advances allow for genetic, genomic, and developmental investigations across a wider array of species, many more examples of developmental genetic bases for evolutionary change will come from teleosts and their spectacular pigment patterns. Already, the field has provided insights into fundamental questions about neural crest lineages, roles for cellular interactions in pattern formation and pattern variants, genetic underpinnings of homoplasy and regressive evolution, and potential roles for gene duplication in the elaboration of morphology. Though not explicitly highlighted above, studies of pigment pattern evolution also contribute to debates over the relative roles of gene regulatory and coding sequence changes, and the numbers of loci responsible for adaptive trait differences. Perhaps most importantly, teleost pigment patterns and examples of their diversification offer tangible phenotypes with wide appeal and opportunities to teach a broad audience both the facts of evolution and the importance of conserving natural diversity in a rapidly changing environment.

Acknowledgements DMP and research in his laboratory are supported by NIH R35 GM122471. Images not attributed in figure legends were stock photographs or taken by D. Parichy.

References

Alberch P, Gould SJ, Oster GF, Wake DB (1979) Size and shape in ontogeny and phylogeny. Paleobiology 5:296–317

Baldwin CC (2013) The phylogenetic significance of colour patterns in marine teleost larvae. Zool J Linnean Soc 168(3):496–563. https://doi.org/10.1111/zoj.12033

Besmer P, Manova K, Duttlinger R, Huang EJ, Packer A, Gyssler C, Bachvarova RF (1993) The kit-ligand (steel factor) and its receptor c-kit/W: pleiotropic roles in gametogenesis and melanogenesis. Dev Suppl 1993:125–137

Braasch I (2006) Asymmetric evolution in two fish-specifically duplicated receptor tyrosine kinase paralogons involved in teleost coloration. Mol Biol Evol 23(6):1192–1202. https://doi.org/10.1093/molbev/msk003

Braasch I, Postlethwait JH (2011) The teleost agouti-related protein 2 gene is an ohnolog gone missing from the tetrapod genome. Proc Natl Acad Sci U S A 108(13):E47–E48. https://doi.org/10.1073/pnas.1101594108

Braasch I, Schartl M (2014) Evolution of endothelin receptors in vertebrates. Gen Comp Endocrinol 209:21–34. https://doi.org/10.1016/j.ygcen.2014.06.028

Braasch I, Salzburger W, Meyer A (2006) Asymmetric evolution in two fish-specifically duplicated receptor tyrosine kinase paralogons involved in teleost coloration. Mol Biol Evol 23 (6):1192–1202. https://doi.org/10.1093/molbev/msk003

Braasch I, Schartl M, Volff JN (2007) Evolution of pigment synthesis pathways by gene and genome duplication in fish. BMC Evol Biol 7:74. https://doi.org/10.1186/1471-2148-7-74

Braasch I, Volff JN, Schartl M (2008) The evolution of teleost pigmentation and the fish-specific genome duplication. J Fish Biol 73(8):1891–1918. https://doi.org/10.1111/j.1095-8649.2008.02011.x

Braasch I, Brunet F, Volff JN, Schartl M (2009a) Pigmentation pathway evolution after whole-genome duplication in fish. Genome Biol Evol 1:479–493. https://doi.org/10.1093/gbe/evp050

Braasch I, Liedtke D, Volff JN, Schartl M (2009b) Pigmentary function and evolution of tyrp1 gene duplicates in fish. Pigment Cell Melanoma Res 22(6):839–850. https://doi.org/10.1111/j.1755-148X.2009.00614.x

Braasch I, Volff JN, Schartl M (2009c) The endothelin system: evolution of vertebrate-specific ligand-receptor interactions by three rounds of genome duplication. Mol Biol Evol 26 (4):783–799

Braasch I, Peterson SM, Desvignes T, McCluskey BM, Batzel P, Postlethwait JH (2015) A new model army: emerging fish models to study the genomics of vertebrate evo-devo. J Exp Zool B Mol Dev Evol 324(4):316–341. https://doi.org/10.1002/jez.b.22589

Braasch I, Gehrke AR, Smith JJ, Kawasaki K, Manousaki T, Pasquier J, Amores A, Desvignes T, Batzel P, Catchen J, Berlin AM, Campbell MS, Barrell D, Martin KJ, Mulley JF, Ravi V, Lee AP, Nakamura T, Chalopin D, Fan S, Wcisel D, Canestro C, Sydes J, Beaudry FE, Sun Y,

Hertel J, Beam MJ, Fasold M, Ishiyama M, Johnson J, Kehr S, Lara M, Letaw JH, Litman GW, Litman RT, Mikami M, Ota T, Saha NR, Williams L, Stadler PF, Wang H, Taylor JS, Fontenot Q, Ferrara A, Searle SM, Aken B, Yandell M, Schneider I, Yoder JA, Volff JN, Meyer A, Amemiya CT, Venkatesh B, Holland PW, Guiguen Y, Bobe J, Shubin NH, Di Palma F, Alfoldi J, Lindblad-Toh K, Postlethwait JH (2016) The spotted gar genome illuminates vertebrate evolution and facilitates human-teleost comparisons. Nat Genet 48(4):427–437. https://doi.org/10.1038/ng.3526

Brawand D, Wagner CE, Li YI, Malinsky M, Keller I, Fan S, Simakov O, Ng AY, Lim ZW, Bezault E, Turner-Maier J, Johnson J, Alcazar R, Noh HJ, Russell P, Aken B, Alfoldi J, Amemiya C, Azzouzi N, Baroiller JF, Barloy-Hubler F, Berlin A, Bloomquist R, Carleton KL, Conte MA, D'Cotta H, Eshel O, Gaffney L, Galibert F, Gante HF, Gnerre S, Greuter L, Guyon R, Haddad NS, Haerty W, Harris RM, Hofmann HA, Hourlier T, Hulata G, Jaffe DB, Lara M, Lee AP, MacCallum I, Mwaiko S, Nikaido M, Nishihara H, Ozouf-Costaz C, Penman DJ, Przybylski D, Rakotomanga M, Renn SC, Ribeiro FJ, Ron M, Salzburger W, Sanchez-Pulido L, Santos ME, Searle S, Sharpe T, Swofford R, Tan FJ, Williams L, Young S, Yin S, Okada N, Kocher TD, Miska EA, Lander ES, Venkatesh B, Fernald RD, Meyer A, Ponting CP, Streelman JT, Lindblad-Toh K, Seehausen O, Di Palma F (2014) The genomic substrate for adaptive radiation in African cichlid fish. Nature 513(7518):375–381. https://doi.org/10.1038/nature13726

Budd PL (1940) Development of the eggs and early larvae of six california fishes. Hopkins Marine Station Stanford University, fish bulletin No. 56

Budi EH, Patterson LB, Parichy DM (2008) Embryonic requirements for ErbB signaling in neural crest development and adult pigment pattern formation. Development 135(15):2603–2614. https://doi.org/10.1242/dev.019299

Budi EH, Patterson LB, Parichy DM (2011) Post-embryonic nerve-associated precursors to adult pigment cells: genetic requirements and dynamics of morphogenesis and differentiation. PLoS Genet 7(5):e1002044. https://doi.org/10.1371/journal.pgen.1002044

Bullara D, De Decker Y (2015) Pigment cell movement is not required for generation of Turing patterns in zebrafish skin. Nat Commun 6:7971. https://doi.org/10.1038/ncomms7971

Caicedo-Carvajal CE, Shinbrot T (2008) In silico zebrafish pattern formation. Dev Biol 315 (2):397–403. https://doi.org/10.1016/j.ydbio.2007.12.036

Cal L, Meglas M, Cerda-Reverter JM, Postlethwait JH, Braasch I, Rotllant J (2017a) BAC recombineering of the agouti loci from spotted gar and zebrafish reveals the evolutionary ancestry of dorsal-ventral pigment asymmetry in fish. J Exp Zool B Mol Dev Evol 328 (7):697–708. https://doi.org/10.1002/jez.b.22748

Cal L, Suarez-Bregua P, Cerda-Reverter JM, Braasch I, Rotllant J (2017b) Fish pigmentation and the melanocortin system. Comp Biochem Physiol A Mol Integr Physiol 211:26–33. https://doi.org/10.1016/j.cbpa.2017.06.001

Cal L, Suarez-Bregua P, Comesana P, Owen J, Braasch I, Kelsh R, Cerda-Reverter JM, Rotllant J (2019) Countershading in zebrafish results from an Asip1 controlled dorsoventral gradient of pigment cell differentiation. Sci Rep 9(1):3449. https://doi.org/10.1038/s41598-019-40251-z

Camargo Sosa K, Colanesi S, Muller J, Schulte-Merker S, Stemple D, Patton EE, Kelsh RN (2019) Endothelin receptor Aa regulates proliferation and differentiation of Erb-dependant pigment progenitors in zebrafish. PLoS Genet. https://doi.org/10.1101/308221

Ceinos RM, Guillot R, Kelsh RN, Cerda-Reverter JM, Rotllant J (2015) Pigment patterns in adult fish result from superimposition of two largely independent pigmentation mechanisms. Pigment Cell Melanoma Res 28(2):196–209. https://doi.org/10.1111/pcmr.12335

Cerda-Reverter JM, Haitina T, Schioth HB, Peter RE (2005) Gene structure of the goldfish agouti-signaling protein: a putative role in the dorsal-ventral pigment pattern of fish. Endocrinology 146(3):1597–1610. https://doi.org/10.1210/en.2004-1346

Crawford NG, Kelly DE, Hansen MEB, Beltrame MH, Fan S, Bowman SL, Jewett E, Ranciaro A, Thompson S, Lo Y, Pfeifer SP, Jensen JD, Campbell MC, Beggs W, Hormozdiari F, Mpoloka SW, Mokone GG, Nyambo T, Meskel DW, Belay G, Haut J, Program NCS, Rothschild H,

Zon L, Zhou Y, Kovacs MA, Xu M, Zhang T, Bishop K, Sinclair J, Rivas C, Elliot E, Choi J, Li SA, Hicks B, Burgess S, Abnet C, Watkins-Chow DE, Oceana E, Song YS, Eskin E, Brown KM, Marks MS, Loftus SK, Pavan WJ, Yeager M, Chanock S, Tishkoff SA (2017) Loci associated with skin pigmentation identified in African populations. Science 358(6365). https://doi.org/10.1126/science.aan8433

Cronk QC (2009) Evolution in reverse gear: the molecular basis of loss and reversal. Cold Spring Harb Symp Quant Biol 74:259–266. https://doi.org/10.1101/sqb.2009.74.034

Curran K, Raible DW, Lister JA (2009) Foxd3 controls melanophore specification in the zebrafish neural crest by regulation of Mitf. Dev Biol 332(2):408–417

Curran K, Lister JA, Kunkel GR, Prendergast A, Parichy DM, Raible DW (2010) Interplay between Foxd3 and Mitf regulates cell fate plasticity in the zebrafish neural crest. Dev Biol 344:107–118

Davis AL, Thomas KN, Goetz FE, Robison BH, Johnsen S, Osborn KJ (2020) Ultra-black camouflage in deep-sea fishes. Curr Biol 30(17):3470–3476. https://doi.org/10.1016/j.cub.2020.06.044

Dehal P, Boore JL (2005) Two rounds of whole genome duplication in the ancestral vertebrate. PLoS Biol 3(10):e314. https://doi.org/10.1371/journal.pbio.0030314

Dooley CM, Mongera A, Walderich B, Nusslein-Volhard C (2013) On the embryonic origin of adult melanophores: the role of ErbB and Kit signalling in establishing melanophore stem cells in zebrafish. Development 140(5):1003–1013. https://doi.org/10.1242/dev.087007

Dupin E, Calloni GW, Coelho-Aguiar JM, Le Douarin NM (2018) The issue of the multipotency of the neural crest cells. Dev Biol 444(Suppl 1):S47–S59. https://doi.org/10.1016/j.ydbio.2018.03.024

Dushane GP (1934) The origin of pigment cells in Amphibia. Science 80(2087):620–621. https://doi.org/10.1126/science.80.2087.620-a

Dutton KA, Pauliny A, Lopes SS, Elworthy S, Carney TJ, Rauch J, Geisler R, Haffter P, Kelsh RN (2001) Zebrafish colourless encodes sox10 and specifies non-ectomesenchymal neural crest fates. Development 128(21):4113–4125

Elworthy S, Lister JA, Carney TJ, Raible DW, Kelsh RN (2003) Transcriptional regulation of mitfa accounts for the sox10 requirement in zebrafish melanophore development. Development 130 (12):2809–2818. https://doi.org/10.1242/dev.00461

Endler JA (1978) A predator's view of animal color patterns. In: Hecht MK, Steere WC, Wallace B (eds) Evolutionary biology. Springer, Boston, MA, pp 319–364. https://doi.org/10.1007/978-1-4615-6956-5_5

Endler JA (1983) Natural and sexual selection on color patterns in Poeciliid fishes. Environ Biol Fishes 9:173–190

Eom DS, Parichy DM (2017) A macrophage relay for long-distance signaling during postembryonic tissue remodeling. Science 355(6331):1317–1320. https://doi.org/10.1126/science.aal2745

Eom DS, Inoue S, Patterson LB, Gordon TN, Slingwine R, Kondo S, Watanabe M, Parichy DM (2012) Melanophore migration and survival during zebrafish adult pigment stripe development require the immunoglobulin superfamily adhesion molecule Igsf11. PLoS Genet 8(8): e1002899. https://doi.org/10.1371/journal.pgen.1002899

Eom DS, Bain EJ, Patterson LB, Grout ME, Parichy DM (2015) Long-distance communication by specialized cellular projections during pigment pattern development and evolution. elife 4: e12401. https://doi.org/10.7554/eLife.12401

Eskova A, Frohnhofer HG, Nusslein-Volhard C, Irion U (2020) Galanin signaling in the brain regulates color pattern formation in zebrafish. Curr Biol 30(2):298–303. https://doi.org/10.1016/j.cub.2019.11.033

Espinasa L, Robinson J, Espinasa M (2018) Mc1r gene in Astroblepus pholeter and Astyanax mexicanus: convergent regressive evolution of pigmentation across cavefish species. Dev Biol 441(2):305–310. https://doi.org/10.1016/j.ydbio.2018.07.016

Fadeev A, Krauss J, Frohnhofer HG, Irion U, Nusslein-Volhard C (2015) Tight junction protein 1a regulates pigment cell organisation during zebrafish colour patterning. elife 4:e06545. https://doi.org/10.7554/eLife.06545

Fadeev A, Krauss J, Singh AP, Nusslein-Volhard C (2016) Zebrafish Leucocyte tyrosine kinase controls iridophore establishment, proliferation and survival. Pigment Cell Melanoma Res 29 (3):284–296. https://doi.org/10.1111/pcmr.12454

Fadeev A, Mendoza-Garcia P, Irion U, Guan J, Pfeifer K, Wiessner S, Serluca F, Singh AP, Nusslein-Volhard C, Palmer RH (2018) ALKALs are in vivo ligands for ALK family receptor tyrosine kinases in the neural crest and derived cells. Proc Natl Acad Sci U S A 115(4):E630–E638. https://doi.org/10.1073/pnas.1719137115

Fitch KR, McGowan KA, van Raamsdonk CD, Fuchs H, Lee D, Puech A, Herault Y, Threadgill DW, Hrabe de Angelis M, Barsh GS (2003) Genetics of dark skin in mice. Genes Dev 17 (2):214–228

Force A, Lynch M, Pickett FB, Amores A, Yan YL, Postlethwait J (1999) Preservation of duplicate genes by complementary, degenerative mutations. Genetics 151(4):1531–1545

Frohnhofer HG, Krauss J, Maischein HM, Nusslein-Volhard C (2013) Iridophores and their interactions with other chromatophores are required for stripe formation in zebrafish. Development 140(14):2997–3007. https://doi.org/10.1242/dev.096719

Fujii R (1993) Cytophysiology of fish chromatophores. Int Rev Cytol 143:191–255

Fukamachi S, Asakawa S, Wakamatsu Y, Shimizu N, Mitani H, Shima A (2004) Conserved function of medaka pink-eyed dilution in melanin synthesis and its divergent transcriptional regulation in gonads among vertebrates. Genetics 168(3):1519–1527. https://doi.org/10.1534/genetics.104.030494

Fukuzawa T (2010) Unusual development of light-reflecting pigment cells in intact and regenerating tail in the periodic albino mutant of Xenopus laevis. Cell Tissue Res 342 (1):53–66. https://doi.org/10.1007/s00441-010-1042-0

Gould SJ (1977) Ontogeny and phylogeny. Harvard University Press, Cambridge, MA

Granneman JG, Kimler VA, Zhang H, Ye X, Luo X, Postlethwait JH, Thummel R (2017) Lipid droplet biology and evolution illuminated by the characterization of a novel perilipin in teleost fish. elife 6:21771. https://doi.org/10.7554/eLife.21771

Greenhill ER, Rocco A, Vibert L, Nikaido M, Kelsh RN (2011) An iterative genetic and dynamical modelling approach identifies novel features of the gene regulatory network underlying melanocyte development. PLoS Genet 7(9):e1002265. https://doi.org/10.1371/journal.pgen.1002265

Gross JB, Wilkens H (2013) Albinism in phylogenetically and geographically distinct populations of Astyanax cavefish arises through the same loss-of-function Oca2 allele. Heredity 111 (2):122–130. https://doi.org/10.1038/hdy.2013.26

Gross JB, Borowsky R, Tabin CJ (2009) A novel role for Mc1r in the parallel evolution of depigmentation in independent populations of the cavefish Astyanax mexicanus. PLoS Genet 5(1):e1000326. https://doi.org/10.1371/journal.pgen.1000326

Gross JB, Weagley J, Stahl BA, Ma L, Espinasa L, McGaugh SE (2018) A local duplication of the Melanocortin receptor 1 locus in Astyanax. Genome 61(4):254–265. https://doi.org/10.1139/gen-2017-0049

Guillot R, Ceinos RM, Cal R, Rotllant J, Cerda-Reverter JM (2012) Transient ectopic overexpression of agouti-signalling protein 1 (asip1) induces pigment anomalies in flatfish. PLoS One 7(12):e48526. https://doi.org/10.1371/journal.pone.0048526

Gur D, Palmer BA, Weiner S, Addadi L (2016) Light manipulation by guanine crystals in organisms: biogenic scatterers, mirrors, multilayer reflectors and photonic crystals. Adv Funct Mater 2016:1603514

Gur D, Bain E, Johnson K, Aman AJ, Pasoili A, Flynn JD, Allen MC, Deheyn DD, Lee JC, Lippincott Schwartz J, Parichy D (2020) In situ differentiation of iridophore crystallotypes underlies zebrafish stripe patterning. Nat Commun 11:6391

Hall BK (2003) Descent with modification: the unity underlying homology and homoplasy as seen through an analysis of development and evolution. Biol Rev 78(3):409–433. https://doi.org/10.1017/S1464793102006097

Hama T (1975) Chromatophores and iridocytes. In: Yamamoto T (ed) Medaka (killifish) biology and strains. Keigaku Publishing Co., Tokyo, pp 138–153

Hamada H, Watanabe M, Lau HE, Nishida T, Hasegawa T, Parichy DM, Kondo S (2014) Involvement of delta/notch signaling in zebrafish adult pigment stripe patterning. Development 141(2):318–324. https://doi.org/10.1242/dev.099804

Hemingson CR, Cowman PF, Hodge JR, Bellwood DR (2019) Colour pattern divergence in reef fish species is rapid and driven by both range overlap and symmetry. Ecol Lett 22(1):190–199. https://doi.org/10.1111/ele.13180

Hendrick LA, Carter GA, Hilbrands EH, Heubel BP, Schilling TF, Le Pabic P (2019) Bar, stripe and spot development in sand-dwelling cichlids from Lake Malawi. EvoDevo 10:18. https://doi.org/10.1186/s13227-019-0132-7

Hoekstra HE, Hirschmann RJ, Bundey RA, Insel PA, Crossland JP (2006) A single amino acid mutation contributes to adaptive beach mouse color pattern. Science 313(5783):101–104

Holzer G, Besson M, Lambert A, Francois L, Barth P, Gillet B, Hughes S, Piganeau G, Leulier F, Viriot L, Lecchini D, Laudet V (2017) Fish larval recruitment to reefs is a thyroid hormone-mediated metamorphosis sensitive to the pesticide chlorpyrifos. elife 6:e27595. https://doi.org/10.7554/eLife.27595

Hörstadius S (1950) The neural crest: its properties and derivatives in light of experimental research. Oxford University Press, London

Houde AE (1997) Sex, color, and mate choice in guppies. Princeton University Press, Princeton

Howe K, Clark MD, Torroja CF, Torrance J, Berthelot C, Muffato M, Collins JE, Humphray S, McLaren K, Matthews L, McLaren S, Sealy I, Caccamo M, Churcher C, Scott C, Barrett JC, Koch R, Rauch GJ, White S, Chow W, Kilian B, Quintais LT, Guerra-Assuncao JA, Zhou Y, Gu Y, Yen J, Vogel JH, Eyre T, Redmond S, Banerjee R, Chi J, Fu B, Langley E, Maguire SF, Laird GK, Lloyd D, Kenyon E, Donaldson S, Sehra H, Almeida-King J, Loveland J, Trevanion S, Jones M, Quail M, Willey D, Hunt A, Burton J, Sims S, McLay K, Plumb B, Davis J, Clee C, Oliver K, Clark R, Riddle C, Elliot D, Threadgold G, Harden G, Ware D, Begum S, Mortimore B, Kerry G, Heath P, Phillimore B, Tracey A, Corby N, Dunn M, Johnson C, Wood J, Clark S, Pelan S, Griffiths G, Smith M, Glithero R, Howden P, Barker N, Lloyd C, Stevens C, Harley J, Holt K, Panagiotidis G, Lovell J, Beasley H, Henderson C, Gordon D, Auger K, Wright D, Collins J, Raisen C, Dyer L, Leung K, Robertson L, Ambridge K, Leongamornlert D, McGuire S, Gilderthorp R, Griffiths C, Manthravadi D, Nichol S, Barker G, Whitehead S, Kay M, Brown J, Murnane C, Gray E, Humphries M, Sycamore N, Barker D, Saunders D, Wallis J, Babbage A, Hammond S, Mashreghi-Mohammadi M, Barr L, Martin S, Wray P, Ellington A, Matthews N, Ellwood M, Woodmansey R, Clark G, Cooper J, Tromans A, Grafham D, Skuce C, Pandian R, Andrews R, Harrison E, Kimberley A, Garnett J, Fosker N, Hall R, Garner P, Kelly D, Bird C, Palmer S, Gehring I, Berger A, Dooley CM, Ersan-Urun Z, Eser C, Geiger H, Geisler M, Karotki L, Kirn A, Konantz J, Konantz M, Oberlander M, Rudolph-Geiger S, Teucke M, Lanz C, Raddatz G, Osoegawa K, Zhu B, Rapp A, Widaa S, Langford C, Yang F, Schuster SC, Carter NP, Harrow J, Ning Z, Herrero J, Searle SM, Enright A, Geisler R, Plasterk RH, Lee C, Westerfield M, de Jong PJ, Zon LI, Postlethwait JH, Nusslein-Volhard C, Hubbard TJ, Roest Crollius H, Rogers J, Stemple DL (2013) The zebrafish reference genome sequence and its relationship to the human genome. Nature 496(7446):498–503. https://doi.org/10.1038/nature12111

Hultman KA, Bahary N, Zon LI, Johnson SL (2007) Gene Duplication of the zebrafish kit ligand and partitioning of melanocyte development functions to kit ligand. PLoS Genet 3(1):e17. https://doi.org/10.1371/journal.pgen.0030017

Inaba M, Yamanaka H, Kondo S (2012) Pigment pattern formation by contact-dependent depolarization. Science 335(6069):677. https://doi.org/10.1126/science.1212821

Inaba M, Jiang TX, Liang YC, Tsai S, Lai YC, Widelitz RB, Chuong CM (2019) Instructive role of melanocytes during pigment pattern formation of the avian skin. Proc Natl Acad Sci U S A 116 (14):6884–6890. https://doi.org/10.1073/pnas.1816107116

Inoue S, Kondo S, Parichy DM, Watanabe M (2014) Tetraspanin 3c requirement for pigment cell interactions and boundary formation in zebrafish adult pigment stripes. Pigment Cell Melanoma Res 27(2):190–200. https://doi.org/10.1111/pcmr.12192

Irion U, Frohnhofer HG, Krauss J, Colak Champollion T, Maischein HM, Geiger-Rudolph S, Weiler C, Nusslein-Volhard C (2014) Gap junctions composed of connexins 41.8 and 39.4 are essential for colour pattern formation in zebrafish. elife 3:e05125. https://doi.org/10.7554/eLife.05125

Irion U, Singh AP, Nusslein-Volhard C (2016) The developmental genetics of vertebrate color pattern formation: lessons from zebrafish. Curr Top Dev Biol 117:141–169. https://doi.org/10.1016/bs.ctdb.2015.12.012

Iwashita M, Watanabe M, Ishii M, Chen T, Johnson SL, Kurachi Y, Okada N, Kondo S (2006) Pigment pattern in jaguar/obelix zebrafish is caused by a Kir7.1 mutation: Implications for the regulation of melanosome movement. PLoS Genet 2(11):1861–1870. https://doi.org/10.1371/journal.pgen.0020197

Iyengar S, Kasheta M, Ceol CJ (2015) Poised regeneration of zebrafish melanocytes involves direct differentiation and concurrent replenishment of tissue-resident progenitor cells. Dev Cell 33 (6):631–643. https://doi.org/10.1016/j.devcel.2015.04.025

Jacobs LC, Wollstein A, Lao O, Hofman A, Klaver CC, Uitterlinden AG, Nijsten T, Kayser M, Liu F (2013) Comprehensive candidate gene study highlights UGT1A and BNC2 as new genes determining continuous skin color variation in Europeans. Hum Genet 132(2):147–158. https://doi.org/10.1007/s00439-012-1232-9

Jeffery WR (2009) Regressive evolution in Astyanax cavefish. Annu Rev Genet 43:25–47. https://doi.org/10.1146/annurev-genet-102108-134216

Johnson SL, Africa D, Walker C, Weston JA (1995) Genetic control of adult pigment stripe development in zebrafish. Dev Biol 167(1):27–33. https://doi.org/10.1006/dbio.1995.1004

Kapp FG, Perlin JR, Hagedorn EJ, Gansner JM, Schwarz DE, O'Connell LA, Johnson NS, Amemiya C, Fisher DE, Wolfle U, Trompouki E, Niemeyer CM, Driever W, Zon LI (2018) Protection from UV light is an evolutionarily conserved feature of the haematopoietic niche. Nature 558(7710):445–448. https://doi.org/10.1038/s41586-018-0213-0

Kelsh RN, Brand M, Jiang Y-J, Heisenberg C-P, Lin S, Haffter P, Odenthal J, Mullins MC, van Eeden FJM, Furutani-Seiki M, Granato M, Hammerschmidt M, Kane DA, Warga RM, Beuchle D, Vogelsang L, Nusslein-Volhard C (1996) Zebrafish pigmentation mutations and the processes of neural crest development. Development 123:369–389

Kelsh RN, Harris ML, Colanesi S, Erickson CA (2009) Stripes and belly-spots-A review of pigment cell morphogenesis in vertebrates. Semin Cell Dev Biol 20:90–104

Kimmel CB, Ballard WW, Kimmel SR, Ullmann B, Schilling TF (1995) Stages of embryonic development of the zebrafish. Dev Dyn 203(3):253–310. https://doi.org/10.1002/aja.1002030302

Kimura T, Nagao Y, Hashimoto H, Yamamoto-Shiraishi Y, Yamamoto S, Yabe T, Takada S, Kinoshita M, Kuroiwa A, Naruse K (2014) Leucophores are similar to xanthophores in their specification and differentiation processes in medaka. Proc Natl Acad Sci U S A 111 (20):7343–7348. https://doi.org/10.1073/pnas.1311254111

Kimura T, Takehana Y, Naruse K (2017) pnp4a is the causal gene of the medaka iridophore mutant guanineless. G3 (Bethesda) 7(4):1357–1363. https://doi.org/10.1534/g3.117.040675

Klaassen H, Wang Y, Adamski K, Rohner N, Kowalko JE (2018) CRISPR mutagenesis confirms the role of oca2 in melanin pigmentation in Astyanax mexicanus. Dev Biol 441(2):313–318. https://doi.org/10.1016/j.ydbio.2018.03.014

Kleinjan DA, Bancewicz RM, Gautier P, Dahm R, Schonthaler HB, Damante G, Seawright A, Hever AM, Yeyati PL, van Heyningen V, Coutinho P (2008) Subfunctionalization of duplicated

zebrafish pax6 genes by cis-regulatory divergence. PLoS Genet 4(2):e29. https://doi.org/10.1371/journal.pgen.0040029
Kottler VA, Fadeev A, Weigel D, Dreyer C (2013) Pigment pattern formation in the guppy, Poecilia reticulata, involves the Kita and Csf1ra receptor tyrosine kinases. Genetics 194(3):631–646. https://doi.org/10.1534/genetics.113.151738
Kratochwil CF, Liang Y, Gerwin J, Woltering JM, Urban S, Henning F, Machado-Schiaffino G, Hulsey CD, Meyer A (2018) Agouti-related peptide 2 facilitates convergent evolution of stripe patterns across cichlid fish radiations. Science 362(6413):457–460. https://doi.org/10.1126/science.aao6809
Kratochwil CF, Urban S, Meyer A (2019) Genome of the Malawi golden cichlid fish (Melanochromis auratus) reveals exon loss of oca2 in an amelanistic morph. Pigment Cell Melanoma Res 32(5):719–723. https://doi.org/10.1111/pcmr.12799
Kurokawa T, Murashita K, Uji S (2006) Characterization and tissue distribution of multiple agouti-family genes in pufferfish, Takifugu rubripes. Peptides 27(12):3165–3175. https://doi.org/10.1016/j.peptides.2006.09.013
Lamason RL, Mohideen MA, Mest JR, Wong AC, Norton HL, Aros MC, Jurynec MJ, Mao X, Humphreville VR, Humbert JE, Sinha S, Moore JL, Jagadeeswaran P, Zhao W, Ning G, Makalowska I, McKeigue PM, O'Donnell D, Kittles R, Parra EJ, Mangini NJ, Grunwald DJ, Shriver MD, Canfield VA, Cheng KC (2005) SLC24A5, a putative cation exchanger, affects pigmentation in zebrafish and humans. Science 310(5755):1782–1786. https://doi.org/10.1126/science.1116238
Lambert MJ, Cochran WO, Wilde BM, Olsen KG, Cooper CD (2015) Evidence for widespread subfunctionalization of splice forms in vertebrate genomes. Genome Res 25(5):624–632. https://doi.org/10.1101/gr.184473.114
Lamoreux ML, Kelsh RN, Wakamatsu Y, Ozato K (2005) Pigment pattern formation in the medaka embryo. Pigment Cell Res 18(2):64–73. https://doi.org/10.1111/j.1600-0749.2005.00216.x
Lang MR, Patterson LB, Gordon TN, Johnson SL, Parichy DM (2009) Basonuclin-2 requirements for zebrafish adult pigment pattern development and female fertility. PLoS Genet 5(11):e1000744. https://doi.org/10.1371/journal.pgen.1000744
Le Douarin NM, Dupin E (2018) The beginnings of the neural crest. Dev Biol. https://doi.org/10.1016/j.ydbio.2018.07.019
Lewis VM, Saunders LM, Larson TA, Bain EJ, Sturiale SL, Gur D, Chowdhury S, Flynn JD, Allen MC, Deheyn DD, Lee JC, Simon JA, Lippincott-Schwartz J, Raible DW, Parichy DM (2019) Fate plasticity and reprogramming in genetically distinct populations of Danio leucophores. Proc Natl Acad Sci 16(24):11806–11811
Li IJ, Chang CJ, Liu SC, Abe G, Ota KG (2015) Postembryonic staging of wild-type goldfish, with brief reference to skeletal systems. Dev Dyn 244(12):1485–1518. https://doi.org/10.1002/dvdy.24340
Liang Y, Gerwin J, Meyer A, Kratochwil CF (2020) Developmental and cellular basis of vertical bar color patterns in the east african cichlid fish Haplochromis latifasciatus. Front Cell Dev Biol 8:62. https://doi.org/10.3389/fcell.2020.00062
Lister JA (2019) Larval but not adult xanthophore pigmentation in zebrafish requires GTP cyclohydrolase 2 (gch2) function. Pigment Cell Melanoma Res 32:724–727. https://doi.org/10.1111/pcmr.12783
Lister JA, Robertson CP, Lepage T, Johnson SL, Raible DW (1999) nacre encodes a zebrafish microphthalmia-related protein that regulates neural-crest-derived pigment cell fate. Development 126(17):3757–3767
Lopes SS, Yang X, Muller J, Carney TJ, McAdow AR, Rauch GJ, Jacoby AS, Hurst LD, Delfino-Machin M, Haffter P, Geisler R, Johnson SL, Ward A, Kelsh RN (2008) Leukocyte tyrosine kinase functions in pigment cell development. PLoS Genet 4(3):e1000026. https://doi.org/10.1371/journal.pgen.1000026

Lorin T, Brunet FG, Laudet V, Volff JN (2018) Teleost fish-specific preferential retention of pigmentation gene-containing families after whole genome duplications in vertebrates. G3 (Bethesda) 8(5):1795–1806. https://doi.org/10.1534/g3.118.200201

Losey GS, Cronin TW, Goldsmith TH, Hyde D, Marshall NJ, McFarland WN (1999) The UV visual world of fishes: a review. J Fish Biol 54(5):921–943. https://doi.org/10.1111/j.1095-8649.1999.tb00848.x

Maan ME, Sefc KM (2013) Colour variation in cichlid fish: developmental mechanisms, selective pressures and evolutionary consequences. Semin Cell Dev Biol 24(6-7):516–528. https://doi.org/10.1016/j.semcdb.2013.05.003

Mahalwar P, Walderich B, Singh AP, Nusslein-Volhard C (2014) Local reorganization of xanthophores fine-tunes and colors the striped pattern of zebrafish. Science 345 (6202):1362–1364. https://doi.org/10.1126/science.1254837

Mahalwar P, Singh AP, Fadeev A, Nusslein-Volhard C, Irion U (2016) Heterotypic interactions regulate cell shape and density during color pattern formation in zebrafish. Biol Open 5 (11):1680–1690. https://doi.org/10.1242/bio.022251

Manceau M, Domingues VS, Mallarino R, Hoekstra HE (2011) The developmental role of Agouti in color pattern evolution. Science 331(6020):1062–1065. https://doi.org/10.1126/science.1200684

Marletaz F, Firbas PN, Maeso I, Tena JJ, Bogdanovic O, Perry M, Wyatt CDR, de la Calle-Mustienes E, Bertrand S, Burguera D, Acemel RD, van Heeringen SJ, Naranjo S, Herrera-Ubeda C, Skvortsova K, Jimenez-Gancedo S, Aldea D, Marquez Y, Buono L, Kozmikova I, Permanyer J, Louis A, Albuixech-Crespo B, Le Petillon Y, Leon A, Subirana L, Balwierz PJ, Duckett PE, Farahani E, Aury JM, Mangenot S, Wincker P, Albalat R, Benito-Gutierrez E, Canestro C, Castro F, D'Aniello S, Ferrier DEK, Huang S, Laudet V, Marais GAB, Pontarotti P, Schubert M, Seitz H, Somorjai I, Takahashi T, Mirabeau O, Xu A, Yu JK, Carninci P, Martinez-Morales JR, Crollius HR, Kozmik Z, Weirauch MT, Garcia-Fernandez J, Lister R, Lenhard B, Holland PWH, Escriva H, Gomez-Skarmeta JL, Irimia M (2018) Amphioxus functional genomics and the origins of vertebrate gene regulation. Nature 564(7734):64–70. https://doi.org/10.1038/s41586-018-0734-6

Marshall NJ, Cortesi F, de Busserolles F, Siebeck UE, Cheney KL (2018) Colours and colour vision in reef fishes: past, present and future research directions. J Fish Biol. https://doi.org/10.1111/jfb.13849

McClintock JM, Carlson R, Mann DM, Prince VE (2001) Consequences of Hox gene duplication in the vertebrates: an investigation of the zebrafish Hox paralogue group 1 genes. Development 128(13):2471–2484

McGuirl MR, Volkening A, Sandstede B (2020) Topological data analysis of zebrafish patterns. Proc Natl Acad Sci U S A 117(10):5113–5124. https://doi.org/10.1073/pnas.1917763117

McKinney ML, McNamara KJ (1991) Heterochrony: the evolution of ontogeny. Plenum Press, New York

McMenamin SK, Parichy DM (2013) Metamorphosis in teleosts. Curr Top Dev Biol 103:127–165. https://doi.org/10.1016/B978-0-12-385979-2.00005-8

McMenamin SK, Bain EJ, McCann AE, Patterson LB, Eom DS, Waller ZP, Hamill JC, Kuhlman JA, Eisen JS, Parichy DM (2014) Thyroid hormone-dependent adult pigment cell lineage and pattern in zebrafish. Science 345(6202):1358–1361. https://doi.org/10.1126/science.1256251

Miller CT, Beleza S, Pollen AA, Schluter D, Kittles RA, Shriver MD, Kingsley DM (2007) cis-Regulatory changes in Kit ligand expression and parallel evolution of pigmentation in sticklebacks and humans. Cell 131(6):1179–1189. https://doi.org/10.1016/j.cell.2007.10.055

Mills MG, Nuckels RJ, Parichy DM (2007) Deconstructing evolution of adult phenotypes: genetic analyses of kit reveal homology and evolutionary novelty during adult pigment pattern development of Danio fishes. Development 134(6):1081–1090. https://doi.org/10.1242/dev.02799

Minchin JE, Hughes SM (2008) Sequential actions of Pax3 and Pax7 drive xanthophore development in zebrafish neural crest. Dev Biol 317(2):508–522. https://doi.org/10.1016/j.ydbio.2008.02.058

Mo ES, Cheng Q, Reshetnyak AV, Schlessinger J, Nicoli S (2017) Alk and Ltk ligands are essential for iridophore development in zebrafish mediated by the receptor tyrosine kinase Ltk. Proc Natl Acad Sci U S A 114(45):12027–12032. https://doi.org/10.1073/pnas.1710254114
Mundy NI, Kelly J, Theron E, Hawkins K (2003) Evolutionary genetics of the melanocortin-1 receptor in vertebrates. Ann N Y Acad Sci 994:307–312
Nagao Y, Suzuki T, Shimizu A, Kimura T, Seki R, Adachi T, Inoue C, Omae Y, Kamei Y, Hara I, Taniguchi Y, Naruse K, Wakamatsu Y, Kelsh RN, Hibi M, Hashimoto H (2014) Sox5 functions as a fate switch in medaka pigment cell development. PLoS Genet 10(4):e1004246. https://doi.org/10.1371/journal.pgen.1004246
Nagao Y, Takada H, Miyadai M, Adachi T, Seki R, Kamei Y, Hara I, Taniguchi Y, Naruse K, Hibi M, Kelsh RN, Hashimoto H (2018) Distinct interactions of Sox5 and Sox10 in fate specification of pigment cells in medaka and zebrafish. PLoS Genet 14(4):e1007260. https://doi.org/10.1371/journal.pgen.1007260
Nakamasu A, Takahashi G, Kanbe A, Kondo S (2009) Interactions between zebrafish pigment cells responsible for the generation of Turing patterns. Proc Natl Acad Sci U S A 106(21):8429–8434
Nakayama T, Nakajima K, Cox A, Fisher M, Howell M, Fish MB, Yaoita Y, Grainger RM (2017) No privacy, a Xenopus tropicalis mutant, is a model of human Hermansky-Pudlak Syndrome and allows visualization of internal organogenesis during tadpole development. Dev Biol 426 (2):472–486. https://doi.org/10.1016/j.ydbio.2016.08.020
Nord H, Dennhag N, Muck J, von Hofsten J (2016) Pax7 is required for establishment of the xanthophore lineage in zebrafish embryos. Mol Biol Cell 27(11):1853–1862. https://doi.org/10.1091/mbc.E15-12-0821
O'Reilly-Pol T, Johnson SL (2008) Melanocyte regeneration reveals mechanisms of adult stem cell regulation. Semin Cell Dev Biol 20(1):117–124
O'Reilly-Pol T, Johnson SL (2013) Kit signaling is involved in melanocyte stem cell fate decisions in zebrafish embryos. Development 140(5):996–1002. https://doi.org/10.1242/dev.088112
Odenthal J, Rossnagel K, Haffter P, Kelsh RN, Vogelsang E, Brand M, van Eeden FJ, Furutani-Seiki M, Granato M, Hammerschmidt M, Heisenberg CP, Jiang YJ, Kane DA, Mullins MC, Nusslein-Volhard C (1996a) Mutations affecting xanthophore pigmentation in the zebrafish, Danio rerio. Development 123:391–398
Odenthal J, Rossnagel K, Haffter P, Kelsh RN, Vogelsang E, Brand M, van Eeden FJM, Furutani-Seiki M, Granato M, Hammerschmidt C-P, Jiang Y-J, Kane DA, Kelsh RN, Mullins MC, Nusslein-Volhard C (1996b) Mutations affecting xanthophore pigmentation in the zebrafish, *Danio rerio*. Development 123:391–398
Ohno S (1970) Duplication of regulatory genes and receptors. In: Evolution by gene duplication. Springer, Berlin, pp 82–88
Oliphant LW, Hudon J (1993) Pteridines as reflecting pigments and components of reflecting organelles in vertebrates. Pigment Cell Res 6(4 Pt 1):205–208
Orton GL (1953) Development and migration of pigment cells in some teleost fishes. J Morphol 93:69–99
Owen JP, Kelsh RN, Yates CA (2020) A quantitative modelling approach to zebrafish pigment pattern formation. elife 9:e52998. https://doi.org/10.7554/eLife.52998
Parichy DM (1996) Salamander pigment patterns: how can they be used to study developmental mechanisms and their evolutionary transformation? Int J Dev Biol 40(4):871–884
Parichy DM (1998) Experimental analysis of character coupling across a complex life cycle: pigment pattern metamorphosis in the tiger salamander, Ambystoma tigrinum tigrinum. J Morphol 237(1):53–67
Parichy DM, Spiewak JE (2015) Origins of adult pigmentation: diversity in pigment stem cell lineages and implications for pattern evolution. Pigment Cell Melanoma Res 28(1):31–50. https://doi.org/10.1111/pcmr.12332
Parichy DM, Turner JM (2003a) Cellular interactions during adult pigment stripe development in zebrafish. Academic, New York, p 486

Parichy DM, Turner JM (2003b) Temporal and cellular requirements for Fms signaling during zebrafish adult pigment pattern development. Development 130(5):817–833. https://doi.org/10.1242/dev.00307

Parichy DM, Turner JM (2003c) Zebrafish puma mutant decouples pigment pattern and somatic metamorphosis. Dev Biol 256(2):242–257. https://doi.org/10.1016/s0012-1606(03)00015-0

Parichy DM, Rawls JF, Pratt SJ, Whitfield TT, Johnson SL (1999) Zebrafish sparse corresponds to an orthologue of c-kit and is required for the morphogenesis of a subpopulation of melanocytes, but is not essential for hematopoiesis or primordial germ cell development. Development 126 (15):3425–3436

Parichy DM, Mellgren EM, Rawls JF, Lopes SS, Kelsh RN, Johnson SL (2000a) Mutational analysis of endothelin receptor b1 (rose) during neural crest and pigment pattern development in the zebrafish Danio rerio. Dev Biol 227(2):294–306. https://doi.org/10.1006/dbio.2000.9899

Parichy DM, Ransom DG, Paw B, Zon LI, Johnson SL (2000b) An orthologue of the kit-related gene fms is required for development of neural crest-derived xanthophores and a subpopulation of adult melanocytes in the zebrafish, Danio rerio. Development 127(14):3031–3044

Parichy DM, Reedy MV, Erickson CA (2006) Chapter 5. Regulation of melanoblast migration and differentiation. In: Nordland JJ, Boissy RE, Hearing VJ, King RA, Oetting WS, Ortonne JP (eds) The pigmentary system: physiology and pathophysiology, 2nd edn. Oxford University Press, New York

Parichy DM, Elizondo MR, Mills MG, Gordon TN, Engeszer RE (2009) Normal table of postembryonic zebrafish development: staging by externally visible anatomy of the living fish. Dev Dyn 238:2975–3015

Patterson LB, Parichy DM (2013) Interactions with iridophores and the tissue environment required for patterning melanophores and xanthophores during zebrafish adult pigment stripe formation. PLoS Genet 9(5):e1003561. https://doi.org/10.1371/journal.pgen.1003561

Patterson LB, Parichy DM (2019) Zebrafish pigment pattern formation: insights into the development and evolution of adult form. Annu Rev Genet 53:043741. https://doi.org/10.1146/annurev-genet-112618-043741

Patterson LB, Bain EJ, Parichy DM (2014) Pigment cell interactions and differential xanthophore recruitment underlying zebrafish stripe reiteration and Danio pattern evolution. Nat Commun 5:5299. https://doi.org/10.1038/ncomms6299

Petratou K, Subkhankulova T, Lister JA, Rocco A, Schwetlick H, Kelsh RN (2018) A systems biology approach uncovers the core gene regulatory network governing iridophore fate choice from the neural crest. PLoS Genet 14(10):e1007402. https://doi.org/10.1371/journal.pgen.1007402

Prazdnikov DV, Shkil FN (2019) Experimental evidence of the role of heterochrony in evolution of the Mesoamerican cichlids pigment patterns. Evol Dev 21(1):3–15

Price AC, Weadick CJ, Shim J, Rodd FH (2008) Pigments, patterns, and fish behavior. Zebrafish 5 (4):297–307. https://doi.org/10.1089/zeb.2008.0551

Prince VE, Pickett FB (2002) Splitting pairs: the diverging fates of duplicated genes. Nat Rev Genet 3(11):827–837. https://doi.org/10.1038/nrg928

Protas ME, Hersey C, Kochanek D, Zhou Y, Wilkens H, Jeffery WR, Zon LI, Borowsky R, Tabin CJ (2006) Genetic analysis of cavefish reveals molecular convergence in the evolution of albinism. Nat Genet 38(1):107–111

Quigley IK, Turner JM, Nuckels RJ, Manuel JL, Budi EH, MacDonald EL, Parichy DM (2004) Pigment pattern evolution by differential deployment of neural crest and post-embryonic melanophore lineages in Danio fishes. Development 131(24):6053–6069. https://doi.org/10.1242/dev.01526

Quigley IK, Manuel JL, Roberts RA, Nuckels RJ, Herrington ER, MacDonald EL, Parichy DM (2005) Evolutionary diversification of pigment pattern in Danio fishes: differential fms dependence and stripe loss in D. albolineatus. Development 132(1):89–104. https://doi.org/10.1242/dev.01547

Rawles ME (1948) Origin of melanophores and their role in development of color patterns in vertebrates. Physiol Rev 28:383–408
Rawls JF, Johnson SL (2003) Temporal and molecular separation of the kit receptor tyrosine kinase's roles in zebrafish melanocyte migration and survival. Dev Biol 262(1):152–161
Ré P, Meneses I (2008) Early stages of marine fishes occuring in the Iberian peninsula. IPIMAR, Isabel Meneses
Rican O, Musilova Z, Muska M, Novak J (2005) Development of coloration patterns in neotropical cichlids (Teleostei: Cichlidae: Cichlasomatinae). Folia Zool 54:1–46
Rican O, Pialek L, Dragova K, Novak J (2016) Diversity and evolution of the Middle American cichlid fishes (Teleostei: Cichlidae) with revised classification. Vertebr Zool 66(1):3–102
Riddle MR, Aspiras AC, Gaudenz K, Peuss R, Sung JY, Martineau B, Peavey M, Box AC, Tabin JA, McGaugh S, Borowsky R, Tabin CJ, Rohner N (2018) Insulin resistance in cavefish as an adaptation to a nutrient-limited environment. Nature 555(7698):647–651. https://doi.org/10.1038/nature26136
Roberts RB, Ser JR, Kocher TD (2009) Sexual conflict resolved by invasion of a novel sex determiner in Lake Malawi cichlid fishes. Science 326(5955):998–1001. https://doi.org/10.1126/science.1174705
Roberts RB, Moore EC, Kocher TD (2017) An allelic series at pax7a is associated with colour polymorphism diversity in Lake Malawi cichlid fish. Mol Ecol 26(10):2625–2639. https://doi.org/10.1111/mec.13975
Rosenthal GG (2007) Spatiotemporal dimensions of visual signals in animal communication. Annu Rev Ecol Evol Syst 38:155–178. https://doi.org/10.1146/annurev.ecolsys.38.091206.095745
Roux N, Salis P, Lambert A, Logeux V, Soulat O, Romans P, Frederich B, Lecchini D, Laudet V (2019) Staging and normal table of postembryonic development of the clownfish (Amphiprion ocellaris). Dev Dyn 248(7):545–568. https://doi.org/10.1002/dvdy.46
Saenko SV, Teyssier J, van der Marel D, Milinkovitch MC (2013) Precise colocalization of interacting structural and pigmentary elements generates extensive color pattern variation in Phelsuma lizards. BMC Biol 11:105. https://doi.org/10.1186/1741-7007-11-105
Saldana-Caboverde A, Kos L (2010) Roles of endothelin signaling in melanocyte development and melanoma. Pigment Cell Melanoma Res 23(2):160–170. https://doi.org/10.1111/j.1755-148X.2010.00678.x
Salis P, Roux N, Soulat O, Lecchini D, Laudet V, Frederich B (2018) Ontogenetic and phylogenetic simplification during white stripe evolution in clownfishes. BMC Biol 16(1):90. https://doi.org/10.1186/s12915-018-0559-7
Salis P, Lorin T, Lewis V, Rey C, Marcionetti A, Escande ML, Roux N, Besseau L, Salamin N, Semon M, Parichy D, Volff JN, Laudet V (2019) Developmental and comparative transcriptomic identification of iridophore contribution to white barring in clownfish. Pigment Cell Melanoma Res 32:391–402. https://doi.org/10.1111/pcmr.12766
Salzburger W, Braasch I, Meyer A (2007) Adaptive sequence evolution in a color gene involved in the formation of the characteristic egg-dummies of male haplochromine cichlid fishes. BMC Biol 5:51. https://doi.org/10.1186/1741-7007-5-51
Santos ME, Braasch I, Boileau N, Meyer BS, Sauteur L, Bohne A, Belting HG, Affolter M, Salzburger W (2014) The evolution of cichlid fish egg-spots is linked with a cis-regulatory change. Nat Commun 5:5149. https://doi.org/10.1038/ncomms6149
Saunders LM, Aman AJ, Mishra AK, Lewis VM, Toomey MB, Packer JS, Qiu X, McFaline-Figueroa JL, Corbo JC, Trapnell C, Parichy DM (2019) Thyroid hormone regulates distinct paths to maturation in pigment cell lineages. elife 8:e45181
Sawada R, Aramaki T, Kondo S (2018) Flexibility of pigment cell behavior permits the robustness of skin pattern formation. Genes Cells 23(7):537–545. https://doi.org/10.1111/gtc.12596
Schartl M, Larue L, Goda M, Bosenberg MW, Hashimoto H, Kelsh RN (2016) What is a vertebrate pigment cell? Pigment Cell Melanoma Res 29(1):8–14. https://doi.org/10.1111/pcmr.12409
Schneider RA, Helms JA (2003) The cellular and molecular origins of beak morphology. Science 299(5606):565–568

Seehausen O, Mayhew PJ, Van Alphen JJM (1999) Evolution of colour patterns in East African cichlid fish. J Evol Biol 12(3):514–534

Seehausen O, Terai Y, Magalhaes IS, Carleton KL, Mrosso HDJ, Miyagi R, van der Sluijs I, Schneider MV, Maan ME, Tachida H, Imai H, Okada N (2008) Speciation through sensory drive in cichlid fish. Nature 455(7213):620–626. https://doi.org/10.1038/nature07285

Selz Y, Braasch I, Hoffmann C, Schmidt C, Schultheis C, Schartl M, Volff JN (2007) Evolution of melanocortin receptors in teleost fish: the melanocortin type 1 receptor. Gene 401 (1-2):114–122. https://doi.org/10.1016/j.gene.2007.07.005

Shainer I, Michel M, Marquart GD, Bhandiwad AA, Zmora N, Ben-Moshe Livne Z, Zohar Y, Hazak A, Mazon Y, Forster D, Hollander-Cohen L, Cone RD, Burgess HA, Gothilf Y (2019) Agouti-related protein 2 is a new player in the teleost stress response system. Curr Biol 29 (12):2009–2019. https://doi.org/10.1016/j.cub.2019.05.021

Singh AP, Schach U, Nusslein-Volhard C (2014) Proliferation, dispersal and patterned aggregation of iridophores in the skin prefigure striped colouration of zebrafish. Nat Cell Biol 16 (6):607–614. https://doi.org/10.1038/ncb2955

Singh AP, Dinwiddie A, Mahalwar P, Schach U, Linker C, Irion U, Nusslein-Volhard C (2016) Pigment cell progenitors in zebrafish remain multipotent through metamorphosis. Dev Cell 38 (3):316–330. https://doi.org/10.1016/j.devcel.2016.06.020

Spiewak JE, Bain EJ, Liu J, Kou K, Sturiale SL, Patterson LB, Diba P, Eisen JS, Braasch I, Ganz J, Parichy DM (2018) Evolution of endothelin signaling and diversification of adult pigment pattern in Danio fishes. PLoS Genet 14(9):e1007538. https://doi.org/10.1371/journal.pgen.1007538

Streelman JT, Albertson RC, Kocher TD (2003) Genome mapping of the orange blotch colour pattern in cichlid fishes. Mol Ecol 12(9):2465–2471. https://doi.org/10.1046/j.1365-294x.2003.01920.x

Sturm RA, Duffy DL, Box NF, Newton RA, Shepherd AG, Chen W, Marks LH, Leonard JH, Martin NG (2003) Genetic association and cellular function of MC1R variant alleles in human pigmentation. Ann N Y Acad Sci 994:348–358

Svetic V, Hollway GE, Elworthy S, Chipperfield TR, Davison C, Adams RJ, Eisen JS, Ingham PW, Currie PD, Kelsh RN (2007) Sdf1a patterns zebrafish melanophores and links the somite and melanophore pattern defects in choker mutants. Development 134(5):1011–1022. https://doi.org/10.1242/dev.02789

Tadokoro R, Shikaya Y, Takahashi Y (2019) Wide coverage of the body surface by melanocyte-mediated skin pigmentation. Dev Biol 449(2):83–89. https://doi.org/10.1016/j.ydbio.2018.04.016

Takahashi G, Kondo S (2008) Melanophores in the stripes of adult zebrafish do not have the nature to gather, but disperse when they have the space to move. Pigment Cell Melanoma Res 21 (6):677–686. https://doi.org/10.1111/j.1755-148X.2008.00504.x

Tryon RC, Higdon CW, Johnson SL (2011) Lineage relationship of direct-developing melanocytes and melanocyte stem cells in the zebrafish. PLoS ONE 6(6):e21010. https://doi.org/10.1371/journal.pone.0021010

Twitty VC, Bodenstein D (1939) Correlated genetic and embryological experiments on Triturus. J Exp Zool 81:357–398

Urban S, Nater A, Meyer A, Kratochwil CF (2020) Different sources of allelic variation drove repeated color pattern divergence in cichlid fishes. Mol Biol Evol. https://doi.org/10.1093/molbev/msaa237

Usui Y, Aramaki T, Kondo S, Watanabe M (2019) The minimal gap-junction network among melanophores and xanthophores required for stripe pattern formation in zebrafish. Development 146(22). https://doi.org/10.1242/dev.181065

Volkening A, Sandstede B (2015) Modelling stripe formation in zebrafish: an agent-based approach. J R Soc 12:112. https://doi.org/10.1098/rsif.2015.0812

Volkening A, Sandstede B (2018) Iridophores as a source of robustness in zebrafish stripes and variability in Danio patterns. Nat Commun 9(1):3231. https://doi.org/10.1038/s41467-018-05629-z

Volkening A, Abbott MR, Chandra N, Dubois B, Lim F, Sexton D, Sandstede B (2020) Modeling stripe formation on growing zebrafish tailfins. Bull Math Biol 82(5):56. https://doi.org/10.1007/s11538-020-00731-0

Wake DB, Wake MH, Specht CD (2011) Homoplasy: from detecting pattern to determining process and mechanism of evolution. Science 331(6020):1032. https://doi.org/10.1126/science.1188545

Walderich B, Singh AP, Mahalwar P, Nusslein-Volhard C (2016) Homotypic cell competition regulates proliferation and tiling of zebrafish pigment cells during colour pattern formation. Nat Commun 7:11462. https://doi.org/10.1038/ncomms11462

Watanabe M, Kondo S (2015) Is pigment patterning in fish skin determined by the Turing mechanism? Trends Genet 31(2):88–96. https://doi.org/10.1016/j.tig.2014.11.005

Watanabe M, Iwashita M, Ishii M, Kurachi Y, Kawakami A, Kondo S, Okada N (2006) Spot pattern of *leopard Danio* is caused by mutation in the zebrafish connexin41.8 gene. EMBO Rep 7 (9):893–897. https://doi.org/10.1038/sj.embor.7400757

Watanabe K, Washio Y, Fujinami Y, Aritaki M, Uji S, Suzuki T (2008) Adult-type pigment cells, which color the ocular sides of flounders at metamorphosis, localize as precursor cells at the proximal parts of the dorsal and anal fins in early larvae. Develop Growth Differ 50(9):731–741

Watanabe M, Watanabe D, Kondo S (2012) Polyamine sensitivity of gap junctions is required for skin pattern formation in zebrafish. Sci Rep 2:473. https://doi.org/10.1038/srep00473

Woodcock MR, Vaughn-Wolfe J, Elias A, Kump DK, Kendall KD, Timoshevskaya N, Timoshevskiy V, Perry DW, Smith JJ, Spiewak JE, Parichy DM, Voss SR (2017) Identification of mutant genes and introgressed tiger salamander DNA in the laboratory axolotl, Ambystoma mexicanum. Sci Rep 7(1):6. https://doi.org/10.1038/s41598-017-00059-1

Yamada T, Okauchi M, Araki K (2010) Origin of adult-type pigment cells forming the asymmetric pigment pattern, in Japanese flounder (Paralichthys olivaceus). Dev Dyn 239(12):3147–3162. https://doi.org/10.1002/dvdy.22440

Yamaguchi M, Yoshimoto E, Kondo S (2007) Pattern regulation in the stripe of zebrafish suggests an underlying dynamic and autonomous mechanism. Proc Natl Acad Sci U S A 104 (12):4790–4793

Yamanaka H, Kondo S (2014) In vitro analysis suggests that difference in cell movement during direct interaction can generate various pigment patterns in vivo. Proc Natl Acad Sci U S A 111 (5):1867–1872. https://doi.org/10.1073/pnas.1315416111

Zhang C, Song Y, Thompson DA, Madonna MA, Millhauser GL, Toro S, Varga Z, Westerfield M, Gamse J, Chen W, Cone RD (2010) Pineal-specific agouti protein regulates teleost background adaptation. Proc Natl Acad Sci U S A 107(47):20164–20171. https://doi.org/10.1073/pnas.1014941107

Zhang YM, Zimmer MA, Guardia T, Callahan SJ, Mondal C, Di Martino J, Takagi T, Fennell M, Garippa R, Campbell NR, Bravo-Cordero JJ, White RM (2018) Distant insulin signaling regulates vertebrate pigmentation through the sheddase Bace2. Dev Cell 45(5):580. https://doi.org/10.1016/j.devcel.2018.04.025

Ziegler I (2003) The pteridine pathway in zebrafish: regulation and specification during the determination of neural crest cell-fate. Pigment Cell Res 16(3):172–182

Ziegler I, McDonald T, Hesslinger C, Pelletier I, Boyle P (2000) Development of the pteridine pathway in the zebrafish, Danio rerio. J Biol Chem 275(25):18926–18932. https://doi.org/10.1074/jbc.M910307199

Chapter 11
Mechanisms of Feather Structural Coloration and Pattern Formation in Birds

Shinya Yoshioka and Toyoko Akiyama

Abstract Many bird species have microstructures that produce diverse feather colors. The morphology of these color-generating microstructures differs largely depending on the avian species. We classified the structural coloration of bird feathers into three types: thin-layer interference, melanin granules, and amorphous networks, and provide examples for each in avian coloration mechanisms. Bird body colors and patterns can depend on the body part, age, sex, and season; however, body patterns and each feather color combination are essentially determined in the feather bud during the feather developmental process. Although birds do not have the colorful erythrophore, xanthophore, and iridophore cells found in lower vertebrates, they obtain dietary pigments from foods and have structural colors within their feathers. With these colorations, feathers in birds often show amazing complex patterns and the diversity of their body colors and patterns are characteristics is very important for their living strategy.

Keywords Birds · Feather color · Structural color · Pattern formation · Microstructure

11.1 Mechanisms of Structural Coloration in Birds

11.1.1 Introduction

Some animals possess surprisingly brilliant colors. The most famous example may be the Morpho butterfly, which has strikingly blue wings (Fig. 11.1a). These colors are called structural colors because they essentially originate from microstructures

S. Yoshioka
Department of Science and Technology, Tokyo University of Science, Noda, Japan

T. Akiyama (✉)
Department of Biology, Keio University, Yokohama, Japan
e-mail: akiyama@a3.keio.jp

H. Hashimoto et al. (eds.), *Pigments, Pigment Cells and Pigment Patterns*,
https://doi.org/10.1007/978-981-16-1490-3_11

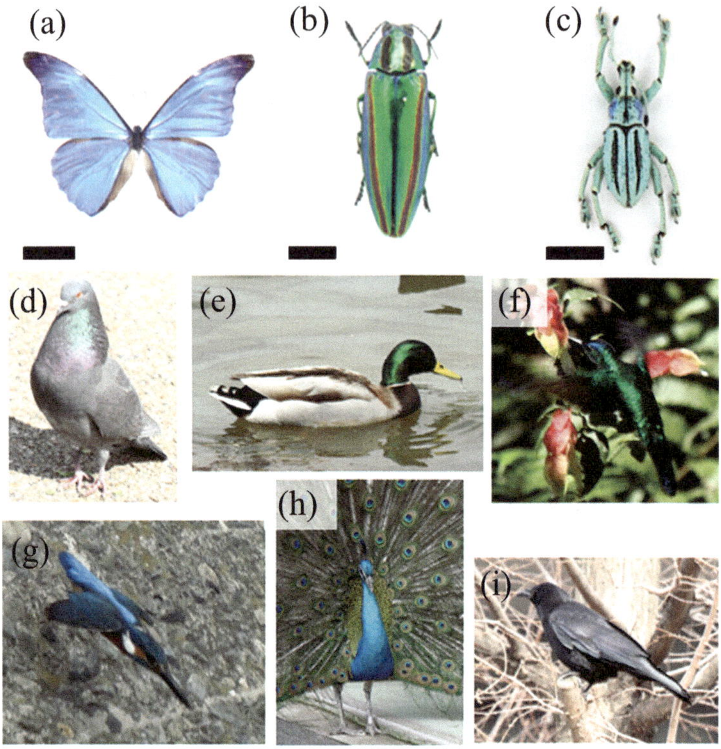

Fig. 11.1 Animals with structural feather coloration. (**a**) Morpho butterfly, (**b**) jewel beetle, (**c**) weevil, (**d**) rock dove, (**e**) mallard, (**f**) hummingbird, (**g**) common kingfisher, (**h**) peacock, and (**i**) crow. Scale bar: (**a**) 2 cm, (**b**) and (**c**) 1 cm

that have a size comparable to the wavelength of light and are highly reflective for a certain wavelength range. Butterflies, jewel beetles, and weevils are some of the best known examples of insects with structural colors (Fig. 11.1a–c). However, structural colors are not only found in insects but also in a wide range of animals, including bird species (Fig. 11.1d–i). The peacock is the most well-known avian example, with upper tail covert that have a characteristic iridescent eye pattern (i.e., the color changes depending on the angle of observation and illumination) and represents one of the major characteristics of structural colors.

In this section, we describe the structural coloration of bird feathers. For the process in other animals, readers are referred to other good review papers and monographs on this subject (Srinivasarao 1999; Vukusic and Sambles 2003; Kinoshita 2008; Kinoshita et al. 2008). First, we begin with a brief explanation of the physical mechanisms behind the coloration. Second, the colored structures found in bird feathers are classified into three types and a few examples of each are provided. Third, factors other than these microstructures that are important for coloration are reported. These latter factors largely affect the vividness of the color and the angular range of reflection. Finally, we present some comments for future research directions.

11.1.2 Basic Mechanisms of Structural Coloration

Figure 11.2 shows four fundamental types of the color-causing microstructures. The simplest type is the single thin layer shown in Fig. 11.2, which despite its simplicity, can exhibit a beautiful iridescence similar to that often seen in soap bubbles (Fig. 11.3a). When light is incident on the thin layer, some of it is reflected off the top surface, while the remaining part transmits and propagates inside the layer and then impinges on the bottom surface where some of it is again reflected. Thus, the observer sees the superposition of the two light waves reflected off the top and bottom surfaces, which can alter the perceived light reflection. If the two light waves overlap *in phase* (constructive interference), then the reflection is enhanced and reflectance increases. Conversely, if the light waves overlap *out of phase* (destructive interference), then the reflection is suppressed and reflectance decreases. The way in which the two light waves overlap depends on the difference in the optical path length, which is expressed using several physical parameters: the light wavelength, the refractive index (n) and layer thickness (d), and the angle of incidence. In the case of the soap bubble, where the refractive index of the layer is higher than that of the surrounding medium, the wavelength at which constructive interference occurs (λ_m) is given as

$$2nd\cos\theta_l = \left(m + \frac{1}{2}\right)\lambda_m \tag{11.1}$$

where m is an integer called the order of the optical interference. The angle θ_l is the refraction angle inside the layer, which is related to the angle of incidence according to Snell's law. Eq. (11.1) was constructed by considering the optical interference of

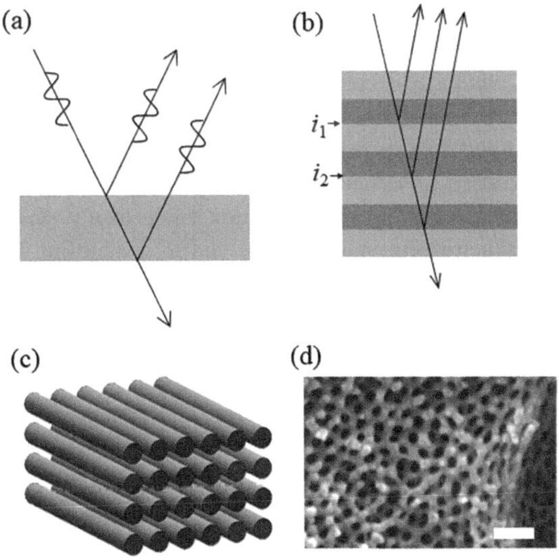

Fig. 11.2 Basic structures that produce structural color. (**a**) A single thin layer. Two light waves reflected at the top and bottom surfaces interfere with each other. (**b**) A periodic multilayer structure. i_1 and i_2 indicate interfaces that are separated by one period. (**c**) A two-dimensional photonic crystal structure. (**d**) An amorphous network structure of a common kingfisher (*Alcedo atthis*) feather barb cross-section observed by scanning electron microscopy. Scale bar: (**d**) 0.5 μm

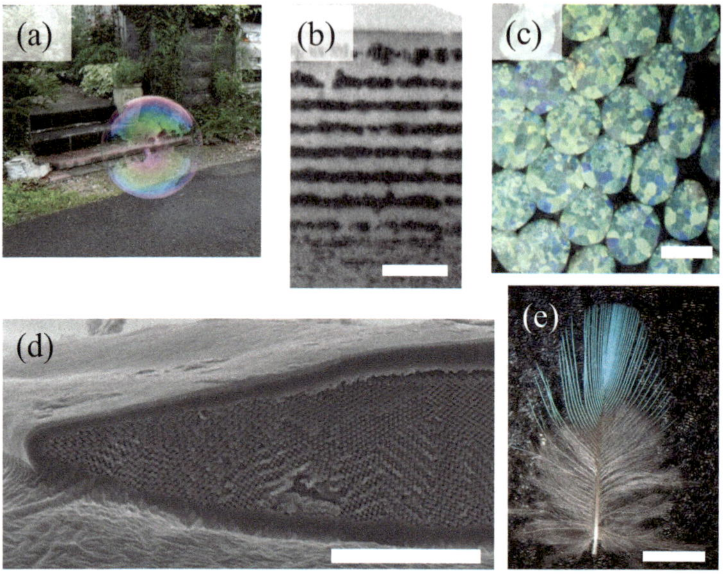

Fig. 11.3 Examples of structural colors and the color-causing microstructures. (**a**) The iridescence of a soap bubble caused by thin-layer interference. (**b**) The multilayer structure of the jewel beetle (*Chrysochroa fulgidissima*) observed by transmission electron microscopy. (**c**) Scales from the weevil (*Eupholus cuvieri*, Fig. 11.1c) and (**d**) its cross-section observed by scanning electron microscopy. (**e**) A feather of the common kingfisher (*Alcedo atthis*). Scale bar: (**b**) 500 nm, (**c**) 50μm, (**d**) 5μm, and (**e**) 3 cm

the two light waves only. However, there are many other light waves that follow optical paths called multiple reflections, in which the light wave reflects many times within the thin layer. However, the multiple reflections do not usually contribute largely to the overall reflection. Thus, Eq. (11.1) is based on the two-wave approximation and usually works well for estimating the wavelength of reflection. However, this also means that the reflectance is not very high except for a layer with an extraordinarily high refractive index because only the two waves are assumed to be involved in the reflection mechanism. In the case of a soap bubble in air, the maximum reflectance is calculated to be less than 8% under the normal incidence when using the refractive index value of $n = 1.33$ (water). The two-wave approximation results in a spectral shape resembling a sinusoidal curve that oscillates smoothly between the maximum and minimum and thus, the wavelength range of the reflection band is not very narrow, implying that it is difficult to produce a highly saturated color via the optical interference of a single thin layer.

In contrast to the single layer, a stack of layers (i.e., the multilayer system) can have a very high reflectance only in a narrow wavelength range, which produces a brilliant and saturated color. One of the most famous natural examples is the jewel beetle (*Chrysochroa fulgidissima*), shown in Fig. 11.1b, which has a multilayer structure beneath the surface of the elytrum that consists of electron-dense and electron-lucent cuticle layers (Fig. 11.3b). To understand the physical mechanism of reflection, we consider the periodic system shown in Fig. 11.2b, where two types of layers, denoted with the subscripts a and b, are periodically stacked at thicknesses d_a and d_b, respectively. The refractive indices of these layers are denoted by n_a and n_b, respectively. In this system, we can expect enhanced reflection when constructive interference occurs between the light waves that are reflected from interfaces separated by one period (e.g., i_1 and i_2 in Fig. 11.2b). The constructive interference occurs when the round trip of the optical path for one period is equal to an integer multiple of the wavelength, as is expressed by the following equation:

$$2(n_a d_a + n_b d_b) = m\lambda_m \tag{11.2}$$

In this equation, we assume that the incident light is normal to the layers and can correctly estimate the wavelength of reflection, except for some special cases. Reflectance rapidly increases as the number of layers increases. For example, reflectance can reach nearly 100% for a system of 20 layers with an $n_a = 1.55$ and $n_b = 1.0$. The exact reflectance value can be calculated by several computational methods, such as the transfer matrix and iteration of Fresnel's equation (Kinoshita 2013). When the layers are infinitely stacked, the reflectance is 100% for a certain wavelength range, including λ_m. This wavelength range depends on the ratio of the refractive indices n_a and n_b; when the ratio is more different from unity, the reflection band widens.

Some animals have more complicated microstructures, such as a photonic crystal structure (Fig. 11.2c). In a typical crystal, the atoms or molecules are arranged periodically with the Å scale; in a photonic crystal, the structure repeats with the size of the light wavelength. A photonic crystal is often classified by the dimensions along which the structure is periodic. As such, the multilayer system is called a one-dimensional photonic crystal because the system is periodic only along the direction perpendicular to the layers (Fig. 11.2b). In contrast, the two-dimensional photonic crystal structure is periodic into two orthogonal directions (Fig. 11.2c). In some insect species, the structural color is caused by three-dimensional photonic crystals (Seago et al. 2009), which are periodic in three orthogonal directions. The weevil (*Eupholus cuvieri*) (Fig. 11.1c) has elytrum covered with brightly colored scales (Fig. 11.3c) that have a complicated cuticle network structure (Fig. 11.4d).

The periodic structure initiates a fixed relationship between the scattered light waves. Thus, the optical interference between the scattered waves results in an enhanced reflection, similar to when X-rays are diffracted by a typical crystal according to Bragg's law. The wavelength of reflection is mainly determined by the period of the structure and therefore, the wavelength becomes longer or shorter when the period is larger or smaller. To estimate the wavelength range of reflection exactly, the frequency range of the photonic band gap in the photonic band diagram must be known because electromagnetic waves with a frequency within the photonic band gap cannot propagate inside the photonic crystal. Therefore, when a light wave with this frequency is incident on the surface of a photonic crystal, the light is perfectly reflected because it cannot propagate in the crystal. Although the photonic

Fig. 11.4 Schematic illustration of the neck feather of the rock dove (Nakamura et al. 2008). The general bird feather has the structures called rachis, barb and barbule. Detailed structures inside barbules are drawn for the iridescent part of rock dove's neck feather

band calculation may be mathematically tedious, a simple estimation can often be calculated using Eq. (11.2) because the multilayer system is usually a good model for the reflection from a photonic crystal (Stavenga et al. 2017).

An amorphous network (Fig. 11.2d) is another type of color-causing microstructure whose size is in the approximate order of the wavelength of light. However, it is quite different from the aforementioned types in that the structure is not periodic. An example of this type of microstructure is the blue feathers of the common kingfisher (*Alcedo atthis*; Figs. 11.1g and 11.3e). Because the network appears random, Rayleigh scattering was thought to be the origin of its coloration, as this scattering process causes the blue color of the sky. However, the reflectance spectra of blue feathers with the amorphous network have a distinctive peak in the visible wavelength range and contradict the Rayleigh scattering theory, which predicts a monotonic increase in reflectance with a decrease in wavelength. However, Prum et al. (1999) revealed that the amorphous network was not actually random but quasi-ordered. Thus, the scattered light could constructively interfere in a wavelength range to produce color. Fourier analysis is often employed to estimate the wavelength of reflection.

11.1.3 Color-Causing Microstructures in Bird Feathers

The color-causing structures in bird feathers can be classified into three types. The first type is a single thin layer on the surface of the feather barbule. The soap bubble can be a good model for this type. In the second type, the structural color is caused by the periodic arrangement of melanin granules inside the barbule. Many examples of this type can be modeled as photonic crystals. The third type is the amorphous network structure of keratin, which is found in feather barbs.

Thin-Layer Interference Microstructures

The rock dove (*Columba livia*) has iridescent neck feathers, with structures schematically illustrated in Fig. 11.4 (Nakamura et al. 2008). The feather structure is hierarchical, with many barbs protruding from the rachis and many small barbules extending from both sides of the barb at an angle of 50–60 °. In the cross-section of the barbule, we can observe melanin granules enveloped by a thin layer of cortex (Fig. 11.5). Because the melanin granules are irregularly arranged and do not have a uniform size, the cortex layer is expected to cause the structural color. Indeed, experimentally determined reflectance spectra can be reproduced by theoretically calculated spectra, thus confirming that the thin-layer interference as the origin of the color (Yin et al. 2006; Yoshioka et al. 2007).

The coloration mechanism in the neck feathers of rock doves is physically simple. However, the iridescence has a peculiar feature in that the color change is limited to only two colors—green and purple—and the change between these colors occurs suddenly according to slight changes in the viewing angle (Yoshioka et al. 2007). This peculiar feature can be understood when we consider several orders of optical interference. Using the thickness (650 nm) of the cortex layer, the interference

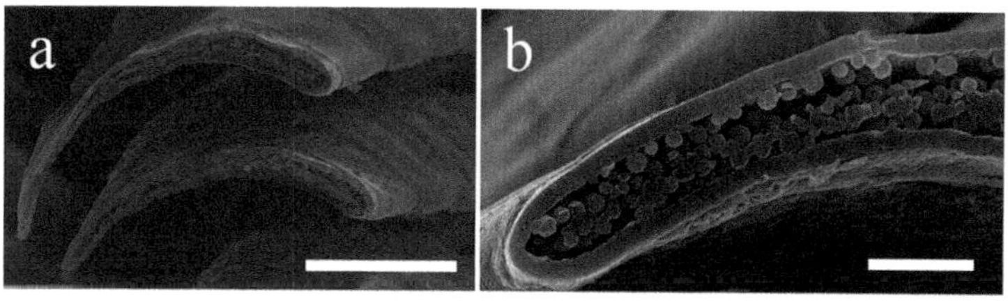

Fig. 11.5 Scanning electron microscopic images of the cross-section of iridescent barbules of the neck feather of the rock dove (Yoshioka et al., 2007). Scale bar: (**a**) 20 μm and (**b**) 3 μm

condition of Eq. (11.1) is satisfied for several wavelengths with different orders in the range of visible light. Among these wavelengths, two reflection bands of two adjacent interference orders overlap in the blue and red sensitivities of the human color vision at a certain viewing angle. At this angle, the feather appears purple owing to the mixture of the blue and red colors. When the viewing angle is changed, the two reflection bands shift gradually according to Eq. (11.1), deviating from the blue and red sensitivities and overlapping the green sensitivity of the human color vision. Thus, this physically simple but chromatically sophisticated mechanism causes the two-color iridescence of the structural color in rock dove neck feathers. The sudden change between the two colors has been quantitatively understood as an effect of the change in the refraction angle (Yoshioka et al. 2007). We may extend the above discussion on the color perception of the rock dove by considering the effective maximum sensitivities of the visual pigments, which have been determined by microspectrophotometry (Bowmaker 1977). Because the wavelengths of the maximum sensitivities are located almost at the peaks of the reflection bands, a similar color effect is expected at least at the level of absorption of the visual pigment.

Not all structural colors are brilliant and in fact, some of them are quite inconspicuous. One such example is the jungle crow (*Corvus macrorhynchos*), whose black feathers exhibit a weak iridescence that originates from thin-layer interference that occurs in the epicuticle layer located on the outermost surface of the barbule (Lee et al. 2012). The refractive index of this layer is not very different from that of the underlying keratin layer and therefore, the reflectance is low even at the wavelength of constructive interference, which directly explains the inconspicuous iridescence. Lee et al. (2012) reported that the thin epicuticle layer may cause weak iridescence in other bird species with black feathers, such as the northern raven (*Corvus corax*), carrion crow (*Corvus corone*), cowbird, and grackles (*Icteridae*). The iridescence of hadeda ibis (*Bostrychia hagedash*) has also been reported to originate from thin-layer interference (Brink and van der Berg 2004).

Melanin Granule Microstructures

The common peacock (Indian peafowl, *Pavo cristatus*) is a good example of a bird species with melanin granule type microstructures. When the feather is observed under an optical microscope, the barbules have a jewel-like appearance (Fig. 11.6a). In the barbule cross-section, the melanin granules form a rectangular lattice beneath the surface (Fig. 11.6b). When the barbule is sectioned longitudinally, the granules appear long (Fig. 11.6c). Thus, the granule has a cylindrical shape with a diameter of approximately 100 nm and length of a few micrometers (Fig. 11.6d). Zi et al. (2003) modeled the structure as a two-dimensional photonic crystal and calculated the photonic band diagram, thereby confirming that the lattice arrangement of the granules was the origin of the structural color. Additionally, a simpler analysis can be used to correctly estimate the wavelength of the reflection by applying the interference condition of the multilayer interference; because the left-hand side of

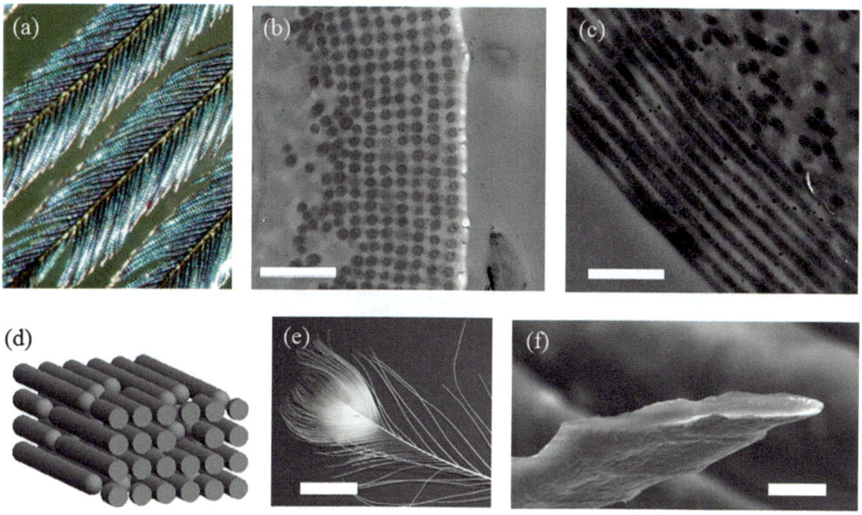

Fig. 11.6 Feather barbule of a peacock. (**a**) An optical micrograph of the blue feather with reflective barbules. Transmission electron micrographs of the (**b**) cross-section and (**c**) longitudinal section of the barbule. (**d**) Schematic of the arrangement of the melanin granules in peacock feather barbules. (**e**) A feather from an albino peacock. (**f**) A cross-section of the barbule of an albino feather that shows the absence pigment granules. Scale bar: (**b**) and (**c**) 1 μm, (**e**) 3 cm, and (**f**) 10 μm

Eq. (11.2) describes the round trip of the optical path length of one period of the structure, it is reasonable to approximate it by the product of the period of the photonic crystal structure and the mean refractive index. The estimated wavelengths were in reasonable agreement with the experimentally determined reflection peaks (Yoshioka and Kinoshita 2002). In a recent study, the adequacy of approximating the photonic crystal as a multilayer system was studied in the feather of the mallard (*Anas platyrhynchos*) by comparing the results of a reflectance calculation that assumed realistic and simplified model structures (Stavenga et al. 2017). This study found that the reflection from the photonic crystal was well explained by simple multilayer modeling.

The different colors of peacock feathers are realized through changes in the period of the photonic crystal lattice. In the case of the characteristic eye pattern of the upper tail covert, the distance between the melanin granules is accurately controlled according to the position of the barbule, forming the eye pattern of the feather. It is interesting to consider what physical mechanisms that regulate pattern formation. It has been reported that the reaction–diffusion system can reproduce various patterns in bird feathers by choosing the appropriate parameter values for the system (Prum and Williamson 2002).

The lattice type of the photonic crystal differs depending on the bird species. In the dabbling duck feathers (Eliason and Shawkey 2012; Stavenga et al. 2017), melanin granules elongated and formed a hexagonal rather than a rectangular lattice (Fig. 11.7a). The shape and composition of these granules also vary depending on

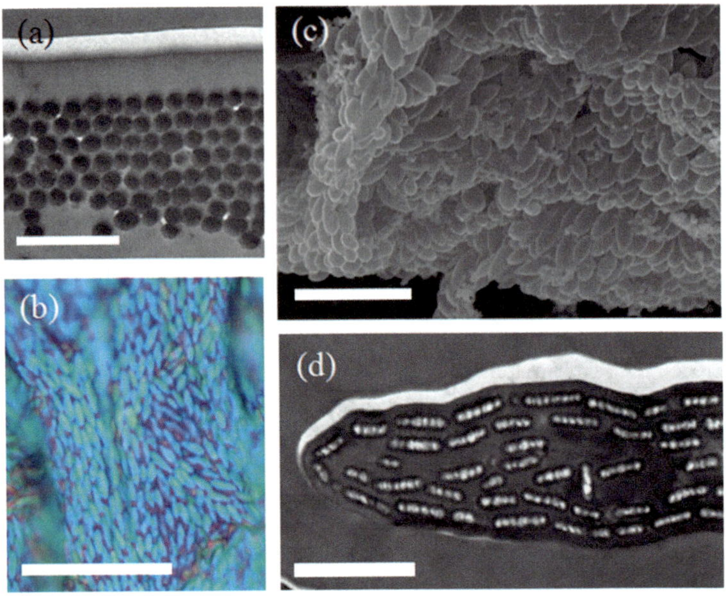

Fig. 11.7 Feather barbules of the mallard (*Anas platyrhynchos*) and hummingbird (*Colibri coruscans*). (**a**) A cross-section of the barbule from a green feather of the mallard head. (**b**) An optical micrograph of a hummingbird barbule. (**c**) A scanning electron micrograph of a fractured barbule of a hummingbird. Melanin granules with a plate shape are observed. (**d**) A transmission electron micrograph of the cross-section of the barbule in which the air cavities within the melanin granules are visible. Scale bar: (**a**) 1μm, (**b** and **c**) 5μm, and (**d**) 2μm

the species. For example, hummingbirds (*Colibri coruscans*) have hollow melanin granules that are shaped like oval plates (Greenewalt et al. 1960; Eliason et al. 2013; Fig. 11.7b–d), which are largely different from cylindrical granules in the peacock feather. Optical micrography showed many small reflection patches on the barbule, which correspond to the electron micrographs that revealed the barbule contains many small oval plate-shaped granules (Fig. 11.7c). Interestingly, transmission electron microscopy showed that these plates are hollow (Fig. 11.7d). The stacked granule plates are thought to cause optical interference and result in structural color. Because the refractive index of air is much smaller than that of keratin, a hollow structure can increase reflectance. The variations in the crystal lattice type and shape of the melanin granule that constitute the microstructure have been thoroughly studied by Durrer (1977).

Now that we know melanin granules are related to the coloration mechanism in bird feathers, one natural question arises: what is the microstructure like in albino individuals? We investigated the white feathers of the albino peacock and found no melanin granules when feathers were observed by electron microscopy (Fig. 11.6e). Instead, the cross-section of the barbule appeared flat and structureless (Fig. 11.6f).

Amorphous Network Microstructures

Structural color is usually iridescent because the interference condition relates the wavelength of light to the angle of observation. However, some bird species have blue feathers that appear less or non-iridescent. That is, the feathers retain the same color even when observed from a highly oblique angle. This unique optical property is caused by the amorphous structure that is located in the feather barb (Fig. 11.8). Indeed, the barbs rather than the barbules look blue in common kingfisher (*A. atthis*) feathers (Fig. 11.3e). A simple experiment clearly demonstrated that this blue color has a structural origin. When the barb was soaked in liquid toluene, the color vanished as the liquid penetrated into the tissue inside the barb (Fig. 11.8c–e). This was because the refractive index of toluene was very close to that of keratin, which suppressed the scattering of light.

When the cross-section of the barb was observed under an optical microscope, the outer region appeared reflective (Fig. 11.8a), which was consistent with the spongy network of keratin observed in the outer part of the barb using scanning electron microscopy (Figs. 11.8b and 11.2d). Around the center of the barb, cavities with melanin granules on the cavity surface were also observed. Similar amorphous network structures have been reported in the blue feathers of *Agapornis roseicollis*, *Cotinga maynana*, and *Cotinga cotinga* (Dyck 1971; Noh et al. 2010a).

The exact optical analysis of the amorphous structure is difficult because there are no analytical solutions. Therefore, Fourier analysis is often employed to estimate the wavelength of reflection (Prum and Torres 2003). The transmission electron

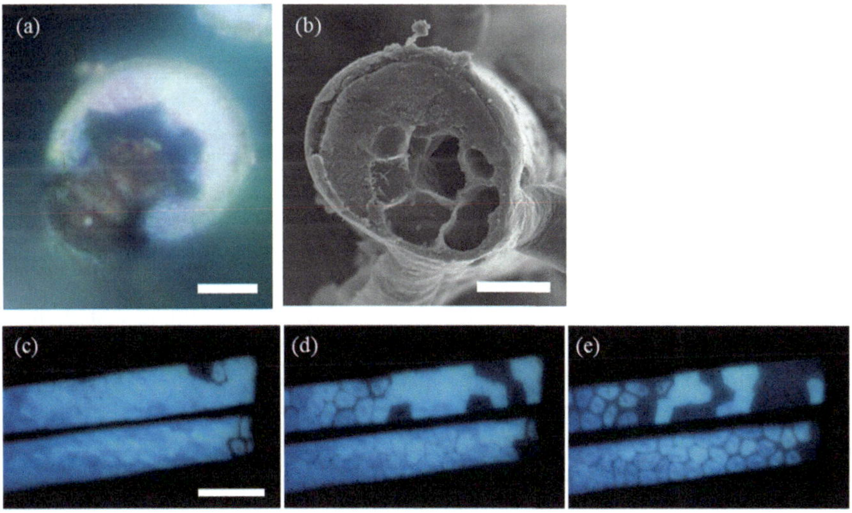

Fig. 11.8 Feather barb of the common kingfisher. A barb cross-section observed under (**a**) an optical microscope and (**b**) a scanning electron microscope. (**c–e**) Time lapse images during a liquid penetration experiment. Two barb tips were cut and soaked in toluene. As the liquid penetrated the barb, the blue color was removed. Scale bar: (**a**) 4μm, (**b**) 6μm, and (**c**) 10μm

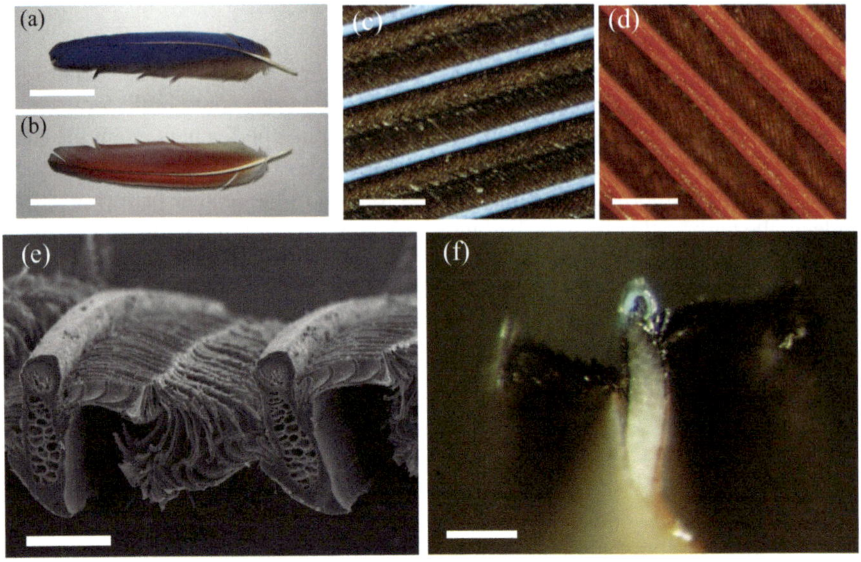

Fig. 11.9 Feather of the red-and-green macaw (*Ara chloropterus*). (**a** and **b**) The same feather observed from the (**a**) outer and (**b**) inner sides. (**c** and **d**) Optical micrographs of the two sides shown in (**a**) and (**b**), respectively. (**e**) Electron and (**f**) optical micrographs of the cross-section of the barb. Scale bar: (**a** and **b**) 5 cm, (**c** and **d**) 500μm, and (**e** and **f**) 100μm

micrograph of the amorphous network is digitally represented as a two-dimensional array of pixel values. This dataset can be analyzed by a two-dimensional Fourier transformation using mathematical packages such as Mathematica (developed by Wolfram Research). Prum et al. (1998, 1999) initially reported the analysis for the feather barb of *Cotinga maynana* using this method. The analysis gave a ring-like pattern in Fourier space, indicating that the structure was isotropic; that is, there was no special direction in the network structure. It also indicated that the network structure was modulated mainly by the spatial frequency at the radius of the ring. The inverse of the radius corresponded approximately to the mean period of the amorphous network, which could cause constructive optical interference. From an optical viewpoint, Fourier analysis is based on the single-scattering approximation, where the incident light is scattered only once within the sample and multiple scattering is ignored. This approximation seems adequate, at least for a simple analysis such as the estimation of the main reflection wavelength. However, the effects of multiple scattering cannot be completely ignored, as the double-scattering process causes a broad peak at a wavelength shorter than the peak of the single-scattering process (Noh et al. 2010b).

In some bird species, the amorphous network structure is combined with color pigments. Figure 11.9a, b shows the feather of a colorful parrot, the red-and-green macaw (*Ara chloropterus*). The two photographs are images of both sides of the same feather. It is interesting that one side looks blue, while the other side appears red, which is due to the barbs (Fig. 11.9c, d); however, the barbules appear dark on both sides. How could the color of the barb be so different between the two sides? The answer can be found in the cross-section. As shown in Fig. 11.9e, the shape of the barb is elongated in the direction perpendicular to the plane of the barbules (vane). On the blue side, a round-shaped structure is observed that is an amorphous network under higher magnification. Below this round part, there are many cavities that are covered with a cortex that appears red on an optical micrograph (Fig. 11.9f). Therefore, both the structural and pigmentary coloration methods combine to produce differently colored feather surfaces in this species.

11.1.4 Other Important Factors in Avian Feather Coloration

The chroma of the structural color (i.e., the wavelength of reflection) is determined mainly by the period of the microstructures. However, there are factors other than microstructures that largely affect the color and/or their appearance and we will provide two such examples. The first is the optical absorption of the melanin pigment. The color-causing microstructures in animals are very different from the theoretical models that physicists usually assume in that the structures are not regular but can have many irregularities and defects. Such imperfections inevitably cause wavelength-independent scattering that produces a white color. Therefore, it is important to reduce the wavelength-independent component in reflectance by optical absorption and melanin granules play a role in this process. For example, in the rock dove, granules behind the thin cortex layer absorb the light that transmits the layer. Melanin granules have dual functions in feather structural color: one is to produce color through optical interference and the other is to reduce the background white light through absorption. Using light-absorbing pigments as a building block of the periodic structure is a very efficient strategy for producing vivid color.

The refractive index is an important parameter for optical analysis because it characterizes the interaction between the material and electromagnetic wave. The melanin granules should have a complex refractive index value because they absorb light. Stavenga et al. (2015) recently utilized Jamin–Lebedeff interference microscopy to determine the complex wavelength-dependent refractive index value. They reported that for Lawes's Parotia (*Parotia lawesii*), the real part of the refractive index of the melanin granules was approximately 1.7 at 500 nm, whereas the imaginary part was 0.11i at the same wavelength. Because the real part is rather high, it indicates that the granules are efficient at scattering and reflecting light.

Another important factor for coloration is the large-sized feather structure. As seen with the rock dove (Fig. 11.5), the cross-section of the barbule usually has a crescent shape. Because of this surface curvature, the reflected light spreads in a wide angle, even when it is illuminated by directional illumination such as sunlight. In the rock dove, the angular range of reflection is more than 120°. The reflection range is important because it is directly related to the direction in which the color can be recognized. In addition to the crescent shape, the barbules have other large-sized structures. The curvature along the barbule length, twists, and the presence of barbules on both sides of the barb (Nakamura et al. 2008) all broaden the range of reflection to make the structural color of the feather more effective as a tool of communication. The structural color of melanin granules includes the arrangement of the granules, which can be treated locally as a small, flat photonic crystal mirror. However, the mirror surfaces are widely distributed owing to their large structures, just like a mirror ball, although one mirror is too small to be identified with the naked eye.

11.1.5 Concluding Remarks

Since the electron microscope was invented, the microstructures inside bird feathers have been studied in a wide variety of species. To date, we believe that all microscopic coloration mechanisms have been identified. However, when it comes to the color effect that is realized through a combination with other factors, such as a large-sized feather structures, there seems to be plenty of opportunity for further investigation. For example, it was reported that the barbule of Lawes's Parotia has a boomerang-shaped cross-section filled with regularly stacked melanin granules (Wilts et al. 2014). The stacking and the peculiarly shaped structures cause reflections in three directions, producing different colors. The study of such combined effects in structural coloration is a needed direction for future research.

There has been rapid development in the research field of biomimetics, where the designs observed in nature are evaluated and mimicked in order to produce novel products. Because structural colors have anti-fading properties, many researchers have been trying to produce structurally colored pigments. In particular, the amorphous network type is thought to be commercially interesting because the color is less iridescent. This feature is desirable for replacing ordinary pigments with structural ones, especially for materials that need to appear the same color independent of viewing angle. Attempts to produce an amorphous aggregation of colloidal particles that exhibit non-iridescent structural colors have been successful (Takeoka et al. 2013). Additionally, by carefully mimicking the design of the barbule structure, it has been demonstrated that the light absorption of a black background is important for generating more vivid structural colors (Iwata et al. 2017). These studies indicate that we can learn more from bird feathers to produce new pigments that are vivid, anti-fading, and environmentally friendly.

11.2 Feather Pattern Formation in Birds

The feather colors and patterns in birds often differ based on age and sex. Additionally, many birds exhibit various colors in different body parts (Fig. 3.1). Their feathers also exhibit various patterns (Crawford 1990), which include single lacing, double lacing, penciling, autosomal barring or parallel penciling, butter

Fig. 11.10 Second Patterns in contour feathers. Single lacing (**a**), double lacing (**b**), penciling (**c**), autosomal barring or parallel penciling (**d**), stippling (**e**), butter cup-autosomal barring modification (**f**), mottling (**g**), and spangling with asymmetric pattern (**h**). bar=10mm. Feathers are provided by Dr. Keiji Kinoshita, Mr. Isao Akaike and Mr. Hiroto Hayashi

cup-autosomal barring modification, stippling, spangling, mottling, and tricolor (Fig. 11.10). These patterns are formed mainly during the feather development process (Figs. 3.5, 3.6, 3.7, 3.8), and is based on the ratio of eumelanin and pheomelanin, and, in some cases, it is based on the combination of these two types of melanin as well as the structural coloration. For example, if melanin production occurs once and then discontinues or is inhibited in the melanocytes of the feather root, a single ring pattern is formed on the feather (Fig. 11.10a). If the melanin is produced again, a double-ring pattern is formed. If the melanin production switches from eumelanin to pheomelanin, the color of that particular part of the feather changes from black to brown or orange. The melanization pathways are similar to that of mammals, including melanocortin 1 receptor (MC1R)-, cAMP-dependent, WNT-, and mitogen-activated protein kinase (MAPK)-signaling pathways, all of which regulate the gene expression of microphthalmia-associated transcription factor (MITF) for tyrosinase, 5,6-hydroxyindole-2-carboxylic acid oxidase (DHICA oxidase; Tyrosinase-related protein-1,Trp-1), and Dopachrome tautomerase (DCT; Tyrp-2). Endothelins and endothelin receptors are also known to be involved in melanization in birds; however, birds possess endothelin receptor B2 in addition to B (see Chap. 3) (Poelstra et al. 2014). Therefore, melanocortin 1 receptor (MC1R), which is the switching factor between eumelanin and pheomelanin in α-melanocyte-stimulating hormone (α-MSH) system, is the first candidate known to be involved in pattern formation (Mundy 2005). Differences in feather color between male and female birds are suggested to be driven by the sex hormones (Siefferman et al. 2013; Lindsay et al. 2016). In addition to different feather colors, patterns, such as dots, stripes, and eyespots, and sometimes a combination of individual patterns (Crawford 1990) is also crucial for camouflage, conspecific recognition, and threatening and mating behaviors. Particularly, appealing patterns in males for females are sometimes surprisingly complicated and beautiful. Despite the elucidation of genes related to melanization, the mechanisms regulating pattern formation are not disclosed completely. In this section, some notable examples of body color patterns are introduced along with certain cues to understand the underlying mechanisms.

11.2.1 Examples of Complicated Feather Patterns: Pheasant

The common pheasant is known for its beautiful and complicated feather patterns, and that the male spreads and displays its stunning tail feathers to attract the females. An example of decorative feathers of pheasant is shown in Fig. 3.1. The color of these feathers is derived from the nanoscale structure of keratin, and the mechanism has already been explained in details in the first section of this chapter. To form this structural color pattern, it is important that the fine structure of feathers consists of keratin and regulatory shaped and compactly arranged melanosomes packed in barbules, as described in Sect. 11.1. To form the feather structure, the differences in the function of α- and β-keratin may be committed. Genes related to the constitution and transfer of melanosomes to feathers are also suggested to be involved in color pattern formation.

Another example of feathers with complicated patterns is of the pheasant species, *Argusianus argus*, which displays its feathers to attract the females similar to the common pheasant. This bird has long tail feathers with a row of eyespots that are like tactile buttons and dots with stripe patterns (Fig. 11.11). Additionally, structural colors are also present in certain parts of the feathers (Figs. 11.12, 11.13). As described before, the contour feather patterns appear as an assembly image of each barb (see Chap. 3, Figs. 3.5, 3.6, 3.7, 3.8). At feather roots, each barb receives melanin produced from the melanocytes located at the ridge of the roots (Fig. 3.8). At this time, each melanocyte produces melanin through complicated regulation such that the pattern-forming is computationally programmed. Surprisingly, these birds have acquired such complicated and beautiful feathers in a long evolutionary process, even though such conspicuous color patterns are often easy to detect, and males can be attacked by their natural predators.

Fig. 11.11 Tale feather display by the Great Argus pheasant. Great Argus males (background) appeals to females (foreground) (**a**). Detail of an upper tail feather (**b**). Scale bar = 15 cm. The photo in (**a**) was provided by the Fukuoka City zoo in Japan, and the feather was gifted by Tamaki Shimosaka

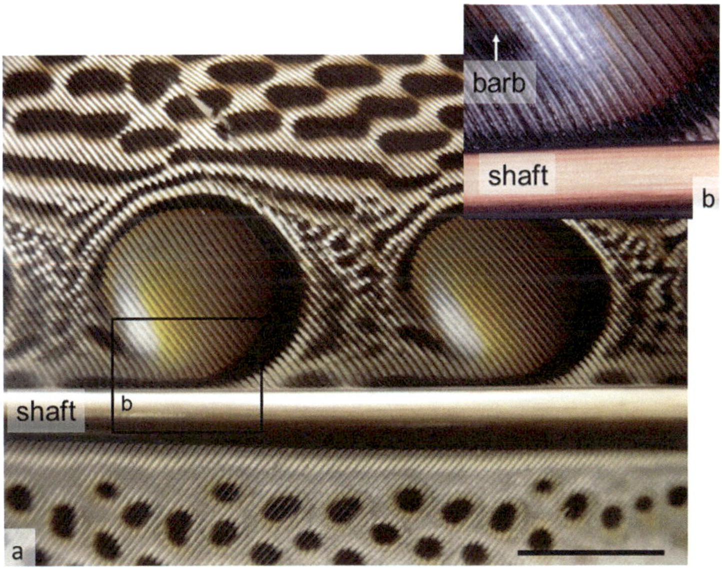

Fig. 11.12 Microscopic image of a feather of Great Argus pheasant. Image of the eye spot pattern and the surrounding area of the Great Argus feather (**a**) shown in Fig. 11.11 a high-magnification image of the square region of (**a**) (**b**). A barb shows the gradation or pattern from the melanin content. Scale bar = 10 mm

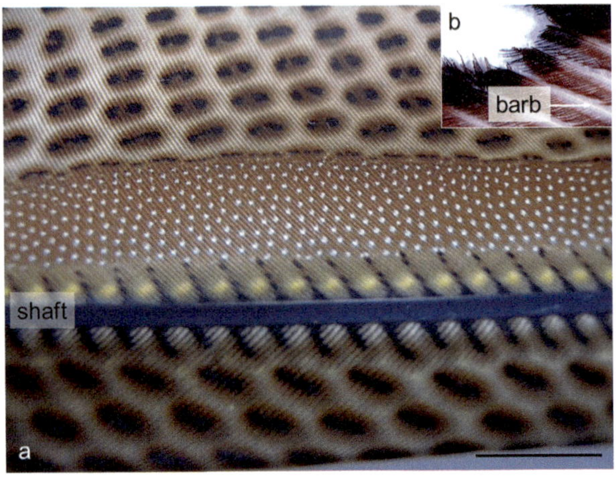

Fig. 11.13 Variable spot patterns in the long feather from a Great Argus pheasant (**a**) shown in Fig. 11.11(a), and a higher magnification of the area surrounding a black spot (**b**). White and black parts of the barbs and barbules are shown. Scale bar = 10 mm

Fig. 11.14 Jaybird and the feather with a characteristic blue stripe. An image of the jaybird provided by Dr. Gen Morimoto, from the Yamashina Bird Laboratory in Japan (**a**). A jaybird feather showing the characteristic blue stripe pattern of the wing (**b**). The feather was provided by the Ishikawa Zoo in Japan. Scale bar = 4.3 cm

11.2.2 Another Example of Complicated Feather Patterns: Jaybird

Jaybird (*Garrulus glandarius*) possesses beautiful feathers with blue, white, and black stripes in a part of remex feathers (Fig. 11.14). There is a repeated melanin

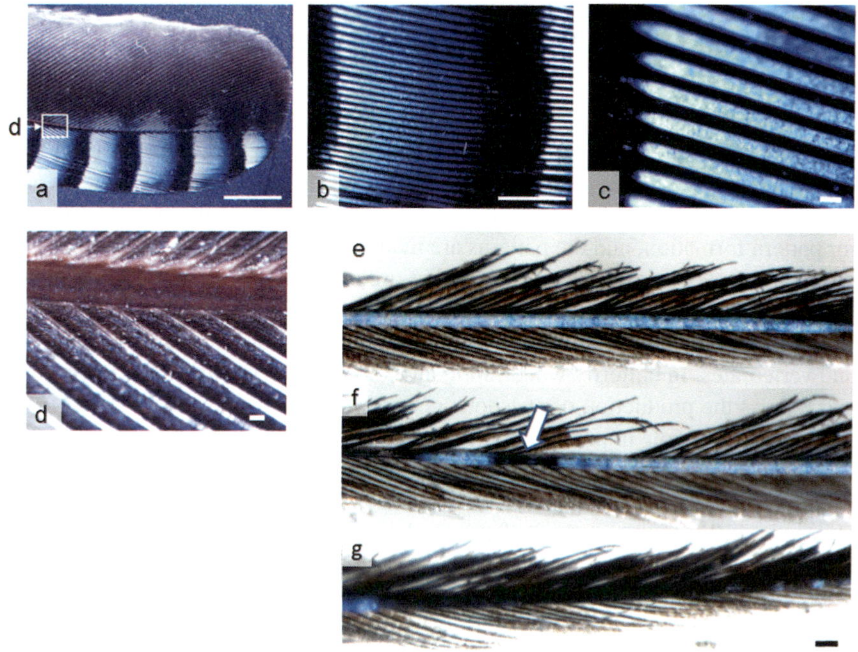

Fig. 11.15 High-magnification profiles of the blue stripes of the jaybird wing feather. The jaybird feather has a repeated stripe pattern of white, blue, and black zones in the barbs (**a**). The stripe unit consists of a gradation of melanin from black to white (**b**). In the middle zone of the stripe, a blue color is emitted (**c**), and in the zones with less melanin, a white color is emitted (**d**). A high magnification of a barb from root to tip (**e–g**). The structure of the barb is similar (**c**, **d**, **e**) regardless of melanin content. However, when the upper part of the barb is removed, the blue color is lost (**f**, arrow). The barb in the black zone is black throughout (g). Scale bar; (**a**) 3 mm, (**b**) 1 mm, (**c–g**) 100 μm

dose gradation observed with a certain distance unit, and the intermediate zone between the black and white colors of the unit is composed of blue color. The blue color in the stripes is due to the structural color of the feather barbs (Fig. 11.15). The blue color is considered to be derived from a photonic crystal structure, one type of structural colors. It depends on melanin content of the barb shaft of the feather that contributes to the structural coloration. The matte coloration is similar to the red-and-green macaw (*Ara chloropterus*) provoding a photonic crystal structure (see Sect. 11.1, Fig. 11.9). As described before, the patterns are created at the roots in the first developmental process of feather formation, and then, they are almost fixed (Fig. 3.8). In this case, it is notable that the feathers with blue stripes exist in a limited narrow region. Therefore, region-specific gene expression for pattern formation, such as stripe, and/or a special keratin formation for structural color might occur. Interestingly, both male and female birds exhibit similar colors and patterns, including the blue patterns. Therefore, the stripe is not expected to attract the females, but possibly help recognize the individuals of the same species.

11.2.3 Mechanism of Feather and Body Pattern Formation with Dietary Pigments and Structural Color

The body pattern in animals is explained in Chap. 10 by Kondo's group using the modified Alan Turing model and their new model (Ball 2015; Kondo et al. 2009; Watanabe et al. 2012; Watanabe and Kondo 2014; Yamanaka and Kondo 2014; Kondo 2017). In birds, melanocytes and keratinocytes are the effective cells for pattern formation, and the patterns are mainly formed in keratin structure formed by keratinocytes receiving of melanin, but are alternatively supported through the addition of dietary pigments and structural colors.

Eyespot patterns are found to be present in many animals, including invertebrates and vertebrates. In butterfly wings, *spalt* and *distal-less* genes are known to promote and repress the production of eyespots, respectively (Zhang and Reed 2016). These genes play a key role in producing the eyespot in many cells at a specific position on the wings. Vertebrates, like fish or amphibians, form integumental color patterns depending on the chromatophores present in the flat, sheet-like epidermal tissue. However, in birds, the pattern of the feathers is formed through the assembly of many barbs with individual pigmented pattern. Additionally, the entire body of birds often exhibits multiple colors in different body parts, especially in wings or tails (see Chap. 3, Fig. 3.1). In a case of Japanese quail, chick has characteristic longitudinal stripes with periodic coloration on the dorsal side. The genetic variants of the quails have slightly different stripe patterns. The regulation mechanism to produce the patterns and function of the one of the related genes, *agouti* is recently elucidated (Haupaix et al. 2018). This report is introduced and discussed in Chap. 3 in details. There are presumably different regulatory mechanisms to express coloration-related genes in each part (Ng and Li 2018). Therefore, it is important to conduct further research to analyze the expression mechanisms in each part of the body separately.

11.2.4 Body Colors and Patterns in Birds as a Survival Strategy

Various body colors and complicated patterns observed in birds contribute to their survival. Birds camouflage themselves using similar colors as that of the surrounding environment to hide and escape from their natural predators. Conversely, some species also display conspicuous feather colors and patterns that help them to frighten their natural predators. Additionally, body colors and patterns are used to recognize or differentiate between various species or sexes. Male-specific body colors and patterns may be efficient tools to attract females other than singing, food presentation, or nest building in birds. Examples of such sexual dimorphism are shown in Fig. 11.16 (*Aix galericulata*) and Fig. 11.17 (*Lagopus muta*). Also, the exhibition of body colors or patterns by peacock (*Pavo cristatus*) (Fig. 11.1h) (Yoshioka and Kinoshita 2002), Great Argus (*Argusianus argus*) (Figs. 11.11,

Fig. 11.16 Male and female mandarin ducks have distinctive coloration differences. During the breeding season, males have specific and colorful feathers (**a**), while females have brownish, modest plumages (**b**). After the breeding season, male plumage becomes closer to that of the female coloration after a molting period

Fig. 11.17 Seasonal variation of plumage colors in the snow grouse (*Lagopus muta*). Plumages of a male and female coloration in the winter (**a**, **d**), breeding season (**b**, **e**), and autumn (**c**, **f**), respectively. Grouse molt their feathers after each season. In the breeding season from spring to summer, the plumage colors are clearly different between males and females, while in the autumn and winter, the plumage can act as camouflage with the surroundings. Photos were provided by a naturalist, Tsutomu Matsuda

11.12, 11.13), and Mandarin duck (*Aix galericulata*) (Fig. 11.16) are the well-known examples of appealing the females.

Another example of sexual dimorphism is the snow grouse (*Lagopus mutu*), which is a high mountain bird known to seasonally change the color of its plumage (Fig. 11.17). This bird changes its feather color three times per year, for instance, brown in summers, dark brown and black for females and males, respectively, during the mating season, and white for both the sexes in winters. These color changes are efficient for camouflage during the summer and winter seasons, and also for the males to attract the females during the mating season. Temperature-sensitive factor(s) might be involved in controlling the molting process and changes in melanin production during the molting season, but this has not been established yet. Despite these effects of body coloration, there is a sharp decline in the number of snow grouses due to the modulation in their natural predators or habitats as a consequence of global warming. Therefore, protecting these birds is now one of the most urgent worldwide missions in bird ecology.

Finally, it is suggested that the body colors and patterns in birds contribute to their biological success. Alternatively, the multiple functions associated with body colors and patterns are strategically very important and useful for their survival. To date, the expression mechanisms of the body colors and feather patterns in birds have not been completely elucidated. Such diverse body colors and complicated feather patterns in birds are considered to be genetically selected as effective and suitable ways to support their evolutionary survival. Therefore, analyses of the diverse body colors and feather patterns can help elucidate the route of evolution and species differentiation in birds (Poelstra et al. 2014). Further, research on the coloration in birds will be very useful for researchers to investigate the gene expression, ecology, and evolution.

11.2.5 Concluding Remarks

In birds, the body color pattern is determined through the following three steps: (1) in each barb at feather roots, (2) in each feather where assembly of barbs occurs, and (3) in every part of the body. Therefore, it is certain that feather patterns in birds are determined by melanization at feather roots during the developmental process. Additionally, coloration in some cases occurs through pigments, such as carotenoids, pteridine, and psittacofulvin that are incorporated from foods, which are also derived from keratin structure in feathers without melanin. Conversely, structural colors generally appear as nanoscale structures consisting of mainly keratin and melanotic melanosomes. In the entire body, birds often show multiple colors and patterns suggesting that the expression of genes is under differential regulation. Birds possess melanocytes, which are exclusively the effective pigment cell types, but often exhibit notably diverse colors and patterns with the support of two systems other than melanization.

Acknowledgements The authors deeply thank Dr. Keiji Kinoshita, Dr. Gen Morimoto, Mr. Tsutomu Matsuda, Mr. Hiroto Hayashi, Ms. Tamaki Shimosaka, Mr. Isao Akaike, Fukuoka City Zoo, Ishikawa Zoo, and Nogeyama Zoo in Japan for providing, donating or supplying photos and feathers. We are also grateful to the Avian Bioscience Research Center in Nagoya University for supplying chickens and other valuable information. The work in Sect. 11.1 was partly supported by a Grant-in-Aid for Scientific Research (No. 18H01191) from the Ministry of Education, Culture, Sports, Science, and Technology (MEXT, Japan). The work in Sect. 11.2 was supported in part by grants from Keio University.

References

Ball P (2015) Forging patterns and making waves from biology to geology: a commentary on Turing (1952) 'The chemical basis of morphogenesis'. Philos Trans R Soc Lond B Biol Sci 370 (1666):20140218. https://doi.org/10.1098/rstb.2014.0218

Bowmaker JK (1977) The visual pigments, oil droplets and spectral sensitivity of the pigeon. Vision Res 17:1129–1138

Brink DJ, van der Berg NG (2004) Structural colours from the feathers of the bird Bostrychia hagedash. J Phys D Appl Phys 37:813–818

Crawford RD (1990) Poultry breeding and genetics. Elsevier Science Publishers. Amsterdam.

Durrer H (1977) Schillerfarben der vogelfeder als evolutions problem. Denkschr Schweiz Naturforsch Ges 91:1–127

Dyck J (1971) Structure and colour-production of the blue barbs of agapornis roseicollis and cotinga maynana. Z Zellforsch 115(1):17–29

Eliason CM, Shawkey MD (2012) A photonic heterostructure produces diverse iridescent colours in duck wing patches. J R Soc Interface 9(74):2279–2289

Eliason CM, Bitton P-P, Shawkey MD (2013) How hollow melanosomes affect iridescent colour production in birds. Proc R Soc Lond B Biol Sci 280:1767

Greenewalt CH, Brandt W, Friel DD (1960) Iridescent colors of hummingbird feathers. J Opt Soc Am 50(10):1005–1013

Haupaix N, Curantz C, Bailleul R, Beck S, Robic A, Manceau M (2018) The periodic coloration in birds forms through a prepattern of somite origin. Science 361(6408):eaar4777. https://doi.org/10.1126/science.aar4777

Iwata M, Teshima M, Seki T, Yoshioka S, Takeoka Y (2017) Bio-inspired bright structurally colored colloidal amorphous array enhanced by controlling thickness and black background. Adv Mater 29(26):1605050

Kinoshita S (2008) Structural colors in the realm of nature. World Scientific, Singapore

Kinoshita S (2013) Bionanophotonics: an introductory textbook. Pan Stanford, Boca Raton

Kinoshita S, Yoshioka S, Miyazaki J (2008) Physics of structural colors. Rep Prog Phys 71:076401

Kondo S (2017) An updated kernel-based Turing model for studying the mechanisms of biological pattern formation. J Theor Biol 414:120–127

Kondo S, Iwashita M, Yamaguchi M (2009) How animals get their skin patterns: fish pigment pattern as a live Turing wave. Int J Dev Biol 53:851–856. https://doi.org/10.1387/ijdb.072502sk

Lee E, Miyazaki J, Yoshioka S, Lee H, Sugita S (2012) The weak iridescent feather color in the jungle crow *corvus macrorhynchos*. Ornithol Sci 11:59–64

Lindsay WR, Barron DG, Webster MS, Schwabl H (2016) Testosterone activates sexual dimorphism including male-typical carotenoid but not melanin plumage pigmentation in a female bird. J Exp Biol 219(Pt 19):3091–3099

Mundy NI (2005) A window on the genetics of evolution: MC1R and plumage colouration in birds. Proc Biol Sci 272(1573):1633–1640. https://doi.org/10.1098/rspb.2005.3107

Nakamura E, Yoshioka S, Kinoshita S (2008) Structural color of rock dove's neck feather. J Physical Soc Japan 77(12):124801

Ng CS, Li W-H (2018) Genetic and molecular basis of feather diversity in birds. Genome Biol Evol 10(10):2572–2586. https://doi.org/10.1093/gbe/evy180

Noh H, Liew SF, Saranathan V, Mochrie SGJ, Prum RO, Dufresne ER, Cao H (2010a) Structural color: how noniridescent colors are generated by quasi-ordered structures of bird feathers. Adv Mater 22(26–27):2871–2880

Noh H, Liew SF, Saranathan V, Prum RO, Mochrie SGJ, Dufresne ER, Cao H (2010b) Contribution of double scattering to structural coloration in quasiordered nanostructures of bird feathers. Phys Rev E 81:051923

Poelstra JW, Vijay N, Bossu CM, Lantz H, Ryll B, Müller I, Baglione V, Unneberg P, Wikelski M, M. G. GrabherrM.G., and Wolf, J. B. W. (2014) The genomic landscape underlying phenotypic

integrity in the face of gene flow in crows. Science 344(6190):1410–1414. https://doi.org/10.1126/science.1253226
Prum RO, Torres RH (2003) A fourier tool for the analysis of coherent light scattering by bio-optical nanostructures. Integr Comp Biol 43(4):591–602
Prum RO, Williamson S (2002) Reaction–diffusion models of within-feather pigmentation patterning. Proc R Soc Lond B Biol Sci 269(1493):781–792
Prum RO, Torres RH, Williamson S, Dyck J (1998) Coherent light scattering by blue feather barbs. Nature 396:28
Prum RO, Torres R, Williamson S, Dyck J (1999) Two-dimensional fourier analysis of the spongy medullary keratin of structurally coloured feather barbs. Proc R Soc Lond B Biol Sci 266 (1414):13–22
Seago AE, Brady P, Vigneron JP, Schultz TD (2009) Gold bugs and beyond: a review of iridescence and structural colour mechanisms in beetles (Coleoptera). J R Soc Interface 6 (Suppl 2):S165
Siefferman L, Liu M, Navara KJ, Mendonça MT, Hill GE (2013) Effect of prenatal and natal administration of testosterone on production of structurally based plumage coloration. Physiol Biochem Zool 86(3):323–332. https://doi.org/10.1086/670383
Srinivasarao M (1999) Nano-optics in the biological world: beetles, butterflies, birds, and moths. Chem Rev 99(7):1935–1961
Stavenga DG, Leertouwer HL, Osorio DC, Wilts BD (2015) High refractive index of melanin in shiny occipital feathers of a bird of paradise. Light: Sci Appl 4:e243
Stavenga DG, van der Kooi CJ, Wilts BD (2017) Structural coloured feathers of mallards act by simple multilayer photonics. J R Soc Interface 14:20170407
Takeoka Y, Yoshioka S, Takano A, Arai S, Nueangnoraj K, Nishihara H, Teshima M, Ohtsuka Y, Seki T (2013) Production of colored pigments with amorphous arrays of black and white colloidal particles. Angew Chem Int Ed 52(28):7261–7265
Vukusic P, Sambles J (2003) Photonic structures in biology. Nature 424(6950):852–855
Watanabe M, Kondo S (2014) Is pigment patterning in fish skin determined by the Turing mechanism? Trends Genet 31(2):88–96. https://doi.org/10.1016/j.tig.2014.11.005
Watanabe M, Watanabe D, Kondo S (2012) Polyamine sensitivity of gap junctions is required for skin pattern formation in zebrafish. Sci Rep 2:473. https://doi.org/10.1038/srep00473
Wilts BD, Michielsen K, De Raedt H, Stavenga DG (2014) Sparkling feather reflections of a bird-of-paradise explained by finite-difference time-domain modeling. Proc Natl Acad Sci 111 (12):4363–4368
Yamanaka H, Kondo S (2014) In vitro analysis suggests that difference in cell movement during direct interaction can generate various pigment patterns in vivo. Proc Natl Acad Sci USA 111 (5):1867–1872. https://doi.org/10.1073/pnas.1315416111
Yin H, Shi L, Sha J, Li Y, Qin Y, Dong B, Meyer S, Liu X, Zhao L, Zi J (2006) Iridescence in the neck feathers of domestic pigeons. Phys Rev E 74:051916
Yoshioka S, Kinoshita S (2002) Effect of macroscopic structure in iridescent color of the peacock feathers. FORMA 17:169–181
Yoshioka S, Nakamura E, Kinoshita S (2007) Origin of two-color iridescence in rock dove's feather. J Physical Soc Japan 76(1):013801
Zhang L, Reeda RD (2016) Genome editing in butterflies reveals that *spalt* promotes and *Distal-less* represses eyespot colour patterns. Nat Commun 7:11769. https://doi.org/10.1038/ncomms11769
Zi J, Yu X, Li Y, Hu X, Xu X, Wang X, Liu X, Fu R (2003) Coloration strategies in peacock feathers. Proc Natl Acad Sci 100(22):12576–12578

Chapter 12
Mechanism of Color Pattern Formation in Insects

Yuichi Fukutomi and Shigeyuki Koshikawa

Abstract Insects have various color patterns on their bodies and have played major roles in elucidating the mechanisms of color pattern formation because of their suitability as experimental models. In particular, studies of *Drosophila* (fruit flies) and butterflies have produced a number of new findings. The logic of generation of a color pattern from the combination of the spatial information of body parts, the genes responsible for those processes, and the mechanism of generating a novel color pattern by *cis*-regulatory evolution have been elucidated by *Drosophila* studies. In butterfly studies, attempts to find color pattern genes through genome-wide analysis and functional analysis of color pattern genes using genome editing technology are producing new results. Theoretical models to explain complex eyespot patterning in butterflies have been developed and are awaiting verification with experimental results. A comprehensive model to explain general color pattern formation in insects and validation with empirical data is required.

Keywords *Drosophila* · Butterfly · *Cis*-regulation · Mimicry · Genome editing · Eyespot

---

Y. Fukutomi
Graduate School of Environmental Science, Hokkaido University, Sapporo, Japan

Present address: Department of Biological Sciences, Tokyo Metropolitan University, Hachioji, Japan

S. Koshikawa (✉)
Graduate School of Environmental Science, Hokkaido University, Sapporo, Japan

Faculty of Environmental Earth Science, Hokkaido University, Sapporo, Japan
e-mail: koshi@ees.hokudai.ac.jp

H. Hashimoto et al. (eds.), *Pigments, Pigment Cells and Pigment Patterns*,
https://doi.org/10.1007/978-981-16-1490-3_12

12.1 Introduction

Like other animals, a diversity of color patterns is observed on the surfaces of insect bodies. Some of these patterns have important functions such as courtship display, camouflage, mimicry, warning coloration, and threats to predators (Cott 1940). For example, the black and yellow stripes of bees and the color pattern on the elytra (forewings) of ladybeetles are thought to function as warning coloration, informing predators that they are dangerous or unpalatable. Color patterns of venomless insects mimic those of venomous or dangerous insects. The color pattern on the body of the tribe Clytini (venomless longhorn beetles) resembles that of venomous wasps (Linsley 1959). Some female individuals of *Papilio polytes* (a venomless butterfly) have a similar color pattern on their wings to that of *Pachliopta aristolochiae* (a poisonous butterfly, Clarke and Sheppard 1972). These are thought to function as Batesian mimicry. In other cases, the color patterns of hornets and paper wasps are hard to distinguish, and the color patterns observed on the wings of *Heliconius* butterflies resemble each other. These patterns are thought to operate as Müllerian mimicry (Sherratt 2008). Insects in various taxa such as butterflies and praying mantises have eyespots on their wings, and some of these eyespots might have the role of threatening predators (Cott 1940; Ruxton et al. 2004). Research on insects has contributed much to establishing concepts about the functions of animal color patterns.

Researchers are attempting to reveal how these diverse patterns have evolved. To answer this question, the mechanism of color pattern formation has been studied using various insects. Insects are prolific and relatively easy to maintain in the laboratory. Their color patterns are easy to analyze because they are often formed in a single plane. These advantages mean that insect color patterns have a central role in the study of the proximate cause of morphological evolution. To understand mechanisms of color pattern formation, color patterns on the abdomens of *Drosophila*, the wings of *Drosophila*, and the wings of butterflies have been especially well studied. As a result, genes that encode transcription factors or signal ligands have been identified as regulators of color pattern formation. Analysis of the *cis*-regulatory regions of these genes has led to the prevalence of the idea that *cis*-regulatory mutations cause differences in gene expression and eventually, morphological diversity. To date, researchers have proposed theoretical models to explain the determination of the position and the area of color patterns. In the future, the molecular mechanisms of these determinations will have to be clarified by experiments.

The mechanisms of color pattern formation can be separated into two steps. The first is the step of (pre)pattern formation, which is an upstream developmental event. The second is the step of pigmentation (coloration). In this chapter, we mainly focus on the first step, pattern formation. The details of pigmentation in insects are reviewed in Futahashi and Osanai-Futahashi (2021, see Chap. 1).

12.2 Genes That Control Color Patterns of Drosophila

Drosophilid flies (fruit flies of the family Drosophilidae, which includes more than 4000 species) have diverse color patterns on their wings and abdomens (Patterson et al. 1943; Carson et al. 1970; Markow and O'Grady 2006; Setoguchi et al. 2014; Werner et al. 2018). Some species can be utilized for experiments such as gene knockout, gene knockdown, gene overexpression, and fluorescent reporter assays, by which the activity of enhancers can be visualized with fluorescent proteins. These species are useful model organisms for exploring genes that control color pattern formation (Wittkopp et al. 2003; Koshikawa et al. 2017; Rebeiz and Williams 2017; Koshikawa 2020).

One of the fruit fly species, *Drosophila melanogaster*, which was utilized as a genetic model by Thomas Hunt Morgan in the 1900s, has now been developed as the best model of developmental genetics with abundant information on epithelial patterning and development. Therefore, it provides a strong basis for studying pigmentation pattern formation. Both males and females have black stripes on their abdomens, while only males have black pigmentation at the posterior part of the dorsal side (tergites) of their abdomens (Fig. 12.1a). The formation of common black stripes is controlled by *optomotor-blind* (*omb,* also called *bi*), which encodes a transcription factor. In loss-of-function mutants of *omb*, common black stripes were not formed. The formation of male-specific pigmentation is positively regulated by *Abdominal-B* (*Abd-B*), one of the *Hox* genes, and negatively regulated by the *bric-à-brac* (*bab*) locus, which encodes transcription factors. Ectopic expression of *Abd-B* at abdominal segment A3-A4 (a place where male-specific pigmentation does not occur) induced ectopic pigmentation. Mutants in which Abd-B does not function at abdominal segment A5-A6 (the place where male-specific pigmentation occurs) did not form male-specific pigmentation. In wildtype, *bab* was expressed in the entire dorsal side of the abdomen in females, but in males, it was not expressed in abdominal segment A5-A6 (the place where male-specific pigmentation occurs). When *bab* was expressed in the abdominal segment A5-A6 (the location of male-specific pigmentation), the pigmentation was lost (Kopp et al. 2000). Abd-B is known to activate the *yellow* gene, which is necessary for melanin synthesis (but probably does not encode an enzyme, Han et al. 2002; Hinaux et al. 2018), by direct binding to an enhancer of *yellow* (Jeong et al. 2006), while Bab proteins are known to suppress the *yellow* gene by direct binding to an enhancer of *yellow* (Roeske et al. 2018; Fig. 12.1b). These results indicate that pigmentation in every abdominal segment and male-specific pigmentation are controlled by different transcription factors in the abdomen of *D. melanogaster*. Moreover, multiple transcription factors affecting pigmentation were identified by using a collection of RNAi (RNA interference) lines (Rogers et al. 2014) and by genetic mapping in polymorphic species and subsequent validation in *D. melanogaster* (Yassin et al. 2016).

Males of the fruit fly species *Drosophila biarmipes* have black pigmentations on the anterior distal side of the wings (Fig. 12.2a). The expression of the Yellow protein, which is necessary for melanin synthesis, is correlated with the pattern of

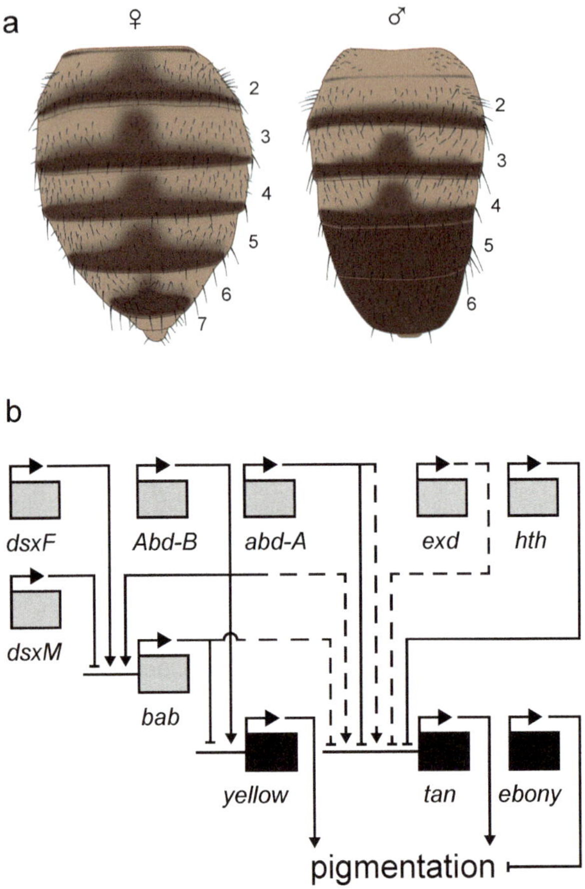

Fig. 12.1 (**a**) The abdomen of adults of *Drosophila melanogaster*. The numbers indicate the abdominal tergites (dorsal plates of the abdomen, A2-A7). In the female (left), the posterior side of each tergite is black (or dark brown). The sixth tergite of the female shows diverse pigmentation patterns, depending on the lineage and the environment. In the male (right), the fifth and sixth tergite always become black, in addition to the posterior side of each tergite. (**b**) The regulatory network controlling abdominal pigmentation in *D. melanogaster*. Gray boxes indicate transcription factors. Black boxes indicate effector genes. Large arrows indicate gene expression, small arrows indicate activation, and T-shapes (tacks) indicate repression. Dashed lines indicate indirect interactions or interactions not yet shown to be direct. The sex-specific isoforms of *doublesex* gene, *dsxF* (female isoform of *doublesex*), and *dsxM* (male isoform of *doublesex*), are coded in the same locus but shown as different genes in this panel for simplicity. The *bab* (*bric-à-brac*) locus contains two partially redundant transcription factor genes (*bab1* and *bab2*), but shown as a single gene for simplicity. Other abbreviations are as follows: *Abd-B* (*Abdominal-B*), *abd-A* (*abdominal-A*), *exd* (*extradenticle*), and *hth* (*homothorax*). Modified from Roeske et al. (2018)

wing pigmentation. Ebony protein, a beta-alanyl-dopamine synthase that suppresses melanin synthesis, is expressed where wing pigmentation does not occur (Wittkopp et al. 2002; Gompel et al. 2005). Using RNAi lines and a transgenic line of

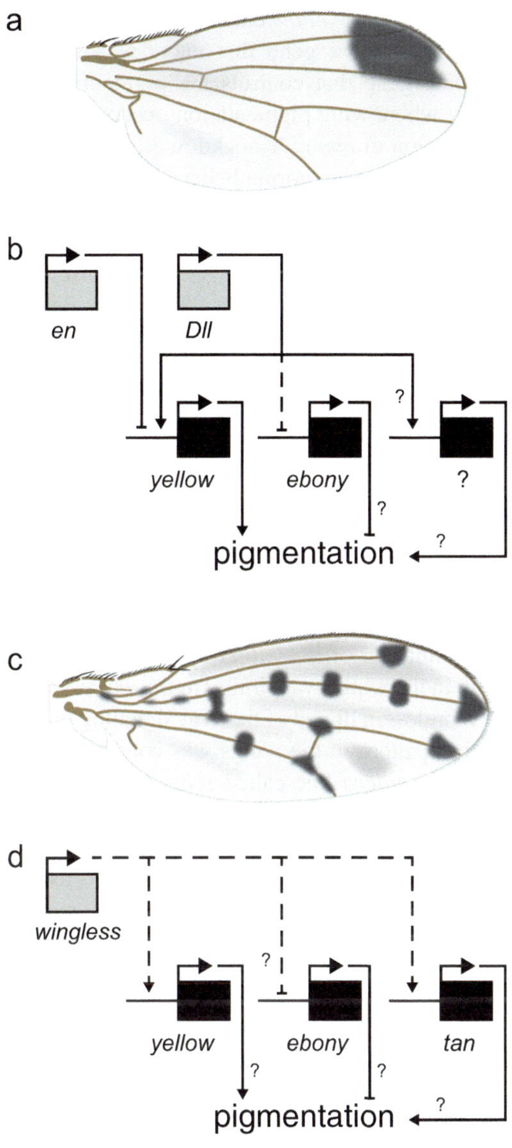

Fig. 12.2 (**a**) The male wing of *Drosophila biarmipes*. The female wing does not have much pigmentation. (**b**) The regulatory network controlling wing pigmentation in *D. biarmipes*. Gray boxes indicate transcription factors. Black boxes indicate effector genes. Large arrows indicate gene expression, small arrows indicate activation, and T-shapes (tacks) indicate repression. Dashed lines indicate indirect interactions or interactions not yet shown to be direct. Abbreviations are as follows: *en* (*engrailed*) and *Dll* (*Distal-less*). Modified from Arnoult et al. (2013). (**c**) The wing of *Drosophila guttifera*. In this species, there are almost no differences in patterns between females and males. The pigmentation is formed around the end points and the intersection points of the wing veins, and around campaniform sensilla, which are mechanical sensory organs. (**d**) The regulatory network controlling wing pigmentation in *D. guttifera*. A gray box indicates a regulatory gene. Black boxes indicate effector genes. Large arrows indicate gene expression, small arrows indicate

D. melanogaster, which carries a *yellow* enhancer of *D. biarmipes* and a fluorescent protein gene, *Distal-less* (*Dll*), a gene that encodes a transcription factor, was identified as a candidate gene that controls the pigmentation pattern. *Distal-less* was expressed strongly where wing pigmentation occurs. The ectopic expression of *Dll* repressed the expression of *ebony*. Knockdown of *Dll* repressed the expression of *yellow*, where wing pigmentation normally occurs and promotes the expression of *ebony*. Distal-less protein was also found to bind directly to a *yellow* enhancer. Furthermore, the expression of Distal-less protein was correlated with the wing pigmentation pattern of three other fruit fly species, *Drosophila pulchrella*, *Drosophila elegans*, and *Drosophila prolongata*. This indicates that changes in the expression pattern of *Dll* can be the cause of the diversification of wing pigmentation patterns (Arnoult et al. 2013; Fig. 12.2b). Comparing the color pattern formation on the wings and abdomens of fruit flies has revealed that some of the genes involved in melanin synthesis are the same, but transcription factors that control the area of color patterns are different.

Another fruit fly species, *Drosophila guttifera*, has a polka-dotted pigmentation pattern on the wings (Fig. 12.2c). Yellow protein is expressed where wing pigmentation occurs, and Ebony protein is expressed where wing pigmentation does not occur (Gompel et al. 2005). To identify the gene that controls the pigmentation pattern, a transgenic line of *D. melanogaster* that carries a fluorescent protein gene connected to a *yellow* enhancer of *D. guttifera* was used. The expression pattern of the fluorescent protein in the wings was similar to the expression pattern of *wingless*, a gene that encodes Wingless, a ligand of the Wnt signaling pathway. In *D. guttifera*, the *wingless* expression pattern in the wings was correlated with the pigmentation pattern. In a mutant *D. guttifera* line called *schwarzvier*, which has ectopic wing pigmentation around the fourth longitudinal vein, ectopic *wingless* expression was observed in the fourth longitudinal vein. When *wingless* was ectopically expressed in the wing, ectopic wing pigmentation was induced. Collectively, these findings led to the conclusion that the wing pigmentation pattern of *D. guttifera* is controlled by *wingless* (Werner et al. 2010, Fig. 12.2d). Although further comparisons among *Drosophila* species will be required to draw a complete picture, the current evidence indicates that the pigmentation positions seem to be determined by different regulatory genes in *D. guttifera* and *D. biarmipes*, which is not surprising because these species are distantly related within the genus *Drosophila* (Markow and O'Grady 2006; Izumitani et al. 2016; O'Grady and DeSalle 2018). In addition, downstream of the Wingless signaling of *D. guttifera*, effector genes involved in melanin synthesis, such as *yellow*, must work, however, their regulation has not been investigated in detail. Wing pigmentation starts from a late pupal stage and continues even after eclosion (Fukutomi et al. 2017, 2018). In a different *Drosophila* species, it is thought that transport of melanin precursors might determine the area of wing pigmentation

---

Fig. 12.2 (continued) activation, and T-shapes (tacks) indicate repression. Dashed lines indicate indirect interactions or interactions not yet shown to be direct. Based on Werner et al. (2010) and Fukutomi et al. (2021)

(True et al. 1999; Hinaux et al. 2018). Understanding the shape, area, and density of pigmentation patterns requires the identification of signal ligands and transcriptional factors (Fukutomi et al. 2021; Dufour et al. 2020), as well as the construction of a comprehensive model that includes melanin synthesis and precursor transport.

To understand how the molecular mechanisms of color pattern formation have evolved, many questions remain to be answered. In the fourth section of this chapter, we introduce the idea that changes in *cis*-regulatory regions play an important role in color pattern evolution.

12.3 Color Pattern Formation on Butterfly Wings

Identification of the genes correlated with the wing color patterns of butterflies started in the 1990s using *Junonia coenia* (common buckeye) and *Bicyclus anynana* (squinting bush brown, Fig. 12.3a). It has been shown that the expression patterns of genes such as *Distal-less*, *wingless*, *hedgehog*, *engrailed/invected*, *patched*, and *cubitus interruptus* prefigured future color patterns (Carroll et al. 1994; Keys et al. 1999). Because the correlation between gene expression patterns and color patterns is maintained in phenotypic variations caused by seasonal polymorphism, mutation, and artificial selection, these genes are thought to be involved in color pattern formation (Brakefield et al. 1996; Brunetti et al. 2001, Fig. 12.3b, c). At the time, functional analysis of genes in butterflies was difficult, and consequently, gene expression analysis, such as in situ hybridization and immunohistochemistry, played a central role in butterfly research. Various new experimental approaches have revealed causal relationships between regulatory genes and phenotypes (Tong et al. 2012, 2014; Monteiro et al. 2013; Monteiro 2015; Dhungel et al. 2016).

In addition, recent progress in genomics has enabled the emergence of new findings. Amplified fragment length polymorphism (AFLP) mapping and gene expression profiles with tiling arrays found that a transcription factor, *optix*, controlled the formation of the red pigmentation pattern on the wings of *Heliconius* butterflies, and the *optix* region showed highly significant genotype–phenotype associations in the hybrid zones (Reed et al. 2011, Fig. 12.3d, e). Formation of a part of the black areas (related to the "Central Symmetry System" in butterfly terminology, Nijhout 2001) on the wings of *Heliconius* butterflies is controlled by *WntA*, a gene encoding a signal ligand and a paralog of *wingless/Wnt1*. This was found by RAD (restriction site associated DNA) mapping, high-resolution linkage analysis, and in situ hybridization (Martin et al. 2012). In another example, genes that regulate the wing color pattern of *Papilio polytes* were identified. It is known that the wing color pattern of female *P. polytes* shows polymorphism. One of the morphs looks like *Pachliopta aristolochiae*, and this morph is known as an example of Batesian mimicry. The gene that controls the formation of the morph is *doublesex* (*dsx*), a gene known for sex determination in various animals, and *dsx* was located in a supergene, a structure in which mutation accumulated as a consequence of a chromosomal inversion, which was revealed by linkage, expression, and functional

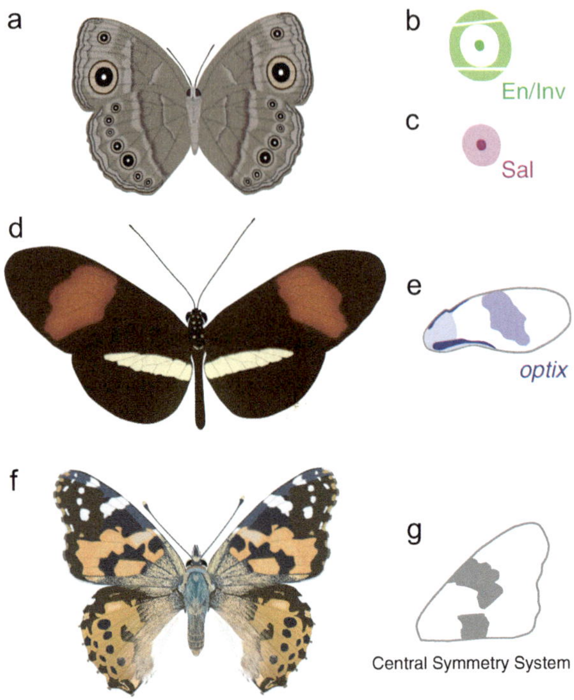

Fig. 12.3 (**a**) *Bicyclus anynana* (squinting bush brown). The figure shows the ventral side. It has been used for the study of eyespot formation (Brakefield et al. 1996; Keys et al. 1999; Brunetti et al. 2001; Marcus et al. 2004; Reed and Serfas 2004; Özsu et al. 2017). It is known that several transcription factors and signal ligands are expressed in the region where eyespots are formed. (**b**) Expression pattern of engrailed/invected (En/Inv, paralogs of NK-like homeobox transcription factors) on a future eyespot in a pupal wing of *B. anynana* (Brunetti et al. 2001). (**c**) Expression pattern of Spalt (Sal, a C2H2 Zinc finger transcription factor) on a future eyespot in a pupal wing of *B. anynana* (Brunetti et al. 2001). (**d**) *Heliconius erato* (red postman or red passion flower butterfly), dorsal side. It has been used for the study of Müllerian mimicry. This butterfly is distributed in Central and South America, and has various patterns depending on the area, each of which mimics other poisonous butterflies in the same area. In recent years, with the advance of genomics, a number of genes controlling the pattern have been identified (Joron et al. 2006; Reed et al. 2011; Martin et al. 2012; Nadeau et al. 2016). (**e**) Expression pattern of *optix* (a gene encoding a *sine oculis* homeobox transcription factor) in a pupal forewing of *H. erato petiverana* (Reed et al. 2011). (**f**) *Vanessa cardui* (painted lady), dorsal side. (**g**) A part of the pattern called the Central Symmetry System (marked in gray, Nijhout 2001) is regulated by WntA, a signaling ligand (Martin and Reed 2014; Mazo-Vargas et al. 2017). The figure shows a forewing

analysis (Kunte et al. 2014; Nishikawa et al. 2015). In addition, in the silkworm *Bombyx mori*, a gene that controls the formation of color patterns was identified. On the dorsal side of larvae of *B. mori*, there are pairs of spots in some of the body segments. Linkage analysis and genetic functional analyses revealed that *Wnt1*, an ortholog of *wingless*, was found to control the formation of the spots (Yamaguchi et al. 2013).

A gene that encodes neither a transcription factor nor a signal ligand was also identified as a controller of color pattern formation. Sequence variations around the *cortex* gene were identified as the cause of polymorphism observed in the color patterns of *Heliconius* butterflies; *cortex* is a member of a gene family known as regulator(s) of the cell cycle. Mutations near the *cortex* gene were also identified as the cause of the increase in the size of eyespots on the wing of *B. anynana*, the famous industrial melanism of *Biston betularia* (peppered moth), and color mutants of *Bombyx mori* (silkmoth) (Ito et al. 2016; Nadeau et al. 2016; van't Hof et al. 2016). Collectively, these findings showed that the *cortex* gene regulates various color patterns and pigmentations of Lepidoptera wings.

Recently, advances in genome editing technology have offered the possibility of a simple test of the gene functions that potentially regulate color pattern formation. Using the CRISPR/Cas9 genome editing system, G0 (generation zero, RNA-injected generation) individuals, which often show mosaically knocked-out phenotypes, can be analyzed for the phenotype of interest. The mosaic phenotypes of various butterflies, such as *J. coenia* and *Vanessa cardui,* were analyzed for *spalt*, *Dll*, *optix*, and *WntA*, which were previously identified as being responsible for or related to color pattern formation (Zhang and Reed 2016; Zhang et al. 2017; Mazo-Vargas et al. 2017, Fig. 12.3f, g). Zhang and Reed (2016) analyzed the *Dll* function in *V. cardui* and concluded that *Dll* acts as a repressor of eyespots, contrary to the prediction made from previous expression analysis. A subsequent study showed that the CRISPR/Cas9 disruption of *Dll* in *B. anynana* led to both ectopic and missing eyespots (Connahs et al. 2019). They successfully explained these apparently contradictory results by a reaction–diffusion model assuming that *Dll* is an activator of eyespots.

12.4 *Cis*-Regulatory Elements That Control Color Pattern Formation

To understand the proximate cause of the evolution of diversity observed in insect color patterns, identifying and analyzing *cis*-regulatory elements of transcriptional factors and signal ligands have been conducted. In the case of the male-specific pigmentation pattern on the abdominal tergites of *D. melanogaster*, *cis*-regulatory regions of the *bab* transcription factor gene locus were analyzed. Tandemly duplicated *bab1* and *bab2* are located at the *bab* locus, and these two genes, collectively called *bab,* are thought to have redundant functions and overlapping expression patterns. A *cis*-regulatory element called "the dimorphic element" was identified to drive sexually dimorphic expression patterns of *bab* genes. Two transcription factors, Doublesex (Dsx) and Abdominal-B (Abd-B), were found to have inputs to the dimorphic element, probably through direct binding. In males, Abd-B and the male isoform of Dsx give input to the dimorphic element and inhibit *bab* expression at the posterior part of the abdomen (A5-A6). In females, Abd-B and the female isoform of Dsx give input to the dimorphic element and activate *bab* expression at the posterior part of the abdomen (A5-A7). To understand how the male-specific pigmentation has arisen, the dimorphic element of *D. melanogaster* and the homologous *cis*-regulatory element of *Drosophila willistoni*, a species with monomorphic pigmentation, were compared (Williams et al. 2008). The dimorphic element of *D. willistoni*, which is believed to retain an ancestral state, has an activity in the female abdominal tip (A7). This activity could be involved in sexually dimorphic traits other than pigmentation, and the area of dimorphic expression may have been expanded by sequence evolution, including changes in polarity (direction) and spacing of transcription factor binding sites in the dimorphic element, resulting in the evolution of sexually dimorphic pigmentation (Williams et al. 2008). This study clarified the regulatory mechanism of sex-specific pigmentation patterns on the abdomen of *Drosophila* as a clear example of the evolution of sexual dimorphism by a *cis*-regulatory change with nucleotide level of resolution. Furthermore, a recent study showed that the activity of dimorphic elements of *D. melanogaster* has intraspecific variation among strains and correlates with their actual pigmentation levels (Rogers et al. 2013).

In *Drosophila guttifera*, the *cis*-regulatory elements of *wingless*, which control the pigmentation pattern on the wings, were identified and analyzed. In *Drosophila*, only *Drosophila guttifera* has pigmentation associated with the campaniform sensilla on the wings. Clarifying the mechanism of the gain of expression of *wingless* in the campaniform sensilla leads to an understanding of the proximate cause of the evolution of the novel wing pigmentation pattern of *D. guttifera*. To this end, we identified a novel enhancer driving expression at the campaniform sensilla, and its activity seemed to be gained in the lineage leading to *D. guttifera* (Koshikawa et al. 2015). This result showed the possibility that a novel color pattern could evolve when a gene responsible for color pattern formation gained novel enhancer activity.

Putative *cis*-regulatory elements for color patterning genes have also been identified in butterflies. The genomic comparison of subspecies of *Heliconius erato* and hybrids between subspecies revealed that genomic variations responsible for phenotypes resided around *optix*, *WntA*, and *cortex*. These variations are thought to be parts of *cis*-regulatory elements and are responsible for the diversity of wing patterns among the subspecies of *H. erato* (Van Belleghem et al. 2017). Another study on *Heliconius* butterflies suggested the mechanism of convergence of color patterns. The genomes of various species of *Heliconius* butterflies were sequenced and compared, and single-nucleotide polymorphisms (SNPs) correlated with wing color patterns were identified. As a result, the region presumed to be the *cis*-regulatory element of *optix* was found to have been shuffled among different lineages (including *Heliconius elevatus*, *Heliconius melpomene*, and *Heliconius timareta*) through genetic introgression. Müllerian mimicry of the different *Heliconius* species is thought to have arisen by this phenomenon, called enhancer shuffling (Wallbank et al. 2016). Lewis et al. (2019) further analyzed the *cis*-regulatory region of *optix* by ATAC-seq, ChIP-seq, and genome editing in *H. erato*, and compared it with that of other species. They found that enhancers were somewhat interdependent and pleiotropic, and proposed a new model to reconcile their data with previous studies.

As described above, changes in the *cis*-regulatory elements of the genes controlling color pattern formation play an important role as the proximate cause of color pattern evolution. This is probably because transcription factors and signal ligands are often used for the development of various tissues (pleiotropy), and changes in coding regions of the genes may have a harmful effect on organism development. In contrast, because *cis*-regulatory elements are modular, the addition or deletion of a specific functional unit is less likely to affect functions other than color pattern formation (Carroll 2008; Stern and Orgogozo 2008; Koshikawa 2015).

12.5 Theoretical Models for Insect Color Pattern Formation

In the study of mechanisms of color pattern formation in insects, progress has been made in theoretical studies using mathematical models as well as experimental studies. Models aim to explain the determination of the place and the area where color patterns are formed. Eyespots on the wings of butterflies are a favored target for constructing a theoretical model. Eyespot formation is thought to consist of two processes. The first is the process of determining the position of the organizer (the source of a signal) called the "focus." The second is the process of signal emission from the focus and subsequent determination of the pattern of the distribution of pigments.

In the first process, the "reaction–diffusion model" (Turing 1952; Meinhardt 1982) was adopted assuming a diffusing activator and an inhibitor that diffuses for longer than the activator (Nijhout 1990). In this model, the process starts with the introduction of a small amount of activator (a virtual factor to determine the position) from a wing vein at a steady state (the state when nothing has occurred). As a result, a stable dot-like pattern is formed on the midline between the wing veins, and it determines the position of the focus. This model accurately predicted the change in the expression pattern of the *Dll* gene in the eyespot on the wing. However, this model is susceptible to the size of the region, and in a different region, a significantly different pattern is formed. When considering a biological system that is robust against variations in parameters and region size caused by genetic and environmental changes, this model is unrealistic (Nijhout 2002; Otaki 2011).

A simple and more robust model was sought, and the "grass fire model" was devised (Nijhout 2017). In this model, virtual molecules that function as "fuel" were assumed to be widely distributed in the area. Product P1 is produced by a reaction using "fuel" as a substrate, and product P2 is generated by a reaction using P1 as a substrate. After that, the same reaction is repeated, and the process will progress like a grass fire. In this model, the entire process is started by introducing P1 at some point in the region. It is also assumed that all chemicals can diffuse from high to low concentration regions. The grass fire model can be used to explain the focus determination process if the distribution of P1 and P2 changes through development, and the positions of P1 and P2 can be read out at arbitrary times (Nijhout 2017). Although the molecular entities of "fuel," P1 and P2 are unknown, the transition of the distribution pattern of P1 and P2 is similar to the transition of the expression pattern of Notch (a signal receptor) and Distal-less (a transcription factor) on the wing (Reed and Serfas 2004; Zhang and Reed 2016; Nijhout 2017).

In the second process, the eyespot pattern is assumed to be formed by the concentration gradient of a morphogen diffusing from the focus (Nijhout 1980). In fact, in butterflies and a fruit fly, Wingless is assumed to be the morphogen that diffuses from one point to determine the pattern of pigment distribution. Wingless is believed to diffuse from the focus and form the pattern of eyespots in *B. anynana* (Özsu et al. 2017). It is also believed that, in *D. guttifera*, Wingless diffuses from the campaniform sensilla and determines the area of the polka-dotted pigmentation pattern (Werner et al. 2010). In developmental biology, the pattern formation of morphology by Wingless diffusion has long been a standard concept; however, the necessity of diffusion has begun to be questioned in recent studies. In *D. melanogaster*, Wingless diffusion was believed to be necessary for pattern formation of the wing shape, but in individuals that express Wingless fusion protein that is tethered to the cell membrane and not allowed to diffuse away from the membrane, and do not have wild-type Wingless protein, the wing shape was normally formed (Alexandre et al. 2014). The mode of Wingless transport remains largely unresolved questions (Parchure et al. 2018). Therefore, even in butterflies, it is necessary to reconsider whether a model assuming a concentration gradient of morphogen emitted from a focus is appropriate. There is a criticism that models based on morphogen diffusion have inconsistency with detailed observations of butterfly wing patterns (Otaki 2011). The process of determining the distribution of pigments can be explained by the above-described grass fire model. Setting a focus as the starting point of a reaction, the formation of the various shapes of the parafocal elements (color patterns formed at the distal side relative to eyespots) can be well explained as well as the formation of various shapes of eyespots. This suggests that both the eyespots and parafocal elements are formed from the focus (Nijhout 2017).

Taking another approach, the "induction model" has also been proposed to explain the second process. This model assumes that eyespot formation involves a physical force that arises by distortion of the epithelia and signal transduction by calcium waves. In the induction model, the following five stages (and self-similar repetition of the processes) are assumed (Otaki 2017).

1. After the position of the focus is determined cells of the focus cause polyploidization, and the size of the cells increases.
2. As a result of the cell size increase, the epithelium undergoes physical distortion, and this distortion propagates as a continuous wave.
3. Calcium waves are emitted as activation signals and self-enhanced by positive feedback.

4. An unknown inhibitory signal is produced at the margin of the calcium wave, and this limits the size of spots by negative feedback.
5. The gene expression for the color pattern formation is induced.

The phenomena mentioned in this model, such as the cell size increase, the calcium wave, and the tissue distortion at the focus, are actually observed in developing butterflies (Ohno and Otaki 2015; Iwata and Otaki 2016; Taira and Otaki 2016). The validation of the consistency of this unique model with experimental data is awaited.

12.6 The Future Directions of the Study of Insect Color Pattern Formation

To date, genes and their *cis*-regulatory elements that control the formation of insect color patterns have been identified and analyzed. The proximate cause of how various insect color patterns have evolved was shown to be attributable mainly to mutations that occurred in *cis*-regulatory elements. Although identification of such genes and their *cis*-regulatory elements provide important information for unraveling the mechanism of color pattern formation, it does not mean the complete elucidation of the color pattern formation. Currently, an unknown part of the mechanism of color pattern formation is how the regions of color patterns are determined. As described above, multiple theoretical models have been proposed for the formation of eyespots on butterfly wings. Studies on the color pattern formation of insects will be directed towards experimentally proving which one of these models is correct and building improved models that can provide more complete explanations. During this process, we expect new findings that will help explain the proximate cause and trajectory of color pattern diversification. So far, experimental models of color pattern formation are largely biased toward flies and butterflies. Recently, a transcription factor gene *pannier* was identified as the color pattern regulator of ladybird beetles (Ando et al. 2018; Gautier et al. 2018). Many other insects that have independently evolved color patterns should be studied to obtain a comprehensive understanding of color pattern formation in insects. In particular, observation of the process of color pattern formation in other insects will provide important information to augment current knowledge.

In this chapter, we divided the process of color pattern formation of insects into two major steps: upstream pattern formation and downstream pigmentation (coloration), and we focused mainly on the former, pattern formation. This division is convenient because color pattern formation is a complicated system involving various factors, such as synthesis and transport of proteins and precursors of pigments. We should aim for a comprehensive understanding of color pattern formation by using both top-down (theoretical and conceptual) approaches and bottom-up (observation-based) approaches.

Acknowledgements We thank Ryo Futahashi for giving us an opportunity to write this chapter, Toshiya Ando for helpful comments, and Elizabeth Nakajima for comments and English editing. This chapter is partially based on our review article published in the Japanese language (Fukutomi and Koshikawa 2018). Part of the writing was supported by KAKENHI (17 K19427, 18H02486, 18 J20452).

References

Alexandre C, Baena-Lopez A, Vincent JP (2014) Patterning and growth control by membrane-tethered wingless. Nature 505:180–185

Ando T, Matsuda T, Goto K, Hara K, Ito A, Hirata J, Yatomi J, Kajitani R, Okuno M, Yamaguchi K, Kobayashi M, Takano T, Minakuchi Y, Seki M, Suzuki Y, Yano K, Itoh T, Shigenobu S, Toyoda A, Niimi T (2018) Repeated inversions within a *pannier* intron drive diversification of intraspecific colour patterns of ladybird beetles. Nat Commun 9:3893

Arnoult L, Su KFY, Manoel D, Minervino C, Magriña J, Gompel N, Prud'homme B (2013) Emergence and diversification of fly pigmentation through evolution of a gene regulatory module. Science 339:1423–1426

Brakefield PM, Gates J, Keys D, Kesbeke F, Wijngaarden PJ, Montelro A, French V, Carroll SB (1996) Development, plasticity and evolution of butterfly eyespot patterns. Nature 384:236–242

Brunetti CR, Selegue JE, Monteiro A, French V, Brakefield PM, Carroll SB (2001) The generation and diversification of butterfly eyespot color patterns. Curr Biol 11:1578–1585

Carroll SB (2008) Evo-devo and an expanding evolutionary synthesis: a genetic theory of morphological evolution. Cell 134:25–36

Carroll SB, Gates J, Keys DN, Paddock SW, Panganiban GE, Selegue JE, Williams JA (1994) Pattern formation and eyespot determination in butterfly wings. Science 265:109–114

Carson HL, Hardy DE, Spieth HT, Stone WS (1970) The evolutionary biology of the Hawaiian Drosophilidae. In: Essays in evolution and genetics in honor of Theodosius Dobzhansky. North-Holland Publishing, Amsterdam, pp 437–543

Clarke CA, Sheppard PM (1972) The genetics of the mimetic butterfly *Papilio polytes* L. Phil Trans R Soc Lond B 263:431–458

Connahs H, Tlili S, van Creij J, Loo TY, Banerjee T, Saunders TE, Monteiro A (2019) Activation of butterfly eyespots by distal-less is consistent with a reaction-diffusion process. Development 146:dev169367

Cott HB (1940) Adaptive coloration in animals. Methuen, London

Dhungel B, Ohno Y, Matayoshi R, Iwasaki M, Taira W, Adhikari K, Gurung R, Otaki JM (2016) Distal-less induces elemental color patterns in *Junonia* butterfly wings. Zool Lett 2:4

Dufour HD, Koshikawa S, Finet C (2020) Temporal flexibility of gene regulatory network underlies a novel wing pattern in flies. Proc Natl Acad Sci USA 117:11589–11596

Fukutomi Y, Koshikawa S (2018) Present and future of research on pattern formation of insects (in Japanese). Sanshi-Konchu Biotech 87:95–102

Fukutomi Y, Matsumoto K, Agata K, Funayama N, Koshikawa S (2017) Pupal development and pigmentation process of a polka-dotted fruit fly, *Drosophila guttifera* (Insecta, Diptera). Dev Genes Evol 227:171–180

Fukutomi Y, Matsumoto K, Funayama N, Koshikawa S (2018) Methods for staging pupal periods and measurement of wing pigmentation of *Drosophila guttifera*. J Vis Exp 131:e56935

Fukutomi Y, Kondo S, Toyoda A, Shigenobu S, Koshikawa S (2021) Transcriptome analysis reveals *wingless* regulates neural development and signaling genes in the region of wing pigmentation of the polka-dotted fruit fly. FEBS J 288:115–126

Futahashi R, Osanai-Futahashi M (2021) Pigments in insects. In: Pigments, pigment cells and pigment patterns. Springer, Singapore, pp 3–43

Gautier M, Yamaguchi J, Foucaud J, Loiseau A, Ausset A, Facon B, Gschloessl B, Lagnel J, Loire E, Parrinello H, Severac D, Lopez-Roques C, Donnadieu C, Manno M, Berges H, Gharbi K, Lawson-Handley L, Zang LS, Vogel H, Estoup A, Prud'homme B (2018) The genomic basis of colour pattern polymorphism in the harlequin ladybird. Curr Biol 28:3296–3302

Gompel N, Prud'homme B, Wittkopp PJ, Kassner VA, Carroll SB (2005) Chance caught on the wing: *cis*-regulatory evolution and the origin of pigment patterns in *Drosophila*. Nature 433:481–487

Han Q, Fang J, Ding H, Johnson JK, Christensen BM, Li J (2002) Identification of *Drosophila melanogaster* yellow-f and yellow-f2 proteins as dopachrome-conversion enzymes. Biochem J 368:333–340

Hinaux H, Bachem K, Battistara M, Rossi M, Xin Y, Jaenichen R, Le Poul Y, Arnoult L, Kobler JM, Grunwald Kadow IC, Rodermund L, Prud'homme B, Gompel N (2018) Revisiting the developmental and cellular role of the pigmentation gene *yellow* in *Drosophila* using a tagged allele. Dev Biol 438:111–123

Ito K, Katsuma S, Kuwazaki S, Jouraku A, Fujimoto T, Sahara K, Yasukochi Y, Yamamoto K, Tabunoki H, Yokoyama T, Kadono-Okuda K (2016) Mapping and recombination analysis of two moth colour mutations, black moth and wild wing spot, in the silkworm *Bombyx mori*. Heredity 116:52–59

Iwata M, Otaki JM (2016) Spatial patterns of correlated scale size and scale color in relation to color pattern elements in butterfly wings. J Insect Physiol 85:32–45

Izumitani HF, Kusaka Y, Koshikawa S, Toda MJ, Katoh T (2016) Phylogeography of the subgenus *Drosophila* (Diptera: Drosophilidae): evolutionary history of faunal divergence between the old and the new worlds. PLoS One 11:e0160051

Jeong S, Rokas A, Carroll SB (2006) Regulation of body pigmentation by the abdominal-B Hox protein and its gain and loss in *Drosophila* evolution. Cell 125:1387–1399

Joron M, Papa R, Beltrán M, Chamberlain N, Mavárez J, Baxter S, Abanto M, Bermingham E, Humphray SJ, Rogers J, Beasley H (2006) A conserved supergene locus controls colour pattern diversity in *Heliconius* butterflies. PLoS Biol 4:e303

Keys DN, Lewis DL, Selegue JE, Pearson BJ, Goodrich LV, Johnson RL, Gates J, Scott MP, Carroll SB (1999) Recruitment of a hedgehog regulatory circuit in butterfly eyespot evolution. Science 283:532–534

Kopp A, Duncun I, Carroll SB (2000) Genetic control and evolution of sexually dimorphic characters in *Drosophila*. Nature 408:553–559

Koshikawa S (2015) Enhancer modularity and the evolution of new traits. Fly (Austin) 9:155–159

Koshikawa S (2020) Evolution of wing pigmentation in *Drosophila*: diversity, physiological regulation, and *cis*-regulatory evolution. Dev Growth Diff 62:269–278

Koshikawa S, Giorgianni MW, Vaccaro K, Kassner VA, Yoder JH, Werner T, Carroll SB (2015) Gain of *cis*-regulatory activities underlies novel domains of *wingless* gene expression in *Drosophila*. Proc Natl Acad Sci USA 112:7524–7529

Koshikawa S, Matsumoto K, Fukutomi Y (2017) *Drosophila guttifera* as a model system for unraveling color pattern formation. In: Sekimura T, Nijhout HF (eds) Diversity and evolution of butterfly wing patterns. Springer, Singapore, pp 287–301

Kunte K, Zhang W, Tenger-Trolander A, Palmer DH, Martin A, Reed RD, Mullen SP, Kronforst MR (2014) *doublesex* is a mimicry supergene. Nature 507:229–232

Lewis JJ, Geltman RC, Pollak PC, Rondem KE, Van Belleghem SM, Hubisz MJ, Munn PR, Zhang L, Benson C, Mazo-Vargas A, Danko CG, Counterman BA, Papa R, Reed RD (2019) Parallel evolution of ancient, pleiotropic enhancers underlies butterfly wing pattern mimicry. Proc Natl Acad Sci USA 116:24174–24183

Linsley EG (1959) Mimetic form and coloration in the Cerambycidae (Coleoptera). Ann Entomol Soc Am 52:125–131

Marcus JM, Ramos DM, Monteiro A (2004) Germline transformation of the butterfly *Bicyclus anynana*. Proc R Soc Lond B 271(Suppl 5):S263–S265

Markow TA, O'Grady P (2006) Drosophila: a guide to species identification and use. Academic Press, New York
Martin A, Reed RD (2014) Wnt signaling underlies evolution and development of the butterfly wing pattern symmetry systems. Dev Biol 395:367–378
Martin A, Papa R, Nadeau NJ, Hill RI, Counterman BA, Halder G, Jiggins CD, Kronforst MR, Long AD, McMillan WO, Reed RD (2012) Diversification of complex butterfly wing patterns by repeated regulatory evolution of a *Wnt* ligand. Proc Natl Acad Sci USA 109:12632–12637
Mazo-Vargas A, Concha C, Livraghi L, Massardo D, Wallbank RWR, Zhang L, Papador JD, Martinez-Najera D, Jiggins CD, Kronforst MR, Breuker CJ, Reed RD, Patel NH, McMillan WO, Martin A (2017) Macroevolutionary shifts of *WntA* function potentiate butterfly wing-pattern diversity. Proc Natl Acad Sci USA 114:10701–10706
Meinhardt H (1982) Models of biological pattern formation. Academic Press, London
Monteiro A (2015) Origin, development, and evolution of butterfly eyespots. Annu Rev Entomol 60:253–271
Monteiro A, Chen B, Ramos DM, Oliver JC, Tong X, Guo M, Wang WK, Fazzino L, Kamal F (2013) *Distal-less* regulates eyespot patterns and melanization in *Bicyclus* butterflies. J Exp Zool B 320:321–331
Nadeau NJ, Pardo-Diaz C, Whibley A, Supple MA, Saenko SV, Wallbank RW, Wu GC, Maroja L, Ferguson L, Hanly JJ, Hines H, Salazar C, Merrill RM, Dowling AJ, Llaurens V, Joron M, WO MM, Jiggins CD (2016) The gene *cortex* controls mimicry and crypsis in butterflies and moths. Nature 534:106–110
Nijhout HF (1980) Pattern formation on lepidopteran wings: determination of an eyespot. Dev Biol 80:267–274
Nijhout HF (1990) A comprehensive model for colour pattern formation in butterflies. Proc R Soc Lond B 239:81–113
Nijhout HF (2001) Elements of butterfly wing patterns. J Exp Zool (Mol Dev Evol) 291:213–225
Nijhout HF (2002) The nature of robustness in development. BioEssays 24:553–563
Nijhout HF (2017) The common developmental origin of eyespots and parafocal elements and a new model mechanism for color pattern formation. In: Sekimura T, Nijhout HF (eds) Diversity and evolution of butterfly wing patterns. Springer, Singapore, pp 3–9
Nishikawa H, Iijima T, Kajitani R, Yamaguchi J, Ando T, Suzuki Y, Sugano S, Fujiyama A, Kosugi S, Hirakawa H, Tabata S, Ozaki K, Morimoto H, Ihara K, Obara M, Hori H, Itoh T, Fujiwara H (2015) A genetic mechanism for female-limited Batesian mimicry in *Papilio* butterfly. Nat Genet 47:405–409
O'Grady PM, DeSalle R (2018) Phylogeny of the genus *Drosophila*. Genetics 209:1–25
Ohno Y, Otaki JM (2015) Spontaneous long-range calcium waves in developing butterfly wings. BMC Dev Biol 15:17
Otaki JM (2011) Color-pattern analysis of eyespots in butterfly wings: a critical examination of morphogen gradient models. Zool Sci 28:403–413
Otaki JM (2017) Self-similarity, distortion waves, and the essence of morphogenesis: a generalized view of color pattern formation in butterfly wings. In: Sekimura T, Nijhout HF (eds) Diversity and evolution of butterfly wing patterns. Springer, Singapore, pp 119–152
Özsu N, Chan QY, Chen B, Gupta MD, Monteiro A (2017) *Wingless* is a positive regulator of eyespot color patterns in *Bicyclus anynana* butterflies. Dev Biol 429:177–185
Parchure A, Vyas N, Mayor S (2018) Wnt and hedgehog: secretion of lipid-modified morphogens. Trends Cell Biol 28:157–170
Patterson JT, Wagner RP, Wharton LT (1943) The Drosophilidae of the southwest. University of Texas, Texas
Rebeiz M, Williams TM (2017) Using *Drosophila* pigmentation traits to study the mechanisms of *cis*-regulatory evolution. Curr Opin Insect Sci 19:1–7
Reed RD, Serfas MS (2004) Butterfly wing pattern evolution is associated with changes in a notch/distal-less temporal pattern formation process. Curr Biol 14:1159–1166

Reed RD, Papa R, Martin A, Hines HM, Counterman BA, Pardo-Diaz C, Jiggins CD, Chamberlain NL, Kronforst MR, Chen R, Halder G, Nijhout HF, McMillan WO (2011) *optix* drives the repeated convergent evolution of butterfly wing pattern mimicry. Science 333:1137–1141

Roeske MJ, Camino EM, Grover S, Rebeiz M, Williams TM (2018) *Cis*-regulatory evolution integrated the Bric-à-brac transcription factors into a novel fruit fly gene regulatory network. elife 7:e32273

Rogers WA, Salomone JR, Tacy DJ, Camino EM, Davis KA, Rebeiz M, Williams TM (2013) Recurrent modification of a conserved *cis*-regulatory element underlies fruit fly pigmentation diversity. PLoS Genet 9:e1003740

Rogers WA, Grover S, Stringer SJ, Parks J, Rebeiz M, Williams TM (2014) A survey of the *trans*-regulatory landscape for *Drosophila melanogaster* abdominal pigmentation. Dev Biol 385:417–432

Ruxton GD, Sherratt TN, Speed MP (2004) Avoiding attack: the evolutionary ecology of crypsis, warning signals and mimicry. Oxford University Press, Oxford

Setoguchi S, Takamori H, Aotsuka T, Sese J, Ishikawa Y, Matsuo T (2014) Sexual dimorphism and courtship behavior in *Drosophila prolongata*. J Ethol 32:91–102

Sherratt TN (2008) The evolution of Müllerian mimicry. Naturwissenschaften 95:681–695

Stern DL, Orgogozo V (2008) The loci of evolution: how predictable is genetic evolution? Evolution 62:2155–2177

Taira W, Otaki JM (2016) Butterfly wings are three-dimensional: pupal cuticle focal spots and their associated structures in *Junonia* butterflies. PLoS One 11:e0146348

Tong X, Lindemann A, Monteiro A (2012) Differential involvement of hedgehog signaling in butterfly wing and eyespot development. PLoS One 7:e51087

Tong X, Hrycaj S, Podlaha O, Popadic A, Monteiro A (2014) Over-expression of *Ultrabithorax* alters embryonic body plan and wing patterns in the butterfly *Bicyclus anynana*. Dev Biol 394:357–366

True JR, Edwards KA, Yamamoto D, Carroll SB (1999) *Drosophila* wing melanin patterns form by vein-dependent elaboration of enzymatic prepatterns. Curr Biol 9:1382–1391

Turing AM (1952) The chemical basis of morphogenesis. Phil Trans R Soc Lond B 237:37–72

Van Belleghem SM, Rastas P, Papanicolaou A, Martin SH, Arias CF, Supple MA, Hanly JJ, Mallet J, Lewis JJ, Hines HM, Ruiz M, Salazar C, Linares M, Moreira GRP, Jiggins CD, Counterman BA, McMillan WO, Papa R (2017) Complex modular architecture around a simple toolkit of wing pattern genes. Nat Ecol Evol 1:52

van't Hof AE, Campagne P, Rigden DJ, Yung CJ, Lingley J, Quail MA, Hall N, Darby AC, Saccheri IJ (2016) The industrial melanism mutation in British peppered moths is a transposable element. Nature 534:102–105

Wallbank RW, Baxter SW, Pardo-Diaz C, Hanly JJ, Martin SH, Mallet J, Dasmahapatra KK, Salazar C, Joron M, Nadeau N, McMillan WO, Jiggins CD (2016) Evolutionary novelty in a butterfly wing pattern through enhancer shuffling. PLoS Biol 14:e1002353

Werner T, Koshikawa S, Williams TM, Carroll SB (2010) Generation of a novel wing colour pattern by the wingless morphogen. Nature 464:1143–1148

Werner T, Steenwinkel T, Jaenike J (2018) The encyclopedia of north American Drosophilids volume 1: Drosophilids of the Midwest and Northeast. Michigan Technological University, Houghton

Williams TM, Selegue JE, Werner T, Gompel N, Kopp A, Carroll SB (2008) The regulation and evolution of a genetic switch controlling sexually dimorphic traits in *Drosophila*. Cell 134:610–623

Wittkopp PJ, True JR, Carroll SB (2002) Reciprocal functions of the *Drosophila* yellow and ebony proteins in the development and evolution of pigment patterns. Development 129:1849–1858

Wittkopp PJ, Carroll SB, Kopp A (2003) Evolution in black and white: genetic control of pigment patterns in *Drosophila*. Trends Genet 19:495–504

Yamaguchi J, Banno Y, Mita K, Yamamoto K, Ando T, Fujiwara H (2013) Periodic *Wnt1* expression in response to ecdysteroid generates twin-spot markings on caterpillars. Nat Commun 4:1857

Yassin A, Delaney EK, Reddiex AJ, Seher TD, Bastide H, Appleton NC, Lack JB, David JR, Chenoweth SF, Pool JE, Kopp A (2016) The *pdm3* locus is a hotspot for recurrent evolution of female-limited color dimorphism in *Drosophila*. Curr Biol 26:2412–2422

Zhang L, Reed RD (2016) Genome editing in butterflies reveals that *spalt* promotes and *Distal-less* represses eyespot colour patterns. Nat Commun 15:11769

Zhang L, Mazo-Vargas A, Reed RD (2017) Single master regulatory gene coordinates the evolution and development of butterfly color and iridescence. Proc Natl Acad Sci USA 114:10707–10712

Part III
Color Changes

Chapter 13
Physiological and Morphological Color Changes in Teleosts and in Reptiles

Makoto Goda and Takeo Kuriyama

Abstract The spectacular changes in the beautiful coloration and conspicuous patterns in animals have commanded the attention of a wide range of people, including hobbyists, scuba divers, and even zoologists since Aristotle's time. For the animals themselves, these color changes and pattern formations can represent strategies of the utmost importance for the survival of individuals or of species. For instance, cryptic coloration and pattern formation is useful for avoiding attacks by predatory animals. On poikilothermic vertebrates, slow changes in integumental coloration and patterns can be attributed to increases and decreases in the number of chromatophores which possess pigmentary substances and/or high refractive index materials in cytoplasm. These slow changes are generally referred to as the "morphological color changes." Conversely, the more rapid changes in color and patterns are called "physiological color changes" and are attributed to the motile activities of chromatophores, to the aggregation of chromatosomes in the perikaryon and their dispersion throughout the cytoplasm or to changes in the arrangement of the high refractive index materials in the cytoplasm. This chapter explores the various studies on the morphological and/or physiological color changes in poikilothermic vertebrates, fish, and reptiles.

Keywords Color changes · Chromatophore · Fish · Reptile

M. Goda (✉)
Hamamatsu University School of Medicine, Hamamatsu, Shizuoka, Japan
e-mail: goda@hama-med.ac.jp

T. Kuriyama
Institute of Natural and Environmental Sciences, University of Hyogo, Hyogo, Japan

H. Hashimoto et al. (eds.), *Pigments, Pigment Cells and Pigment Patterns*,
https://doi.org/10.1007/978-981-16-1490-3_13

13.1 Color Changes in Teleosts

Serious scientific study of the details of animal coloration began only in the middle of the nineteenth century, when zoologists conducted the first investigations into the remarkable color changes of the African chameleon. Subsequently, various important queries were raised and investigated regarding the physiology and cell biology of integumental color changes in poikilothermic animals. Teleosts, and especially those of smaller size, that change their integument color and pattern provide excellent material for these types of investigations, since the morphological features of the colored cells of the skin are easy to observe. This ease of observation is a quite important for cell biological studies; hence, many researchers continue to study these teleosts today.

13.1.1 Varieties of Chromatophores

Integumental colorations arise due to the presence of pigments and/or the high refractive index materials in dermis and epidermis (Fujii 1993a, b; Needham 1974). Specifically, the colors of the integuments are generated by the absorption, reflection, and scattering of certain light wavelength by the pigments and by the presence of high refractive index materials. In poikilothermic animals, including fish, the materials are packed within specialized organelles called chromatosomes, which are localized in colored cells called "chromatophores."

Various colorations and patterns are displayed in the integument of fish due to the presence of dermal and epidermal chromatophores. These chromatophores can appropriately be classified into eight categories. Those that assume their characteristic colors by absorbing light include the melanophores (black), the xanthophores (yellow), the erythrophores (red), the cyanophores (blue), and the erythrocyanophores (red and blue), while those that reflect light are known as leucophores (white), iridophores (iridescent), and erythro-iridophores (reddish-violet) (Bagnara and Hadley 1973; Fujii 1993a, b; Goda and Fujii 1995; Goda et al. 2011, 2013; Schartl et al. 2016) (Figs. 13.1 and 13.2).

Melanophores are pigmentary chromatophores of the dendritic type and containing black chromatosomes, the melanosomes, in the cytoplasm. These are the most common type of chromatophore in fish. In many fish species, melanophores are mostly found in the dermis, so they are also called "dermal melanophores". The epidermis of some species, however, contains melanophores with extremely fine dendritic processes projecting from small cell bodies (Fujii 1993b). Melanophores play a principal part in integumental color changes; however, other kinds of dendritic pigmentary chromatophores also exist in the skin. The melanosomes effectively absorb light rays within the entire range of the visible spectrum, whereas the chromatosomes of the other pigmentary chromatophores absorb light rays within limited ranges of the visible spectrum to give rise to a specific coloration. The

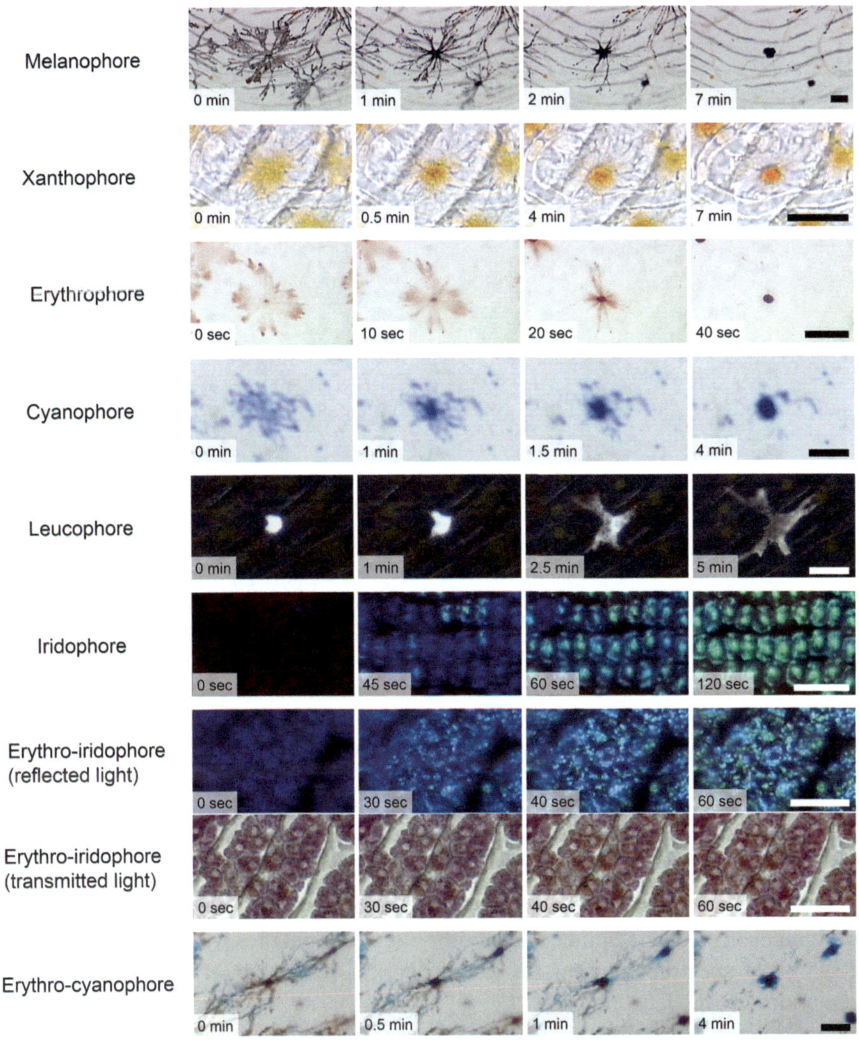

Fig. 13.1 Photomicrographs showing the various chromatophore types in teleosts and their motile responses in four states. The motile responses of the chromatophores were induced using K^+-rich saline as an inducer of neuronal stimulation. The elapsed time after application of K^+-rich saline is superimposed on each photomicrograph. Coloration of the pigmentary chromatophores and the reddish color of the erythro-iridophores are photographed under the transmission optics of a light microscope. The photomicrographs of interference light or scattered light from light-reflecting chromatophores, leucophores, iridophores and erythro-iridophores are taken using epi-illumination optics. The melanophores, xanthophores, and leucophores are from a scale of the medaka, *Oryzias latipes*. The erythrophores are from a scale of the biotcheyes soldierfish, *Myripristis berndti*. Cyanophores and erythro-cyanophores are from the pectoral fin of the mandarin fish, *Synchiropus splendidus*. Erythro-iridophores are from a scale of the diadem dottyback, *Pseudochromis diadema*. Bar = 30 μm

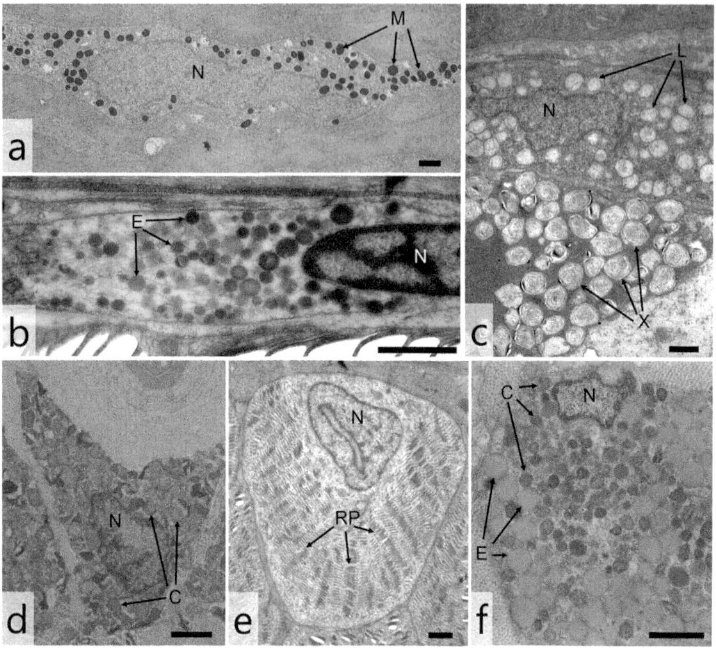

Fig. 13.2 Electron micrographs showing the various chromatophore types in teleosts. (**a**) Melanophore of the ice goby, *Leucopsarion petersii*. (**b**) Erythrophore of the biotcheyes soldierfish, *Myripristis berndti*. (**c**) Xanthophore and leucophore of the *Oryzias latipes*. (**d**) Cyanophore of the mandarin fish *Synchiropus splendidus*. (**e**) Iridophore of the palette surgeonfish, *Paracanthurus hepatus*. (**f**) Erythro-cyanophore of the mandarin fish *Synchiropus splendidus*. *N* nucleus, *M* melanosomes, *E* erythrosomes, *X* xanthosomes, *C* cyanosomes, *L* leucosomes, *RP* reflecting platelets. Bars = 1 μm

chromatosomes of leucophores, by contrast, scatter light rays within a wider range of the visible spectrum, giving rise to the optical phenomenon that gives a white coloration by the leucophores.

Iridophores are commonly found wherever silvery or whitish parts of the skin are observed. They are usually non-dendritic and do not possess pigment granules, but instead contain stacks of transparent flat guanine crystals, called "reflecting platelets," in their cytoplasm (Denton and Nicol 1966; Land 1972). The reflecting platelets are responsible for the reflection of light from the skin, because they are strongly light-reflecting owing to their high refractive index (no less than 1.83). In the iridophore cytoplasm, the spacing between adjacent reflecting platelets is strikingly uniform within a stack, and higher light reflection can be achieved as a result of the thin-film interference phenomenon (Huxley 1968; Land 1972). Generally, the iridophores in colored parts of the skin are static cells and do not exhibit integumental color changes in response to any environmental cues.

In several teleostean species, the iridophores have cellular motility and play a predominant role in the peculiar color changes of these fish (Fujii 1993a, b). As in the static iridophores, the platelets in the motile iridophores are piled up to form stacks within the cytoplasm, but their thickness is always very thin—no thicker than the plasma membrane—and the spacing between adjacent platelets can vary in response to controlling signals. The morphological features and the motility of the platelets are responsible for the exhibition of higher chroma colors and changes in hue (i.e., shifts in spectral peaks), respectively. Fish iridophores are usually non-dendritic; however, dendritic iridophores were reported in 1986 in the integument of some gobiid fishes (Iga and Matsuno 1986). The reflecting platelets of the dendritic iridophores have similar motile activities, aggregation, and dispersion to those in the other motile pigmentary chromatophores.

13.1.2 *Motile Mechanisms of Chromatophores*

To date, a number of investigations have revealed the motile mechanism of fish chromatophores. For the dendritic chromatophores, the microtubules that radiate from the cell body, with minus-ends toward the periphery of the dendritic processes, play an active role (Euteneuer and McIntosh 1981; McNiven and Porter 1986). The fish melanophores also contain a centrosome with two centrioles located at the center of the radiation of the microtubules (Skold et al. 2002). The general consensus is that the movement of the chromatosomes in a centripetal or centrifugal direction is caused by imposition of a sliding force between the surface of the chromatosomes and the microtubules (Murphy and Tilney 1974; Nilsson et al. 1996; Obika 1978; Obika et al. 1986; Rodionov et al. 1998; Schliwa 1984). The use of specific antibodies for dynein and kinesin confirmed that motor proteins are distributed on fish melanosomes for both the centripetal and centrifugal translocation of pigments (Nilsson et al. 1996; Rodionov et al. 1991).

In fish melanophores treated to remove microtubules in the cytoplasm, melanosome movement was not eliminated; instead, slow movements similar to the "shuttle" movements of untreated cells were observed (Schliwa and Euteneuer 1978; Schliwa et al. 1981). Western blot experiments and immunoelectron microscopy indicated the presence of the actin-based motor myosin Va on the melanosomes of the fish melanophores (Rodionov et al. 1998, 2003). Thus, the activation of myosin Va on melanosomes appears to be a principal cause of the "shuttle" movements. Pharmacological experiments using actin disassembling drugs suggested that the random location of dispersed chromatosomes requires a dynamic actin network on the fish melanophores (Ohta 1974; Rodionov et al. 1998; Skold et al. 2002).

Conversely, in teleosts, non-dendritic chromatophores (i.e., iridophores) are concentrated in the whitish and silvery belly surface and are responsible for the light reflectance in the coloration. The light reflectivity of the iridophores is undoubtedly due to the presence of light-reflecting materials within it. Previous morphological studies of fish iridophores by electron microscopy have revealed the presence of many high-density platelets that tend to form piles within the cytoplasm (Kawaguti 1965; Kawaguti and Kamishima 1966). The iridophores found on pale skin usually shown no variation in the distance between the light-reflecting platelets in a pile. However, in the bluish skin of some tropical fish, the distance between the reflecting platelets in the iridophores changes in response to various stimuli (Fujii 2000; Oshima and Fujii 1987).

13.1.3 Physiological Color Changes

In poikilothermic vertebrates, including teleosts, the integumental color changes due to motile activities of chromatophores are called "physiological color changes." The motile activities are regulated by the endocrine and/or nervous system, and the changes are rather rapid and are involved in faster chromatic adaptations (Figs. 13.3 and 13.4). The coordinating systems for the motile activities of chromatophores found in fish show rich diversity. In some teleosts, endocrine hormones are predominantly responsible for the chromatophore movements, while in others, a neural mechanism seems to dominate. Between these systems, many examples are found of teleosts that use both hormonal and neural mechanisms in various proportions.

Fig. 13.3 Pigmentary patterns of the brown pencilfish, *Nannostomus eques*, in daytime (**a**) and nighttime (**b**)

Fig. 13.4 Typical integumental colorations of the blue damselfish, *Chrysiptera cyanea*. (**a**) Cobalt-blue coloration commonly observable among most fish in a school. (**b**) Dark violet coloration observable under nervous conditions

In order to facilitate the understanding of the system that regulates the motile activities of fish chromatophores, a diagram of the system is shown in Fig. 13.5. This diagram was drawn primarily to show the control by the dendritic chromatophores of the light-absorbing type, namely the melanophores, erythrophores, xanthophores, and cyanophores. The different optical properties naturally result in somewhat different regulatory systems for the light-reflecting chromatophores than for the light-absorbing chromatophores. Therefore, although some parts are analogous, the diagram cannot be applied as it stands to the control of leucophores or motile iridophores.

Effects of Hormonal Substances

The effects of hormonal substances on the motile activities of chromatophores have been the focus of considerable interest for many years. Among several hormonal substances confirmed to control fish chromatophores, the melanophore-stimulating hormone (MSH) secreted by the intermediate lobe of the pituitary is the best known. Numerous investigators have reported that MSH induces the dispersion of melanosomes in a number of teleost species. Adrenocorticotropic hormone (ACTH), which contains the amino acid sequence of MSH, has also been shown to be melanosome dispersing in teleosts (Fujii 1969, 1993a, b, 2000). In 1982, Fujii and Miyashita used a synthetic, purified sample of MSH to show that both the dermal and epidermal melanophores of a siluroid catfish, *Siluris asotus*, responded to MSH by the dispersion of melanosomes (Fujii and Miyashita 1982). Working on the medaka, *Oryzias*

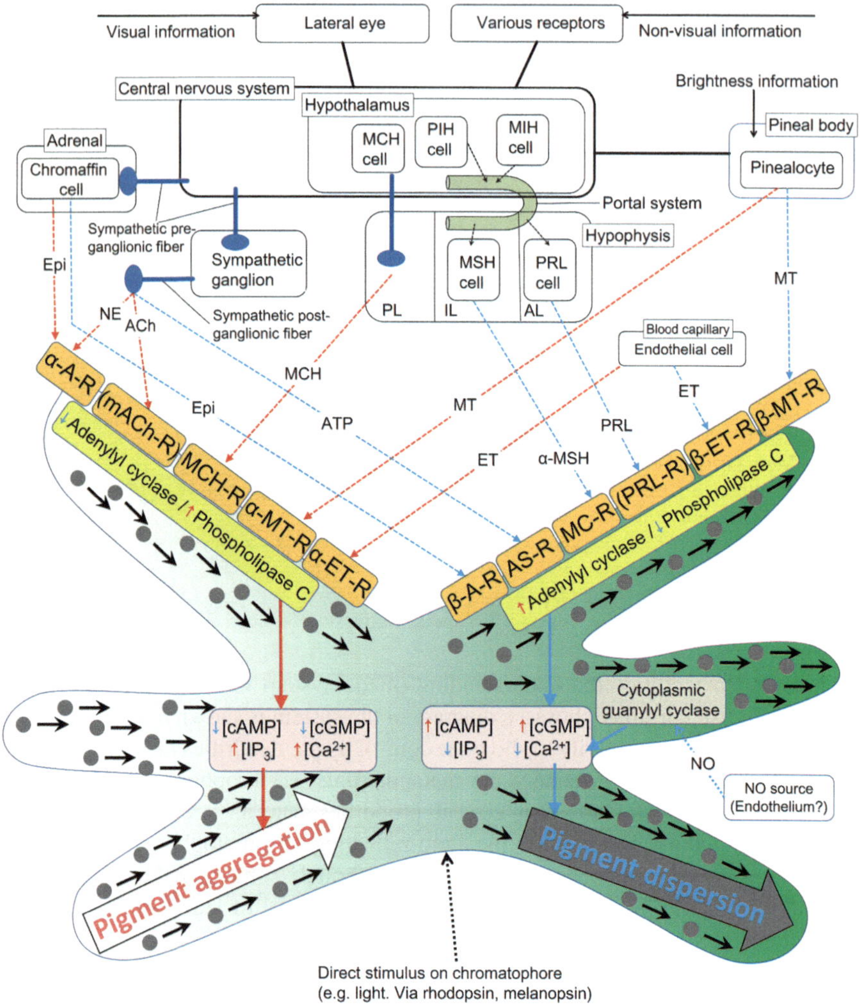

Fig. 13.5 Diagrammatic representation of the system for the control of motile activities of pigmentary chromatophores in teleosts. *α-A-R* α-Adrenoceptor, *NE* norepinephrine, *mACh-R* muscarinic acetylcholine receptor, *ACh* acetylcholine, *MCH-R* MCH receptor, *Epi* epinephrine, *ATP* adenosine 5′-triphosphate, *α-MT-R* α-MT receptor, *PRL cell* prolactin-producing cell, *MCH* melanin-concentrating hormone, *α-ET-R* α-ET receptor, *AL* anterior lobe of hypophysis, *cAMP* cyclic adenosine 3′,5′-monophosphate, *cGMP* cyclic guanosine 3′,5′-monophosphate, *IP3* inositol-1,4,5-triphosphate, *PRL* prolactin, *AS-R* adenosine receptor, *MSH cell* MSH-producing cell, *PIH cell* PRL-release inhibiting hormone-secreting cell, *IL* intermediate lobe of hypophysis, *PRL-R* PRL receptor, *MIH cell* MSH release-inhibiting hormone, *MC-R* melanocortin receptor, *α-MSH* α-melanophore-stimulating hormone, *β-A-R* β-adrenoceptor, *PL* posterior lobe of hypophysis, *ET* endothelin, *MT* melatonin, *MCH cell* MCH-producing neuron in hypothalamus, *β-ET-R* β-ET receptor, *β-MT-R* β-melatonin receptor, *NO* nitric oxide

latipes, Sugimoto et al. found that MSH induces dispersion of colorless premelanosomes which lack mature melanin in amelanotic melanophores (Sugimoto et al. 1985). In 1978, Matsumoto et al. showed clearly that primary cultured erythrophores isolated from swordtail, *Xiphophorus helleri*, responded to MSH (Matsumoto et al. 1978). Recently, the fluorescent chromatophores of the neon pygmy goby, *Eviota pellucida*, were show to react to alpha-MSH by exhibiting the dispersion of fluorescent chromatosomes (Wucherer and Michiels 2012). The leucophores of medakas also responded to MSH by the dispersion of leucosomes (Negishi and Obika 1980; Oshima and Fujii 1985). This was a rather unexpected response, since the direction of a leucophore response is normally opposite to that of the light-absorbing chromatophores. Dendritic motile iridophores of the dark sleeper goby, *Odontobutis obscura*, responded to MSH by aggregation of their reflecting platelets (Fujii et al. 1991a; Iga and Matsuno 1986). The round motile iridophores of the damselfish and neon tetra type showed a shift in the spectral peak of their reflected light toward shorter wavelength in response to alpha-MSH, but only when very strong solutions were applied (Nagaishi and Oshima 1989; Oshima et al. 1989).

Fujii and Miyashita revealed that the effect of MSH in melanophores of catfish may be mediated by specific receptors (Fujii and Miyashita 1982). Extracellular Ca^{2+} ions are required for the action of this hormone on melanophores (Fujii and Miyashita 1980; Iga and Takabatake 1982). The dependence on Ca^{2+} for the action of this hormone was demonstrated in xanthophores and leucophores of the medaka and in erythrophores of the platyfish (Oshima and Fujii 1984). Of the known hormonal and neural substances that signal motile responses in fish chromatophores, only MSH requires extracellular Ca^{2+} ions for its effect.

Prolactin (PRL) is a peptide hormone secreted by the anterior lobe of the pituitary. Sage, in studies on the mudsucker *Gillichthys mirabilis*, was the first to show that PRL has ability to disperse pigment in xanthophores (Sage 1970). Oshima and her associates investigated the effects of PRL on chromatophores of the Nile tilapia, *Oreochromis niloticus* (Kitta et al. 1993). They used two molecular species of PRL, $tPRL_{177}$ and $tPRL_{188}$, which had been purified from the Mozambique tilapia, *O. mossanbicus* (Kitta et al. 1993). They reported little, if any, effect of the peptides on pigment aggregation in the melanophores, whereas $tPRL_{177}$ caused a clear pigment dispersal in xanthophores. Their further pharmacological investigations of the pigment-dispersing effects of PRL with some teleosts revealed that cAMP is a second messenger in these effects (Oshima and Goto 2000). In addition, the response to the peptide showed a seasonal pattern in the xanthophores of the rose bitterling, *Rhodeus ocellatus ocellatus*, and in the erythrophores of the Nile tilapia (Oshima et al. 1996).

In many poikilothermic vertebrates, including teleosts, the integumental color changes are under dual pituitary hormonal control (Baker and Ball 1975; Pickford and Atz 1957). One principal hormone is MSH, as mentioned above, which gives rise to the darkening of animal coloration. The other principal hormone, which is antagonistic in function (i.e., it brightens the skin color or promotes aggregation of pigmentary organelles in the chromatophores) was demonstrated in the pituitary as well as in the hypothalamus of the siluroid catfish, *Silurus asotus* (Enami 1955). Enami named it the melanophore-concentrating hormone (MCH). In 1983, Kawauchi et al. succeeded in isolating MCH from the pituitary glands of the chum salmon, *Oncorhynchus keta* (Kawauchi et al. 1983). It has since been renamed melanin-concentrating hormone, because what concentrates are not melanophores but melanin-carrying organelles. In 1986, Nagai et al. revealed that MCH induces the aggregation of melanosomes in melanophores of all teleostean species (Nagai et al. 1986).

Pigmentary chromatophores other than melanophores show similar responses to MCH (Mizusawa et al. 2011; Oshima et al. 1986a; Wucherer and Michiels 2012). By contrast, leucosomes, which are light-scattering organelles in leucophores, are dispersed in response to MCH, but high concentrations of the hormone were needed for motile activity (Oshima et al. 1986a). Previous investigations of intracellular signaling in response to MCH showed that the hormone binds to a G-protein-coupled receptor on the plasma membrane of chromatophores and reduces cAMP levels (Kawauchi 2006; Logan et al. 2006; Oshima et al. 1986a). Abrao et al. suggested that activation of the MCH receptor induces the production of the second messenger diacylglycerol and activation of protein kinase C (PKC) via activation of phospholipase C (Abrao et al. 1991). Oshima and her associates suggested the existence of two types of receptors on chromatophores that respond to MCH because the application of high concentrations of MCH caused the dispersion of chromatosomes in melanophores and leucophores in some teleosts (Castrucci et al. 1988; Oshima et al. 1986a, 2001). Further analyses are clearly warranted to explain this issue.

In 1958, a potent principal compound that caused blanching of frog skin was isolated from the bovine pineal gland and identified as N-acetyl-5-methoxytryptamine (Lerner et al. 1958). These researchers named it "melatonin." Circadian changes are now well known to occur in the rate of production of melatonin and its release. Melatonin functions in the circadian changes of body patterns of the pencilfish, *Nannostomus beckfordi* (Reed 1968). Immediate aggregation of melanosomes in fish melanophores in response to melatonin was first demonstrated by Fujii (Fujii 1961), and a number of studies with fish melanophores have since been conducted on the pigment-aggregation action of this hormone (Aspengren et al. 2003; Fujii 1993a, b, 2000; Fujii and Oshima 1986, 1994; Fujishige et al. 2000; Skold et al. 2008). The pigment-aggregating action of melatonin is mediated via a melatonin receptor and is accompanied by an increase and a decrease in the cytosolic level of Ca^{2+} and of cAMP, respectively (Aspengren et al. 2003).

In pigmentary chromatophores other than melanophores, a discernible aggregation of chromatosomes by melatonin has been reported. For example, Matsumoto et al. observed that cultured erythrophores from the swordtail, *Xiphophorus helleri*, responded to melatonin by pigment aggregation (Matsumoto et al. 1978). Erythrophores of the neon tetra, *Cheirodon innesi*, and the cardinal tetra, *C. axelrodi*, are also responsive to melatonin (Fujii 2000). By contrast, the effects of melatonin on light-reflecting chromatophores have been investigated. Obika et al. observed that leucosomes in the leucophores of medaka aggregated in response to melatonin (Obika 1976, 1988). Melatonin at high concentrations also induced a shift in the spectral peak of reflected light toward longer wavelengths in the motile iridophores of the neon tetra, *C. innesi*, and in the blue-green damselfish, *Chromis viridis* (Nagaishi and Oshima 1989; Oshima et al. 1989). Although the melanophores in many teleosts are very sensitive to melatonin, those in other teleosts are frequently reported as refractory (Fujii and Oshima 1986). Furthermore, differential responsiveness to melatonin was observed even among chromatophores in limited areas of the fish skin (Fujii and Oshima 1986; Fujii and Taguchi 1969; Grundstrom et al. 1985; Martensson and Andersson 2000). Recently, Ovais et al. reported seasonal variations in the responsiveness of melanophores to melatonin in the slender rasbora, *Rasbora daniconius* (Ovais et al. 2015). Nishi and Fujii reported that certain parts of the body pattern of the pencilfish, *Nannostomus beckfordi*, darkened during the night, and they investigated the motile activities of the melanophores present in the integument of those parts (Nishi and Fujii 1992). They concluded that those melanophores possess melatonin receptors that mediate the dispersion of pigment, and they designated these as "beta-melatonin receptors." The conventional pigment-aggregate receptor was named "alpha-melatonin receptor." The beta-melatonin receptor was found in melanophores of the three-lined pencilfish, *N. trifasciatus*, and the slender lasbora (Masagaki and Fujii 1999; Ovais et al. 2015).

Endothelin (ET) was originally found in the culture medium of porcine aortic endothelial cells (Yanagisawa et al. 1988). Imokawa et al. reported that human keratinocytes produce ET and that it acts as a strong mitogen, as well as a melanogen, for melanocytes (Imokawa et al. 1992). Work on teleostean fish has revealed that ET is secreted from epidermal cells in goldfish (Akimoto et al. 2000), and Fujii and his associates found that ET induced motile responses in most chromatophores (Fujii et al. 1993). The pharmacological properties of the ET receptors of melanophores (Hayashi et al. 1996), erythrophores, and xanthophores (Murata and Fujii 2000) resemble those of the ET_B described in mammalian tissues. The direction of the responses to ET of these chromatophores is toward the aggregation of the chromatosomes. By contrast, the leucosomes in the leucophores of the medaka, *Oryzias latipes*, are dispersed under the influence of ET (Fujita and Fujii 1997).

Nitric oxide (NO), an unstable gaseous radical, has been shown to have various effects on different biological processes in animals. In 2001, NO was reported to play an active role in the elaborate control of color changes in teleosts by dispersing the pigment in melanophores via activation of intracellular guanylyl cyclase and an increase in cytosolic cGMP levels (Hayashi and Fujii 2001).

Effects of Neuronal Substances

Many species belonging to the Osteichthyes show regulation of the motile activities of chromatophores by both nervous and endocrine signals. The peripheral chromatic nerves regulating chromatophore motility belong to the sympathetic division of the autonomic nervous system (Fujii and Oshima 1986). In teleosts, the neuronal substances of the nerve fibers that signal chromatophores have been investigated for many years (Fujii 1961, 1969; Fujii and Novales 1972), and the peripheral neurotransmitter was naturally expected to be adrenergic. In 1961, Fujii observed that dibenamine, a blocker of alpha-adrenoceptors, interfered with the pigment-aggregation action of sympathetic stimuli on the melanophores of bony fish (Fujii 1961). Similarly, Kumazawa and Fujii used radiolabeling methods to show that norepinephrine is released from the nerves in response to nervous system stimulation (Kumazawa and Fujii 1984). Several researchers have since investigated the subtype of alpha-adrenoceptors on fish chromatophores. Some of these studies reported that alpha-2-agonists were more effective than alpha-1-agonists (Andersson et al. 1984; Morishita 1987). These researchers concluded that the pigment-aggregating action of chromatophores in response to norepinephrine was mediated by alpha-2-type adrenoceptors, and that cAMP functions as a second messenger in this action. In some species, the pigment-aggregating action may be triggered by an increase in the level of intracellular Ca^{2+} (Aspengren et al. 2003; Oshima et al. 1988; Thaler and Haimo 1990, 1992).

Fujii et al. also reported that inositol 1,4,5-triphosphate (IP_3) is involved in the pigment-aggregating action of melanophores in tilapia (Fujii et al. 1991b). Many different cells are known to respond to IP_3 by the release of Ca^{2+} from intracellular storage compartments into the cytosol, and that alpha-1-adrenergic stimuli activate phospholipase C, which catalyzes the production of IP_3. These observations indicate that, in addition to the alpha-2-adrenoceptors, alpha-1-adrenoceptors are also functional, at least in some cases.

The motile activities of pigmentary chromatophores other than melanophores are also under the control of the sympathetic nervous system. For example, the erythrophores of the swordtail, *Xiphophorus helleri* and of the squirrelfish *Holocentrus ascensionis* have been shown to be under the influence of the nervous system (Luby-Phelps and Porter 1982; Matsumoto et al. 1978). A comparison of the characteristics of the motile activities of xanthophores with those of melanophores in the scales of the medaka, *O. latipes*, by Iwata et al. indicated that the xanthophores responded in the same way as the melanophores (Iwata et al. 1981). These observations suggest that the nervous mechanisms controlling erythrophores and xanthophores are analogous to those of melanophores.

In light-reflecting chromatophores, the nervous mechanism that controls their motile activities has been investigated by several researchers. Studies on the leucophores of medaka showed that dispersal of leucosomes takes place when the controlling nerves are stimulated (Fujii and Miyashita 1979; Iga 1983). The receptors concerned with the action were of the beta-1-adrenergic type (Iga et al. 1977; Obika 1976; Yamada 1980). Blockage of beta-adrenoceptors caused an aggregation of the leucosomes in the medaka leucophores under the influence of catecholamines (Iga 1979; Morishita and Yamada 1989). The results of pharmacological investigations suggested that this response was mediated by adrenoceptor of the alpha-2 type.

Motile iridophores of the non-dendritic type in the damselfish and in the neon tetra responded to nervous stimulation by a shift in the spectral peak of their reflected light toward shorter wavelengths (Kasukawa et al. 1987; Nagaishi and Oshima 1989; Oshima et al. 1989). Dendritic iridophores in the integument of the dark sleeper goby, *Odontobutis obscura*, showed dispersal of the reflecting platelets in the cytoplasm into the nerve processes upon nervous stimulation (Fujii et al. 1991a; Iga and Matsuno 1986).

Peripheral nerve transmission to melanophores in two species of siluroid catfish (the Japanese common catfish, *Silurus asotus*, and the translucent glass catfish, *Kryptopterus bicirrhis*) is cholinergic, even though the postganglionic fibers to the effector cells have the usual sympathetic character (Fujii and Miyashita 1976; Fujii et al. 1982; Hayashi and Fujii 1994). The M3 subtype of muscarinic cholinoceptors replaces the alpha-adrenoceptors for the transduction of nerve signals to the melanophores. Fish species other than those in the order Siluriformes have muscarinic cholinoceptors on their melanophores, as found in two species of the genus *Zacco* (Hayashi and Fujii 1993). Until the present time, the existence of these cholinoceptors has not been reported for chromatophores other than melanophores.

Fujii and Miyashita showed that the melanophores of guppies may utilize adenosine or adenine nucleotides in the control of the dispersion of melanosomes (Fujii and Miyashita 1976). They found that non-cyclic adenyl compounds were even more effective than cyclic adenosine 3′, 5′-monophosphate (cAMP) for the dispersal response of the melanophores. Work with the melanophores of guppies and catfish revealed that the dispersal response induced by these nucleotides was mediated by adenosine receptors (Miyashita et al. 1984). Studies on tilapia by Kumazawa et al. revealed that ATP is released as a co-transmitter from postganglionic chromatic fibers together with the true transmitter, norepinephrine (NE) (Kumazawa and Fujii 1984).

Norepinephrine induces a rapid pigment aggregation in pigmentary chromatophores via alpha-adrenoceptors on the cytoplasmic membrane. Most of this NE is quickly removed by re-uptake into the fibers. The residue is either taken away in the general circulation or is inactivated by catecholamine O-methyltransferase (COMT) and monoamine oxidase (MAO). ATP, after its concurrent release with the true transmitter, NE, is dephosphorylated by ATPase and then by 5′-nucleotidase in the synaptic cleft. The resultant nucleoside, adenosine, remains in the cleft for a long time, where it functions in the re-dispersion of pigment via the adenosine receptors. Finally, most of the NE true transmitter and ATP co-transmitter are removed by re-uptake into the presynaptic nerves or are carried away by the circulation.

The light-reflecting chromatophores, like medaka leucophores, respond to adenosine by dispersion of leucosomes (Oshima et al. 1986b) in a response mediated by A2 type adenosine receptors. The motile iridophores in damselfish species (Kasukawa et al. 1986; Oshima et al. 1989) and neon tetras (Nagaishi and Oshima 1989) respond to adenine derivatives with a shift in the spectral peak of their reflected light toward shorter wavelengths, which is opposite to the shift in the spectral peak elicited by alpha-adrenergic stimuli.

Direct Effects of Light

Direct effects of incident light on chromatophores are mainly observable during the embryonic and larval stages, when physiological color changes are not yet under the control of endocrine and/or nervous systems. In 1978, Wakamatsu reported that some cultured melanophores from embryos, larvae, or young black platyfish, *Xiphophorus maculatus*, responded to light by aggregating their pigments, although all the melanophores had sensitivity to light initially (Wakamatsu 1978; Wakamatsu et al. 1980). By contrast, melanophores from larvae of the rose bittering, *Rhodeus ocellatus*, responded to light by dispersing their melanosomes, whereas the melanosomes aggregated in the dark (Ohta 1983; Ohta and Muramatsu 1988).

Chromatophores other than melanophores have also been investigated for their responsiveness to light. The xanthophores of medaka were found to respond to light by xanthosome aggregation, and the effective wavelength was around 410–420 nm (Kawai 1989; Oshima et al. 1998). The response was not mediated through alpha-adrenoceptors, and Ca^{2+} ions and calmodulin were not involved in the intracellular signaling of the response. Oshima and Yokozaki reported that erythrophores of the Nile tilapia, *Oreochromis niloticus*, responded to light of a defined wavelength by aggregation or dispersion of erythrosomes (Oshima and Yokozeki 1999). The spectral peak of the response to illuminated light was between 400 and 440 nm and between 550 and 600 nm, where aggregation of erythrosomes was observed, whereas their dispersal was accelerated at wavelengths around 470–530 nm. They concluded that three kinds of visual pigments coexist in the erythrophores of the Nile tilapia. Motile iridophores of the neon tetra, *Paracheirodon innesi*, showed a shift in the spectral peak of the reflected light toward a shorter wavelength than the illuminated light (Lythgoe and Shand 1982; Nagaishi and Oshima 1989). Using medaka, Ohta and Sugimoto reported that leucophores responded to illuminated light by dispersing leucosomes (Ohta and Sugimoto 1980).

13.1.4 Morphological Color Changes

Changes in integumental hue or pattern are crucial strategies for fish for survival under various situations in Nature. Slow changes can be attributed to the increase or decrease in the number of chromatophores in the skin and/or to the net amount of pigmentary material contained in their cytoplasm. These changes are usually referred to as "morphological color changes" (Fig. 13.6). Sugimoto, working with medaka,

Fig. 13.6 Color changes in medaka following long-term adaptation to an illuminated white or black background. Upper and lower individuals dwelled for 9 weeks in a white-colored tank and a black-colored tank respectively

counted the number of chromatophores after long-term adaptation to a dark or a white background (Sugimoto 1993). These experiments showed that adaptation to a dark background gave rise to a significant increase in the numbers of melanophores in the dermis and a concomitant decrease in the number of leucophores, whereas adaptation to a white background resulted in an opposite pattern.

In many teleosts, such as the Koran angelfish, *Pomacanthus semicirculatus*, and the African coris, *Coris gaimard*, the integumental coloration and patterns change remarkably during growth. For example, a nuptial coloration in integument can be observed in many fish during their breeding seasons. These ontogenic integumental color changes are attributed to morphological color changes.

As mentioned above, concurrent changes in the amount of pigmentary substance were also confirmed in the chromatophores. The morphological color changes in the adult rainbow trout are modulated by MCH, which inhibits the upregulation of plasma MSH levels and prevents melanogenesis in the skin (Baker et al. 1986). Similarly, melatonin also reduces the activity of MSH in the cells in the cichlid teleost *O. mossambicus* (Van Der Salm et al. 2005), and long-term administration of MCH decreases the melanophore size in medaka (Sugimoto et al. 2000). Alpha-MSH stimulates the proliferation of melanophores and melanization within them (Fujii 2000; Sugimoto 2002). The developmental effects induced by MSH are mediated by activation of intracellular protein kinase A (PKA) signaling (Fujii 2000). Sugimoto reported that apoptosis is responsible for the decrease in melanophore number during long-term adaptation to a white background, and that this apoptosis is induced by a downregulation of intracellular cAMP-PKA signaling by the binding of NE to the alpha-2-adrenoceptor (Sugimoto et al. 2000; Uchida-Oka and Sugimoto 2001). Previous histological studies with freshwater fish have suggested that the balance of apoptosis and differentiation in chromatophores was important for the morphological color changes (Sugimoto 2002; Sugimoto et al. 2000).

13.2 Chromatophores of Reptiles

The color pattern of animals is a traditional yet important topic in modern evolutionary biology. Reptiles, in particular, exhibit a wide variety of color patterns as visual signals to their conspecific individuals, predators, or sometimes even prey (Allen et al. 2013; Cooper and Greenberg 1992; Olsson et al. 2013; Stuart-Fox et al. 2008). Numerous studies have therefore attempted to measure color fitness; however, few histological studies of chromatophores have been conducted to determine the pigmentation mechanisms by which particular color patterns are imparted among different species or among different populations of the same species (Morrison 1995; Morrison et al. 1995; Saenko et al. 2013). The phylogenetic position of reptiles as a stem lineage of terrestrial vertebrates and the rich diversity in body coloration also provide rich opportunities to study functionally convergent characteristics among distantly related lineages (Fig. 13.7) (e.g., Kronforst et al. 2012; Kuriyama et al. 2020; McKinnon and Pierotti 2010; Olsson et al. 2013; Roy et al. 2020).

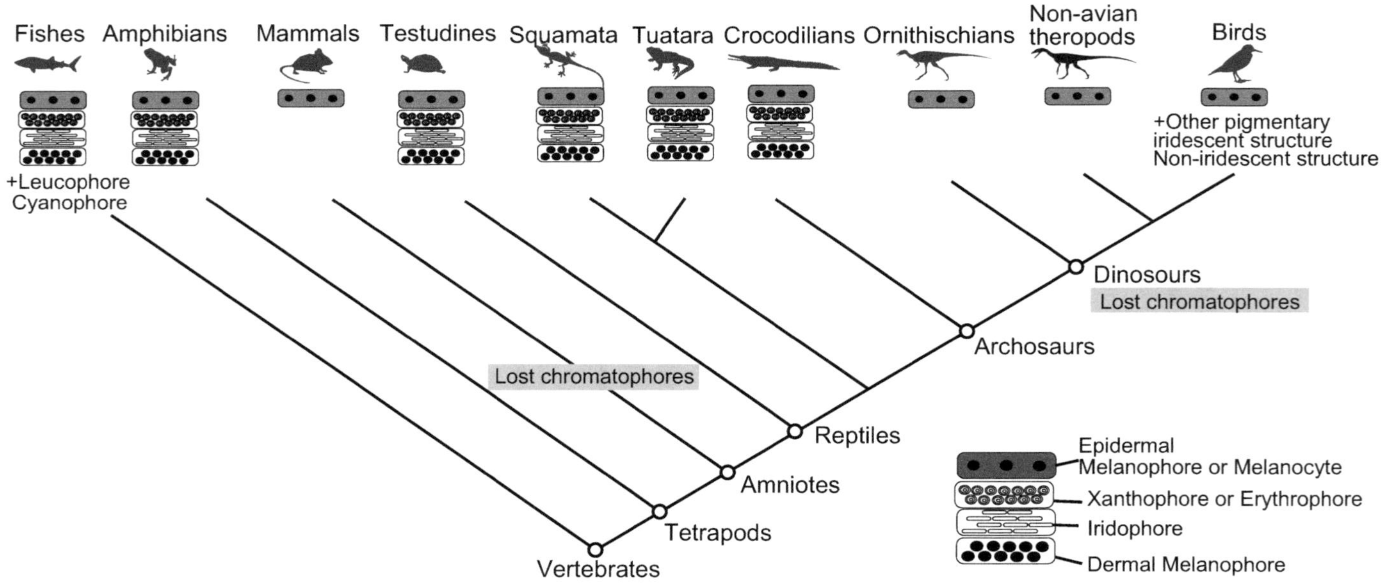

Fig. 13.7 Phylogeny showing the distribution of color producing pigments in vertebrates. Figure from Roy et al. (2020), used with permission. Silhouettes were downloaded from www.phylopic.org, free of copyrights (CG Public Domain Dedication 1.0 license)

A mechanistic understanding of how a color pattern has been formed requires an elucidation of the spatial and vertical architecture of the chromatophores within the skin tissue as a fundamental step (Kuriyama et al. 2006; Saenko et al. 2013). This necessitates more than just a description of the composition spatial arrangement of chromatophores in adults, but also a characterization of the developmental processes of chromatophore specification, migration, and spatial organization during the embryonic and postembryonic stages (Kelsh et al. 2009; Kuriyama et al. 2013; Murakami et al. 2016, 2018), as this knowledge then helps to link the understanding of the gene regulatory networks (GRNs) responsible for color pattern formation. However, our current knowledge about the genetic mechanisms of color pattern formation in vertebrates with multiple chromatophores is primarily based on model organisms, such as the zebrafish (e.g., Greenhill et al. 2011; Kelsh 2004; Kelsh et al. 2009; Parichy 2006; Parichy and Spiewak 2015; Petratou et al. 2018; Ziegler 2003), and the developmental mechanisms underlying color pattern formation and its evolution remain unclear in reptiles.

In this section, we review recent advances in our understanding of the roles of chromatophores in production of various skin colorations and patterning in reptilian sauropsids (e.g., Alibardi 2011, 2012, 2013, 2015; Kuriyama and Hasegawa 2017; Kuriyama et al. 2006; Kuriyama et al. 2013; Kuriyama et al. 2016a, b, 2020; Murakami et al. 2016, 2018; Saenko et al. 2013; Szydlowski et al. 2017) and genetic studies to identify genes and gene systems that generate color patterns (Irizarry and Bryden 2016; Iwanishi et al. 2018; Manceau et al. 2010; Rosenblum et al. 2004; Saenko et al. 2015). Our focus in reviewing the cellular mechanisms producing different coloration is an initial, but important, step in identifying the genetic mechanism underlying adaptive coloration in reptiles. It represents a key research strategy for revealing the genetic mechanisms of coloration in lower vertebrates where spatial and vertical combinations of multiple chromatophores determine skin coloration (Kuriyama et al. 2006, 2013, 2020; Murakami et al. 2016; Saenko et al. 2013). Chromatophores are differentiated from the neural crest (NC), a multipotent and migratory cell population, along with other diverse derivatives, such as the facial skeleton and the peripheral nervous system (Bagnara et al. 1979). The candidate GRNs for color pattern formation are therefore not single genes but multiple genes responsible for cell differentiation, proliferation, death and localization (Chang et al. 2009; Olsson et al. 2013; Parichy and Spiewak 2015).

13.2.1 Chromatophore-Based Understanding of Color and Pattern Formation

The various skin color patterns of reptiles are produced by the epidermal and dermal chromatophore architectures. Four basic types of chromatophores have been identified in the dermal skin of reptiles: xanthophores, erythrophores, iridophores, and melanophores (Bagnara and Matsumoto 2006; Cooper and Greenberg 1992). The

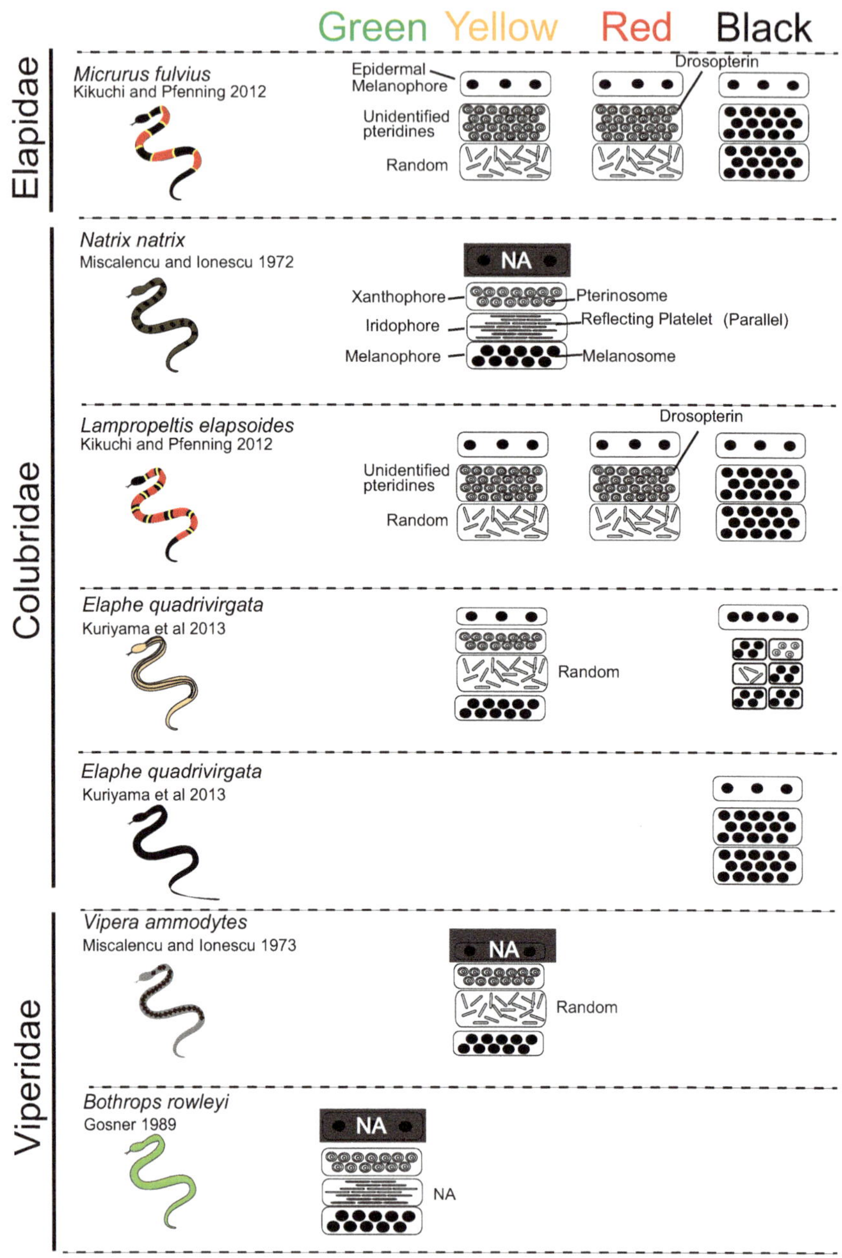

Fig. 13.8 A schematic drawing of the vertical combination and relative thickness of epidermal and dermal chromatophores producing different skin coloration in snakes (Kuriyama et al. 2020)

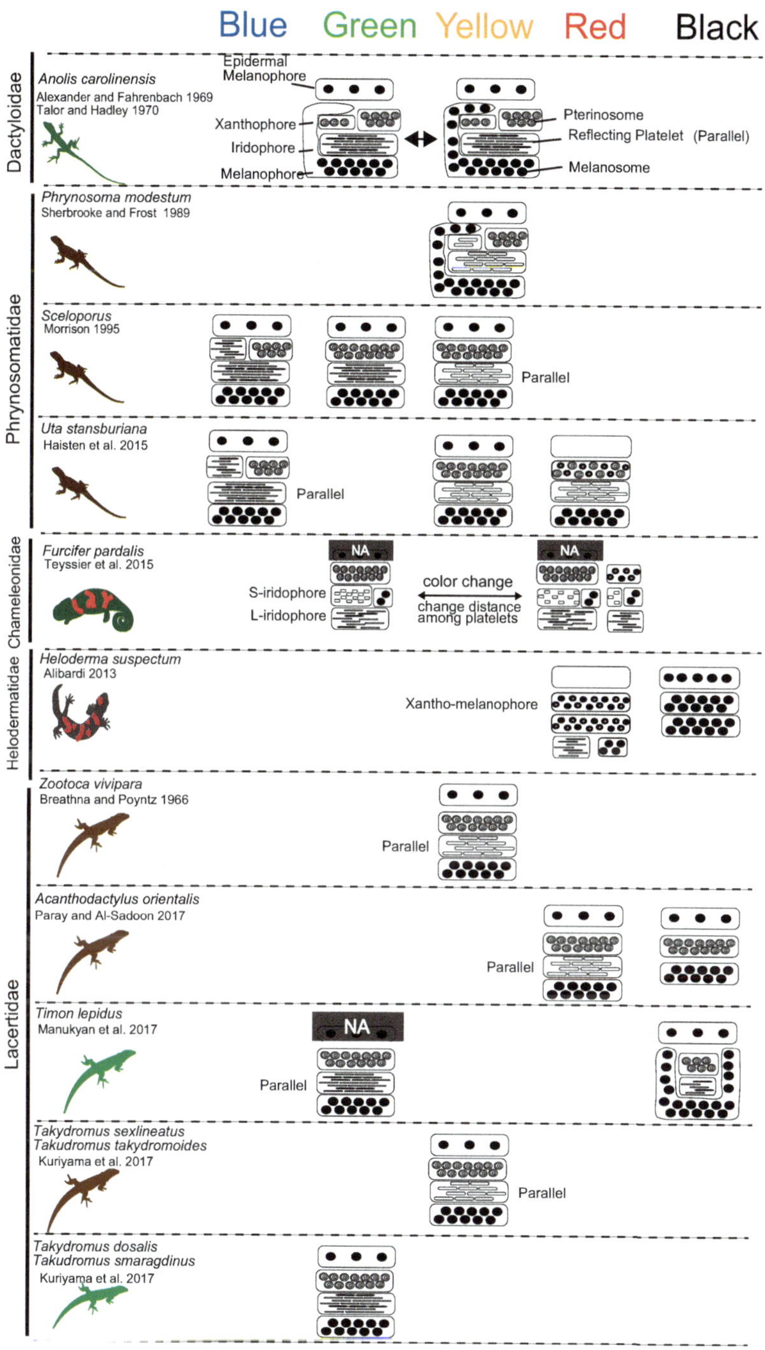

Fig. 13.9 A schematic drawing of the vertical combination and relative thickness of epidermal and dermal chromatophores producing different skin coloration in lizards. Silhouettes were downloaded from www.phylopic.org, free of copyrights (CG Public Domain Dedication 1.0 license)

spatial arrangement and architectural combination of these chromatophores can produce a multitude of skin colors in reptiles such as lizards and snakes (Figs. 13.8, 13.9, 13.10; e.g., Kuriyama et al. 2013, 2017; Murakami et al. 2016; Saenko et al. 2013), tuataras (Fig. 13.10; Alibardi 2012), crocodiles (Fig. 13.10; Alibardi 2011), and chelonians (Fig. 13.10; Alibardi 2013; Brejcha et al. 2019).

In lizards and snakes, the final color of a given patch of skin is strongly influenced by the static structural combination of chromatophores (Kuriyama et al. 2006) and by the dynamic interaction of chemical and physical parameters (Saenko et al. 2013; Teyssier et al. 2015). Xanthophores and erythrophores are light-absorbing chromatophores that impart yellow to red coloration, depending on the wavelengths absorbed by various compositions of pterinosomes and carotenoid granules in the cells. Saenko et al. revealed that the pH or redox states reversibly change yellow xanthophores to red erythrophores, and vice versa (Saenko et al. 2013).

Iridophores are light-reflecting chromatophores that contain light-reflecting platelets made of crystalline purines and pteridines and give rise to structural colors from thin-layer interference and scattering or diffraction of light from the stacks of reflecting platelets (Bagnara and Matsumoto 2006). The size, shape, orientation, number, and conformation of the reflecting platelets and the distance between them determines what structural colors the iridophores produce (Kuriyama et al. 2017, Kuriyama et al. 2016a; Morrison 1995; Morrison et al. 1995). The peak wavelength of light reflected by iridophores is determined by the depth and orientation of the platelets and the cytoplasmic spacing between them (e.g., Kuriyama et al. 2006; Morrison 1995; Rohrlich and Porter 1972; Saenko et al. 2013). Rohrlich noted that extensive filament networks in passive iridophores of the *Anolis* lizard played a cytoskeletal role by holding crystal sheets in their strict parallel array, whereas the motile filament system appeared responsible for mediating cellular changes by altering the array, spacing, or tilt of cellular crystals (Rohrlich 1974). Recently, Teyssier et al. showed that the geometric parameters of the structural elements, including the size, shape, and orientation of the iridophore nanocrystals and the geometry and level of order of the lattice of these elements, enable the rapid color changes observed in the chameleon (Teyssier et al. 2015).

Melanophores are light-absorbing chromatophores that intensely absorb light to produce black or brown colors. The spatial arrangement and architectural combination of these chromatophores can produce a multitude of skin colors in reptiles (Morrison 1995; Morrison et al. 1995).

13.2.2 *Blue Tail Coloration and Stripes of Lizards*

A combination of longitudinal yellowish white-colored stripes on the dark background of the body trunk and the vivid blue tail coloration evolved independently among the different lizards families (Fig. 13.11; Bateman and Fleming 2009; Bateman et al. 2014; Kuriyama et al. 2016a; Murali et al. 2018; Pianka and Vitt 2003; Watson et al. 2012), as well as in parallel among the different species of

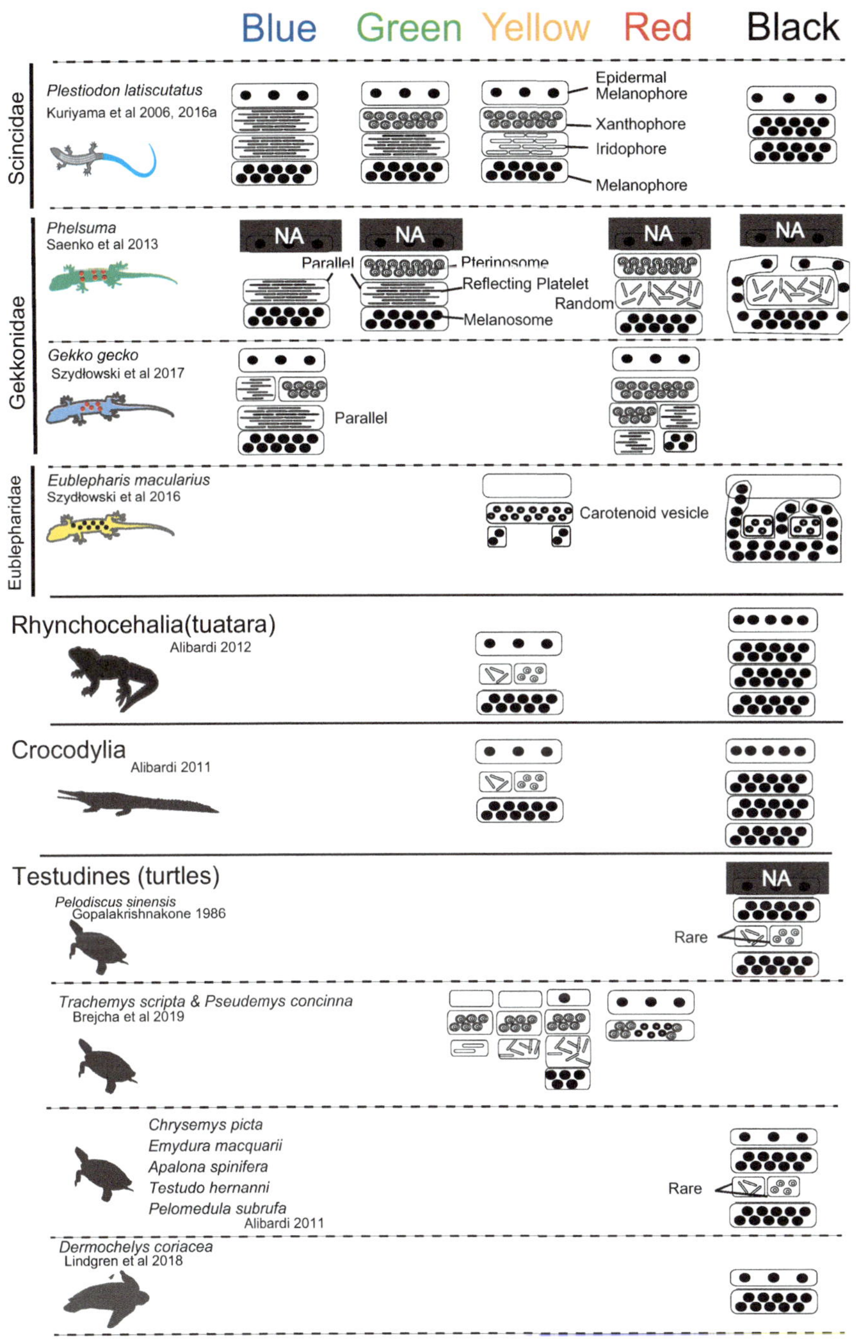

Fig. 13.10 A schematic drawing of the vertical combination and relative thickness of epidermal and dermal chromatophores producing different skin coloration in lizards, tuataras, crocodiles, and turtles. Silhouettes were downloaded from www.phylopic.org, free of copyrights (CG Public Domain Dedication 1.0 license)

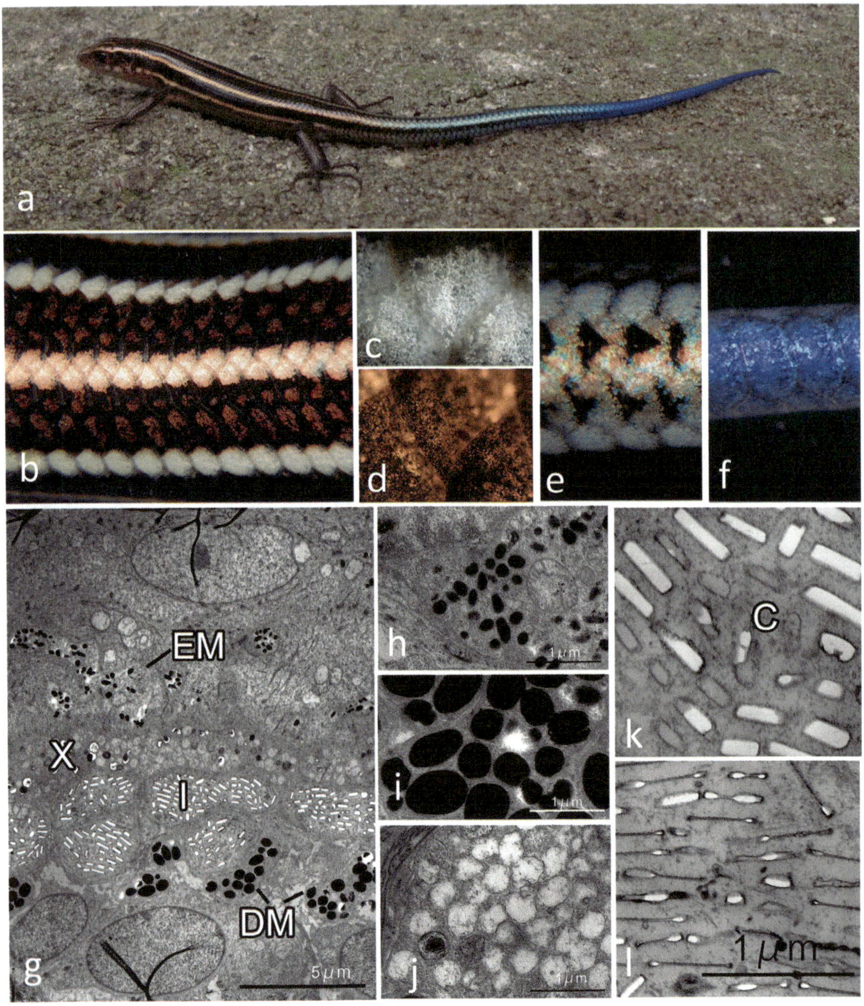

Fig. 13.11 Blue tail coloration and stripes of lizards. (**a**) Dorso-lateral view of a juvenile lizard (*Plestiodon latiscutatus*). The juvenile lizard has yellowish-white longitudinal stripes on a black or dark-brown background of the trunk (**b**). Micrograph of iridophores (**c**) and melanophores (**d**). The tail is colored green in the anterior half (**e**) to deep blue in the posterior half (**f**). (**g**) Electron microphotographs showing the vertical combination of chromatophores on a yellow-colored skin. *EM* epidermal melanophore, *X* xanthophore, *I* iridophore, *DM* melanophore. Electron microphotographs of an epidermal melanophore (**h**), dermal melanophore (**i**), and xanthophore (**j**). Reflecting platelets of an iridophore in yellow skin (**k**) and blue skin (**l**). **c**, light-reflecting platelets with crystals

Plestiodon (Richmond 2006; Richmond and Reeder 2002), thereby providing an excellent opportunity for studying independent, parallel, and divergent color pattern evolution (Clark and Hall 1970; Cooper and Vitt 1985; Hawlena et al. 2006; Ortega et al. 2014), as well as the ability to examine intraspecific geographical variations in the coloration of juveniles (Brandley et al. 2014; Kuriyama et al. 2016a).

The blue coloration of the lizard tail comes from iridophores in the uppermost layer and melanophores in the basal layer (Fig. 13.11; Bagnara et al. 2007, Kuriyama et al. 2006). Iridophores with thinner platelets are found only in the skin tissues taken from the tail, and this localization of iridophores with thinner reflecting platelets in tail skin is highly consistent with the spatial coincidence between the blue coloration and the thinner reflecting platelets (Kuriyama et al. 2006, 2016a). Although mechanism that determines the actual border between iridophores with thin or thick platelets has not yet been specified, a sharp shift may occur in the iridophore types along the longitudinal axis of the trunk and tail. Five yellow longitudinal lines on a black background form a vivid stripe in the trunk region of a juvenile *Plestiodon* lizard (Kuriyama et al. 2006). The combination of xanthophores, iridophores, and melanophores from top to bottom in the dermal layer causes the yellowish-white line, whereas the black inter-stripe region consists of only melanophores (Kuriyama et al. 2006). Thus, a sharp melanophore density gradient exists between the whitish-yellow stripes and the dark black lines and results in clear yellow stripes adjacent to a black background in the juvenile lizard. During ontogeny after hatching, the juvenile with the vivid striping and vivid blue tail assumes the adult coloration with a uniform brown back and tail. This predicts that the loss of the stripe pattern would involve a decay in the sharp density gradient of melanophores built up during embryogenesis and that almost all dermis would consist of a layer of xanthophores, iridophores with thick platelets, and melanophores. Similarly, the clear boundary between the vivid blue tail due to iridophores with thin platelets and the anterior portion of the green or brown tail would fade away during ontogeny, allowing the tail color to change from blue to brown due to the presence of iridophores with thicker platelets and the appearance of xanthophores (Kuriyama and Hasegawa 2017).

The stripe pattern and blue tail formation during embryogenesis were revealed by Kuriyama and Hasegawa (Kuriyama and Hasegawa 2017). The blue tail pattern forms with the appearance of iridophores at a specific region in the tail, and this position on the tail coincides with the boundary area between the posterior portion of the tail with blue coloration and the anterior portion of the tail with green or brown coloration after hatching (Fig. 13.12). Thereafter, iridophores with thicker platelets appear in the anterior part of the tail, and iridophores with thinner or thicker platelets co-occur to form a boundary area in the middle part of the tail.

These findings suggested that controlling factors induce the regional appearance of iridophores with thinner or thicker platelets in different populations of the same species and in different species within the genus (Kuriyama et al. 2016a). The stripe pattern was established during embryogenesis by the initial appearance of melanophores that form a clear pattern, prior to the appearance of iridophores and xanthophores (Kuriyama and Hasegawa 2017). Iridophores then occupied the area above the melanophore layers. The density gradient of melanophores between yellow lines and the dark inter-stripe region differed markedly between the island populations of *P. latiscutatus* (Kuriyama et al. 2016a). The melanophore density in the black background region was higher in lizards with vivid stripes than in lizards

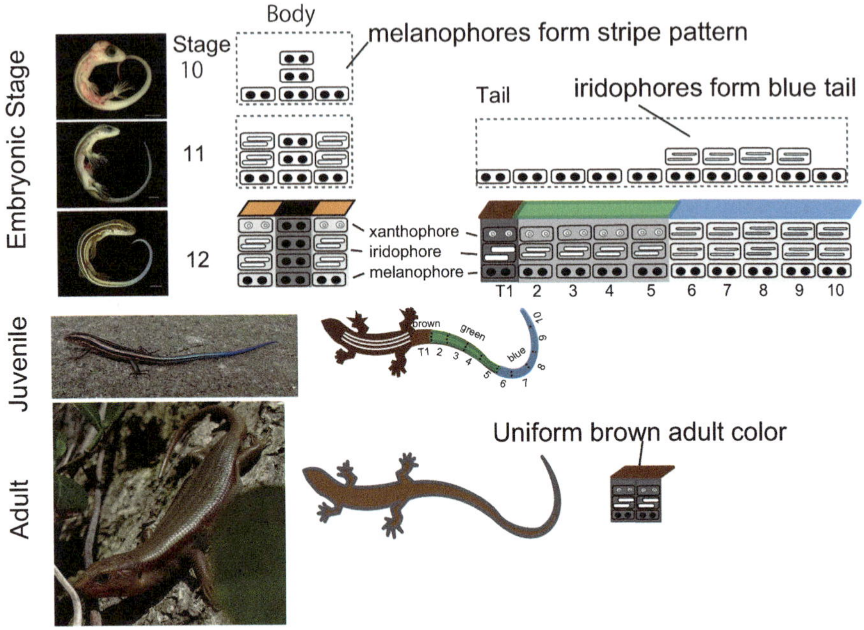

Fig. 13.12 A schematic drawing of the embryonic stripe and tail pattern formation and ontogenetic color changes in *Plestiodon latiscutatus*. The vividness of the dorsal stripes was determined by the high-density melanophores that developed before the appearance of iridophores and xanthophores. In addition, iridophores with thin platelets that reflect blue light appeared at particular positions in the tail during embryonic development, and these positions coincided with the boundary area between the distal part of tail, showing vivid blue posterior coloration, and the basal part of the tail, showing green or brown coloration, after hatching. The uniform brown color of the adult lizard is formed by the same vertical chromatophore combination

with drab stripes. Iridophores started to fill in the dermal space not yet occupied by melanophores, resulting in a higher iridophore density in the yellowish-white stripes than in the inter-stripe region. This observation suggested that the mechanisms controlling the density gradient of melanophores during embryogenesis are crucial for understanding the GRNs responsible for the naturally occurring variation in stripe pattern formation.

13.2.3 Stripe Pattern Formation in Snakes

The Japanese four-lined snake, *Elaphe quadrivirgata,* has a naturally occurring body color polymorphism that includes striped, pale-striped, non-striped, banded, and melanistic morphs (Kuriyama et al. 2013; Mori et al. 2005). The vividly striped and non-striped morphs possess the same set of epidermal melanophores and dermal

Fig. 13.13 Dorso-lateral views and vertical combination of chromatophores of three color morphs of *Elaphe quadrivirgata*. (**a**) The striped morph with four vivid dark-brown longitudinal stripes over a yellowish-brown back ground. (**b**) The non-striped morph with a uniform brown or light-brown ground color

chromatophores (Fig. 13.13, xanthophores at the top, iridophores in the middle, and melanophores at the bottom), but the spatial aggregation and concentrations of epidermal and dermal melanophores differ between the two morphs in terms of the sharpness of the boundary between the dark-colored and light-colored scales (Kuriyama et al. 2013). Even at hatching, the striped and non-striped patterns are detectable (Murakami et al. 2014, 2016), indicating that the basic stripe pattern could be established during embryonic development (Murakami et al. 2016).

Studies on embryonic development have revealed that the positions where melanophores initially appeared did not correspond exactly to the stripe pattern (Murakami et al. 2018). Melanophores that first appeared in the epidermis played a role in forming a precursor stripe pattern, and the region of high-density melanophores in hatchlings followed these precursor stripes. An increase in the density gradient of the epidermal melanophores preceded the development of the stripe pattern due to dermal melanophores. A clear stripe pattern was detected during late embryonic stages, in association with the appearance of the dermal melanophores. However, the higher density and a double stratum of dermal melanophores in the stripe region, which is characteristics of adult snakes (Kuriyama et al. 2013), did not develop in the embryos by the time of hatching. Contrary to the stripe pattern of the *Plestiodon* lizard, in which the stripe pattern is completely developed at the time of hatching but then decays and is lost with ontogeny, the snake stripe pattern continues to develop during ontogeny after hatching (Murakami et al. 2016). The density of the dermal melanophores increases, presumably to replace xanthophores and to form the double stratum in the dark-brown scales during postembryonic growth.

13.2.4 Green and Brown Colors for Background Color Matching

Animal coloration that resembles the background substrate, known as "background matching," is the most widespread strategy for predator evasion (Stevens and Merilaita 2011). Brown and green are the most commonly imitated colors for prey animals because both occur in a wide range of habitats (e.g., due to plant leaves, tree trunks, and soil). Many researchers have evaluated the survival value with respect to

---

Fig. 13.13 (continued) without a trace of the stripe pattern except in the region just behind the head. (**c**) The uniformly jet black melanistic morph, with no sign of a stripe pattern. (**d**) Magnified view of the dorsal surface of a striped morph. (**e**) Magnified view of the dorsal surface of a non-striped morph. (**f**) Magnified view of the dorsal surface of a melanistic morph. A schematic drawing of the vertical combination and relative thickness of the epidermal and dermal chromatophores producing different skin colorations and color patterns. The number, shape, and size of the chromatophores and organelles are simplified and exaggerated. *EM* epidermal melanophore, *X* xanthophore, *I* iridophore, *DM* dermal melanophore. Figure from Kuriyama et al. (2013) with permissions

background color matching (e.g., Stuart-Fox et al. 2008), but the chromatophore characteristics underlying this coloration are not known.

Green and brown dorsal background matching has been observed in many reptilian species, resulting in inter- and/or intraspecific variation. The coloration of grass lizards in the genus *Takydromus*, which inhabit East Asia, varies among species (Arnold 1997; Schlüter 2003). A molecular phylogeny of *Takydromus* lizards has shown that their dorsal coloration has shifted from green to brown and brown to green on multiple occasions during the diversification of the genus (Fig. 13.14; Arnold 1997; Lin et al. 2002; Ota et al. 2002). Thus, the interspecific color variation in *Takydromus* lizards provides an opportunity to study the chromatophore characteristics that underlie these two colorations for background color matching.

Brown and green skin were found to differ with respect to the morphological characteristics of the iridophores, with differences observed in the thicknesses of the reflecting platelets and in the cytoplasmic spacing between platelets, despite a shared vertical arrangement of chromatophores (i.e., xanthophores in the upper layer, iridophores in the middle layer, and melanophores at the bottom of the dermal layer) among the different *Takydromus* lizards (Fig. 13.9). The iridophores of brown skin reflect light of longer wavelengths than do those in green skin, corresponding to the thicker platelets and greater distances between platelets observed in brown skin. The predicted peak wavelengths of light reflected by the iridophores in green and brown skin were also shorter than those actually measured (Fig. 13.14), suggesting a filtering effect of the xanthophores and a subsequent shift in the reflected wavelengths. The number of reflecting platelets was higher in the iridophores of green skin than in brown skin, causing a higher relative reflectance.

13.2.5 Genetic Mechanisms of Color Pattern Formation in Reptiles

Seanko et al. reviewed the strategies available for investigating the molecular genetic basis of color variation in squamates (Saenko et al. 2015). The lack of extensive genomic resources limited the early studies, which had to rely on sequence analyses of only a few candidate genes, such as *MC1R*. The *MC1R* gene encodes a G-protein coupled receptor called the melanocortin-1 receptor, which is located on the surface of melanocytes. This receptor is a key switch in a signal transduction pathway and is involved in color variations in mammals and birds (San-Jose and Roulin 2017). Color variation in some species has been successful in some reptiles (Rosenblum et al. 2004) but ineffective in many others (e.g., Cox et al. 2013).

Positional cloning to identify mutations or genomic intervals harboring causal mutations requires both controlled pedigrees to generate recombinant mapping populations and a dense and large set of genetic markers that co-segregate with the phenotype of interest. Therefore, Saenko et al. adopted an unbiased next-generation

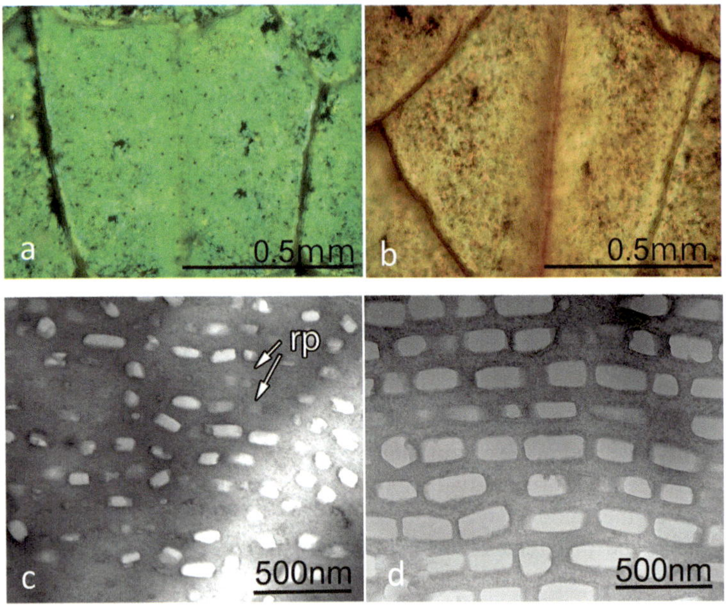

Fig. 13.14 Photographs of the dorsal skin of *Takydromus dorsalis* (**a**) and *T. takydromoides* (**b**). Transmission electron photomicrographs of iridophores in the green skin of *T. smaragdinus* (**c**) and the brown skin of *T. takydromoides* (**d**) containing light-reflecting platelets (rp)

sequencing and exome assembly approach that used extensive genotyping and candidate gene sequencing (Saenko et al. 2015). This approach enabled the identification of a retrotransposon insertion in the *OCA2* gene as the causal mutation determining the amelanistic phenotype in captive-bred corn snakes. The *OCA2* gene provides instructions for making a protein that may regulate the relative acidity of the melanosome and the transport of some molecules into and out of the melanosomes in melanocytes.

Understanding the color pattern formation in reptiles with four types of chromatophores (i.e., epidermal melanophores, dermal melanophores, xanthophores, and iridophores) intuitively suggests that GRNs may control pattern formation and color production in a step-by-step manner. Instead of an exhaustive review of the accumulating research, we prefer instead to begin with a theoretical analysis of pattern-generating mechanisms, such as cell-chemotaxis mechanisms (Murray and Myerscough 1991; Oster and Murray 1989). This is a promising strategy for identifying the molecular components of a biological system by mathematical modeling and can provide a basis for experimental studies and refine and improve our knowledge of the mechanisms underlying pattern formation (Chang et al. 2009).

Acknowledgments We are grateful to K. Miyaji, A. Murakami, M. Ohata, and M. Sugimoto for their observations and discussions about chromatophores and to M.C. Brandley, M. Hasegawa, G. Morimoto for their valuable discussions about the ecology and evolution of reptile coloration.

We also thank the members of the Laboratory of Geographical Ecology of Toho University for their field assistance. It is a pleasure to thank for Risa Goda for her assistance in taking photographs of fish, and Isao Ohta, Hamamatsu University School of Medicine, for his unfailing assistance in taking electron micrographs. This research was partially supported by the Japan Society for the Promotion of Science to M. Hasegawa (19H03307).

References

Abrao MS, Castrucci AM, Hadley ME, Hruby VJ (1991) Protein-kinase C mediates MCH signal transduction in teleost, Synbranchus marmoratus, melanocytes. Pigment Cell Res 4:66–70

Akimoto K, Takaoka T, Sorimachi K (2000) Development of a simple culture method for the tissues contaminated with microorganisms and application to establishment of a fish cell line. Zool Sci 17:61–63

Alibardi L (2011) Histology, ultrastructure, and pigmentation in the horny scales of growing crocodilians. Acta Zool 92:187–200

Alibardi L (2012) Cytology and localization of chromatophores in the skin of the Tuatara (Sphenodon punctaus). Acta Zool 93:330–337

Alibardi L (2013) Observations on the ultrastructure and distribution of chromatophores in the skin of chelonians. Acta Zool 94:222–232

Alibardi L (2015) Ultrastructural features of skin pigmentation in the lizard Heloderma suspectum with emphasis on xantho-melanophores. Acta Zool 96:154–159

Allen WL, Baddeley R, Scott-Samuel NE, Cuthill IC (2013) The evolution and function of pattern diversity in snakes. Behav Ecol 24:1237–1250

Andersson RG, Karlsson JO, Grundstrom N (1984) Adrenergic nerves and the alpha 2-adrenoceptor system regulating melanosome aggregation within fish melanophores. Acta Physiol Scand 121:173–179

Arnold EN (1997) Interrelationships and evolution of the east Asian grass lizards, Takydromus (Squamata: Lacertidae). Zool J Linn Soc 119:267–296

Aspengren S, Skold HN, Quiroga G, Martensson L, Wallin M (2003) Noradrenaline- and melatonin-mediated regulation of pigment aggregation in fish melanophores. Pigment Cell Res 16:59–64

Bagnara JT, Hadley ME (1973) Chromatophores and color change. Prentice-Hall, Englewood Cliffs, NJ

Bagnara JT, Matsumoto J (2006) Comparative anatomy and physiology of pigment cells in nonmammalian tissues. In: Nordlund JJ, Boissy RE, Hearing VJ, King RA, Ortonne J (eds) The pigmentary system, physiology and pathophysiology, 2nd edn. Oxford University Press, New York, pp 11–59

Bagnara JT, Matsumoto J, Ferris W, Frost SK, Turner WA, Tchen TT, Taylor JD (1979) Common origin of pigment cells. Science 203:410–415

Bagnara JT, Fernandez PJ, Fujii R (2007) On the blue coloration of vertebrates. Pigment Cell Res 20:14–26

Baker BI, Ball JN (1975) Evidence for a dual pituitary control of teleost melanophores. Gen Comp Endocrinol 25:147–152

Baker BI, Bird DJ, Buckingham JC (1986) Effects of chronic administration of melanin-concentrating hormone on corticotrophin, melanotrophin, and pigmentation in the trout. Gen Comp Endocrinol 63:62–69

Bateman PW, Fleming PA (2009) To cut a long tail short: a review of lizard caudal autotomy studies carried out over the last 20 years. J Zool 277:1–14

Bateman PW, Fleming PA, Rolek B (2014) Bite me: blue tails as a "risky-decoy" defense tactic for lizards. Curr Zool 60:333–337

Brandley MC, Kuriyama T, Hasegawa M (2014) Snake and bird predation drive the repeated convergent evolution of correlated life history traits and phenotype in the Izu Island Scincid lizard (Plestiodon latiscutatus). PLoS One 9:e92233
Brejcha J, Bataller JV, Bosakova Z, Geryk J, Havlikova M, Kleisner K, Marsik P, Font E (2019) Body coloration and mechanisms of colour production in Archelosauria: the case of deirocheline turtles. R Soc Open Sci 6:190319
Castrucci AM, Visconti MA, Hadley ME, Hruby VJ, Oshima N, Fujii R (1988) Melanin concentrating hormone (MCH) control of chromatophores. Prog Clin Biol Res 256:547–557
Chang C, Wu P, Baker RE, Maini PK, Alibardi L, Chuong CM (2009) Reptile scale paradigm: evo-devo, pattern formation and regeneration. Int J Dev Biol 53:813–826
Clark DR, Hall RJ (1970) Function of the blue tail coloration of the five-lined skink (Eumeces fasciatus). Herpetologica 26:271–274
Cooper WE, Greenberg N (1992) Reptilian coloration and behavior. In: Gans C, Crews D (eds) Biology of the reptilia. University of Chicago Press, Chicago, pp 498–422
Cooper WE, Vitt LJ (1985) Blue tails and autotomy: enhancement of predation avoidance in juvenile skinks. Z Tierpsychol 70:265–276
Cox CL, Rabosky AR, Chippindale PT (2013) Sequence variation in the Mc1r gene for a group of polymorphic snakes. Gene 513:282–286
Denton EJ, Nicol JAC (1966) A survey of reflectivity in silvery teleosts. J Mar Biol Assoc UK 46:685–722
Enami M (1955) Melanophore-contracting hormone (MCH) of possible hypothalamic origin in the catfish, Parasilurus. Science 121:36–37
Euteneuer U, Mcintosh JR (1981) Polarity of some motility-related microtubules. Proc Natl Acad Sci USA 78:372–376
Fujii R (1961) Demonstration of the adrenergic nature of transmission at the junction between melanophore-concentrating nerve and melanophore in bony fish. J Fac Sci Univ Tokyo Sec IV 9:171–196
Fujii R (1969) Chromatophores and pigments. In: Hoar WS, Randall DJ (eds) Fish physiology. Academic Press, New York, pp 307–353
Fujii R (1993a) Coloration and chromatophores. In: The physiology of fishes. CRC Press, Boca Raton, pp 535–562
Fujii R (1993b) Cytophysiology of fish chromatophores. Int Rev. Cytol 143:191–255
Fujii R (2000) The regulation of motile activity in fish chromatophores. Pigment Cell Res 13:300–319
Fujii R, Miyashita Y (1976) Receptor mechanisms in fish chromatophores--III. Neurally controlled melanosome aggregation in a siluroid (Parasilurus asotus) is strangely mediated by cholinoceptors. Comp Biochem Physiol C 55:43–49
Fujii R, Miyashita Y (1979) Photo-electric recording of motile responses of fish leucophores. Annot Zool Jpn 52:87–94
Fujii R, Miyashita Y (1980) Action of melanophore-stimulating hormone on dermal and epidermal melanophores of siluroid, Parasilurus asotus. Yale J Biol Med 53:422
Fujii R, Miyashita Y (1982) Receptor mechanisms in fish chromatophores--V. MSH disperses melanosomes in both dermal and epidermal melanophores of a catfish (Parasilurus asotus). Comp Biochem Physiol C 71C:1–6
Fujii R, Novales R (1972) Nervous control of melanosome movements in vertebrate melanophores. In: Riley V (ed) Pigmentation: its genesis and biologic control. Appleton-Century-Crofts, New York, pp 315–326
Fujii R, Oshima N (1986) Control of chromatophore movements in teleost fishes. Zool Sci 3:13–47
Fujii R, Oshima N (1994) Factors influencing motile activities of fish chromatophores. In: Gilles R (ed) Advances in comparative and environmental physiology. Springer, Berlin, pp 1–54
Fujii R, Taguchi S (1969) The response of fish melanophores to some melanin-aggregating and dispersing agents in potassium-rich medium. Annot Zool Jpn 42:176–182

Fujii R, Miyashita Y, Fujii Y (1982) Muscarinic cholinoceptors mediate neurally evoked pigment aggregation in glass catfish melanophores. J Neural Transm 54:29–39
Fujii R, Hayashi H, Toyohara J, Nishi H (1991a) Analysis of the light reflection from motile iridophores of the dark sleeper, Odontobutis obscura obscura. Zool Sci 8:461–470
Fujii R, Wakatabi H, Oshima N (1991b) Inositol 1,4,5-triphosphate signals the motile response of fish chromatophores I. Aggregation of pigment in the tilapia melanophore. J Exp Zool 259:9–17
Fujii R, Tanaka Y, Hayashi H (1993) Endothelin-1 causes aggregation of pigment in teleostean melanophores. Zool Sci 10:763–772
Fujishige A, Moriwake T, Ono A, Ishii Y, Tsuchiya T (2000) Control of melanosome movement in intact and cultured melanophores in the bitterling, Acheilognathus lanceolatus. Comp Biochem Physiol A 127:167–175
Fujita T, Fujii R (1997) Endothelins disperse light-scattering organelles in leucophores of the Medaka, Oryzias latipes. Zool Sci 14:559–569
Goda M, Fujii R (1995) Blue chromatophores in two species of callionymid fish. Zool Sci 12:811–813
Goda M, Ohata M, Ikoma H, Fujiyoshi Y, Sugimoto M, Fujii R (2011) Integumental reddish-violet coloration owing to novel dichromatic chromatophores in the teleost fish, Pseudochromis diadema. Pigment Cell Melanoma Res 24:614–617
Goda M, Fujiyoshi Y, Sugimoto M, Fujii R (2013) Novel dichromatic chromatophores in the integument of the mandarin fish Synchiropus splendidus. Biol Bull 224:14–17
Greenhill ER, Rocco A, Vibert L, Nikaido M, Kelsh RN (2011) An iterative genetic and dynamical modelling approach identifies novel features of the gene regulatory network underlying melanocyte development. PLoS Genet 7:e1002265
Grundstrom N, Karlsson JO, Andersson RG (1985) The control of granule movement in fish melanophores. Acta Physiol Scand 125:415–421
Hawlena D, Boochnik R, Abramsky Z, Bouskila A (2006) Blue tail and striped body: why do lizards change their infant costume when growing up? Behav Ecol 17:889–896
Hayashi H, Fujii R (1993) Muscarinic cholinoceptors that mediate pigment aggregation are present in the melanophores of cyprinids (Zacco spp). Pigment Cell Res 6:37–44
Hayashi H, Fujii R (1994) Pharmacological profiles of the subtypes of muscarinic cholinoceptors that mediate aggregation of pigment in the melanophores of two species of catfish. Pigment Cell Res 7:175–183
Hayashi H, Fujii R (2001) Possible involvement of nitric oxide in signaling pigment dispersion in teleostean melanophores. Zool Sci 18:1207–1215
Hayashi H, Nakamura S, Fujii R (1996) The endothelin receptors that mediate aggregation of pigment in fish melanophores. Comp Biochem Physiol B 115:143–152
Huxley A (1968) A theoretical treatment of the reflexion of light by multilayer structures. J Exp Biol 48:227–245
Iga T (1979) Alpha adrenoceptors: pigment aggregation in Oryzias leucophores. Mem Fac Lit Sci Shimane Univ 13:87–95
Iga T (1983) Electric stimulation experiments on leucophores of a freshwater teleost, Oryzias latipes. Comp Biochem Physiol C 74:103–108
Iga T, Matsuno A (1986) Motile iridophores of a freshwater goby, Odontobutis obscura. Cell Tissue Res 244:165–171
Iga T, Takabatake I (1982) Action of melanophore-stimulating hormone on melanophores of the cyprinid fish Zacco temmincki. Comp Biochem Physiol C 73:51–55
Iga T, Yamada K, Iwakiri M (1977) Adrenergic receptors mediating pigment dispersion in leucophores of a teleost, Oryzias latipes. Mem Fac Lit Sci Shimane Univ Nat Sci 11:63–72
Imokawa G, Yada Y, Miyagishi M (1992) Endothelins secreted from human keratinocytes are intrinsic mitogens for human melanocytes. J Biol Chem 267:24675–24,680
Irizarry KJ, Bryden RL (2016) In silico analysis of gene expression network components underlying pigmentation phenotypes in the python identified evolutionarily conserved clusters of transcription factor binding sites. Adv Bioinformatics 2016:1286510

Iwanishi S, Zaitsu S, Shibata H, Nitasaka E (2018) An albino mutant of the Japanese rat snake (Elaphe climacophora) carries a nonsense mutation in the tyrosinase gene. Genes Genet Syst 93:163–167
Iwata K, Takahashi T, Okada Y (1981) Nervous control in chromatophores of the medaka. In: Seiji M (ed) Pigment cell 1981: phenotypic expression in pigment cells. University Tokyo Press, Tokyo, pp 369–372
Kasukawa H, Oshima N, Fujii R (1986) Control of chromatophore movements in dermal chromatic units of blue damselfish--II. The motile iridophore. Comp Biochem Physiol C 83:1–7
Kasukawa H, Oshima N, Fujii R (1987) Mechanism of light reflection in blue damselfish motile iridophore. Zool Sci 4:243–257
Kawaguti S (1965) Electron microscopy on iridophores in the scale of the blue wrasse. Proc Jpn Acad 41:610–613
Kawaguti S, Kamishima Y (1966) Electron microscopy on the blue black of a clupeoid fish, Harengula zunasi. Proc Jpn Acad 42:389–393
Kawai I (1989) Light sensitive response of the scale xanthophores of a teleost. Oryzias latipes. Med. Biol (Tokyo) 118:93–97
Kawauchi H (2006) Functions of melanin-concentrating hormone in fish. J Exp Zool 305:751–760
Kawauchi H, Kawazoe I, Tsubokawa M, Kishida M, Baker BI (1983) Characterization of melanin-concentrating hormone in chum salmon pituitaries. Nature 305:321–323
Kelsh RN (2004) Genetics and evolution of pigment patterns in fish. Pigment Cell Res 17:326–336
Kelsh RN, Harris ML, Colanesi S, Erickson CA (2009) Stripes and belly-spots -- a review of pigment cell morphogenesis in vertebrates. Semin Cell Dev Biol 20:90–104
Kitta K, Makino M, Oshima N, Bern HA (1993) Effects of prolactins on the chromatophores of the tilapia, Oreochromis niloticus. Gen Comp Endocrinol 92:355–365
Kronforst MR, Barsh GS, Kopp A, Mallet J, Monteiro A, Mullen SP, Protas M, Rosenblum EB, Schneider CJ, Hoekstra HE (2012) Unraveling the thread of nature"s tapestry: the genetics of diversity and convergence in animal pigmentation. Pigment Cell Melanoma Res 25:411–433
Kumazawa T, Fujii R (1984) Concurrent releases of norepinephrine and purines by potassium from adrenergic melanosome-aggregating nerve in Tilapia. Comp Biochem Physiol C 78:263–236
Kuriyama T, Hasegawa M (2017) Embryonic developmental process governing the conspicuousness of body stripes and blue tail coloration in the lizard Plestiodon latiscutatus. Evol Dev 19:29–39
Kuriyama T, Miyaji K, Sugimoto M, Hasegawa M (2006) Ultrastructure of the dermal chromatophores in a lizard (Scincidae: Plestiodon latiscutatus) with conspicuous body and tail coloration. Zool Sci 23:793–799
Kuriyama T, Misawa H, Miyaji K, Sugimoto M, Hasegawa M (2013) Pigment cell mechanisms underlying dorsal color-pattern polymorphism in the Japanese four-lined snake. J Morphol 274:1353–1364
Kuriyama T, Morimoto G, Miyaji K, Hasegawa M (2016a) Cellular basis of anti-predator adaptation in a lizard with autotomizable blue tail against specific predators with different colour vision. J Zool 300:89–98
Kuriyama T, Okamoto T, Miyaji K, Hasegawa M (2016b) Iridophore- and xanthophore-deficient melanistic color variant of the lizard Plestiodon latiscutatus. Herpetologica 72:189–195
Kuriyama T, Esashi J, Hasegawa M (2017) Light reflection from crystal platelets in iridophores determines green or brown skin coloration in Takydromus lizards. Zoology (Jena) 121:83–90
Kuriyama T, Murakami A, Brandley M, Hasegawa M (2020) Blue, black, and stripes: evolution and development of color production and pattern formation in lizards and snakes. Front Ecol Evol 8:232
Land MF (1972) The physics and biology of animal reflectors. Prog Biophys Mol Biol 24:75–106
Lerner AB, Case JD, Takahashi Y, Lee TH, Mori W (1958) Isolation of melatonin; the pineal gland factor that lightens. J Am Chem Soc 80:2587
Lin SM, Chen CA, Lue KY (2002) Molecular phylogeny and biogeography of the grass lizards genus Takydromus (Reptilia: Lacertidae) of East Asia. Mol Phylogenet Evol 22:276–288

Logan DW, Burn SF, Jackson IJ (2006) Regulation of pigmentation in zebrafish melanophores. Pigment Cell Res 19:206–213
Luby-Phelps K, Porter KR (1982) The control of pigment migration in isolated erythrophores of Holocentrus ascensionis (Osbeck). II. The role of calcium. Cell 29:441–450
Lythgoe JN, Shand J (1982) Changes in spectral reflexions from the iridophores of the neon tetra. J Physiol 325:23–34
Manceau M, Domingues VS, Linnen CR, Rosenblum EB, Hoekstra HE (2010) Convergence in pigmentation at multiple levels: mutations, genes and function. Philos Trans R Soc Lond B Biol Sci 365:2439–2450
Martensson LG, Andersson RG (2000) Is Ca2+ the second messenger in the response to melatonin in cuckoo wrasse melanophores? Life Sci 66:1003–1010
Masagaki A, Fujii R (1999) Differential action of melatonin on melanophores of the threeline pencilfish, Nannostomus trifasciatus. Zool Sci 16:35–42
Matsumoto J, Watanabe Y, Obika M, Hadley M (1978) Mechanisms controlling pigment movements with in swordtail (Xiphophorus helleri) erythrophores in primary culture. Comp Biochem Physiol A 61:509–517
Mckinnon JS, Pierotti ME (2010) Colour polymorphism and correlated characters: genetic mechanisms and evolution. Mol Ecol 19:5101–5125
Mcniven MA, Porter KR (1986) Microtubule polarity confers direction to pigment transport in chromatophores. J Cell Biol 103:1547–1555
Miyashita Y, Kumazawa T, Fujii R (1984) Receptor mechanisms in fish chromatophores--VI. Adenosine receptors mediate pigment dispersion in guppy and catfish melanophores. Comp Biochem Physiol C 77:205–210
Mizusawa K, Kobayashi Y, Sunuma T, Asahida T, Saito Y, Takahashi A (2011) Inhibiting roles of melanin-concentrating hormone for skin pigment dispersion in barfin flounder, Verasper moseri. Gen Comp Endocrinol 171:75–81
Mori A, Tanaka K, Moriguchi H, Hasegawa M (2005) Color variation in Elaphe quadrivirgata throughout Japan. Jpn J Herpetol 2005:22–38
Morishita F (1987) Responses of the melanophores of the medaka, Oryzias latipes, to adrenergic drugs: evidence for involvement of alpha 2 adrenergic receptors mediating melanin aggregation. Comp Biochem Physiol C 88:69–74
Morishita F, Yamada K (1989) Subtype of alpha adrenoceptors mediating leucosome aggregation in medaka leucophore. J Sci Hiroshima Univ Ser B Div 1(33):99–112
Morrison RL (1995) A transmission electron microscopic (TEM) method for determining structural colors reflected by lizard iridophores. Pigment Cell Res 8:28–36
Morrison RL, Rand MS, Frost-Mason SK (1995) Cellular basis of color differences in three morphs of the lizard Sceloporus undulatus erythrocheilus. Copeia 1995:397–408
Murakami A, Hasegawa M, Kuriyama T (2014) Identification of juvenile color morphs for evaluating heredity model of stripe/non-stripe pattern polymorphism in Japanese four-lined snake Elaphe quadrivirgata. Curr Herpetol 33:68–74
Murakami A, Hasegawa M, Kuriyama T (2016) Pigment cell mechanism of postembryonic stripe pattern formation in the Japanese four-lined snake. J Morphol 277:196–203
Murakami A, Hasegawa M, Kuriyama T (2018) Developmental mechanisms of longitudinal stripes in the Japanese four-lined snake. J Morphol 279:27–36
Murali G, Merilaita S, Kodandaramaiah U (2018) Grab my tail: evolution of dazzle stripes and colourful tails in lizards. J Evol Biol 31:1675–1688
Murata N, Fujii R (2000) Pigment-aggregating action of endothelins on medaka xanthophores. Zool Sci 17:853–862
Murphy DB, Tilney LG (1974) The role of microtubules in the movement of pigment granules in teleost melanophores. J Cell Biol 61:757–779
Murray JD, Myerscough MR (1991) Pigmentation pattern formation on snakes. J Theor Biol 149:339–360

Nagai M, Oshima N, Fujii R (1986) A comparative-study of melanin-concentrating hormone (MCH) action on teleost melanophores. Biol Bull 171:360–370
Nagaishi H, Oshima N (1989) Neural control of motile activity of light-sensitive iridophores in the neon tetra. Pigment Cell Res 2:485–492
Needham A (1974) Significance of zoochromis. Springer, Berlin
Negishi S, Obika M (1980) The effects of melanophore-stimulating hormone and cyclic nucleotides on teleost fish chromatophores. Gen Comp Endocrinol 42:471–476
Nilsson H, Rutberg M, Wallin M (1996) Localization of kinesin and cytoplasmic dynein in cultured melanophores from Atlantic cod, Gadus morhua. Cell Motil Cytoskeleton 33:183–196
Nishi H, Fujii R (1992) Novel receptors for melatonin that mediate pigment dispersion are present in some melanophores of the pencil fish (Nannostomus). Comp Biochem Physiol C 103:263–268
Obika M (1976) An analysis of the mechanism of pigment migration in fish chromatophores. In: Riley V (ed) Pigment cell, Unique properties of melanocytes, vol 3. Karger, Basel, pp 254–265
Obika M (1986) Intracellular transport of pigment granules in fish chromatophores. Zool Sci 3:1–11
Obika M (1988) Ultrastructure and physiological-response of leucophores of the medaka Oryzias latipes. Zool Sci 5:311–321
Obika M, Turner WA, Negishi S, Menter DG, Tchen TT, Taylor JD (1978) The effects of lumicolchicine, colchicine and vinblastine on pigment migration in fish chromatophores. J Exp Zool 205:95–110
Ohta T (1974) Movement of pigment granules within melanophores of an isolated fish scale. Biol Bull 146:258–266
Ohta T (1983) Melanosome dispersion in direct response to light in melanophores of Phodeus ocellatus fry. Annot Zool Jpn 56:155–162
Ohta T, Muramatsu K (1988) Spectral sensitivity of melanophores in the primary color response of the rose bittering, Rhodeus ocellatus ocellatus. Jpn J Ichthyol 34:483–487
Ohta T, Sugimoto S (1980) Leucosome dispersion under light in Medaka leucophores. Jpn J Ichthyol 27:72–76
Olsson M, Stuart-Fox D, Ballen C (2013) Genetics and evolution of colour patterns in reptiles. Semin Cell Dev Biol 24:529–541
Ortega J, Lopez P, Martin J (2014) Conspicuous blue tails, dorsal pattern morphs and escape behaviour in hatchling Iberian wall lizards (Podarcis hispanicus). Biol J Linn Soc 113:1094–1106
Oshima N, Fujii R (1984) A precision photoelectric method for recording chromatophore responses in vitro. Zool Sci 1:545–552
Oshima N, Fujii R (1985) Calcium requirement for MSH action on non-melanophoral chromatophores of some fish. Zool Sci 2:127–129
Oshima N, Fujii R (1987) Motile mechanism of blue damselfish (Chrysiptera cyanea) iridophores. Cell Motil Cytoskeleton 8:85–90
Oshima N, Goto M (2000) Prolactin signaling in erythrophores and xanthophores of teleost fish. Pigment Cell Res 13(Suppl 8):35–40
Oshima N, Yokozeki A (1999) Direct control of pigment aggregation and dispersion in tilapia erythrophores by light. Zool Sci 16:51–54
Oshima N, Kasukawa H, Fujii R, Wilkes BC, Hruby VJ, Hadley ME (1986a) Action of melanin-concentrating hormone (MCH) on teleost chromatophores. Gen Comp Endocrinol 64:381–388
Oshima N, Yamaji N, Fujii R (1986b) Adenosine receptors mediate pigment dispersion in leucophores of the medaka, Oryzias latipes. Comp Biochem Physiol C 85:245–248
Oshima N, Suzuki M, Yamaji N, Fujii R (1988) Pigment aggregation is triggered by an increase in free calcium ions within fish chromatophores. Comp Biochem Physiol A 91:27–32
Oshima N, Kasukawa H, Fujii R (1989) Control of chromatophore movements in the blue-green damselfish, Chromis viridis. Comp Biochem Physiol C 93:239–245
Oshima N, Makino M, Iwamuro S, Bern HA (1996) Pigment dispersion by prolactin in cultured xanthophores and erythrophores of some fish species. J Exp Zool 275:45–52

Oshima N, Nakata E, Ohta M, Kamagata S (1998) Light-induced pigment aggregation in xanthophores of the medaka, Oryzias latipes. Pigment Cell Res 11:362–367

Oshima N, Nakamaru N, Araki S, Sugimoto M (2001) Comparative analyses of the pigment-aggregating and -dispersing actions of MCH on fish chromatophores. Comp Biochem Physiol C 129:75–84

Oster GF, Murray JD (1989) Pattern formation models and developmental constraints. J Exp Zool 251:186–202

Ota H, Honda M, Chen S-L, Hikida T, Panha S, Oh H-S, Matsui A (2002) Phylogenetic relationships, taxonomy, character evolution and biogeography of the lacertid lizards of the genus Takydromus (Reptilia. Squamata): a molecular perspective. Biol J Linn Soc 76:493–509

Ovais M, Srivastava SK, Sumoona S, Mubashshir M (2015) Evidence for the presence of novel beta-melatonin receptors along with classical alpha-melatonin receptors in the fish Rasbora daniconius (Ham.). J Recept Signal Transduct Res 35:238–248

Parichy DM (2006) Evolution of danio pigment pattern development. Heredity (Edinb) 97:200–210

Parichy DM, Spiewak JE (2015) Origins of adult pigmentation: diversity in pigment stem cell lineages and implications for pattern evolution. Pigment Cell Melanoma Res 28:31–50

Petratou K, Subkhankulova T, Lister JA, Rocco A, Schwetlick H, Kelsh RN (2018) A systems biology approach uncovers the core gene regulatory network governing iridophore fate choice from the neural crest. PLoS Genet 14:e1007402

Pianka ER, Vitt LJ (2003) Lizards: windows to the evolution of diversity. University of California Press, Berkeley

Pickford G, Atz JW (1957) The physiology of the pituitary gland of fishes. New York Zoological Society, New York

Reed BL (1968) The control of circadian pigment changes in the pencil fish: a proposed role for melatonin. Life Sci 7:961–973

Richmond JQ (2006) Evolutionary basis of parallelism in North American scincid lizards. Evol Dev 8:477–490

Richmond JQ, Reeder TW (2002) Evidence for parallel ecological speciation in scincid lizards of the Eumeces skiltonianus species group (Squamata: Scincidae). Evolution 56:1498–1513

Rodionov VI, Gyoeva FK, Gelfand VI (1991) Kinesin is responsible for centrifugal movement of pigment granules in melanophores. Proc Natl Acad Sci USA 88:4956–4960

Rodionov VI, Hope AJ, Svitkina TM, Borisy GG (1998) Functional coordination of microtubule-based and actin-based motility in melanophores. Curr Biol 8:165–168

Rodionov V, Yi J, Kashina A, Oladipo A, Gross SP (2003) Switching between microtubule- and actin-based transport systems in melanophores is controlled by cAMP levels. Curr Biol 13:1837–1847

Rohrlich ST (1974) Fine structural demonstration of ordered arrays of cytoplasmic filaments in vertebrate iridophores. A comparative survey. J Cell Biol 62:295–304

Rohrlich ST, Porter KR (1972) Fine structural observations relating to the production of color by the iridophores of a lizard Anolis carolinensis. J Cell Biol 53:38–52

Rosenblum EB, Hoekstra HE, Nachman MW (2004) Adaptive reptile color variation and the evolution of the Mc1r gene. Evolution 58:1794–1808

Roy A, Pittman M, Saitta ET, Kaye TG, Xu X (2020) Recent advances in amniote palaeocolour reconstruction and a framework for future research. Biol Rev. Camb Philos Soc. 95:22–50

Saenko SV, Teyssier J, Van Der Marel D, Milinkovitch MC (2013) Precise colocalization of interacting structural and pigmentary elements generates extensive color pattern variation in Phelsuma lizards. BMC Biol 11:105

Saenko SV, Lamichhaney S, Martinez Barrio A, Rafati N, Andersson L, Milinkovitch MC (2015) Amelanism in the corn snake is associated with the insertion of an LTR-retrotransposon in the OCA2 gene. Sci Rep 5:17118

Sage M (1970) Control of prolactin release and its role in color change in the teleost Gillichthys mirabilis. J Exp Zool 173:121–127

San-Jose LM, Roulin A (2017) Genomics of coloration in natural animal populations. Philos Trans R Soc Lond B Biol Sci 372(1724):20160337372
Schartl M, Larue L, Goda M, Bosenberg MW, Hashimoto H, Kelsh RN (2016) What is a vertebrate pigment cell? Pigment Cell Melanoma Res 29:8–14
Schliwa M (1984) Mechanisms of intracellular organelle transport. In: Shay JW (ed) Cell muscle motility. Plenum, New York, pp 1–80
Schliwa M, Euteneuer U (1978) A microtuble-independent component may be involved in granule transport in pigment cells. Nature 273:556–558
Schliwa M, Weber K, Porter KR (1981) Localization and organization of actin in melanophores. J Cell Biol 89:267–275
Schlüter U (2003) Die langschwanzeidechsen der gattung Takydromus. Pflege, zucht und lebensweise. Kirschner & Seufer Verlag, Keltern-Weiler
Skold HN, Norstrom E, Wallin M (2002) Regulatory control of both microtubule- and actin-dependent fish melanosome movement. Pigment Cell Res 15:357–366
Skold HN, Amundsen T, Svensson PA, Mayer I, Bjelvenmark J, Forsgren E (2008) Hormonal regulation of female nuptial coloration in a fish. Horm Behav 54:549–556
Stevens M, Merilaita S (2011) Animal camouflage: mechanisms and function. Cambridge University Press, Cambridge
Stuart-Fox D, Moussalli A, Whiting MJ (2008) Predator-specific camouflage in chameleons. Biol Lett 4:326–329
Sugimoto M (1993) Morphological color changes in the medaka, Oryzias latipes, after prolonged background adaptation—I. Changes in the population and morphology of melanophores. Comp Biochem Physiol A 104:513–518
Sugimoto M (2002) Morphological color changes in fish: regulation of pigment cell density and morphology. Microsc Res Tech 58:496–503
Sugimoto M, Oshima N, Fujii R (1985) Mechanisms controlling motile responses of amelanotic melanophores in the medaka, Oryzias latipes. Zool Sci 2:317–322
Sugimoto M, Uchida N, Hatayama M (2000) Apoptosis in skin pigment cells of the medaka, Oryzias latipes (Teleostei), during long-term chromatic adaptation: the role of sympathetic innervation. Cell Tissue Res 301:205–216
Szydlowski P, Madej JP, Mazurkiewicz-Kania M (2017) Histology and ultrastructure of the integumental chromatophores in tokay gecko (Gekko gecko) (Linnaeus, 1758) skin. Zoomorphology 136:233–240
Teyssier J, Saenko SV, Van Der Marel D, Milinkovitch MC (2015) Photonic crystals cause active colour change in chameleons. Nat Commun 6:6368
Thaler CD, Haimo LT (1990) Regulation of organelle transport in melanophores by calcineurin. J Cell Biol 111:1939–1948
Thaler CD, Haimo LT (1992) Control of organelle transport in melanophores: regulation of Ca2+ and cAMP levels. Cell Motil Cytoskeleton 22:175–184
Uchida-Oka N, Sugimoto M (2001) Norepinephrine induces apoptosis in skin melanophores by attenuating cAMP-PKA signals via alpha2-adrenoceptors in the medaka, Oryzias latipes. Pigment Cell Res 14:356–361
Van Der Salm AL, Spanings FA, Gresnigt R, Bonga SE, Flik G (2005) Background adaptation and water acidification affect pigmentation and stress physiology of tilapia, Oreochromis mossambicus. Gen Comp Endocrinol 144:51–59
Wakamatsu Y (1978) Light sensitive fish melanophores in culture. J Exp Zool 204:299–304
Wakamatsu Y, Kawamura S, Yoshizawa T (1980) Light-induced pigment aggregation in cultured fish melanophores: spectral sensitivity and inhibitory effects of theophylline and cyclic adenosine-3′",5′"-monophosphate. J Cell Sci 41:65–74
Watson CM, Roelke CE, Pasichnyk PN, Cox CL (2012) The fitness consequences of the autotomous blue tail in lizards: an empirical test of predator response using clay models. Zoology (Jena) 115:339–344

Wucherer MF, Michiels NK (2012) A fluorescent chromatophore changes the level of fluorescence in a reef fish. PLoS One 7:e37913

Yamada K (1980) Actions of sympathomimetic amines on leucophores in isolated scales of a teleost fish with special reference to beta-adrenoceptors mediating pigment dispersion. J Sci Hiroshima Univ Ser B Div 1(28):95–114

Yanagisawa M, Kurihara H, Kimura S, Tomobe Y, Kobayashi M, Mitsui Y, Yazaki Y, Goto K, Masaki T (1988) A novel potent vasoconstrictor peptide produced by vascular endothelial cells. Nature 332:411–415

Ziegler I (2003) The pteridine pathway in zebrafish: regulation and specification during the determination of neural crest cell-fate. Pigment Cell Res 16:172–182

Chapter 14
Color Change in Cephalopods

Yuzuru Ikeda

Abstract Cephalopods, such as octopuses, cuttlefish, and squid, are members of the phylum Mollusca, are the most common marine organisms utilized as fisheries resource by humans, and play a key role in marine food webs. Besides these characteristics, cephalopods have gained attention in biological science owing to their physiological and anatomical features, such as camera-type eye and a large brain, by which they can perform advanced learning and exhibit some intelligent behaviors such as tool use. All these examples indicate cephalopods as intelligent invertebrates or primates of the ocean. Color change is another remarkable feature of cephalopods; it is achieved through specific chromatic and reflecting cells and controlled neural system connected with the brain, thus accomplishing the fastest and the most varied chromatic changes among the animal kingdom. This chapter reviews color change in cephalopods, which includes a unique chromophore system contributing to rapid color change, related physiological mechanism, and unique chromatic behavior, such as body patterning. Ecological significance of color change in cephalopods has also been explained with some challenging hypothesis for possible color perception.

Keywords Body patterning · Cephalopods · Chromatophores · Reflecting cells

14.1 Introduction

Cephalopods are members of the molluscan class, which include ammonites, belemnites (the only two fossil species), nautili, octopuses, cuttlefish, and squid. Among them, octopuses, cuttlefish, and squid are the major living cephalopods, namely, coleoid cephalopods that contain ca. 700 species inhabiting a wide ocean range from coastal to oceanic, shallow to deep, and tropical to cold waters (Nixon

Y. Ikeda (✉)
Faculty of Science, University of the Ryukus, Okinawa, Japan
e-mail: ikeda@sci.u-ryukyu.ac.jp

H. Hashimoto et al. (eds.), *Pigments, Pigment Cells and Pigment Patterns*,
https://doi.org/10.1007/978-981-16-1490-3_14

and Young 2003; Boyle and Rodhouse 2005). In addition, coleoid cephalopods (hereafter referred to as "cephalopods") represent a dominant group that plays an important role in marine food webs; they are often consumed by higher ranked predators such as mammals, birds, and fishes, yet also represent active predators, which makes them key species that mediate predator–prey interactions (Clark 1987; Nixon 1987). Further, due to their high abundance, cephalopods are the major target for fisheries worldwide, which provide huge protein resources for humans (Rathjen and Voss 1987; Boyle and Rodhouse 2005). On the other hand, cephalopods are known as unique living organisms, due to their biological characteristics. They are facilitated with advanced sensory organs and nervous systems, including lens eyes similar to our own, and large brains, with a size that is relatively equivalent to that of the brains of phylogenetically higher vertebrates, such as birds and mammals (Packard 1972; Budelmann 1994, 1995). Therefore, cephalopods have been recognized as intelligent invertebrates, which can shed light on the evolution of brains, recognition, and intelligence among animals. One such example can be seen in their learning and memory abilities, which have been studied in octopuses since the 1930s (Young 1971; Wells 1978). These studies induce comparisons of advanced recognizing abilities between cephalopods and its counterparts (Edelman and Seth 2009). Interestingly, in spite of their highly advanced neural systems and consequential related behaviors, most cephalopods have short life spans, typically a year (Nixon and Young 2003; Boyle and Rodhouse 2005), which is contrary to other animals that possess large brains and advanced abilities of learning and memory as they tend to live longer. The question arises then, as to why and how cephalopods have acquired such large brains. A question which has yet to be answered.

With regard to their neural controlled behaviors, cephalopods are creatures with impressive color changing abilities, that are the most diverse among the animal kingdom (Figs. 14.1, 14.2, 14.3). There are many well-known examples of animal taxa which mimic and/or assimilate their body color or color patterns to their environmental background, namely, camouflage (Alcock 2005). However, cephalopods are the most outstanding among these examples, as they not only have the ability to change their color but also their texture and form, in a process that is highly flexible and rapid, enabling them to camouflage themselves perfectly in their environment (Fig. 14.4). Due to these characteristics, cephalopods are labeled as "masters of camouflage" (Hanlon and Messenger 1996; Hanlon 2007). This directs our attention towards the idea that cephalopods are unique subjects that can be studied to understand the mechanisms and ecological significance of color change in animals. There are multiple excellent, published reviews that fully explain the detail regarding color change in cephalopods (Packard 1995; Hanlon and Messenger 1996; Messenger 2001). By following previous research and touching upon recent findings that are missed in past reviews, this chapter attempts to review color change in cephalopods, and discuss its ecological significance.

Fig. 14.1 Body colors of coleoid cephalopods. (**a**) Broadclub cuttlefish (*Sepia latimanus*), (**b**) Paintpot cuttlefish (*Metasepia tullbergi*), (**c**) Tropical bottletail squid (*Sepiadarium kochi*), (**d**, **e**) Small shallow water octopus (*Octopus laqueus*), (**f**) Starry night octopus (*Callistoctopus luteus*), (**g**) Oval squid (*Sepioteuthis lessoniana*), (**h**) Japanese bobtail squid (*Euprymna morsei*) (Photographs, **a–c**, **e**, **f**, **h** by R. Yanagisawa, **d** by D. Ueno, **g** by C. Sugimoto)

Fig. 14.2 Gradational change of body patterning in Paintpot cuttlefish (*Metasepia tullbergi*). Time elapsed shown in second (Photograph, K. Okamoto)

14.2 Elements and Underlining Mechanisms of Color Change

The pigments that produce color have been identified in mollusks, dominantly in their shell (bivalve and gastropod). These include melanins, porphyrins, bile pigments, and carotenoids (Williams 2017). In non-shelled molluscs such as sea slug,

Fig. 14.3 Gradational change of body patterning in Broadclub cuttlefish (*Sepia latimanus*). Time course **a–c** (exact time not recorded). (Photograph, R. Yanagisawa)

color pigments are derived from diets (algae), which include chlorophyll a, chlorophyll b, zeaxanthin, antheraxanthin, and lutein (Costa et al. 2012). Because most of molluscs have poor vision, their surface color would not contribute intraspecies communication. By contrast, cephalopods that have good vision can detect complex color patterns of their body surface as their own signals (but see the later section describing color blindness of cephalopods).

Cephalopods can change not only their color but also their texture and form, which enables them to camouflage chromatically and morphologically (Fig. 14.4). This behavior facilitated with color change is termed as "body patterning," and is unique to cephalopods. According to the detailed definition, body patterning is a behavior comprised of four components, namely, chromatic (color), textural

Fig. 14.4 Camouflage of coleoid cephalopods. (**a**) Algae octopus (*Abdopus aculeatus*) on coral reef, (**b**) Algae octopus (*A. aculeatus*) on rock, (**c**) Broadclub cuttlefish (*Sepia latimanus*) on rock, (**d**) Pharaoh cuttlefish on fine gravel (Photograph, **a**, **c** by H. Yasumuro, **b** by R. Yanagisawa, **d** by Y. Nakai)

(surface), postural (appearance), and locomotor (movement) (Packard and Sanders 1971; Hanlon 1982; Hanlon and Messenger 1988). Because color is the major component of body patterning, this chromatic factor will be the main topic of focus in this chapter.

The main elements responsible for body color in cephalopods are known as chromatophores and reflecting cells, which are distributed across the skin (Budelmann et al. 1997) (Figs. 14.5, 14.6, 14.7). Chromatophores are cells which contain color pigments of various wavelengths, namely, yellow, orange, red, brown, and black (Figs. 14.1, 14.2, 14.3, 14.4, 14.7). There are no pigmentary blues or greens in cephalopods. Type and density of chromatophores differ among species. There are also differences in the appearance of chromatophores during ontogeny of a single species. In addition, the density of chromatophores varies across different regions of the body, and changes during ontogeny. Reflecting cells include three types (Figs. 14.6, 14.7, 14.8, 14.9). Iridophores mainly reflect pink, yellow, green, and blue light, and produce silvery colors. This reflecting cell type occurs in the skin, ink sac, and iris of octopuses, cuttlefish, and squid. Reflector cells that reflect mainly blue and green light occur in the skin and iris of octopods only. Leucophores mainly reflect white light and are highly concentrated among the white patches/spots of the skin across the head, arms, and mantle of cuttlefish and octopuses (Hanlon 1982; Hanlon and Messenger 1996). Another cell which produces color in cephalopods is known as the photophore, a luminescent organ which is visible in some species usually living in deep water [e.g., firefly squid *Watasenia scintillans* (Fig. 14.10)],

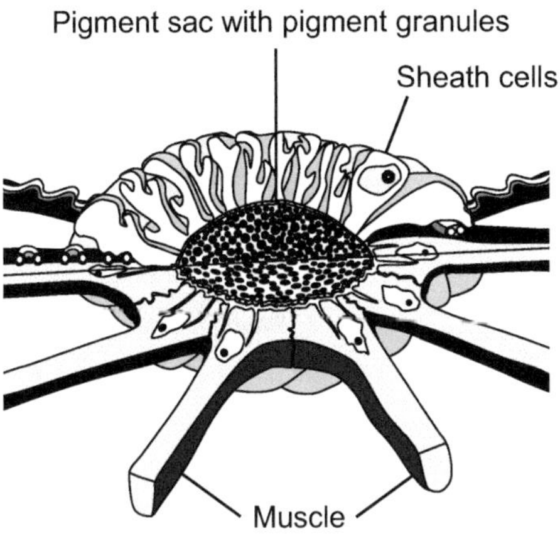

Fig. 14.5 Chromatophore organ of coleoid cephalopods. Chromatophore containing pigment granules are radially connected to muscles that are under neuronal control. Chromatophore is expanded or retracted by the action of radially connected muscles, and according to these activities the inner pigment granules disperse or congregate, causing the pigmented color to appear or disappear (Redrew based on Cloney and Florey 1968; Cloney and Brocco 1983)

and some in shallow waters [e.g., Japanese bobtail squid *Euprymna morsei* (Fig. 14.1)].

A single chromatophore is radially connected with muscle cells that are under neuronal control (Fig. 14.5). This set of cells, known as the chromatophore organ, regulates the expansion and contraction of chromatophores, which then produces various colors on the surface of cephalopods. In detail, the radially connected muscles contract, causing the chromatophore to expand outward, changing its diameter at least 10–20-fold, which then causes the inner pigments to disperse, resulting in pigmented colors becoming dominant. In contrast, when the radially connected muscles relax, this causes the chromatophore to shrink, which leads to the inner pigmented cells becoming centralized, thus resulting in a less pigmented color (i.e., transparent) (Figs. 14.2, 14.3, 14.5, 14.11). Although the ultrastructure of reflecting cells differs between the three cell types, all types physically reflect the outcome light. In iridophores, there are many lamellae of the plasma membrane that contain electron-dense proteinaceous material, namely, reflecting platelets of ca. 160 nm thick (Figs. 14.6 and 14.9). Several neighboring platelets form an iridosome with a 35 nm wide interplatelet space (Budelmann et al. 1997). In squid, there are two types of iridophores, namely, passive iridophores and active iridophores, which are distributed, respectively, in the ventral and dorsal part of the mantle (the dorsal and ventral body wall of a cephalopod, which covers the visceral mass) (Figs. 14.7 and 14.8). The number of platelets and their arrangement differ between these two iridophores (Budelmann et al. 1997). Chromatophores and reflecting cells are located in the dermal layers below the epidermis of the skin (Fig. 14.12); chromatophores are situated among the upper layer, whereas reflecting cells (iridophores, reflector cells, and leucophores) are situated among the lower

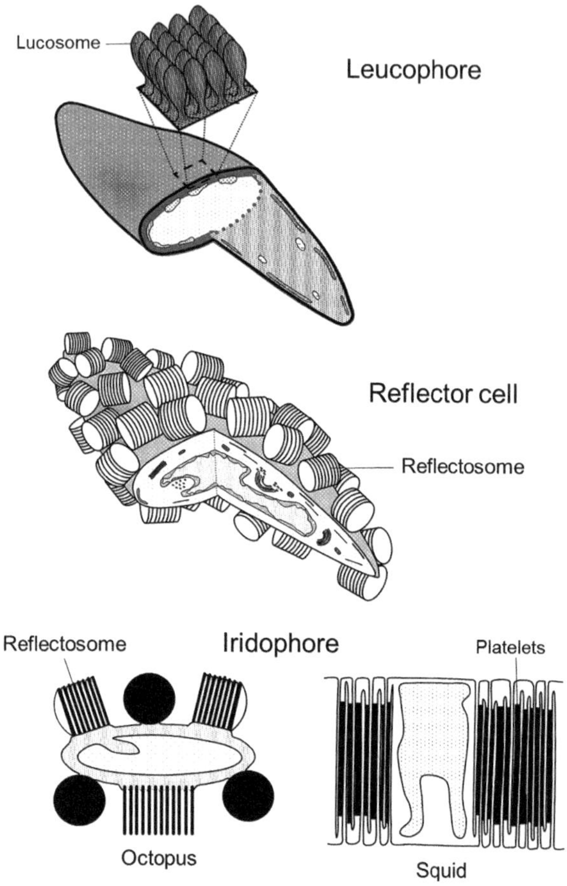

Fig. 14.6 Reflecting cells (leucophores, reflector cells, and iridophores) of coleoid cephalopods (Redrew based on Cloney and Brocco 1983)

layer. The vertical arrangement of these chromatic elements produces various wavelengths of differing strength for body color, as chromatophores can provide a parasol-like coverage on reflecting cells, which actively creates a complex, iridescent appearance (Figs. 14.12 and 14.13). Furthermore, chromatophores are functionally arranged along the horizontal axis, where they are distributed precisely to one another in a "unit" (Hanlon 1982), that also contributes to the complex and various appearance of body color (Figs. 14.11 and 14.13).

Chromatophores are under neuronal control, as the activity of the muscles connecting the chromatophores is mediated by neurons that connect to the brain (Figs. 14.5 and 14.14). Neural pathways in the brain, which mediate color change, are anatomically identified. They are situated within the central part of the brain (in the subesophageal mass, satellite ganglia, and anterior pedal lobes) (Young 1976; Dubas et al. 1986a, b; Novicki et al. 1990; Nixon and Young 2003). Chromatophore oscillation with miniature expansions or retractions at various frequencies, and

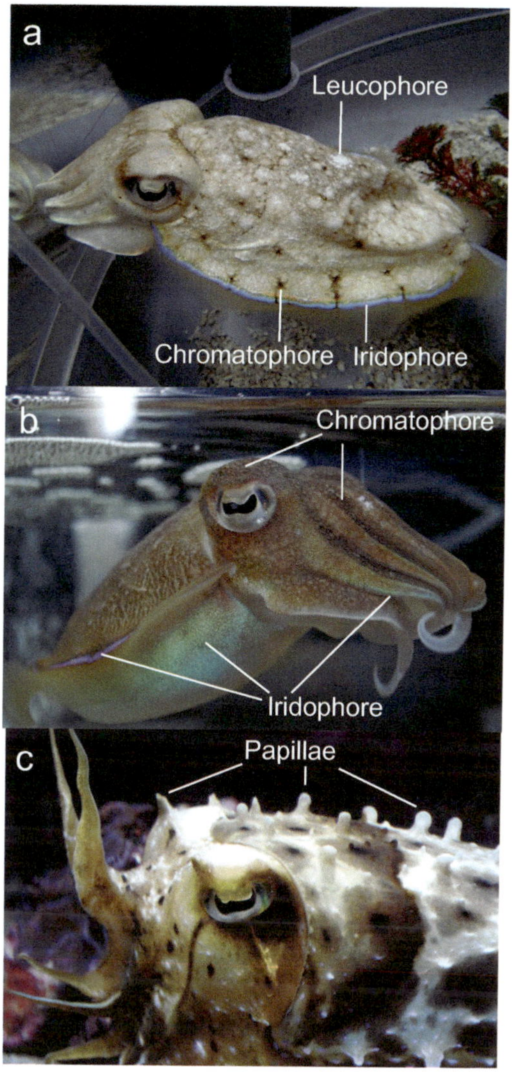

Fig. 14.7 Elements producing body patterning in coleoid cephalopods. (**a**) Pharaoh cuttlefish (*Sepia pharaonis*), (**b**, **c**) Broadclub cuttlefish (*Sepia latimanus*) (Photographs, **a** by Y. Nakai, **b**, **c** by H. Yasumuro)

synchronization of this oscillation were detected by electrophysiological study on semi-intact squid (Suzuki et al. 2011). Marine light, which induces a reaction in chromatophore organs, has previously been considered to be perceived through the eyes. However, recent findings suggest that another light pathway exists in the skin of cephalopods. In the Californian two-spot octopus (*Octopus bimaculoides*), chromatophore expansion can be induced by directing light on to the skin. This behavior is coined as "light-activated chromatophore expansion" (LACE), and this is caused by light that is sensitive to r-opsin, the visual pigment in cephalopods (Ramirez and Oakley 2015). Similarly, in squid (*Doryteuthis pealeii*) and cuttlefish (*Sepia officinalis*, *S. latimanus*), transcripts encoding rhodopsin and retinochrome present

Fig. 14.8 Iridophore of coleoid cephalopods. (**a**) Greater blue-ringed octopus (*Hapalochlaena lunulata*) expressing blue-colored iridophores (arrow), (**b**) Eris (arrow) of oval squid (*Sepioteuthis lessoniana*) eye, (**c**) Ink sac (arrow) of Pharaoh cuttlefish (*Sepia pharaonis*), (**d**) Japanese flying squid (*Todarodes pacificus*) with combination of chromatophores and iridophores, (**e**) Japanese flying squid (*T. pacificus*) with silver-colored iridophores on the whole body, (**f**) Japanese flying squid (*T. pacificus*) with green-colored iridophores on the head and arms (arrows). Note, glittering (silver) at dorsal part (half lower part of body) of Japanese flying squid (**d**, **f**) (Photographs, **a** by courtesy of D. Ueno, **c** by H. Yasumuro, **b**, **d–f** by Y. Ikeda)

in the skin, and $G_{q\alpha}$ proteins are specifically identified in chromatophore components, including pigment cell membranes, radial muscle fibers, and sheath cells in *D. pealeii* (Kingstone et al. 2015). These findings widen our viewpoint regarding color change in cephalopods, in that the process must be more complex than previously assumed, and be under sophisticated control, which is feasible

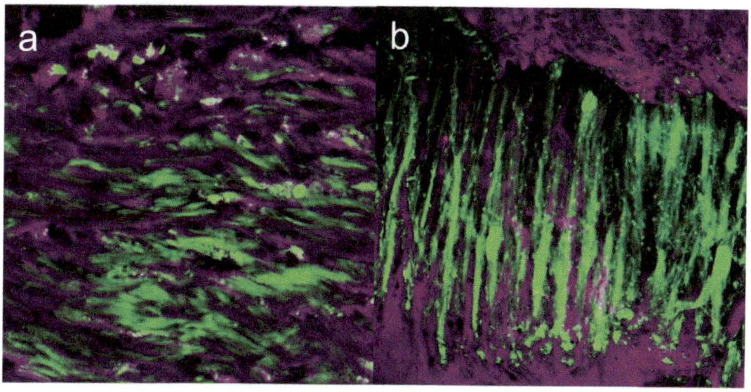

Fig. 14.9 Iridophore of oceanic water squid. (**a**) Histological section of the dermis of the tentacle of a giant squid (*Architeuthis dux*), (**b**) Histological section of the dermis of the ventral mantle of Japanese flying squid (*Todarodes pacificus*). SYTO® green-fluorescent nucleic acid stains. SYTO usually stains only nuclei, but a strong green fluorescence signal can be seen in iridophores. Muscles and connective tissues were stained with Alexa Fluor 594 Phalloidin for counterstaining. Optical sections were taken under a laser confocal microscope (Photographs, courtesy of S. Shigeno)

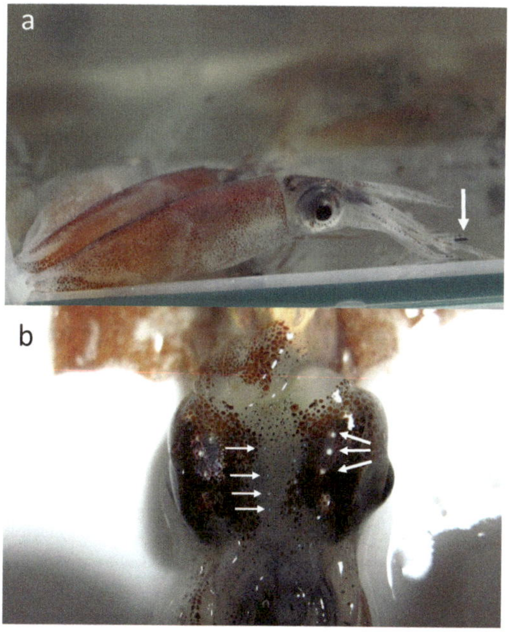

Fig. 14.10 Firefly squid (*Watasenia scintillans*). (**a**) whole appearance, (**b**) head region. Arrows indicate luminescent organs (Photographs, C. Sugimoto)

considering how closely matched they can become to their background (Fig. 14.4). Furthermore, a recent finding in squid (*Doryteuthis pealeii*) revealed that the activation of iridophores is also under neuronal control, and acetylcholine contributes to

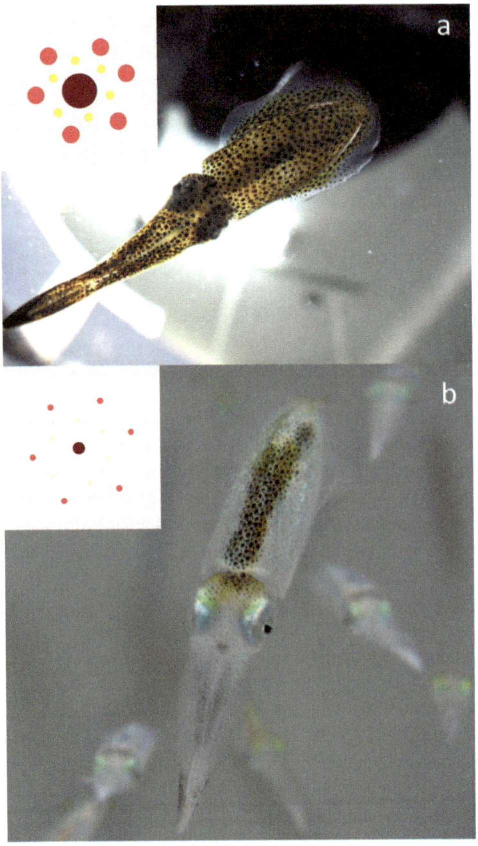

Fig. 14.11 Color change of oval squid (*Sepioteuthis lessoniana*). (**a**) Squid exhibits dark body color with many expanded chromatophores (ventral view). (**b**) Squid appears transparent without central dark area, in which many chromatophores are retracted (dorsal view). Upper left squares illustrate the state of chromatophores and the unit arrangement of chromatophores (Photographs by C. Sugimoto)

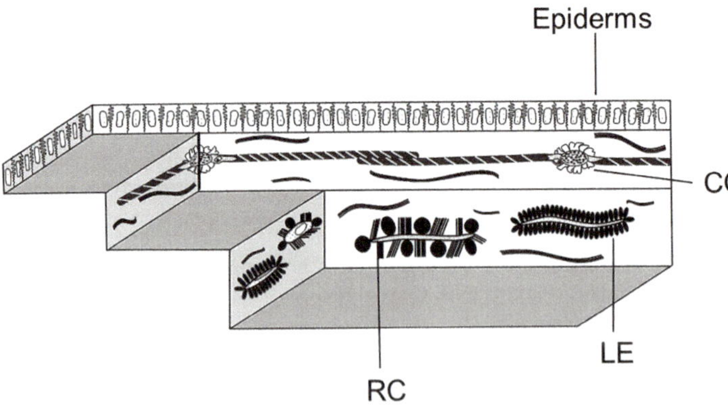

Fig. 14.12 Distribution of the chromatophore organ (CO), reflector cells (RC), and leucophores (LE) in the dermis of coleoid cephalopods (Redrew based on Cloney and Brocco 1983)

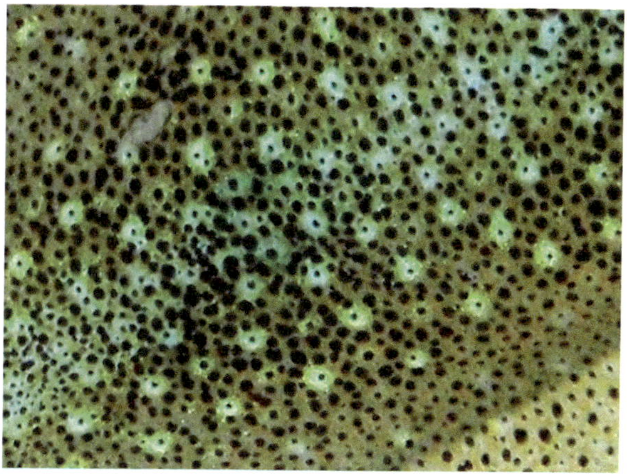

Fig. 14.13 Dorsal skin of Oval squid (*Sepioteuthis lessoniana*). Below the brown chromatophores that expand with various size, iridophores that reflect yellow-green light are visible (Photograph, Y. Oshima)

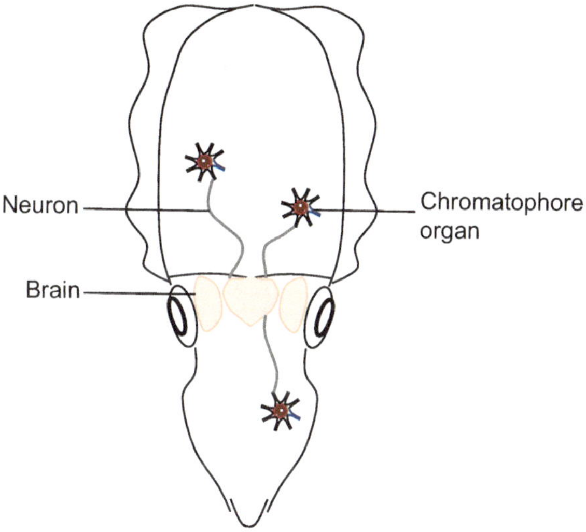

Fig. 14.14 Diagrammatic illustration of the system in coleoid cephalopod brains controlling the chromatophores

this process (Wardill et al. 2015). This is likely to contribute to the rapid and gradational color change, and the production of unique coloration in cephalopods (Figs. 14.1, 14.2, 14.3, 14.7, 14.8). Recent findings also revealed another aspect of coloration in cephalopods. Reflectin (the high-refractive-index protein that is associated with the production of structural coloration) localizes to the sheath cells that surround the chromatophore, and Ω-crystallin (the lens protein) is localized within the chromatocyte (a pigment sac of chromatophore) (Fig. 14.5), which produces structural coloration with seemingly iridescent appearance, similar to that produced by reflecting cells (Deravi et al. 2014; Williams et al. 2019).

Three neuroactive substances regulate the chromatophores of cephalopods: L-glutamate (L-glu) and FMRFamide-related peptides (FaRPs), which expand the chromatophores (Florey et al. 1985; Cornwell and Messenger 1995; Loi et al.

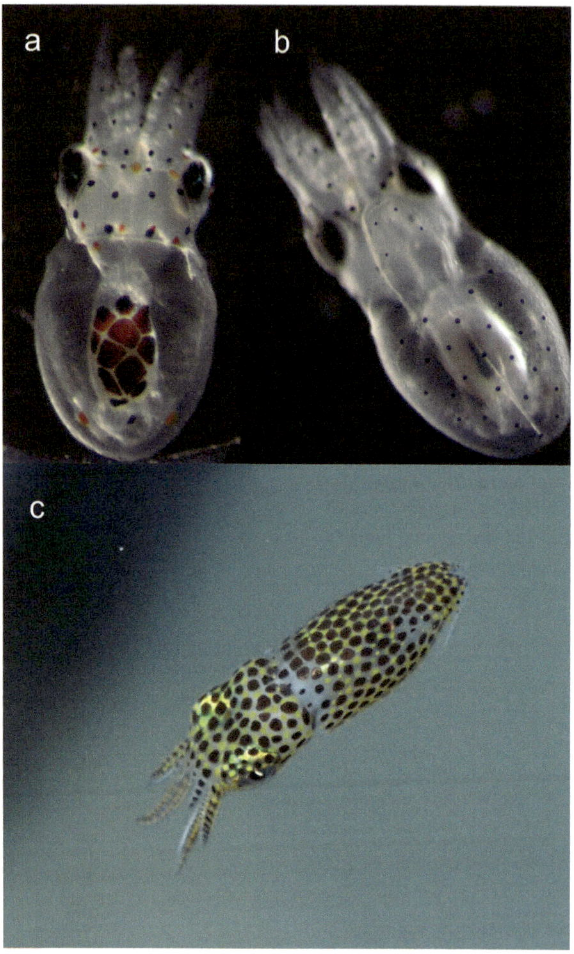

Fig. 14.15 Patterned chromatophores of coleoid cephalopod hatchlings. (**a**) Dorsal view of Algae octopus (*Abdopus aculeatus*), (**b**) Ventral view of Algae octopus (*A. aculeatus*), (**c**) Dorsal view of Oval squid (*Sepioteuthis lessoniana*). Note, pattern arrangement of chromatophores (unit) on the ventral skin of *A. aculeatus* hatchling (**b**) (Photographs, **a** and **b** by K. Shimazoe, **c** by C. Sugimoto)

1996; Loi and Tubliz 2000), and serotonin (5-HT) which retracts the chromatophores (Cornwell and Messenger 1995; Messenger et al. 1997). In the reflecting cell iridophore, acetylcholine (ACh) acts as a neural control agent that produces tunable structural coloration (Wardill et al. 2015). It was revealed recently that nitric oxide acts as a messenger for the long-term maintenance of coloration, which reacts to chromatophores and iridophores (Mattiello et al. 2013).

The embryo and hatchlings of cephalopods are already equipped with chromatophores that can contract and retract; therefore, the necessary neural circuit forms in the early stage of life (Fig. 14.15). The distribution and arrangement of chromatophores is an ontogenetic process. For example, in the common octopus (*Octopus vulgaris*), there are rules for chromatophore development. Older chromatophores retain their position and never disappear, whereas new chromatophores arise in spaces between extant ones; younger chromatophores are smaller than older ones

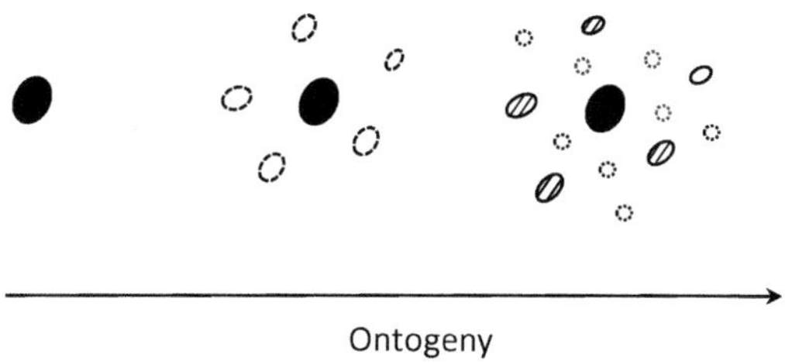

Fig. 14.16 Pattern generation of chromatophores during development (Redrew based on Packard 1988)

and new chromatophores are yellow (Packard 1985, 1988; Messenger 2001) (Fig. 14.16). Details of ontogeny for chromatophore arrangement have previously been investigated in several cephalopods (McConathy et al. 1980; Toyofuku and Wada 2018). Recently, development of chromatophore arrangement was quantitatively analyzed for European cuttlefish (*Sepia officinalis*) by high resolved tracking system and computational method, which revealed skin pattern dynamics at the cellular level (Reiter et al. 2018).

14.3 Function and Ecological Significance of Color Change

The impressive color change of cephalopods is named as body patterning. The body patterning, as previously mentioned, is a unique behavior of cephalopods, which involves chromatic, textural, postural, and locomotor components (Packard and Sanders 1971; Hanlon 1982; Hanlon and Messenger 1988). Among these four components, the chromatic component is the main infrastructure that can produce a variety of body patterning, and can do so rapidly. The number of patterns a single species exhibits can be described in a catalog of body patterning, known as an ethogram. Ethograms include descriptions of cephalopods of many coastal species (e.g., Moynihan 1985; Hanlon and Messenger 1996; Jantzen and Havenhand 2003; Huffard 2007; Nakajima and Ikeda 2017) and a few oceanic species (Trueblood et al. 2015). In general, the number of body patterns is smaller (e.g., 3–17) than the number of chromatic components (e.g., 11–35), which varies among different species (Halon and Messenger 1996). This reflects the hierarchical classification of body patterning (Hanlon and Messenger 1988). For example, the chromatic component consists of a unit with an arrangement of chromatophores and reflecting cells (Figs. 14.13 and 14.15).

The main function of body patterning is for defense and communication. The former is associated with concealment, which includes general background resemblance (i.e., conformance to the appearance of its background in brightness, color, pattern, and texture) (Fig. 14.4a, b, d), countershading (chromatophores with iridophores on the ventral surface are sparse to enhance reflection, whilst chromatophores on the dorsal surface are expanded to produce darkness; animal is indistinguishable from its background from both perspectives) (Fig. 14.8d–f), disruptive coloration (the production of a band and/or patchy pattern to break up the shape of the animal in the landscape) (Fig. 14.4c), and deceptive resemblance (the resemblance of an animal to an inanimate object in the environment, e.g., stones) (Hanlon and Messenger 1996; Messenger 2001). These behavioral traits are incredible, in that they completely camouflage individuals amongst their landscape, and are undetectable by human eyes and probably also from potential predators. This concealment is a chronic behavior that is often exhibited for hours. On the contrary, body patterning can also be an acute process, whereby body color is altered gradually for seconds or minutes, due to the process being under neuronal control (Figs. 14.2 and 14.3). This too represents the uniqueness of color change in cephalopods. The number and type of body patterns vary among species, yet some specific patterns, such as "eye spots" (deimatic pattern) are shared in some species. These shared patterns are thought to be conservative signals that had been evolutionary-acquired in the ancient species, functioning for primitive defense (i.e., same purpose for all species exhibiting these patterns), and therefore, were never lost during speciation (Moynihan 1975).

Communication is another important function of body patterning in cephalopods. It is well documented that body patterning occurs at the reproduction phase in cephalopods, which includes contest among rival males, courtship, and mating (Hanlon 1982; Hanlon and Messenger 1996) (Fig. 14.17). Specific body patterns are exhibited at each of these reproductive opportunities. There are some unique and impressive body patterns that can be invoked under specific social situations. For example, squid and cuttlefish simultaneously exhibit a threatening display across half their body, towards a rival male, while exhibiting a courtship display across the other side of their body, towards the female. Sneaker males (a small satellite male) of the Australian cuttlefish (*Sepia apama*) approach females by mimicking a female (with body color and pattern) and throwing dust in the eyes of larger males that guard females; both of which can lead to successful mating in sneaker males (Norman et al. 1999; Hanlon et al. 2005). Regarding mimicking, cephalopods exhibit a much more outstanding behavior, an example of which is attributed to a defense strategy in interspecific communication, as opposed to intraspecific communication. The mimic octopus (*Thaumoctopus mimicus*) can mimic harmful animals such as sea snakes, Luna lionfish, and tonguefish (all of which are facilitated with poison) (Fig. 14.18). This octopus can mimic not only color pattern, but also the movement of the animals it mimics (Fig. 14.18). Species-specific variation exists in the repertoire of body patterning in cephalopods; those that inhabit shallow waters with a rich landscape (coral reefs, algae, and other marine organisms) express many patterns, whereas those that inhabit oceanic or deep waters with simple landscapes express only a few patterns (Hanlon and Messenger 1996) [cf. coastal species (Fig. 14.1) vs. oceanic species (Fig. 14.8d–f)]. This seems to simply reflect the degree of change that is required to camouflage against their background at different habitats. Whether the repertoire of body patterns also relates to the level of interspecific communication is currently unknown. A recent observation indicates that oceanic water squid communicates more actively via body pattern than had been previously expected (Trueblood et al. 2015).

Fig. 14.17 Broadclub cuttlefish (*Sepia latimanus*) males that initiated agonistic contest. Note, intense zebra pattern of each male. Time course 1–3 (exact time not recorded) (Photograph, Y. Nakai)

Aside from the possible signals associated with reproductive behaviors, the purpose of body patterning has not been thoroughly clarified. It has been assumed for a long time that language is an advanced and specific communication tool limited to humans. However, this has been refuted by findings related to non-human animals; some of which have their own language, advanced with original syntax and the ability to send particular messages [e.g., courtship songs by song birds and the waggle dance of honeybees (Alcock 2005)]. Compared to signals that aim to determine a purpose/object (i.e., male attractiveness, site of food source), body

Fig. 14.18 Mimic octopus (*Thaumoctopus mimicus*). (**a**) Sitting on sandy bottom, (**b**) Swimming with mimicking tonguefish (Photographs, R. Yanagisawa)

patterning in cephalopods is much more complex, with many repertories due to the gradational change of color, therefore providing the flexibility to portray many messages. Based on wide-scale field observations, an idea was presented that cephalopod body patterning acts as language (Moynihan 1985; Moynihan and Rodanichae 1982); although this hypothesis has not yet been tested in detail.

Another possible explanation for the function of cephalopod body patterning is the expression of motivation and/or emotion (Moynihan 1985). Male European common cuttlefish (*Sepia officinalis*) exhibit dark face patterns towards rival males, with darkness being more intense when males are about to use physical contact in a fight (Adamo and Hanlon 1996). Pharaoh cuttlefish (*S. pharaonis*) diminish their camouflage pattern, before gradually exhibiting a dark body pattern as the dangerous subject (predator model) approaches himself/herself (Okamoto et al. 2015). Specific body patterns that reflect emotional states during a defensive situation are also observed in oval squid (*Sepioteuthis lessoniana*) (Kawabata and Ikeda, submitted). The possible function of body pattern for expressing emotion would appear as a gradational change in chromatic pattern, exhibiting a changeable state of individual fear or pleasantness (Figs. 14.2 and 14.3). There is another interesting hypothesis for body patterning in intraspecific communication, namely, the "secret channel" that is accomplished through polarized vision. Cephalopods, like some other animals, such as honey bees, can detect polarized light (Budelmann 1994). European cuttlefish (*S. officinalis*) exhibit a specific body pattern toward their polarized mirror image, organized with chromatophores and iridophores, whereas cuttlefish did not exhibit this pattern when their mirror image was not polarized (through inserting a polarized filter); this observation indicates that cuttlefish communicate with conspecifics via their polarized pattern, which is invisible to potential predators that lack polarized vision (Shashar et al. 1996). This is a presumable hypothesis because, as mentioned in the previous section, a recent finding has revealed that iridophores are also under neuronal control, as in chromatophores (Wardill et al. 2015); the individual can produce various patterns by controlling these two chromatic devices. The possibility of communication through polarized vison in cephalopods has been discussed further with accompanying evidence (Mäthger et al. 2009). Recent findings revealed body patterning also as frequency of exhibition of specific body patterning that gradually changes in octopus (*Abdopus aculeatus*) while it proceeds with the learning task (Kawashima et al. 2020).

The appearance repertoire of body pattern is an ontogenetic process. In the European common cuttlefish (*Sepia officinalis*) the change in body pattern was recorded across its entire life time, and it was revealed that some patterns disappear, while others appear in the repertoire as the cuttlefish grows (Hanlon and Messenger 1988). A similar finding was observed in oval squid (*Sepioteuthis lessoniana*), at the juvenile phase (Iwasaki and Ikeda, unpublished). Development of body pattern has been found to be affected by the raising environment. For example, in European common cuttlefish (*S. officinalis*), young individuals reared in tanks with conspecifics, sand, rocks, and artificial objects (i.e., enriched environment) for 90 days after birth, acquired more advanced cryptic body patterning abilities (matching their appearance to the bottom) than those reared in isolation, in visually poor tanks (Poirier et al. 2005). Similar long-term effects were observed in the expression of successful camouflage by Pharaoh cuttlefish (*S. pharaonis*) juveniles; *S. pharaonis* raised in visually poor environments (lacking a sandy bottom) were delayed in exhibiting correct body patterns on various bottom surfaces, whereas cuttlefish raised in the standard environment (with conspecifics and a sandy bottom) and an enriched environment (with conspecifics, a sandy bottom, and objects) were not delayed (Yasumuro and Ikeda 2016). Further, there is an opinion that cuttlefish can alter their choice of exhibited body pattern based on the experience of learning (Hough et al. 2016). Furthermore, a unique display (possibly for intraspecific communication), termed flushing, that is characterized by a global oscillation (2–4 Hz) of body color between white and red was observed in the oceanic species, Humboldt squid (*Dosidicus gigas*) (Rosen et al. 2015), a species being categorized as a rather simple cephalopod, in terms of their body pattern. Understanding the function of body patterning, other than for camouflage and reproduction-related signals, is one possible direction for future research, along with studies that assess the influence of genetic/environmental factors on the ontogeny of body patterning.

14.4 Mystery of Color Change

Contrary to the fact that cephalopods can exhibit various color patterns, most cephalopods lack color vision, possessing only a single visual pigment, rhodopsin (Hanlon and Messenger 1996). The exception to this is the firefly squid (*Watasenia scintillans*) that possesses three visual pigments in their retina (Matsui et al. 1988; Seido et al. 1990); thus, possessing color vision (Fig. 14.10). One behavioral experiment suggests that an octopus species (*Amphioctopus kagoshimensis*) can discriminate color (Kawamura et al. 2001), although this has not been tested further, and pigment(s) within their retina have not been determined. In other studies, there has been no evidence to indicate that color vision exists for other cephalopods. The way in which cephalopods match their color to their surrounding environment is thought to be based on information relating to contrast and brightness (Marshall and Messenger 1996; Mäthger et al. 2006). Although the peak absorbance of a single visual pigment varies among species (e.g., 471–504 nm, Seido et al. 1990), single species have single or very narrow-ranged peaks that indicates color blindness; except for firefly squid (*W. scintillans*) that have three peaks in their visual pigment (484, 500, and 470 nm) (Matsui et al. 1988; Seido et al. 1990). It seems adaptable for cephalopods to have such peaks in their visual pigment repertoire, as light with long wavelengths (i.e., red in color) is lost as water depth increases, with the main component of light being that of short wave length (i.e., blue-green), which is sensitive to their visual pigment. However, a question still exists as to how cephalopods match their color to the surrounding chromatin-rich environment, such as that seen in shallow waters where many cephalopods live (Mäthger et al. 2006).

In the case of firefly squid (*W. scintillans*) that inhabit deep water, possible color vision may contribute to bioluminescent mediated communication (Matsui et al. 1988). This hypothesis can be related to the fact that animals tend to be facilitated with highly sensitive vision when they inhabit areas with poor sources of light, such as in the deep sea. It is well known that giant squids (*Architeuthis dux*) possess the largest eyes in the animal kingdom; possibly using their eyes to detect large predators, such as sperm whales, at a distance exceeding 120 m in the deep sea (Nilsson et al. 2012). Further, through a theoretical analysis, it has been shown that the deep-water octopus (*Japatella diaphana*) possesses eyes with higher resolution than that of human eyes (Sweeny et al. 2007). Bioluminescence is a shared characteristic among some deep sea and shallow water cephalopods; whether only firefly squid has acquired color vision or not, is yet to be examined. Recently, a new hypothesis for possible color vision in cephalopods states that the ability to recognize color is due to chromatic aberration, mediated by an off-axis pupil (eye slit) (Stubbs and Stubbs 2016). As previously mentioned, general background resemblance might be achieved through the function of photoreceptors in the skin (Rarmirez and Oakley 2015), color vision, if it exists, might also play a role in shallow water. Interestingly, the absorption peak of photoreceptor cells in the skin is similar to that of the visual cells in the retina of the Californian two-spot octopus

Fig. 14.19 Simulated image between color blind (**a**) and color vision (**b**) in an underwater environment (coastal water of Okinawa Island, Ryukyu Archipelago; Photograph, H. Yasumuro)

(*Octopus bimaculoides*), which can be thought to support the matching of skin photoreceptor pattern to the background (Rarmirez and Oakley 2015).

Advantages must exist for cephalopods living in shallow waters with a chromatically rich landscape, if they can detect color in their surroundings, this would provide a more informative picture than the information that can be derived from monochromatic vision (contrast and brightness). For example, because cephalopods can detect depth (Messenger 1977), this ability would be supported by the chromatic character of objects, including biotic and abiotic organisms, by which cephalopods could perceive depth more effectively (Fig. 14.19). Color vision must also be effective in intraspecific communication in cephalopods, in which various complex body patterns are displayed. Recent findings for the function of body pattern at reproduction seem much more complex than was previously understood [e.g., low-risk and high-risk displays in Australian cuttlefish *Sepia apama* (Schnell et al. 2016); grammar in oval squid *Sepioteuthis lessoniana* (Lin et al. 2017)]. There is no phylogenetic relationship among animals to acquire color vision; it is well known that tiny insects, including the honey bee, can discriminate color, which contributes to finding flowers (i.e., food sources) (Alcock 2005), whereas dolphins (possessing large brains) possess a single visual cell, inducing controversy regarding color blindness (Peichl et al. 2001). These variations for the existence of color vision would depend upon the ecological significance for each species. In other words, how it relates to the survival of the animal. The viewpoint regarding the cost vs. benefit to maintain color vision should be kept in mind. Cephalopods already expend a physiological cost to produce color (to activate chromatophores and reflecting cells via neural and physical control). We should consider whether this cost to producing color can balance color blindness, in which cephalopods lose various chromatic information from their environment.

Acknowledgements I give thanks to my laboratory members who offered photographs. Daisuke Ueno (Kagoshima University) and Shuichi Shigeno (Osaka University) are acknowledged for their permission to use photographs and unpublished data. Sumire Kawashima (University of the Ryukyus) is given thanks for drawing Figs. 14.5, 14.6, and 14.12.

References

Adamo SA, Hanlon RT (1996) Do cuttlefish (Cephalopoda) signal their intentions to conspecifics during agonistic encounters? Anim Behav 52:73–81

Alcock J (2005) Animal behavior: an evolutionary approach, 8th edn. Sinauer Associates, Sunderland

Boyle P, Rodhouse P (2005) Cephalopods: ecology and fisheries. Blackwell Science, Oxford

Budelmann BU (1994) Cephalopod sense organs, nerves and the brain: adaptation for high performance and life style. Mar Freshw Behav Physiol 25:13–33

Budelmann BU (1995) The cephalopod nervous system: what evolution has made of the molluscan design. In: Breidbach O, Kutsch W (eds) The nervous system of invertebrates: an evolutionary and comparative approach. Basel, Birkhäuser, pp 115–138

Budelmann BU, Schipp R, Boletzky SV (1997) Cephalopoda. In: Frederick FW, Kohn AJ (eds) Microscopic anatomy of invertebrates, mollusca II. Wiley, New York, pp 119–414

Clark MR (1987) Cephalopod biomass–estimation from predation. In: Boyle PR (ed) Cephalopod life cycles volume II comparative reviews. Academic Press, London, pp 221–237

Cloney RA, Brocco SL (1983) Chromatophore organs, reflector cells, iridocytes and leucophores in cephalopods. Am Zool 23:581–592

Cloney RA, Florey E (1968) Ultrastructure of cephalopod chromatophore organs. Z Zellforsch 89:250–280

Cornwell CJ, Messenger JB (1995) Neurotransmitters of squid chromatophores. In: Abbott NJ, Williamson R, Maddock L (eds) Cephalopod Neurobiology. Oxford University Press, Oxford, pp 369–379

Costa J, Giménez-Casalduero F, Melo R, Jesus B (2012) Colour morphotypes of *Elysia timida* (Sacoglossa, Gastropoda) are determined by light acclimation in food algae. Aquat Biol 17:81–89

Deravi LF, Magyar AP, Sheehy SP, Bell GRR, Mäthger LM, Senft SL, Wardill TJ, Lane ES, Kuzirian AM, Hanlon RT, Hu EL, Parker KK (2014) The structure–function relationships of a natural nanoscale photonic device in cuttlefish chromatophores. J R Soc Interface 11:20130942

Dubas F, Hanlon RT, Ferguson GP, Pinsker HM (1986a) Localization and stimulation of chromatophore motoneurons in the brain of the squid, *Lolliguncula brevis*. J Exp Biol 117:415–431

Dubas F, Leonard RB, Hanlon RT (1986b) Chromatophore motoneurons in the brain of the squid, *Lolliguncula brevis*. Brain Res 374:21–29

Edelman DB, Seth AK (2009) Animal consciousness: a synthetic approach. Trends Neurosci 32 (9):476–484

Florey E, Dubas F, Hanlon RT (1985) Evidence for L-glutamate as a transmitter of motoneurons innervating squid chromatophore muscles. Comp Biochem Physiol 82C:259–268

Hanlon RT (1982) The functional organization of chromatophores and iridescent cells in the body patterning of *Loligo plei* (Cephalopoda: Myopsida). Malacologia 23:89–119

Hanlon RT (2007) Cephalopod dynamic camouflage. Curr Biol 17:400–404

Hanlon RT, Messenger JB (1988) Adaptive coloration in young cuttlefish (*Sepia officinalis* L.): the morphology and development of body patterns and their relation to behaviour. Philos Trans Royal Soc London Ser B 320:437–487

Hanlon RT, Messenger JB (1996) Cephalopod behaviour. Cambridge University Press, Cambridge

Hanlon RT, Naud M-J, Shaw PW, Havenhand JN (2005) Transient sexual mimicry leads to fertilization. Nature 433:212

Hough AR, Case AR, J. and Boal, J. G. (2016) Learned control of body patterning in cuttlefish *Sepia officinalis*. J Moll Stud 82:427–431

Huffard CL (2007) Ethogram of *Abdopus aculeatus* (d'Orbigny, 1834) (Cephalopoda: Octopodidae): can behavioural characters inform octopodid taxonomy and systematics? J Moll Stud 73:185–193

Jantzen TM, Havenhand JN (2003) Reproductive behavior in the squid *Sepioteuthis australis* from South Australia: ethogram of reproductive body patterns. Biol Bull 204:290–304

Kawamura G, Nobutoki K, Anraku K, Tanaka Y, Okamoto M (2001) Color discrimination conditioning in two octopus *Octopus aegina* and *O. vulgaris*. Nippon Suisann Gakkaishi 67 (1):35–39. (in Japanese with English abstract)

Kawashima S, Takei K, Yoshikawa S, Yasumuro H, Ikeda Y (2020) Tropical octopus *Abdopus aculeatus* can learn to recognize real and virtual symbolic objects. Biol Bull 238:12–24

Kingstone ACN, Kuzirian AM, Hanlon RT, Cronin TW (2015) Visual phototransduction components in cephalopod chromatophores suggest dermal photoreception. J Exp Biol 218:1596–1602

Lin C-Y, Tsai Y-C, Chao C-C (2017) Quantitative analysis of dynamic body patterning reveals the grammar of visual signals during reproductive behavior of the oval squid *Sepioteuthis lessoniana*. Front Ecol Evol 5:30

Loi PK, Tublitz NJ (2000) The roles of glutamate and FMRFamide-related peptides at the chromatophore neuro- muscular junction in the cuttlefish *Sepia officinalis*. J Comp Neurol 420:499–511

Loi PK, Saunders RG, Young DC, Tublitz NJ (1996) Peptidergic regulation of chromatophore function in the European cuttlefish *Sepia officinalis*. J Exp Biol 199:1177–1187

Marshall NJ, Messenger JB (1996) Colour-blind camouflage. Nature 382:408–409

Mäthger LM, Barbosa A, Miner S, Hanlon RT (2006) Color blindness and contrast perception in cuttlefish (*Sepia officinalis*) determined by a visual sensorimotor assay. Vision Res 46:1746–1753

Mäthger LM, Shashar N, Hanlon RT (2009) Do cephalopods communicate using polarized light reflections from their skin? J Exp Biol 212:2133–2140

Matsui S, Seido M, Horiuchi S, Uchiyama I, Kito Y (1988) Adaptation of deep-sea cephalopod to the photic environment. Evidence for three visual pigments. J Gen Physiol 92:55–66

Mattiello T, d'Ischia M, Palumbo A (2013) Nitric oxide in chromatic body patterning elements of *Sepia officinalis*. J Exp Mar Biol Ecol 447:128–131

McConathy DA, Hanlon RT, Hixon RF (1980) Chromatophore arrangements of hatchling loliginid squids (Cephalopoda, Myopsida). Malacologia 19:279–288

Messenger JB (1977) Prey-capture and learning in the cuttlefish, *Sepia*. Symp Zool Soc Lond 38:347–376

Messenger JB (2001) Cephalopod chromatophores: neurobiology and natural history. Biol Rev 76:473–528

Messenger JB, Cornwell CJ, Reed CM (1997) L-glutamate and serotonin are endogenous in squid chromatophore nerves. J Exp Biol 200:3043–3054

Moynihan M (1975) Conservatism of displays and comparable stereotyped patterns among cephalopods. In: Baerends G, Beer C, Manning A (eds) Function and evolution in behaviour. Clarendon Press, Oxford, pp 276–291

Moynihan M (1985) Communication and noncommunication by cephalopods. Indiana University Press, Bloomington

Moynihan M, Rodaniche AF (1982) The behavior and natural history of the Caribbean reef squid *Sepioteuthis sepioidea*. Adv Ethol 25:1–151

Nakajima R, Ikeda Y (2017) A catalog of the chromatic, postural, and locomotor behaviors of the pharaoh cuttlefish (*Sepia pharaonis*) from Okinawa Island, Japan. Mar Biodivers 47 (2):735–753

Nilsson D-E, Warrant EJ, Johnsen S, Hanlon R, Shashar N (2012) A unique advantage for giant eyes in giant squid. Curr Biol 22:683–688

Nixon M (1987) Cephalopod diets. In: Boyle PR (ed) Cephalopod life cycles volume II comparative reviews. Academic Press, London, pp 201–219

Nixon M, Young JZ (2003) The brains and lives of cephalopod. Cambridge University Press, Cambridge

Norman MD, Finn J, Tregenza T (1999) Female impersonation as an alternative reproductive strategy in giant cuttlefish. Proc Roy Soc Lond A 266:1347–1349

Novicki A, Budelmann BU, Hanlon RT (1990) Brain pathway of the chromatophore system in the squid, *Lolliguncula brevis*. Brain Res 519:315–323

Okamoto K, Mori A, Ikeda Y (2015) Effects of visual cues of a moving predator on body patterns in cuttlefish *Sepia pharaonis*. Zoolog Sci 32:336–344

Packard A (1972) Cephalopods and fish: the limits of convergence. Biol Rev 47:241–307

Packard A (1985) Sizes and distribution of chromatophores during post-embryonic development in cephalopods. Vie et Milieu 35:285–298

Packard A (1988) The skin of cephalopods (Coleoids): general and special adaptations. In: Trueman ER, Clarke MR (eds) The mollusca, vol. 11: form and function. Academic Press, San Diego, pp 37–67

Packard A (1995) Organization of cephalopod chromatophore systems: a neuromuscular image-generator. In: Abbott NJ, Williamson R, Maddock L (eds) Cephalopod neurobiology. Oxford University Press, Oxford, pp 503–520

Packard A, Sanders GD (1971) Body patterns of *Octopus vulgaris* and maturation of the response to disturbance. Anim Behav 19:780–790

Peichl L, Behrmann G, Kröger RHH (2001) For whales and seals the ocean is not blue: a visual pigment loss in marine mammals. Eur J Neurosci 13:1520–1528

Poirier R, Chichery R, Dickel L (2005) Early experience and postembryonic maturation of body patterns in cuttlefish (*Sepia officinalis*). J Comp Psychol 119:230–237

Ramirez MD, Oakley TH (2015) Eye-independent, light-activated chromatophore expansion (LACE) and expression of phototransduction genes in the skin of *Octopus bimaculoides*. J Exp Biol 218:1513–1520

Rathjen WF, Voss GL (1987) The cephalopod fisheries: a review. In: Boyle PR (ed) Cephalopod life cycles volume II comparative reviews. Academic Press, London, pp 253–275

Reiter S, Hülsdunk P, Woo T, Lauterbach MA, Eberle JS, Akay LA, Longo A, Meier-Credo J, Kretschmer F, Langer JD, Kaschube M, Laurent G (2018) Elucidating the control and development of skin patterning in cuttlefish. Nature 562:361–366

Rosen H, Gilly W, Bell L, Abernathy K, Marshall G (2015) Chromogenic behaviors of the Humboldt squid (*Dosidicus gigas*) studied *in situ* with an animal-borne video package. J Exp Biol 218:265–275

Schnell AK, Smith CL, Hanlon RT, Hall KC, Harcourt R (2016) Cuttlefish perform multiple agonistic display to communicate a hierarchy of threats. Behav Ecol Sociobiol 70:1643–1655

Seidou M, Sugahara M, Uchiyama H, Hiraki K, Hamanaka T, Michinomae M, Yoshihara K, Kito Y (1990) On the three visual pigments of the firefly squid, *Watasenia scintillans*. J Comp Physiol A 166:769–773

Shashar N, Rutledge PS, Cronin TW (1996) Polarization vision in cuttlefish – a concealed communication channel? J Exp Biol 199:2077–2084

Stubbs AL, Stubbs CW (2016) Spectral discrimination in color blind animals via chromatic aberration and pupil shape. Proc Natl Acad Sci 113(29):8206–8211

Suzuki M, Kimura T, Ogawa H, Hotta K, Oka K (2011) Chromatophore activity during natural pattern expression by the squid *Sepioteuthis lessoniana*: contributions of miniature oscillation. PLoS One 6(4):e18244

Sweeny AM, Haddock SH, Johnsen S (2007) Comparative visual acuity of coleoid cephalopods. Integr Comp Biol 47(6):808–814

Toyofuku T, Wada T (2018) Chromatophore arrangement and photophore formation in the early development of swordtip squid *Uroteuthis* (*Photololigo*) *edulis*. Fish Sci 84:9–15

Trueblood LA, Zylinski S, Robison BH, Seibel BA (2015) An ethogram of the Humboldt squid *Dosidicus gigas* Orbigny (1835) as observed from remotely operated vehicles. Behaviour 152:1911–1932

Wardill TJ, Gonzalez-Bellido PT, Crook RJ, Hanlon RT (2015) Neural control of tuneable skin iridescence in squid. Proc R Soc Lond B 279:4243–4252

Wells MJ (1978) Octopus physiology and behaviour of an advanced invertebrate. University Printing House, London

Williams ST (2017) Molluscan shell colour. Biol Rev 92:1039–1058

Williams TL, Senft SL, Yeo J, Martín-Martínez FJ, Kuzirian AM, Martin CA, DiBona CW, Chen C-T, Dinneen SR, Nguyen HT, Gomes CM, Rosenthal JJC, MacManes MD, Chu F, Buehler MJ, Hanlon RT, Deravi LF (2019) Dynamic pigmentary and structural coloration within cephalopod chromatophore organs. Nat Commun 10:1004

Yasumuro H, Ikeda Y (2016) Environmental enrichment accelerates the ontogeny of cryptic behavior in pharaoh cuttlefish (*Sepia pharaonis*). Zoolog Sci 33:255–265

Young JZ (1971) The anatomy and nervous system of *Octopus vulgaris*. Clarendon Press, Oxford

Young JZ (1976) The nervous system of *Loligo*. II Subesophageal centres. Philos Trans Roy Soc London Ser B 274:101–167

Chapter 15
Physiological and Biochemical Mechanisms of Insect Color Change Towards Understanding Molecular Links

Minoru Moriyama

Abstract Insect color patterns are prominently diverse, serving significant physiological and ecological functions. However, the environments surrounding insects are not always constant, complicating attempts to set the most suitable color pattern as a fixed phenotype. While not ubiquitous, many insects possess a sophisticated mechanism that adjusts their surface color in response to external cues to cope with such variable environments. Compared with other well-known color changing-animals like vertebrates and aquatic invertebrates, insects face the challenge of flexible color change because they lack chromatophores like vertebrates but have a melanized exoskeleton. Instead, a variety of unique processes have been adopted in a wide range of insects. This chapter outlines the diverse mechanisms underlying various insect color changes by introducing physiological and biochemical research results and recent progress in molecular studies. Here, the term color change is considered in a broad sense; it covers not only color changes that take place within the life of a single insect, but also alternative color morphs resulting from adaptive developmental plasticity. This review highlights that ecologically relevant color changes are achieved by an elaborate system that requires the coordination of several dedicated pieces of machinery, including variable architectures, perception and integration systems for environmental cues, neuroendocrine signaling systems, and color-generating apparatuses.

Keywords Developmental plasticity · Diapause · Ontogenic color change · Polyphenism · Reversible color change · Seasonal form

M. Moriyama (✉)
Bioproduction Research Institute, National Institute of Advanced Industrial Science and Technology, Tsukuba, Japan
e-mail: m-moriyama@aist.go.jp

H. Hashimoto et al. (eds.), *Pigments, Pigment Cells and Pigment Patterns*,
https://doi.org/10.1007/978-981-16-1490-3_15

15.1 Introduction

Diverse insect color patterns, which are made up of various types of pigment compounds and color-reflecting structures, serve significant physiological and ecological functions such as crypsis, aposematism, mimetism, thermal regulation, and sexual recognition (Booth 1990; Clusella-Trullas et al. 2007; Duarte et al. 2017; Stevens and Ruxton 2012). However, environments surrounding insects are not homogeneous through time and space, and thus the most suitable color for a species may vary through life cycles. A plethora of literature has reported that although not ubiquitous, many insects have the ability to cope with fluctuating environments through sophisticated mechanisms that flexibly adjust their coloration in response to adaptive demands.

In addition to insects, flexible color change is well documented in vertebrates and aquatic invertebrates such as cephalopods and crustaceans. In these animals, aggregation/dispersion of pigment granules in chromatophores enables rapid reversible color changes (Schartl et al. 2016; Sköld et al. 2016). In mammals and birds, alternative colors are recruited after a seasonal exchange of differentially pigmented coats and plumages, respectively (Zimova et al. 2018). The color formation processes of insects are significantly different from these animals due to the presence of an exoskeleton and general lack of chromatophores. At the time of molting, melanin and other pigments are often embedded in the exoskeletal cuticle layer and cuticular scales. If present, the epidermis colors beneath the cuticle can be visible through translucent parts of the cuticle. Cuticle melanization is associated with cuticle sclerotization, plays a protective role against UV light, desiccation, invasion of pathogens, and physical damages. Cuticle melanization is considered a crucial factor assisting insect infestations in terrestrial ecosystems (Andersen 2010; Noh et al. 2016; Sugumaran 2002). The prevalence of the melanized and opaque exoskeleton among insects may be a burden for flexible color change without exoskeleton ecdysis. Nevertheless, a considerable number of insects are known to accomplish color change in a variety of unique ways.

This chapter outlines the diverse mechanisms underlying insect color changes with a special interest in recent progress in molecular understanding, evolving from physiological and biochemical research on representative model species. Ecologically relevant color changes require an organized system that includes the perception of environmental change, a neural and endocrinal regime, and pigment metabolism. In some cases, gaps in our knowledge are compensated by discoveries of other terrestrial arthropods, such as spiders and mites, so I highlight some examples in this chapter. Ecological significances of insect color changes are sometimes apparent, but are often ambiguous or associated with more than one function. In this chapter, the ecological implications of color changes are minimally interpreted.

Temporal color shift can occur between individuals as well as within an individual, and all color variations can be broadly referred to as a color change in insects, probably because of their relatively small body sizes and short generation times among the well-studied animals. Here, I focus on color changes that act as an

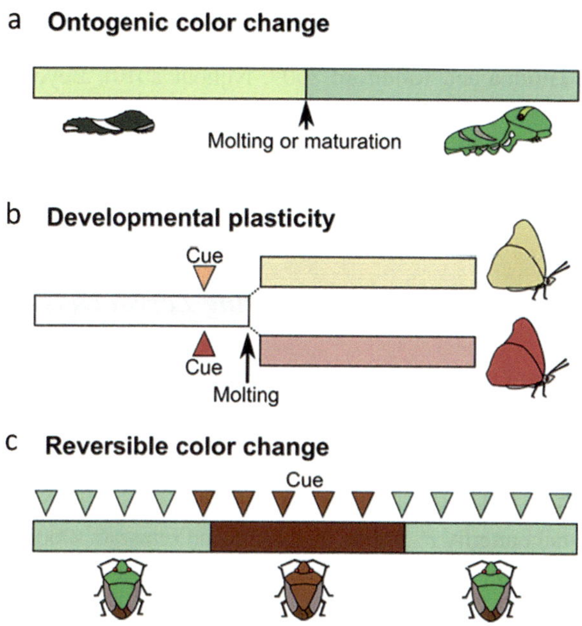

Fig. 15.1 Schematic description of three modes of insect color change. (**a**) Obligate color change induced during the course of juvenile molting or on maturation irrespective of the environment. (**b**) Selecting developmental options with different color phenotypes in response to environmental cues. The changes take place with molting, pupation, or adult eclosion, and the sensitive period for cues usually precedes it. (**c**) Appearance change reversibly induced along with environmental fluctuations. These changes can be accomplished within a single developmental stage without molting

adaptive mechanism with environmental fluctuations. Therefore, I discuss inter-individual phenotypic variations that are expressed as alternative developmental options derived from a single genotype (known as phenotypic plasticity). Meanwhile, it does not cover inter-individual color variations attributed to different genotypes, known as genetic polymorphisms (such as geographic variations and intra-population variations), and sexual dimorphisms. In the following sections, insect color changes are roughly divided into the following three categories: ontogenic color change, developmental plasticity, and reversible color change (Fig. 15.1).

15.2 Ontogenic Color Changes

Insect development progresses in a stepwise manner, which is punctuated by a renewal of the exoskeleton or molting. The color patterns of insects are not necessarily homogeneous through molting, and especially in holometabolous insects, distinct adult color patterns are built at the final molting. At this time, a series of molecular signaling pathways are provoked by coordination of the juvenile hormone (JH) and the ecdysteroid hormone, to allocate appropriate pigments in appropriate positions (e.g. Hiruma and Riddiford 2009; Nijhout 2010). Beyond these normal ontogenic color pattern formations, some insects exhibit drastic color changes during the course of juvenile molting or after reaching the adult stage. These types of color change are called "ontogenic color changes" (Bückmann 1985; Booth 1990).

15.2.1 Color Pattern Switching During Larval Development

Young caterpillars of the swallowtail butterfly species in the genus *Papilio* exhibit a mottled black and white color pattern (which resembles bird droppings) until the third or fourth instar (Prudic et al. 2007). At the next molting, this mimetic color pattern either is switched to a cryptic green color, a conspicuous aposematic color, or remains the white/black mimetic color pattern, depending on the species. Using the Asian swallowtail butterfly *P. xuthus*, Futahashi and Fujiwara (2008a) demonstrated that these drastic changes in color pattern are governed by the JH. In this species, the white/black form lasts until the fourth molting. The fifth (final) instar larvae show a green base coloration decorated with a pair of eyespots with a few black bands (Fig. 15.2). This color pattern is formed from region-specific expression of green- or black-associated genes. An intended green integument region is consistent with the spatial expression pattern of the bilin-binding proteins (*BBP1* and *BBP2*), a yellow-related protein (*YRG*), and putative carotenoid-binding proteins (*PCBP1* and *PCBP2*) (Futahashi and Fujiwara 2008a, b; Futahashi et al. 2012b; Shirataki et al. 2010). Conversely, black regions are characterized by the sequential expression of several melanin synthesis genes, including *MCO2* (*laccase2*), *yellow*, *tyrosine hydroxylase* (*TH*), *dopa decarboxylase* (*DDC*), and *tan* (Futahashi and Fujiwara 2005; Futahashi and Fujiwara 2007; Futahashi and Fujiwara 2008a; Futahashi et al. 2010; Shirataki et al. 2010, Chap. 1). The application of synthetic JH on fourth-instar larvae generates fifth instar larvae exhibiting a white/black coloration similar to the fourth-instar larvae (Futahashi and Fujiwara 2008a). The sensitive period for

Fig. 15.2 Developmental variation in the larval color pattern in the Asian swallowtail butterfly *Papilio xuthus*. (**a**) Mimetic black/white coloration of the fourth-instar larva resembles a bird dropping. (**b**) Cryptic green coloration o the fifth instar larva

external JH application is narrow and restricted to within 20 h from the onset of the fourth instar, when the JH titer normally declines. JH treatment suppresses green-associated gene expression and maintains the mimetic-type distribution pattern of the melanin-related genes. This research demonstrates that the decline in JH titer at the onset of the fourth-instar stage acts as a trigger for the switch in color pattern. Thereafter, three homeobox genes (*clawless*, *abdominal-A*, and *Abdominal-B*) were demonstrated to be indispensable for the formation of the normal cryptic color pattern of the fifth instar larvae (Jin et al. 2019).

15.2.2 Color Changes on Maturation

Some insects undergo a color change after reaching the adult stage, thus these changes are attained without molting. These maturation-associated ontogenic color changes are often observed in dragonflies and damselflies. These insect groups are characterized by large conspicuous compound eyes with highly multiplied visual opsins (Futahashi et al. 2015) and small antennae, suggesting that they rely highly on visual signals for inter-individual recognition (Futahashi 2016). In many red dragonflies in the genera of *Crocothemis* and *Sympetrum*, young adults of both sexes show yellowish coloration, with only males developing the reddish coloration upon sexual maturation (Fig. 15.3). Futahashi et al. (2012a) revealed that these color changes are attributed to the conversion of the redox status of ommochrome pigments in the epidermis layer. They identified that the integument of the dragonflies contains two ommochrome pigments, decarboxylated xanthommatin and xanthommatin, both are yellow in the oxidized form and red in the reduced form. Consistent with the appearance color, the proportion of the reduced form is elevated in mature red males when compared to immature yellow males. They also experimentally reproduced the yellow-to-red color shift in vivo by injecting a chemical reductant (ascorbic acid) into yellow immature males and yellow females. To date, the molecular mechanisms regulating the redox conditions of ommochromes remain undiscovered. Gynandromorph dragonfiles, which are rarely found sexual mosaic individuals, have both yellow and red regions, suggests that the redox status is likely to be regulated in a cell-autonomous manner (Futahashi et al. 2012a). In the case of the damselfly, *Calopteryx japonica*, the male wing color changes from a dull light

Fig. 15.3 Color shift on maturation observed in a red dragonfly, *Crocothemis servilia*. (**a**) Immature yellow male. (**b**) Mature red male. Photos courtesy of Dr. Ryo Futahashi

brown to an iridescent blue-black on maturation (Stavenga et al. 2012). During maturation, a deposition of black melanin is layered into the wing cuticle matrix that creates a multi-layered structure, generating a blue reflectance.

A carotenoid-binding yellow protein of the takeout family (YPT) has been demonstrated to be involved in male-specific mature coloration in the desert locust *Schistocerca gregaria*. In crowded conditions (see below sections for more detail), adult males of this species turn their integument color from brown to yellow on maturation (Wybrandt and Andersen 2001; De Loof et al. 2010). The mature male-specific YPT gene expression is considered positively regulated by JH (Sas et al. 2007; Sugahara and Tanaka 2018). There are two possibilities causing this male-specific color change, male-specific endocrine regulation (Sas et al. 2007), and/or a sex-specific epidermal response (Nishide and Tanaka 2012).

15.2.3 Involvement of Symbiotic Microbes

Recent studies performing molecular diagnoses have revealed significant influences of symbiotic microbes on insect coloration. Experimental removal of mutualistic bacterial symbionts often causes deformation of normal coloration of the host insect because they supply substantial resources for pigment production, for example, tyrosine for melanin production (Anbutsu et al. 2017). These heritable and mutualistic symbionts are often essential for host development and reproduction. Therefore, color changes due to the loss or gain of these symbiotic partners during insect development rarely occur in ecological situations (for an exception, see Kikuchi et al. 2016).

Interestingly, Tsuchida et al. (2010) found that an infection in a facultative symbiotic bacterium modifies the coloration of the pea aphid *Acyrthosiphon pisum*. It has been found that there are genetic green/red variations within a population of pea aphids. This dimorphism is considered to be maintained under two different selective pressures of parasitism and predation. The green morph is advantageous for avoiding predation by ladybird beetles (Losey et al. 1997), whereas the red morph is not as attractive to parasitoid wasps (Libbrecht et al. 2007). Tsuchida et al. (2010) found that some red-morph nymphs collected from several populations had turned green by the adult stage. A diagnostic survey of symbiotic microbiota revealed that these color-changing strains consistently harbor a particular bacterium, *Rickettsiella viridis*. Artificial infection of this bacterial symbiont verified its causal role in the red-green transition. This color change is due to the accumulation of green aphins, which are aphid-specific polycyclic quinone pigments (see Chap. 1). The symbiont genome analysis revealed that *Rickettsiella* is unable to synthesize polycyclic quinones, implying that unidentified effectors may stimulate aphin production by the host (Nikoh et al. 2018).

15.3 Developmental Plasticity of Color Forms

Insects have been shown to possess an ability to adapt to changing environments by selecting alternative developmental pathways that are preprogrammed in the genome and lead to various color phenotypes. The phenotypic responses to environmental cues are continuous or discrete (i.e., polyphenism), allowing them to adapt to the heterogeneity of the season, background conditions, and population density (Beldade et al. 2011; Nijhout 1999, 2003; Simpson et al. 2011). Short-lived insects are unlikely to experience relatively long-term environmental changes within a certain developmental stage. Therefore, selection would have favored a life cycle strategy that more suited alternative phenotypes are recruited generationally or at developmental stages by predicting upcoming environmental conditions, rather than flexibly modulating coloration within a stage.

15.3.1 Seasonal Forms

Insects can produce multiple generations a year; therefore, different generations are exposed to different seasonal environments. Seasonally specialized color morphs or seasonal forms are observed in such insects. In some instances, the alternative seasonal forms can be mistaken for different species because of their distinctly different appearances (Nijhout 2010). Seasonal environmental change is a predictable oscillation, and many biotic and abiotic factors fluctuate in a coordinated fashion. Photoperiod and temperature are frequently used as a cue to predict the time of a year (Nijhout 1999, 2003). Additional factors such as seasonally limited food (Greene 1989) can also be a primary determinant of plastic responses. Expression of a specific seasonal morph can be associated with diapause. Diapause is a particular physiological state involving a programmed developmental arrest, often accompanied by a low metabolic rate and enhanced temperature tolerance, to overcome adverse conditions (Danks 2006; Denlinger et al. 2012).

Seasonal polyphenisms of wing color patterns in butterflies have attracted considerable attention (Clarke 2017; Nijhout 2010). In some model butterflies, the pigment chemistry, external stimuli, endocrine regulation, and molecular mechanisms involved in color polyphenism have been identified. A well-organized research program is progressively investigating the common buckeye butterfly, *Junonia* (*Precis*) *coenia*. The summer form (called the *liena* form) develops light beige ventral sides in the hindwings. Conversely, the autumn form (called the *rosa* form) is a reddish brown (Daniels et al. 2012; Smith 1991). Larvae reared under a long-day photoperiod and high temperature produced the summer form, while a short-day photoperiod and low temperature induced the autumn form. This shift to a darker coloration in the autumn generation is considered a cryptic color on senescent leaves (Daniels et al. 2012). The color forms are under endocrine control by the ecdysteroid hormone. The high ecdysteroid titer within the critical period of the earlier pupal stage induces the summer form and vice versa (Rountree and Nijhout 1995). Ommochromes are the pigments responsible for this color difference (Nijhout 1997). Yellowish xanthommatin is the primary component in the beige hindwings of summer-form butterflies. In contrast, the reduced form of xanthommatin and its sulfated compound (ommatin-D) are the major components in the autumn-form brown hindwing. As previously mentioned, xanthommatin turns red upon reduction. These findings imply that different seasonal coloration of the buckeye butterfly is achieved by specific expression of xanthommatin-reductive enzymes and their sulfotransferases. There may be also a possible function of ommochrome binding proteins to stabilize the reduced red state of xanthommatin (Gilbert et al. 1988; Nijhout 1997). Daniels et al. (2014) attempted to isolate the genes relevant to this seasonal color change using a comprehensive transcriptome analysis by RNA sequencing. They compared the gene expression patterns of the wing primordia extracted from pupae destined to be summer- or autumn-form. They confirmed that the key gene of the ommochrome synthesis pathway, *cinnabar*, is upregulated during wing development in both phenotypes, but at different times. However, the identification remains unknown of the xanthommatin-reductive enzymes or the sulfotransferases that show expression patterns consistent with different color forms. As suggested by transcriptome analyses of other butterfly species (Hines et al. 2012; Connahs et al. 2016), unknown factors may be involved in the ommochrome-based red pigmentation in butterfly wings. In *J. coenia*, genes related to melanin synthesis, such as *tan* and *yellow*, are upregulated in the brown wings, suggesting that melanin accumulation may partly contribute to the darkening of the autumn-form wings in addition to ommochrome reduction (Daniels et al. 2014). Recently, van der Burg et al. (2020) found three genes (*cortex*, *herfst*, and *trehalase*) related to red wing formation by a genome-wide association study using artificially selected red butterflies.

A transcriptome approach for seasonal polyphenism in butterfly wings was also performed in the European map butterfly *Araschnia levana*. This butterfly has two distinct wing phenotypes that are controlled by photoperiod and temperature, similar to the buckeye butterfly. The black summer form is generated from larvae reared under long-day and warmer conditions, whereas the spring form with a mottled orange/black pattern emerges after pupal diapause in larvae exposed to short-day and cooler conditions (Koch and Bückmann 1987). These wing phenotypes are determined by the different timing of the ecdysteroid release during the pupal stage (Koch and Bückmann 1987; Nijhout 2003) in combination with the summer-morph-producing hormone (Yamashita et al. 2014). Vilcinskas and Vogel (2016) explored differentially expressed genes between prepupae derived from two rearing conditions using a subtraction method. A gene encoding the prophenoloxidase-activating protein (presumably involved in melanin formation) was preferentially expressed in the darker summer form. A yellow family gene (also expected to serve for melanin production) was highly expressed in the summer-form destined pupae. Conversely, another *yellow* family gene was upregulated in the autumn-form destined pupae. The form-specific expression patterns of the *yellow* family genes are also observed in the buckeye butterfly (Daniels et al. 2014). Vilcinskas and Vogel (2016) presumed that the regulation of the spatial expression patterns of these specific *yellow* family genes underlies the different wing color patterns in *A. levana*. They also suggested that post-transcriptional regulation by microRNAs plays a substantial role in the expression of seasonal polyphenism (Mukherjee et al. 2020).

A third example addresses dry season-specific color pattern formation in the nymphalid butterfly *Bicyclus anynana*. In this species, high temperature induces the wet-season phenotype exhibiting several eyespots, markings composed of concentric color rings, are developed in the ventral side of hindwings, whereas these eyespots are small or less conspicuous on the low temperature-inducible dry-season wings (Brakefield et al. 1996, 1998). Conversely, rearing condition has limited effect on the large eyespots on the forewings. Monteiro et al. (2015) provided clear evidence explaining the regulatory mechanisms of this condition-dependent pattern formation and the heterogeneous responsibility among eyespots. *Bicyclus anynana* butterflies were most sensitive to temperature in the wandering stage, at the final larval stage just before pupation. In this stage, the ecdysteroid titer increases with the variation in rearing temperature. Interestingly, the ecdysone receptor (EcR) (which receives ecdysteroid signals and provokes a sequential expression of functional genes) is expressed in the central cells of the prospective hindwing eyespots, but not in the forewing eyespots, irrespective of rearing temperatures. It suggested that in the sensitive period, only the eyespots on the hindwing are responsible for ecdysteroid signals that carry environmental information (Monteiro et al. 2015). Subsequently, EcR receiving elevated ecdysteroid signals induce gene expression of pigment metabolism via the regulatory genes involved in eyespot formation (Beldade and Peralta 2017; Monteiro 2015; Nijhout 2010), such as the transcription factor *Distal-less* (Brakefield et al. 1996; Beldade et al. 2002; Monteiro et al. 2013). In addition, different timing of the ecdysteroid peak in the pupal period is considered due to different rearing conditions affecting the size of eyespot color rings (Mateus et al. 2014; Beldade and Peralta 2017).

15.3.2 Background Matching

Some insects are able to develop different color patterns in order to mimic background colors. In addition to seasonal wing color polyphenism, plastic responses in pupal coloration are often observed in butterfly species (Fujiwara and Nishikawa 2016; Mayekar and Kodandaramaiah 2017). Due to the sedentary nature of pupae, developing cryptic pupal coloration in response to heterogeneous environments can confer a potential fitness advantage. In these butterfly species, pupal green/brown phenotypes can be determined by a combination of various types of stimuli, including relative humidity, substrate texture, coloration of the peripheral environment, and seasonal cues such as photoperiod and temperature (Mayekar and Kodandaramaiah 2017). In the case of the swallowtail butterfly *Papilio xuthus*, tactile stimuli play a primary role in pupal color determination in natural settings; smooth and rough substrate surfaces induce green and brown pupae, respectively

(Hiraga 2006). An endocrine factor that regulates pupal coloration is the pupal-cuticle-melanizing hormone (PCMH) that is found in the central nervous system, but the substance and relevant genes involved are yet to be identified (Yamanaka et al. 1999). Interestingly, similar hormones have been reported in other butterfly species, affecting pupal pigmentation in either direction depending on the species (Jones et al. 2007; Yamanaka et al. 2009). Fujiwara and Nishikawa (2016) reported that black and red pigment-related genes are upregulated in the brown pupae of another *Papilio* species, whereas a few bilin-binding proteins are specifically upregulated in green pupae. Yoda et al. (2020) compared gene expression profiles between the two-color form of *P. polytes* pupae, demonstrating that genes in the melanin synthesis pathway are upregulated in the brown pupae, whereas genes related to bilins and carotenoids in the green pupae.

Twig-mimicking caterpillars of the peppered moth, *Biston betularia*, are also able to adjust color in the subsequent larval stage in response to the color of the twig that they rest on (Eacock et al. 2017; Noor et al. 2008). The color can vary from pale and light green to dark brown, but the involved pigments remain unknown. Nymphs of the desert locust *Schistocerca gregaria* in uncrowded conditions (see the next section for details) will background match in a range of off-white, yellow, green, brown, and black (Pener and Simpson 2009; Tanaka et al. 2012, 2016a). High temperature, however, suppresses the development of the black pigmentation (Tanaka et al. 2012; Sugahara et al. 2018). It is quite curious how these insects discriminate light wavelengths, and how they produce desired color patterns by expressing a complicated combination of various pigments in the subsequent developmental stages.

15.3.3 Density-Dependent Phase Change

One representative example of insect polyphenism is the density-dependent phase change observed in locust species. A multitude of studies have addressed this biologically curious and economically important phenomenon, especially using the migratory locust *Locusta migratoria* and the desert locust *Schistocerca gregaria* (Pener and Simpson 2009; Tanaka 2006; Tanaka et al. 2016a; Wang and Kang 2014). In the uncrowded "solitarious" phase, the integuments exhibit various colors depending on the external environmental factors, as mentioned previously. A crowding signal induces the "gregarious" phase, in which black melanization patterns are developed over the yellow or orange base coloration (Fig. 15.4a, b). Younger nymphs possess increased sensitivity to crowding, and the fate of the color phases can vary to some degree in the course of nymphal development (Pener and Simpson 2009; Tanaka and Nishide 2012; Tanaka et al. 2016b). Visual, olfactory, tactile, and auditory stimuli have been considered as potential crowding cues (Pener and Simpson 2009; Tanaka et al. 2016a). Interestingly, Tanaka and Nishide (2012) demonstrated that displaying a video image of moving tadpoles as well as locusts (but not frozen frame) effectively induces gregarious phase-like black

Fig. 15.4 Density-dependent color polyphenism in the desert locust *Schistocerca gregaria*. Final instar nymphs in the solitarious phase (**a**) and gregarious phase (**b**). The black pattern development of the gregarious locust is suppressed in high temperatures (**c**). Photos courtesy of Dr. Seiji Tanaka

pigmentation in isolated locusts, suggesting a primary role of visual stimuli. This plastic black pattern is known to be regulated by the melanization-inducible neuropeptide corazonin (Pener and Simpson 2009; Tanaka 2006; Verlinden et al. 2009). Recently, the genes encoding corazonin (*CRZ*) and its receptor (*CRZR*) have been identified in both of these locusts, and their pivotal roles in the gregarious-form pigmentation have been verified using RNAi gene silencing techniques (Sugahara et al. 2015, 2016, 2017). Expression analyses of *CRZ* and *CRZR* suggest the corazonin release is under neural regulation rather than transcriptomic regulation. The downstream pathway of corazonin signaling was investigated using a comparative transcriptome analysis using normal gregarious nymphs, CRZR-deficient albino nymphs, and *CRZ*-knockdown nymphs (Sugahara et al. 2018). The authors found a novel transcription factor gene, named locust corazonin-related transcription factor (*LOCT*), which is particularly upregulated downstream of normal corazonin signaling. The RNAi-mediated gene silencing experiment indicated that LOCT possibly serves as a link between corazonin signaling and the expression of melanin synthesis genes, such as *yellow* (Sugahara et al. 2018). A microarray analysis also revealed that catecholamine metabolic pathway genes (including the rate-limiting enzyme, tyrosine hydroxylase (*pale*)) are upregulated in gregarious locusts. This result suggests that the facilitated production of dopamine may be effectively utilized for melanin synthesis, although dopamine can also serve as a neurotransmitter in gregarious behavior (Ma et al. 2011).

The development of black pigmentation in the gregarious phase of *desert locusts* is suppressed at high temperatures, and the yellow coloration is induced

(Fig. 15.4c, Pener and Simpson 2009; Tanaka et al. 2012). This color shift is thought to be caused by the elevated expression of the carotenoid-binding yellow protein gene (*YPT*) and the simultaneous suppression of corazonin secretion (Sugahara and Tanaka 2018).

15.4 Reversible Color Changes

To cope with relatively short-term environmental fluctuations, some insects can flexibly exhibit alternative colorations without exchanging their exoskeleton. Although such a reversible color change is observed in a very limited number of insect species, a variety of color change mechanisms have been proposed (Umbers et al. 2014). Similar to other animals, insect reversible color change is roughly divided into two categories based on the absence or presence of pigment metabolism: physiological color change or morphological color change (Bückmann 1985; Umbers et al. 2014; Duarte et al. 2017).

15.4.1 Physiological Color Changes

The physiological color change includes appearance shifts that are independent of pigment metabolism. A handful of insects are known to possess physiological color changes. This change usually takes place in a relatively short time, within minutes to hours (Umbers et al. 2014).

The first method of rapid color change is the migration of pigment granules within the epidermis cell. The stick insect *Carausius morosus* changes color from green to brown in response to visual stimuli. During the darkening process, red-violet ommochrome granules come to the surface, migrating along microtubules from basal (inner) to distal (outer) parts of the epidermal cell (Berthold 1980). Some damselflies can exhibit a rapid color transition between blue-turquoise and black; this is presumably associated with thermoregulatory posturing behavior. The bluish color is a structural color generated by light-scattering spherical nanostructures (Prum 2004; Veron et al. 1974). In the high temperature-inducible bluish phase, these refractive nanospheres are tightly packed in the distal part of the epidermal cells, while the larger granules containing ommochrome pigments, mainly xanthommatin, are withdrawn in the proximal section. In the dark phase, ommochrome granules are dispersed throughout the cells quenching scattering light from the nanospheres. A similar color change mechanism was observed in the thermo-sensitive turquoise/black change of the male chameleon grasshopper *Kosciuscola tristis* (Filshie et al. 1975; Umbers 2011). Efforts to determine the driving mechanism behind this type of physiological color changes have been restricted to histological observations and optical modeling analyses, therefore exploring physiological and molecular mechanisms are desired.

Another type of reversible color change that is recorded in some beetle species is hygrochromic color; hydration or dehydration of the porous cuticle affects the structural color by modifying the refractive index of the multilayer structure (Seago et al. 2009). In the male Hercules beetle, *Dynastes hercules*, the elytron (the sclerotized forewing commonly observed in beetles) is formed by a light-scattering spongy multilayer and normally exhibits a khaki-green appearance (Rassart et al. 2008). Under high humidity conditions, infiltration of water into the spongy layer results in a black appearance. Similarly, the structural color of iridescent scales on the elytron and thorax regions is changed after contact with water from gold to red in the longhorn beetles *Tmesisternus isabellae* (Liu et al. 2009) or from blue to green in males of the scarabaeid beetle Hoplia coerulea (Mouchet et al. 2016). In contrast, a startle hygrochromic color change takes place independently of environmental moisture in a tortoise beetle, *Charidotella egregia*. In the normal beetles at rest, metallic gold is emitted from the humid state of the multilayer reflector in the dorsal cuticle (Vigneron et al. 2007). When disturbed, the golden structural color is quickly (within minutes) abolished by the expulsion of water from the cuticle layers, and the red pigment beneath the cuticle becomes visible. In this case, the moisture of the cuticle layer seems to be manipulated by the insect, but it is an enigma as to what is involved in withdrawing water from the cuticle.

As we have seen above, these physiological color changes are accomplished by a specialized microtubule rail apparatus built up in the epidermal cells or a finely elaborated multilayer structure in the cuticles. Transcriptomic regulation is unlikely to be involved in rapid color changes. Rather, we need to identify the set of molecular components that create these precise structures with variable gimmicks.

15.4.2 Morphological Color Changes

Morphological color change is accompanied by pigment production or degradation and chemical modification of pigment compounds (Bückmann 1985; Umbers et al. 2014; Duarte et al. 2017). Various types of pigments are related to insect morphological color changes including carotenoids, ommochromes, pteridines, melanin, and bile pigments. Generally, morphological color change takes longer than physiological color changes, usually days to weeks. As with physiological color changes, morphological color changes are not widespread among insects.

The stick insect *Carausius morosus*, in addition to the short-term physiological color change mentioned previously, also possesses the ability of morphological color change for background matching (Bückmann 1977). In this insect, continuous production of ommochrome pigments (ommin and xanthommatin) is induced in dark conditions and appears to accumulate dark pigments in the integuments through nymphal development. Conversely, a cessation of pigment production occurs under light conditions resulting in a lighter color, possibly because the ommochrome pigments are diluted as it grows.

Fig. 15.5 Diapause-associated reversible color change of the brown-winged green stinkbug *Plautia stali*. (**a**) Non-diapausing green adult. (**b**) Diapausing brown adult

Although not included in insects, the histochemical mechanisms of morphological color changes are well illustrated in the two crab spiders, *Misumena vatia* and *Thomisus spectabilis*. Within several days, their color reversibly alters between white and yellow in accordance with the color of the flower on which they sit. The camouflage appears to effectively facilitate the hunting of prey or avoidance of predators (Théry and Casas 2002; Théry 2007). The appearance color is made by a combination of the outermost cuticle layer, the hypodermis layer, and the innermost guanocyte layer (Insausti and Casas 2008, 2009; Gawryszewski et al. 2014). In the white phase, light passes through the transparent cuticle and the hypodermis layer and is reflected as white light by guanine crystals stored in the guanocytes (Gawryszewski et al. 2014). In the transient process from white to yellow, colorless pigment progranules in the hypodermis cells develop into the ommochrome-containing pigment granules (Insausti and Casas 2008). It is associated with the metabolism of kynurenine to 3-hydroxy-kynurenine and subsequently to several ommochrome compounds, such as xanthommatin and decarboxylated xanthommatin (Riou and Christidès 2010). The reverse change is accompanied by the degradation of ommochrome granules, regenerating the colorless progranules (Insausti and Casas 2009). The possible involvement of the ecdysteroid hormone in color-phase regulation is proposed, although further confirmation is required (Llandres et al. 2013). Transcriptome analyses using other spider species are starting to accumulate molecular information on pigment metabolism pathways in spiders (Croucher et al. 2013). Considering the wealth of background information, crab spiders are a promising candidate to solve the mystery of molecular mechanisms involved in morphological color change.

Another example of insect morphological color change is a syndrome of diapause. This diapause-associated color change is observed in several phytophagous stinkbugs (Musolin 2012), the physiological mechanisms are addressed in *Nezara viridula* and *Plautia stali* (Fig. 15.5). Their color change appears to serve a cryptic role against seasonal habitat shifts. The adults of these species are a green color on host plants in reproductive seasons, whereas during winter they are in reproductive diapause and turn brown and hibernate in the litter or bark cavities. After overwintering, the green coloration is restored along with reproductive activity (Kotaki 1998). The diapause-associated color darkening is mainly induced by a short-day photoperiod as the cue signally an approaching winter (Kotaki and Yagi 1987; Musolin and Numata 2003). In *N. viridula*, the orange-red pteridine pigment (erythropterin) is thought to be involved in this change (Gogala and Michieli 1962, 1967). Erythropterin is found in both green and brown insects, but the authors suggest that the color change was caused by the change in the pigment state between aqueous yellow and crystalline red. The diapause is regulated under neuroendocrinological control. The external administration of JH and surgical ablation/implantation of JH-producing neurosecretory glands, corpora allata (CA), revealed that both diapause and its associated color darkening of *P. stali* are induced by the absence of JH (Kotaki and Yagi 1989; Kotaki 1999). JH release from the CA is controlled by the brain, where diapause is decided by integrating photoperiodic and other environmental information (Denlinger et al. 2012; Shiga and Numata 2007). Recently it has been identified that brain gene encoding neurosecretory peptides (allatostatins) negatively regulate the JH-releasing activity of CA (Matsumoto et al. 2013, 2017).

In the two-spotted spider mite *Tetranychus urticae*, a comparative transcriptome analysis was performed between diapause-associated color phases. Diapausing red mites accumulate a twofold volume of carotenoid pigments, such as 3-hydroxyechinenone, phoenicoxanthin, and astaxanthin, compared to normal green mites (Veerman 1974, 1977). Although animals generally lack the de novo synthetic ability of carotenoids, the genome of the spider mites contains carotenoid biosynthetic genes, a fused carotenoid cyclase–carotenoid synthase gene, and carotenoid desaturase gene, by horizontal gene transfer from fungi (Altincicek et al. 2012; Grbić et al. 2011). The gene expression of these two genes increased in the diapausing red spider mites (Altincicek et al. 2012; Bryon et al. 2013). In spider mites, ecdysteroids, rather than JH, are thought to be involved in reproductive diapause as an endocrine effector (Goto 2016), but the molecular pathway linking it to carotenoid synthesis remains unknown.

15.5 Conclusions and Perspectives

This review highlights the diverse modes of insect color changes involving a variety of physiological and molecular mechanisms to create alternative color patterns. Insect color changes were considered in three categories. However, these categories are not necessarily exclusive to each other, often sharing regulatory machinery. Notably, a fine adjustment of the titer and the peak timing of JH and ecdysteroid are the two major hormones that govern insect growth, development, and reproduction (Dubrovsky 2005; Gäde et al. 1997; Riddiford et al. 2003). These hormones are shown to be crucial for inducing alternative color patterns in ontogenic color changes (Futahashi and Fujiwara 2008a), color polyphenisms (Koch and Bückmann 1987; Rountree and Nijhout 1995; Monteiro et al. 2015; Suzuki and Nijhout 2006), and morphological color changes (Kotaki and Yagi 1989; Goto 2016). These shared features imply that insects may possess common background mechanisms inducing hidden phenotypes. It also implies that through evolution, genetic assimilation and genetic accommodation processes may shift a mode of color change reciprocally across different categories by fixing or liberating alternative color forms (Rountree

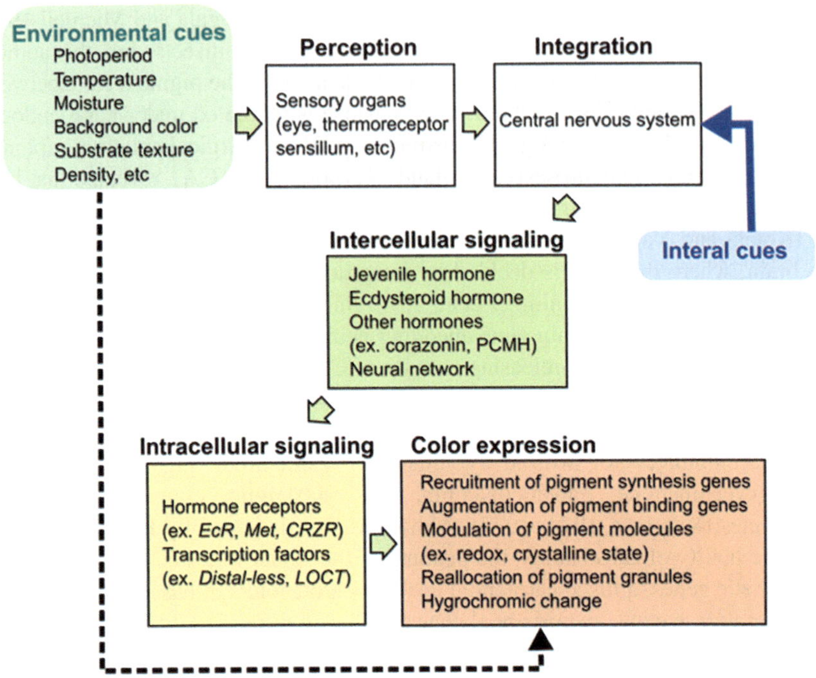

Fig. 15.6 A scheme of coordinated mechanisms of insect color change. In ontogenic color change, internal cues are used instead of environmental cues (blue line). Environmental stimuli can directly evoke insect color change (broken line); e.g. hydrochromic change of structural colors in coleopteran species. Although environmental cues and color expression modes are diverse among species, ecologically relevant color changes share needs for the coordination of perception, integration, signal transduction, and color expression machineries

and Nijhout 1995; Safranek and Riddiford 1975; Suzuki and Nijhout 2006; van der Burg et al. 2020).

Research is in its infancy on the molecular mechanisms underlying insect color change. New technological advances, especially next-generation sequencing, will facilitate the improvement of our fragmented molecular knowledge in prospective non-model insects. In particular, research should unveil the core molecular factors that serve as a switch for alternative color-generating pathways. Color change is an ideal study model system to explore unknown pathways of pigment synthesis and transport or the regulatory mechanisms for color pattern formation. Finally, ecologically relevant color change can be achieved by an elaborate system that requires the coordination of several dedicated pieces of machinery, such as variable architectures, perception and integration systems for environmental cues, neuroendocrine signaling systems, and color-generating apparatuses (Fig. 15.6). Therefore, we need comprehensive studies linking physiological, biochemical, histological, and molecular biological changes associated with color pattern changes.

Acknowledgements I would like to thank Dr. Ryo Futahashi and Dr. Seiji Tanaka for kindly offering photos of insects.

References

Altincicck B, Kovacs JL, Gerardo NM (2012) Horizontally transferred fungal carotenoid genes in the two-spotted spider mite *Tetranychus urticae*. Biol Lett 8:253–257

Anbutsu H, Moriyama M, Nikoh N et al (2017) Small genome symbiont underlies cuticle hardness in beetles. Proc Natl Acad Sci 114:201712857

Andersen SO (2010) Insect cuticular sclerotization: a review. Insect Biochem Mol Biol 40:166–178

Beldade P, Peralta CM (2017) Developmental and evolutionary mechanisms shaping butterfly eyespots. Curr Opin Insect Sci 19:22–29

Beldade P, Brakefield PM, Long AD (2002) Contribution of *Distal-less* to quantitative variation in butterfly eyespots. Nature 415:315–318

Beldade P, Mateus A, Keller R (2011) Evolution and molecular mechanisms of adaptive developmental plasticity. Mol Ecol 20:1347–1363

Berthold G (1980) Microtubules in the epidermal cells of *Carausius morosus* (Br.). Their pattern and relation to pigment migration. J Insect Physiol 26:421–425

Booth C (1990) Evolutionary significance of ontogenetic color change in animals. Biol J Linn Soc 40:125–163

Brakefield PM, Gates J, Keys D et al (1996) Development, plasticity and evolution of butterfly eyespot patterns. Nature 384:236–242

Brakefield PM, Kesbeke F, Koch PB (1998) The regulation of phenotypic plasticity of eyespots in the butterfly *Bicyclus anynana*. Am Nat 152:853–860

Bryon A, Wybouw N, Dermauw W et al (2013) Genome wide gene-expression analysis of facultative reproductive diapause in the two-spotted spider mite *Tetranychus urticae*. BMC Genomics 14:1–20

Bückmann D (1977) Morphological colour change: stage independent, optically induced ommochrome synthesis in larvae of the stick insect, *Carausius morosus* Br. J Comp Physiol B 115:185–193

Bückmann D (1985) Color change in insects. In: Bagnara J, Klaus SN, Paul E (eds) Biological, molecular and clinical aspects of pigmentation. University of Tokyo Press, Tokyo, pp 209–217

Clarke JW (2017) Evolutionary trends in phenotypic elements of seasonal forms of the tribe Junoniini (Lepidoptera: Nymphalidae). In: Diversity and evolution of butterfly wing patterns. Springer, Singapore, pp 239–253

Clusella-Trullas S, van Wyk JH, Spotila JR (2007) Thermal melanism in ectotherms. J Therm Biol 32:235–245

Connahs H, Rhen T, Simmons RB (2016) Transcriptome analysis of the painted lady butterfly, *Vanessa cardui* during wing color pattern development. BMC Genomics 17:1–16

Croucher PJP, Brewer MS, Winchell CJ et al (2013) De novo characterization of the gene-rich transcriptomes of two color-polymorphic spiders, *Theridion grallator* and *T. californicum* (Araneae: Theridiidae), with special reference to pigment genes. BMC Genomics 14:1–18

Daniels EV, Mooney KA, Reed RD (2012) Seasonal wing colour plasticity varies dramatically between buckeye butterfly populations in different climatic zones. Ecol Entomol 37:155–159

Daniels EV, Murad R, Mortazavi A, Reed RD (2014) Extensive transcriptional response associated with seasonal plasticity of butterfly wing patterns. Mol Ecol 23:6123–6134

Danks H (2006) Insect adaptations to cold and changing environments. Can Entomol 23:1 23

De Loof A, Huybrechts J, Geens M et al (2010) Sexual differentiation in adult insects: male-specific cuticular yellowing in *Schistocerca gregaria* as a model for reevaluating some current (neuro)-endocrine concepts. J Insect Physiol 56:919–925

Denlinger DL, Yocum GD, Rinehart JP (2012) Hormonal control of diapause. Elsevier, Amsterdam
Duarte RC, Flores AAV, Stevens M (2017) Camouflage through colour change: mechanisms, adaptive value and ecological significance. Philos Trans R Soc B Biol Sci 372:20160342
Dubrovsky EB (2005) Hormonal cross talk in insect development. Trends Endocrinol Metab 16:6–11
Eacock A, Rowland HM, Edmonds N, Saccheri IJ (2017) Colour change of twig-mimicking peppered moth larvae is a continuous reaction norm that increases camouflage against avian predators. PeerJ 5:e3999
Filshie BK, Day MF, Mercer EH (1975) Colour and colour change in the grasshopper, *Kosciuscola tristis*. J Insect Physiol 21:1763–1770
Fujiwara H, Nishikawa H (2016) Functional analysis of genes involved in color pattern formation in Lepidoptera. Curr Opin Insect Sci 17:16–23
Futahashi R (2016) Color vision and color formation in dragonflies. Curr Opin Insect Sci 17:32–39
Futahashi R, Fujiwara H (2005) Melanin-synthesis enzymes coregulate stage-specific larval cuticular markings in the swallowtail butterfly, *Papilio xuthus*. Dev Genes Evol 215:519–529
Futahashi R, Fujiwara H (2007) Regulation of 20-hydroxyecdysone on the larval pigmentation and the expression of melanin synthesis enzymes and yellow gene of the swallowtail butterfly, *Papilio xuthus*. Insect Biochem Mol Biol 37:855–864
Futahashi R, Fujiwara H (2008a) Juvenile hormone regulates butterfly larval pattern switches. Science 319:1061–1061
Futahashi R, Fujiwara H (2008b) Identification of stage-specific larval camouflage associated genes in the swallowtail butterfly, *Papilio xuthus*. Dev Genes Evol 218:491–504
Futahashi R, Banno Y, Fujiwara H (2010) Caterpillar color patterns are determined by a two-phase melanin gene prepatterning process: new evidence from tan and laccase2. Evol Dev 12:157–167
Futahashi R, Kurita R, Mano H, Fukatsu T (2012a) Redox alters yellow dragonflies into red. Proc Natl Acad Sci U S A 109:12626–12631
Futahashi R, Shirataki H, Narita T et al (2012b) Comprehensive microarray-based analysis for stage-specific larval camouflage pattern-associated genes in the swallowtail butterfly, Papilio xuthus. BMC Biol 10:46
Futahashi R, Kawahara-Miki R, Kinoshita M et al (2015) Extraordinary diversity of visual opsin genes in dragonflies. Proc Natl Acad Sci 112:E1247–E1256
Gäde G, Hoffmann KH, Spring JH (1997) Hormonal regulation in insects: facts, gaps, and future directions. Physiol Rev 77:963–1032
Gawryszewski FM, Birch D, Kemp DJ, Herberstein ME (2014) Dissecting the variation of a visual trait: the proximate basis of UV-Visible reflectance in crab spiders (Thomisidae). Funct Ecol 29:44–54
Gilbert LE, Forrest HS, Schultz TD, Harvey DJ (1988) Correlations of ultrastructure and pigmentation suggest how genes control development of wing scales of *Heliconius* butterflies. J Res Lipid 26:141–160
Gogala M, Michieli Š (1962) Sezonsko prebarvanje pri nekaterih vrstah stenic (Heteroptera). Biološki Vestnik (Ljubljana). Biološki Vestn 10:33–44
Gogala M, Michieli Š (1967) Berichtigung zu unseren Veröffentlichungen über Farbstoffe bei Heteropteren. Bull Sci Cons Acad 12:6
Goto SG (2016) Physiological and molecular mechanisms underlying photoperiodism in the spider mite: comparisons with insects. J Comp Physiol B 186:969–984
Grbić M, Van Leeuwen T, Clark RM et al (2011) The genome of *Tetranychus urticae* reveals herbivorous pest adaptations. Nature 479:487–492
Greene E (1989) A diet-induced developmental polymorphism in a caterpillar. Science 243:643–646
Hines HM, Papa R, Ruiz M et al (2012) Transcriptome analysis reveals novel patterning and pigmentation genes underlying Heliconius butterfly wing pattern variation. BMC Genomics 13:288

Hiraga S (2006) Interactions of environmental factors influencing pupal coloration in swallowtail butterfly *Papilio xuthus*. J Insect Physiol 52:826–838
Hiruma K, Riddiford LM (2009) The molecular mechanisms of cuticular melanization: the ecdysone cascade leading to dopa decarboxylase expression in *Manduca sexta*. Insect Biochem Mol Biol 39:245–253
Insausti TC, Casas J (2008) The functional morphology of color changing in a spider: development of ommochrome pigment granules. J Exp Biol 211:780–789
Insausti TC, Casas J (2009) Turnover of pigment granules: cyclic catabolism and anabolism of ommochromes within epidermal cells. Tissue Cell 41:421–429
Jin H, Seki T, Yamaguchi J, Fujiwara H (2019) Prepatterning of *Papilio xuthus* caterpillar camouflage is controlled by three homeobox genes: *clawless*, *abdominal-A*, and *Abdominal-B*. Sci Adv 5:eaav7569
Jones M, Rakes L, Yochum M et al (2007) The proximate control of pupal color in swallowtail butterflies: Implications for the evolution of environmentally cued pupal color in butterflies (Lepidoptera: Papilionidae). J Insect Physiol 53:40–46
Kikuchi Y, Tada A, Musolin DL et al (2016) Collapse of insect gut symbiosis under simulated climate change. MBio 7:1–8
Koch PB, Bückmann D (1987) Hormonal control of seasonal morphs by the timing of ecdysteroid release in *Araschnia levana* L. (Nymphalidae: Lepidoptera). J Insect Physiol 33:823–829
Kotaki T (1998) Effects of low temperature on diapause termination and body colour change in adults of a stink bug, *Plautia stali*. Physiol Entomol 23:53–61
Kotaki T (1999) Relationships between JH-biosynthetic activity of the corpora allata in vitro, their size and adult diapause in a stink bug, *Plautia crossota stali* Scott. Entomol Sci 2:307–313
Kotaki T, Yagi S (1987) Relationship between diapause development and coloration change in brown-winged green bug, *Plautia stali* Scott (Heteroptera: Pentatomidae). Jpn J Appl Entomol Zool 31:285–290
Kotaki T, Yagi S (1989) Hormonal control of adult diapause in the brown-winged green bug, *Plautia stali* Scott (Heteroptera : Pentatomidae). Appl Entomol Zool 24:42–51
Libbrecht R, Gwynn DM, Fellowes MDE (2007) *Aphidius ervi* preferentially attacks the green morph of the pea aphid, *Acyrthosiphon pisum*. J Insect Behav 20:25–32
Liu F, Dong BQ, Liu XH et al (2009) Structural color change in longhorn beetles *Tmesisternus isabellae*. Opt Express 17:16183
Llandres AL, Figon F, Christides J-P et al (2013) Environmental and hormonal factors controlling reversible colour change in crab spiders. J Exp Biol 216:3886–3895
Losey JE, Ives AR, Harmon J et al (1997) A polymorphism maintained by opposite patterns of parasitism and predation. Nature 388:269–272
Ma Z, Guo W, Guo X et al (2011) Modulation of behavioral phase changes of the migratory locust by the catecholamine metabolic pathway. Proc Natl Acad Sci U S A 108:3882–3887
Mateus ARA, Marques-Pita M, Oostra V et al (2014) Adaptive developmental plasticity: compartmentalized responses to environmental cues and to corresponding internal signals provide phenotypic flexibility. BMC Biol 12:97
Matsumoto K, Numata H, Shiga S (2013) Role of the brain in photoperiodic regulation of juvenile hormone biosynthesis in the brown-winged green bug *Plautia stali*. J Insect Physiol 59:387–393
Matsumoto K, Suetsugu Y, Tanaka Y et al (2017) Identification of allatostatins in the brown-winged green bug *Plautia stali*. J Insect Physiol 96:21–28
Mayekar HV, Kodandaramaiah U (2017) Pupal colour plasticity in a tropical butterfly, Mycalesis mineus (Nymphalidae: Satyrinae). PLoS One 12:1–14
Monteiro A (2015) Origin, development, and evolution of butterfly eyespots. Annu Rev Entomol 60:253–271
Monteiro A, Chen B, Ramos DM et al (2013) *Distal less* regulates eyespot patterns and melanization in *Bicyclus* butterflies. J Exp Zool Part B Mol Dev Evol 320:321–331
Monteiro A, Tong X, Bear A et al (2015) Differential expression of ecdysone receptor leads to variation in phenotypic plasticity across serial homologs. PLoS Genet 11:1–20

Mouchet SR, Van Hooijdonk E, Welch VL et al (2016) Liquid-induced colour change in a beetle: the concept of a photonic cell. Sci Rep 6:1–10

Mukherjee K, Baudach A, Vogel H, Vilcinskas A (2020) Seasonal phenotype-specific expression of microRNAs during metamorphosis in the European map butterfly *Araschnia levana*. Arch Insect Biochem Physiol 104:e21657

Musolin DL (2012) Surviving winter: diapause syndrome in the southern green stink bug *Nezara viridula* in the laboratory, in the field, and under climate change conditions. Physiol Entomol 37:309–322

Musolin DL, Numata H (2003) Photoperiodic and temperature control of diapause induction and colour change in the southern green stink bug *Nezara viridula*. Physiol Entomol 28:65–74

Nijhout HF (1997) Ommochrome pigmentation of the linea and rosa seasonal forms of *Precis coenia* (Lepidoptera: Nymphalidae). Arch Insect Biochem Physiol 36:215–222

Nijhout HF (1999) Control mechanisms of polyphenic development in insects. Bioscience 49:181–192

Nijhout HF (2003) Development and evolution of adaptive polyphenisms. Evol Dev 5:9–18

Nijhout HF (2010) Molecular and physiological basis of colour pattern formation. In: Advances in insect physiology, 1st edn. Elsevier, London, pp 219–265

Nikoh N, Tsuchida T, Maeda T et al (2018) Genomic insight into symbiosis-induced insect color change by a facultative bacterial endosymbiont "*Candidatus* Rickettsiella viridis". MBio 9: e00890

Nishide Y, Tanaka S (2012) Yellowing, morphology and behaviour in sexually mature gynandromorphs of the desert locust *Schistocerca gregaria*. Physiol Entomol 37:379–383

Noh MY, Muthukrishnan S, Kramer KJ, Arakane Y (2016) Cuticle formation and pigmentation in beetles. Curr Opin Insect Sci 17:1–9

Noor MAF, Parnell RS, Grant BS (2008) A reversible color polyphenism in American peppered moth (Biston betularia cognataria) caterpillars. PLoS One 3:e3142

Pener MP, Simpson SJ (2009) Locust phase polyphenism: an update. Elsevier, Amsterdam

Prudic KL, Oliver JC, Sperling FAH (2007) The signal environment is more important than diet or chemical specialization in the evolution of warning coloration. Proc Natl Acad Sci 104:19381–19386

Prum RO (2004) Blue integumentary structural colours in dragonflies (Odonata) are not produced by incoherent Tyndall scattering. J Exp Biol 207:3999–4009

Rassart M, Colomerl JF, Tabarrant T, Vigneron JP (2008) Diffractive hygrochromic effect in the cuticle of the hercules beetle *Dynastes hercules*. New J Phys 10:033014

Riddiford LM, Hiruma K, Zhou X, Nelson CA (2003) Insights into the molecular basis of the hormonal control of molting and metamorphosis from *Manduca sexta* and *Drosophila melanogaster*. Insect Biochem Mol Biol 33:1327–1338

Riou M, Christidès J-P (2010) Cryptic color change in a crab spider (*Misumena vatia*): identification and quantification of precursors and ommochrome pigments by HPLC. J Chem Ecol 36:412–423

Rountree DB, Nijhout HF (1995) Hormonal control of a seasonal polyphenism in *Precis coenia* (Lepidoptera: Nymphalidae). J Insect Physiol 41:987–992

Safranek L, Riddiford LM (1975) The biology of the black larval mutant of the tobacco hornworm, Manduca sexta. J Insect Physiol 21:1931–1938

Sas F, Begum M, Vandersmissen T et al (2007) Development of a real-time PCR assay for measurement of yellow protein mRNA transcription in the desert locust *Schistocerca gregaria*: a basis for isolation of a peptidergic regulatory factor. Peptides 28:38–43

Schartl M, Larue L, Goda M et al (2016) What is a vertebrate pigment cell? Pigment Cell Melanoma Res 29:8–14

Seago AE, Brady P, Vigneron J-P, Schultz TD (2009) Gold bugs and beyond: a review of iridescence and structural colour mechanisms in beetles (Coleoptera). J R Soc Interface 6: S165–S184

Shiga S, Numata H (2007) Neuroanatomical approaches to the study of insect photoperiodism. Photochem Photobiol 83:76–86
Shirataki H, Futahashi R, Fujiwara H (2010) Species-specific coordinated gene expression and trans-regulation of larval color pattern in three swallowtail butterflies. Evol Dev 12:305–314
Simpson SJ, Sword GA, Lo N (2011) Polyphenism in insects. Curr Biol 21:R738–R749
Sköld HN, Aspengren S, Cheney KL, Wallin M (2016) Fish chromatophores–from molecular motors to animal behavior. In: International review of cell and molecular biology. Elsevier Inc., Amsterdam, pp 171–219
Smith KC (1991) The effects of temperature and daylength on the rosa polyphenism in the buckeye butterfly, *Precis coenia* (Lepidoptera: Nymphalidae). J Res Lipid 30:225–236
Stavenga DG, Leertouwer HL, Hariyama T et al (2012) Sexual dichromatism of the damselfly *Calopteryx japonica* caused by a melanin-chitin multilayer in the male wing veins. PLoS One 7:1–7
Stevens M, Ruxton GD (2012) Linking the evolution and form of warning coloration in nature. Proc R Soc B Biol Sci 279:417–426
Sugahara R, Tanaka S (2018) Environmental and hormonal control of body color polyphenism in late-instar desert locust nymphs: role of the yellow protein. Insect Biochem Mol Biol 93:27–36
Sugahara R, Saeki S, Jouraku A et al (2015) Knockdown of the corazonin gene reveals its critical role in the control of gregarious characteristics in the desert locust. J Insect Physiol 79:80–87
Sugahara R, Tanaka S, Jouraku A, Shiotsuki T (2016) Functional characterization of the corazonin-encoding gene in phase polyphenism of the migratory locust, *Locusta migratoria* (Orthoptera: Acrididae). Appl Entomol Zool 51:225–232
Sugahara R, Tanaka S, Jouraku A, Shiotsuki T (2017) Two types of albino mutants in desert and migratory locusts are caused by gene defects in the same signaling pathway. Gene 608:41–48
Sugahara R, Tanaka S, Jouraku A, Shiotsuki T (2018) Identification of a transcription factor that functions downstream of corazonin in the control of desert locust gregarious body color. Insect Biochem Mol Biol 97:10–18
Sugumaran M (2002) Comparative biochemistry of eumelanogenesis and the protective roles of phenoloxidase and melanin in insects. Pigment Cell Res 15:2–9
Suzuki Y, Nijhout HF (2006) Evolution of a polyphenism by genetic accommodation. Science 311:650–652
Tanaka S (2006) Corazonin and locust phase polyphenism. Appl Entomol Zool 41:179–193
Tanaka S, Nishide Y (2012) Do desert locust hoppers develop gregarious characteristics by watching a video? J Insect Physiol 58:1060–1071
Tanaka S, Harano KI, Nishide Y (2012) Re-examination of the roles of environmental factors in the control of body-color polyphenism in solitarious nymphs of the desert locust *Schistocerca gregaria* with special reference to substrate color and humidity. J Insect Physiol 58:89–101
Tanaka S, Harano KI, Nishide Y, Sugahara R (2016a) The mechanism controlling phenotypic plasticity of body color in the desert locust: some recent progress. Curr Opin Insect Sci 17:10–15
Tanaka S, Saeki S, Nishide Y et al (2016b) Body-color and behavioral responses by the mid-instar nymphs of the desert locust, *Schistocerca gregaria* (Orthoptera: Acrididae) to crowding and visual stimuli. Entomol Sci 19:391–400
Théry M (2007) Colours of background reflected light and of the prey's eye affect adaptive coloration in female crab spiders. Anim Behav 73:797–804
Théry M, Casas J (2002) Predator and prey views of spider camouflage. Nature 415:133–133
Tsuchida T, Koga R, Horikawa M et al (2010) Symbiotic bacterium modifies aphid body color. Science 330:1102–1104
Umbers KDL (2011) Turn the temperature to turquoise: Cues for colour change in the male chameleon grasshopper (*Kosciuscola tristis*) (Orthoptera: Acrididae). J Insect Physiol 57:1198–1204
Umbers KDL, Fabricant SA, Gawryszewski FM et al (2014) Reversible colour change in Arthropoda. Biol Rev 89:820–848

van der Burg KRL, Lewis JJ, Brack BJ et al (2020) Genomic architecture of a genetically assimilated seasonal color pattern. Science 370:721–725

Veerman A (1974) Carotenoid metabolism in *Tetranychus urticae* Koch (Acari: Tetranychidae). Comp Biochem Physiol 47:101–116

Veerman A (1977) Photoperiodic termination of diapause in spider mites. Nature 266:526–527

Verlinden H, Badisco L, Marchal E et al (2009) Endocrinology of reproduction and phase transition in locusts. Gen Comp Endocrinol 162:79–92

Veron JEN, O'Farrell AF, Dixon B (1974) The fine structure of odonata chromatophores. Tissue Cell 6:613–626

Vigneron JP, Pasteels JM, Windsor DM et al (2007) Switchable reflector in the Panamanian tortoise beetle *Charidotella egregia* (Chrysomelidae: Cassidinae). Phys Rev E 76:1–10

Vilcinskas A, Vogel H (2016) Seasonal phenotype-specific transcriptional reprogramming during metamorphosis in the European map butterfly *Araschnia levana*. Ecol Evol 6:3476–3485

Wang X, Kang L (2014) Molecular mechanisms of phase change in locusts. Annu Rev Entomol 59:225–244

Wybrandt GB, Andersen SO (2001) Purification and sequence determination of a yellow protein from sexually mature males of the desert locust, *Schistocerca gregaria*. Insect Biochem Mol Biol 31:1183–1189

Yamanaka A, Endo K, Nishida H et al (1999) Extraction and partial characterization of pupal-cuticle-melanizing hormone (PCMH) in the swallowtail butterfly, *Papilio xuthus* L. (Lepidoptera, Papilionidae). Zool Sci 16:261–268

Yamanaka A, Kometani M, Yamamoto K et al (2009) Hormonal control of pupal coloration in the painted lady butterfly *Vanessa cardui*. J Insect Physiol 55:512–517

Yamashita K, Kanzaki K, Hinauchi M et al (2014) Changes of seasonal morph development induced by surgical operations in pupae of the large map butterfly *Araschnia burejana* Bermer (Lepidoptera: Nymphalidae). J Exp Zool Part A Ecol Genet Physiol 321:276–282

Yoda S, Otaguro E, Nobuta M, Fujiwara H (2020) Molecular mechanisms underlying pupal protective color switch in papilio polytes butterflies. Front Ecol Evol 8:51. https://doi.org/10.3389/FEVO.2020.00051

Zimova M, Hackländer K, Good JM et al (2018) Function and underlying mechanisms of seasonal colour moulting in mammals and birds: what keeps them changing in a warming world? Biol Rev 93:1478–1498

GPSR Compliance

The European Union's (EU) General Product Safety Regulation (GPSR) is a set of rules that requires consumer products to be safe and our obligations to ensure this.

If you have any concerns about our products, you can contact us on ProductSafety@springernature.com

In case Publisher is established outside the EU, the EU authorized representative is:

Springer Nature Customer Service Center GmbH
Europaplatz 3
69115 Heidelberg, Germany

Batch number: 10755214

Printed by Printforce, the Netherlands